Michael W. Diest

D0458992

Multiple Regression in Behavioral Research

Fred N. Kerlinger
Elazar J. Pedhazur
New York University

HOLT, RINEHART AND WINSTON, INC.
New York   Chicago   San Francisco   Atlanta
Dallas   Montreal   Toronto   London   Sydney

Copyright © 1973 by Holt, Rinehart and Winston, Inc.
All rights reserved
Library of Congress Catalog Card Number: 73-3936
ISBN: 0-03-086211-6
Printed in the United States of America
3456   038   123456789

To Geula and Betty

Preface

Like many ventures, this book started in a small way: we wanted to write a brief manual for our students. And we started to do this. We soon realized, however, that it did not seem possible to write a brief exposition of multiple regression analysis that students would understand. The brevity we sought is possible only with a mathematical presentation relatively unadorned with numerical examples and verbal explanations. Moreover, the more we tried to work out a reasonably brief manual the clearer it became that it was not possible to do so. We then decided to write a book.

Why write a whole book on multiple regression analysis? There are three main reasons. One, multiple regression is a general data analytic system (Cohen, 1968) that is close to the theoretical and inferential preoccupations and methods of scientific behavioral research. If, as we believe, science's main job is to "explain" natural phenomena by discovering and studying the relations among variables, then multiple regression is a general and efficient method to help do this.

Two, multiple regression and its rationale underlie most other multivariate methods. Once multiple regression is well understood, other multivariate methods are easier to comprehend. More important, their use in actual research becomes clearer. Most behavioral research attempts to explain one dependent variable, one natural phenomenon, at a time. There is of course research in which there are two or more dependent variables. But such research can be more profitably viewed, we think, as an extension of the one dependent variable case. Although we have not entirely neglected other multivariate methods, we have concentrated on multiple regression. In the next decade and beyond, we think it will be seen as the cornerstone of modern data analysis in the behavioral sciences.

Our strongest motivation for devoting a whole book to multiple regression is that the behavioral sciences are at present in the midst of a conceptual and technical revolution. It must be remembered that the empirical behavioral sciences are young, not much more than fifty to seventy years old. Moreover, it is only recently that the empirical aspects of inquiry have been emphasized. Even after psychology, a relatively advanced behavioral science, became strongly empirical, its research operated in the univariate tradition. Now, however, the availability of multivariate methods and the modern computer makes possible theory and empirical research that better reflect the multivariate nature of psychological reality.

The effects of the revolution are becoming apparent, as we will show in the latter part of the book when we describe studies such as Frederiksen et al.'s (1968) study of organizational climate and administrative performance and the now well-known *Equality of Educational Opportunity* (Coleman et al., 1966).

Within the decade we will probably see the virtual demise of one-variable think-ing and the use of analysis of variance with data unsuited to the method. Instead, multivariate methods will be well-accepted tools in the behavioral scientist's and educator's armamentarium.

The structure of the book is fairly simple. There are five parts. Part 1 provides the theoretical foundations of correlation and simple and multiple regression. Basic calculations are illustrated and explained and the results of such calculations tied to rather simple research problems. The major purpose of Part 2 is to explore the relations between multiple regression analysis and analysis of variance and to show the student how to do analysis of variance and covariance with multiple regression. In achieving this purpose, certain techni-cal problems are examined in detail: coding of categorical and experimental variables, interaction of variables, the relative contributions of independent variables to the dependent variable, the analysis of trends, commonality analy-sis, and path analysis. In addition, the general problems of explanation and prediction are attacked.

Part 3 extends the discussion, although not in depth, to other multivariate methods: discriminant analysis, canonical correlation, multivariate analysis of variance, and factor analysis. The basic emphasis on multiple regression as the core method, however, is maintained. The use of multiple regression analysis — and, to a lesser extent, other multivariate methods — in behavioral and educa-tional research is the substance of Part 4. We think that the student will profit greatly by careful study of actual research uses of the method. One of our purposes, indeed, has been to expose the student to cogent uses of multiple regression. We believe strongly in the basic unity of methodology and research substance.

In Part 5, the emphasis on theory and substantive research reaches its climax with a direct attack on the relation between multiple regression and scientific research. To maximize the probability of success, we examine in some detail the logic of scientific inquiry, experimental and nonexperimental research, and, finally, theory and multivariate thinking in behavioral research. All these problems are linked to multiple regression analysis.

In addition to the five parts briefly characterized above, four appendices are included. The first three address themselves to matrix algebra and the compu-ter. After explaining and illustrating elementary matrix algebra — an indispens-able and, happily, not too complex a subject — we discuss the use of the computer in data analysis generally and we give one of our own computer programs in its entirety with instructions for its use. The fourth appendix is a table of the F distribution, 5 percent and 1 percent levels of significance.

Achieving an appropriate level of communication in a technical book is always a difficult problem. If one writes at too low a level, one cannot really explain many important points. Moreover, one may insult the background and intelligence of some readers, as well as bore them. If one writes at too advanced

.. ievel, then one loses most of one's audience. We have tried to write at a fairly elementary level, but have not hesitated to use certain advanced ideas. And we have gone rather deeply into a number of important, even indispensable, concepts and methods. To do this and still keep the discussion within the reach of students whose mathematical and statistical backgrounds are bounded, say, by correlation and analysis of variance, we have sometimes had to be what can be called excessively wordy, although we hope not verbose. To compensate, the assumptions behind multiple regression and related methods have not been emphasized. Indeed, critics may find the book wanting in its lack of discussion of mathematical and statistical assumptions and derivations. This is a price we had to pay, however, for what we hope is comprehensible exposition. In other words, understanding and intelligent practical use of multiple regression are more important in our estimation than rigid adherence to statistical assumptions. On the other hand, we have discussed in detail the weaknesses as well as the strengths of multiple regression.

The student who has had a basic course in statistics, including some work in inferential statistics, correlation, and, say, simple one-way analysis of variance should have little difficulty. The book should be useful as a text in an intermediate analysis or statistics course or in courses in research design and methodology. Or it can be useful as a supplementary text in such courses. Some instructors may wish to use only parts of the book to supplement their work in design and analysis. Such use is feasible because some parts of the books are almost self-sufficient. With instructor help, for example, Part 2 can be used alone. We suggest, however, sequential study since the force of certain points made in later chapters, particularly on theory and research, depends to some extent at least on earlier discussions.

We have an important suggestion to make. Our students in research design courses seem to have benefited greatly from exposure to computer analysis. We have found that students with little or no background in data processing, as well as those with background, develop facility in the use of packaged computer programs rather quickly. Moreover, most of them gain confidence and skill in handling data, and they become fascinated by the immense potential of analysis by computer. Not only has computer analysis helped to illustrate and enhance the subject matter of our courses; it has also relieved students of laborious calculations, thereby enabling them to concentrate on the interpretation and meaning of data. We therefore suggest that instructors with access to computing facilities have their students use the computer to analyze the examples given in the text as well as to do exercises and term projects that require computer analysis.

We wish to acknowledge the help of several individuals. Professors Richard Darlington and Ingram Olkin read the entire manuscript of the book and made many helpful suggestions, most of which we have followed. We are grateful for their help in improving the book. To Professor Ernest Nagel we express our thanks for giving us his time to discuss philosophical aspects of causality. We are indebted to Professor Jacob Cohen for first arousing our curiosity about multiple regression and its relation to analysis of variance and its application to data analysis.

The staff of the Computing Center of the Courant Institute of Mathematical Sciences, New York University, has been consistently cooperative and helpful. We acknowledge, particularly, the capable and kind help of Edward Friedman, Neil Smith, and Robert Malchie of the Center. We wish to thank Elizabeth Taleporos for valuable assistance in proofreading and in checking numerical examples. Geula Pedhazur has given fine typing service with ungrateful material. She knows how much we appreciate her help.

New York University's generous sabbatical leave policy enabled one of us to work consistently on the book. The Courant Institute Computing Center permitted us to use the Center's CDC-6600 computer to solve some of our analytic and computing problems. We are grateful to the university and to the computing center, and, in the latter case, especially to Professor Max Goldstein, associate director of the center.

Finally, but not too apologetically, we appreciate the understanding and tolerance of our wives who often had to undergo the hardships of talking and drinking while we discussed our plans, and who had to put up with, usually cheerfully, our obsession with the subject and the book.

This book has been a completely cooperative venture of its authors. It is not possible, therefore, to speak of a "senior" author. Yet our names must appear on some order on the cover and title page. We have solved the problem by listing the names alphabetically, but would like it understood that the order could just as well have been the other way around.

Amsterdam, The Netherlands   Fred N. Kerlinger
Brooklyn, New York   Elazar J. Pedhazur
March 1973

Contents

Foundations of Multiple Regression Analysis

The Nature of Multiple Regression Analysis

Remarkable advances in the analysis of educational, psychological, and socio-logical data have been made in recent decades. The high-speed computer has made it possible to analyze large quantities of complex data with relative ease. The basic conceptualization of data analysis, too, has advanced, although per-haps not as rapidly as computer technology. Much of the increased understand-ing and mastery of data analysis has come about through the wide propagation and study of statistics and statistical inference and especially from the analysis of variance. The expression "analysis of variance" is well-chosen. It epito-mizes the basic nature of most data analysis: the partitioning, isolation, and identification of variation in a dependent variable due to different independent variables. In any case, analysis of variance has thrived and has become a vital part of the analytic armamentarium of the behavioral scientist.

Another group of analytic-statistical techniques known as multivariate analysis has also thrived, even though its purposes, mechanics, and uses are not as well-understood as those of analysis of variance. Of these methods, two in particular, factor analysis and multiple regression analysis, have been fairly widely used. In this book we concentrate on multiple regression analysis, a most important branch of multivariate analysis.[1] We will find that it is a power-ful analytic tool widely applicable to many different kinds of research problems.

---

[1]Strictly speaking, the expression "multivariate analysis" has meant analysis with more than one dependent variable. A univariate method is one in which there is only one dependent variable. We prefer to consider all analytic methods that have more than one independent variable or more than one dependent variable or both as multivariate methods. Thus, multiple regression is a multi-variate method. Although the point is not all-important, it needs to be clarified early to prevent reader confusion.

It can be used effectively in sociological, psychological, economic, political, and educational research. It can be used equally well in experimental or non-experimental research. It can handle continuous and categorical variables. It can handle two, three, four, or more independent variables. In principle, the analysis is the same. Finally, as we will abundantly show, multiple regression analysis can do anything the analysis of variance does — sums of squares, mean squares, F ratios — and more. Handled with knowledge, understanding, and care, it is indeed a general and potent tool of the behavioral scientist.

Multiple Regression and Scientific Research

Multiple regression is a method of analyzing the collective and separate contributions of two or more independent variables, X_i, to the variation of a dependent variable, Y. The fundamental task of science is to explain phenomena. As Braithwaite (1953) says, its basic aim is to discover or invent general explanations of natural events. The purpose of science, then, is theory. A theory is an interrelated set of constructs or variables "that presents a systematic view of phenomena by specifying relations among variables, with the purpose of explaining . . . the phenomena" (Kerlinger, 1964, p. 11). But this view of science is close to the definition of multiple regression.

Natural phenomena are complex. The phenomena and constructs of the behavioral sciences — learning, achievement, anxiety, conservatism, social class, aggression, reinforcement, authoritarianism, and so on — are especially complex. "Complex" in this context means that a phenomenon has many facets and many causes. In a research-analytic context, "complex" means that a phenomenon has several sources of variation. To study a construct or variable scientifically we must be able to identify the sources of the variable's variation. We say that a variable varies. This means that when we apply an instrument that measures the variable to a sample of individuals we will obtain more or less different measures from each of them. We talk about the variance of Y, or the variance of college grade-point averages (a measure of achievement), or the variance of a scale of ethnocentrism.

It can be asserted that all the scientist has to work with is variance. If variables do not vary, if they do not have variance, the scientist cannot do his work. If in a sample, all individuals get the same score on a test of mathematical aptitude, the variance is zero and it is not possible to "explain" mathematical aptitude. In the behavioral sciences, variability is itself a phenomenon of great scientific curiosity and interest. The large differences in the intelligence and achievement of children, for instance, and the considerable differences between schools and socioeconomic groups in critical educational variables are phenomena of deep interest and concern to behavioral scientists. Because of the analytic and substantive importance of variation, then, the expressions "variance" and "covariance" will be used a great deal in this book.

Multiple regression's task is to help "explain" the variance of a dependent variable. It does this, in part, by estimating the contributions to this variance of two or more independent variables. Educational researchers seek to explain the variance of school achievement by studying various correlates of school achievement: intelligence, aptitude, social class, race, home background, school atmosphere, teacher characteristics, and so on. Political scientists seek to explain voting behavior by studying variables presumed to influence such behavior: sex, age, income, education, party affiliation, motivation, place of residence, and the like. Psychological scientists seek to explain risk-taking behavior by searching for variables that covary with risk taking: communication, group discussion, group norms, type of decision, group interaction, diffusion of responsibility (Kogan & Wallach, 1967).

The traditional view of research amounts to studying the relation between one independent variable and one dependent variable, studying the relation between another independent variable and the dependent variable, and so on, and then trying to put the pieces together. The traditional research design is the so-called classic experimental group and control group setup. While one can hardly say that the traditional view is invalid, one can say that in the behavioral sciences it is obsolescent, even obsolete (Campbell & Stanley, 1963; Kerlinger, 1964, 1969). One simply cannot understand and explain phenomena in this way because of the complex interaction of independent variables as they impinge on dependent variables.

Take a simple example. We wish to study the effects of different methods on reading proficiency. We study the effects of the methods with boys. Then we study their effects with girls, then with middle-class children, then with working-class children, then with children of high, medium, and low intelligence. This is, of course, a travesty that dramatizes the near futility of studying one variable at a time. The job is not only endless; it is self-defeating because methods probably differ in their effectiveness with different kinds of children, and one cannot really study the implied problems without studying the variables together.

Multiple regression analysis is nicely suited to studying the influence of several independent variables, including experimental (manipulated) variables, on a dependent variable. Let us look at three examples of its use in actual research and put research flesh on the rather bare and abstract bones of this discussion.

Research Examples

Before describing the illustrative uses of multiple regression, a word on prediction may be helpful. Many, perhaps most, uses of multiple regression have emphasized prediction from two or more independent variables, X_i, to a dependent variable, Y. The results of multiple regression analysis fit well into a prediction framework. Prediction, however, is really a special case of explanation; it can be subsumed under theory and explanation. Look at it this way: scientific explanation consists in specifying the relations between empirical events. We say: If p, then q, under conditions r, s, and t.[2] This is explanation, of course. It is also prediction, prediction from p (and r, s, and t) to q. In this book, the word prediction will often be used. (See Chapter 11, below, for an extended discussion.) It is to be understood in the larger scientific sense, however, except in largely practical studies in which researchers are interested only in successful prediction to a criterion, say grade-point average or skill performance.

Holtzman and Brown Study: Predicting High School Achievement

Holtzman and Brown (1968), in a study of the prediction of high school grade-point average (GPA), used study habits and attitudes (SHA) and scholastic aptitude (SA) as independent variables. The correlation between high school GPA (Y) and SHA (X_1) was .55; between GPA and SA (X_2) it was .61.[3] By using SHA alone, then, $.55^2 = .30$, which means that 30 percent of the variance of GPA was accounted for by SHA. Using SA alone, $.61^2 = 37$ percent of the variance of GPA was accounted for. Since SHA and SA overlap in their variance (the correlation between them was .32), we cannot simply add the two r^2's together to determine the amount of variance that SHA and SA accounted for *together*. By using multiple regression, Holtzman and Brown found that in their sample of 1684 seventh graders, the correlation between *both* SHA *and* SA, on the one hand, and GPA, on the other hand, was .72, or $.72^2 = .52$. Thus 52 percent of the variance of GPA was accounted for by SHA *and* SA. By adding scholastic aptitude to study habits and attitudes, Holtzman and Brown raised predictive power considerably: from 30 percent to 52 percent.[4] Clearly, using both independent variables and their joint relation to GPA was advantageous.

Coleman Study: Equality of Educational Opportunity

The massive and important Coleman report, *Equality of Educational Opportunity* (Coleman et al., 1966), contains numerous and effective examples of multiple regression analysis. One of the basic purposes of the study was to explain school achievement, or rather, inequality in school achievement. Although we cannot at this point do justice to this study, we can, hopefully, give the reader a feeling for the use of multiple regression in explaining a complex educational, psychological, and sociological phenomenon. Toward the end of the book, we will examine the report in more detail.

---

[2]Whenever we say If p, then q in this book, we always mean: If p, then *probably q*. The insertion of "probably" is consistent with our probabilistic approach and does not affect the basic logic.

[3]The dependent variable is always indicated by Y and the independent variables by X_i, or X_1, $X_2, \ldots, X_k$.

[4]Recall one or two points from elementary statistics. The square of the correlation coefficient, r^2_{xy}, expresses the amount of variance shared in common by two variables, X and Y. This amount of variance is a proportion or percentage. One can say, for instance, if $r_{xy} = .60$, that $.60^2 = .36 = 36$ percent of the variance of Y is accounted for, "explained," predicted by X. Similar expressions are used in multiple regression analysis except that we usually speak of the variance of Y accounted for by two or more X's. See Hays (1963, pp. 501–502).

The researchers chose as their most important dependent variable verbal ability or achievement (VA). Some 60 independent variable measures of different kinds were correlated with VA. By combining certain of these measures in multiple regression analyses, Coleman and his colleagues were able to sort out the relative effects of different kinds of independent variables on the dependent variable, verbal achievement. They found, for instance, that what they called attitudinal variables—students' interest in school, their self-concept (in relation to learning and success in school), and their sense of control of the environment —accounted for more of the variance of verbal achievement than family background variables and school variables (*ibid.*, pp. 319–325). They were able to reach this conclusion by combining and recombining different independent variables in multiple regression analyses, the results of which indicated the relative contributions of the individual variables and sets of variables. This is a case where the data could not have yielded the conclusions with any other method.

Lave and Seskin Study: Air Pollution and Disease Mortality

In an analysis of the effects of air pollution on human health, Lave and Seskin (1970) used bronchitis, lung cancer, pneumonia, and other human disease mortality rates as dependent variables and air pollution and socioeconomic status (population density or social class) as independent variables.[5] The two independent variables were related to the various indices of mortality in a series of multiple regression analyses. If air pollution is a determinant of, say, bronchitis mortality or lung cancer mortality, the correlation should be substantial, especially as compared to the correlation between air pollution and a variable which should probably not be affected by air pollution—stomach cancer, for example.

A number of analyses yielded similar results: air pollution contributed substantially to the variance of mortality indices and population density did not. The authors concluded that air pollution is a significant explanatory variable and the socioeconomic variable of doubtful significance.

Multiple Regression Analysis and Analysis of Variance

The student should be aware early in his study of the almost virtual identity of multiple regression analysis and analysis of variance. As we said in the Preface, Part II of the book will explain how analysis of variance can be done with multiple regression analysis. It will also show that multiple regression analysis not only gives more information about the data; it is also applicable to more kinds of data.

---

[5]The socioeconomic designation is a bit misleading. In the relations reported here the designation really means, in most of the analyses, population density. Although Lave and Seskin do not elaborate the point, presumably the denser the population the greater the mortality rate.

Analysis of variance was designed to analyze data yielded by experiments. If there is more than one experimental variable, one of the conditions that must be satisfied to use analysis of variance is that the experimental variables be independent. In a factorial design, for instance, the researcher takes pains to satisfy this condition by randomly assigning an equal number of subjects to the cells of the design. If, for some reason, he does not have equal numbers in the cells, the independent variables will be correlated and the usual analysis of variance, strictly speaking, is not appropriate.

We have labored the above point as background for an important analytic problem. Many, perhaps most, behavioral research problems are of the ex post facto kind that do not lend themselves to experimental manipulation and to random assignment of equal numbers of subjects to groups. Instead, intact already existing groups must be used. The condition of no correlation between the independent variables is systematically violated, so to speak, by the very nature of the research situation and problem. Is social class an independent variable in a study? To use analysis of variance one would have to have equal numbers of middle-class and working-class subjects in the cells corresponding to the social class partition. (It is assumed that there is at least one other independent variable in the study.) This point is so important, especially for later discussions, that it is necessary to clarify what is meant.

In general, there are two kinds of independent variables, active and attribute (Kerlinger, 1973, Chapter 3). *Active variables* are manipulated variables. If, for example, an experimenter rewards one group of children for some kind of performance and does not reward another group, he has manipulated the variable reward. *Attribute variables* are measured variables. Intelligence, aptitude, social class, and many other similar variables cannot be manipulated; subjects come to research studies, so to speak, with the variables. They can be assigned to groups on the basis of their possession of more or less intelligence or on the basis of being members of middle-class or working-class families. When there are two or more such variables in a study and, for analysis of variance purposes, the subjects are assigned to subgroups on the basis of their status on the variables, it is next to impossible, without using artificial means, to have equal numbers of subjects in the cells of the design. This is because the variables are correlated, are not independent. The variables intelligence and social class, for instance, are correlated. To see what is meant and the basic difficulty involved, suppose we wish to study the relations between intelligence and social class, on the one hand, and school achievement, on the other hand. The higher intelligence scores will tend to be those of middle-class children, and the lower intelligence scores those of working-class children. (There are of course many exceptions, but they do not alter the argument.) The number of subjects, therefore, will be unequal in the cells of the design simply because of the correlation between intelligence and social class.

Figure 1.1 illustrates this in a simple way. Intelligence is dichotomized into high and low intelligence; social class is dichotomized into middle class and working class. Since middle-class subjects tend to have higher intelligence

	High Intelligence	Low Intelligence
Middle Class	*a*	*b*
Working Class	*c*	*d*

FIGURE 1.1

than working-class subjects, there will be more subjects in cell *a* than in cell *b*. Similarly, there will be more working-class subjects in cell *d* than in cell *c*. The inequality of subjects is inherent in the relation between intelligence and social class.

Researchers commonly partition a continuous variable into high and low, or high, medium, and low groups — high intelligence–low intelligence, high authoritarianism–low authoritarianism, and so on — in order to use analysis of variance. Although it is valuable to conceptualize design problems in this manner, it is unwise and inappropriate to analyze them so. Such devices, for one thing, throw away information. When one dichotomizes a variable that can take on a range of values, one loses considerable variance.[6] This can mean lowered correlations with other variables and even nonsignificant results when in fact the tested relations may be significant. In short, researchers can be addicted to an elegant and powerful method like analysis of variance (or factor analysis or multiple regression analysis) and force the analysis of research data on the procrustean bed of the method.

Such problems virtually disappear with multiple regression analysis. In the case of the dichotomous variables, one simply enters them as independent variables using so-called dummy variables, in which 1's and 0's are assigned to subjects depending on whether they possess or do not possess a characteristic in question. If a subject is middle class, assign him a 1; if he is working class, assign him a 0. If a subject is male, assign him a 1; if female, assign a 0 — or the other way around if one is conscious of the Lib Movement. In the case of the continuous variables that in the analysis of variance would be partitioned, simply include them as independent continuous variables. These matters will be clarified later. The main point now is that multiple regression analysis has the fortunate ability to handle different kinds of variables with equal facility. Experimental treatments, too, are handled as variables. Although the problem of unequal *n*'s, when analyzing experimental data with multiple regression

---

[6]Later, we will define types of variables. For the present, a *nominal*, or *categorical*, variable is a partitioned variable — partitions are subsets of sets that are disjoint and exhaustive (Kemeny, Snell, and Thompson, 1966, p. 84) — the separate partitions of which have no quantitative meaning except that of qualitative difference. The numbers assigned to such partitions, in other words, are really labels that do not have number meaning: strictly speaking, they cannot be ordered or added, even though we will later treat them *as though* they had number meaning. A *continuous* variable is one that has some range of values that have number meaning. Actually, "continuous" literally means spread over a linear continuum, that is, capable of having points in a continuum that are arbitrarily small. For convenience, however, we take "continuous" to include scales with discrete values.

analysis, does not disappear, it is so much less a problem as to be almost negligible.

Prolegomenon to Part I

In the present chapter, we have tried to give the reader an intuitive and general feeling for multiple regression analysis and its place in the analysis of behavioral research data. It would be too much to expect any appreciable depth of understanding. We have only been preoccupied with setting the stage for what follows. We should cling to certain ideas as we enter deeper into our study. First, multiple regression is used as is any other analytic method: to help us understand natural phenomena by indicating the nature and magnitude of the relations between the phenomena and other phenomena. Second, the phenomena are named by scientists and are called constructs or variables. It is the relations among these constructs or variables that we seek.

Third, the technical keys to understanding most statistical methods are variance and covariance. Relations are sets of ordered pairs, and if we wish to study relations, we must somehow observe or create such sets. But observation is insufficient. The sets of ordered pairs have to be analyzed to determine the nature of the relations. We can create, for example, a set of ordered pairs of intelligence test scores and achievement test scores by administering the two tests to a suitable sample of children. While the set of ordered pairs is by definition a relation, merely observing the set is not enough. We must know how much the two components of the set covary. Are the two components in approximately the same rank order? Are the high intelligence scores generally paired with the high achievement scores and the lows with the lows?

After determining the direction of the covariation, one must then determine to what extent the components covary. In addition, one must answer the question: If the covariation is not perfect, that is, if the rank orders of the two components are not the same, to what extent are they not the same? Are there other sources of the variation of the achievement scores, assuming it is achievement one wishes to explain? If so, what are they? How do they covary with achievement and with intelligence? Do they contribute as much as, more, or less to the variation of achievement than intelligence does? Such questions are what science and analysis are about. Multiple regression analysis is a strong aid in answering them.

These basic ideas are quite simple. Ways of implementing them are not simple. In the chapters of Part I we lay the foundation for understanding and mastering multiple regression ways of implementing the ideas and answering the questions. We examine and study in detail simple linear regression, regression with only one independent variable, and then extend our study to multiple regression theory and practice and to the theory and practice of statistical control using what are known as partial correlation and semipartial correlation. The constant frame of reference is scientific research and research problems. Wherever possible, contrived and real research problems will be used to illustrate the discussion so that the student "thinks" research problems rather than being solely preoccupied with mathematics, arithmetic, and statistics.

The reader who first approaches the kinds of things we are going to do, even the reader who knows a good deal of statistics, is likely to feel a sense of mystery or magic about the procedures and the results. Of course there is no mystery or magic. We agree that the procedures sometimes seem magical, especially those of Part II, where we try to show how and why analysis of variance is done with multiple regression analysis. But they are quite straightforward; there is almost no aspect of them that cannot be revealed and understood, even though we will occasionally have to take things on faith. We suggest that the reader enter his study of the subject by agreeing to work through the examples. We will try to explain the steps and the reasoning that go into a problem. We will also try to reflect our own puzzlement and difficulties — still fresh in our minds — and thus perhaps help the student by taking his role and working from there.

Study Suggestions

1. It can be said that multiple regression and science are closely interrelated. Justify this statement. Can it be said that multiple regression is more realistic, closer to the "real" world, than simple regression? Why?
2. What does scientific "explanation" mean? How are explanation and multiple regression analysis related? Why was multiple regression generously used in the Coleman report, *Equality of Educational Opportunity*?
3. Why is it difficult to use analysis of variance when there are two or more attribute independent variables? What happens when a continuous variable like intelligence or authoritarianism is converted to a dichotomous or trichotomous variable? That is, what happens to the variance of the variable? Does multiple regression analysis avoid such difficulties? How?

Relations, Correlations, and Simple Linear Regression[1]

Relations and Correlations

A correlation is a relation. Since a relation is a set of ordered pairs, a correlation is a set of ordered pairs. Correlation means more than this, however. It also means the covarying of two variables. In Figure 2.1 a set of ordered pairs is given. According to the definition of relation as a set of ordered pairs, Figure 2.1 depicts a relation. The lines connecting the pairs of numbers in the figure indicate the ordered pairs.

As pointed out in Chapter 1, such a definition of relation, while general and unambiguous, cannot satisfy the scientist's need to "know" the relation. What is the nature of the relation? What is its direction? What is its magnitude? Do the two subsets of numbers, X and Y, show any systematic covariation? How do they "go along" with each other? In this simple example, it is clear that X and Y "go along" with each other. As X gets larger so does Y, with one exception. The usage of the word "correlation" is rather loose. It sometimes means the covarying of the numbers of the two subsets of the set of ordered pairs, as just described. It sometimes means the direction, positive or negative, and the magnitude of the relation, or what is called the coefficient of correlation. In the present case, for instance, the coefficient of correlation is .80.

A coefficient of correlation is an index of the direction and magnitude of a

---

[1]It was said in the Preface that elementary knowledge of statistics is assumed. A complete exposition of correlation would require a whole chapter. We limit ourselves to discussing only those aspects of correlation that throw light on regression theory and analysis. It is suggested that the student study concomitantly elementary correlation theory and the following statistical notions: variance, standard deviation, standard error of estimate, standard scores, and the assumptions behind correlation statistics.

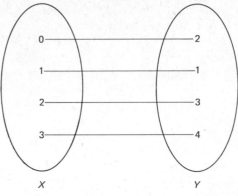

FIGURE 2.1

relation. The product-moment coefficient of correlation, r, is defined by several formulas, all of which are equivalent. Here are three of them, which we will occasionally use:

$$r_{xy} = \frac{\Sigma\, xy}{\sqrt{\Sigma\, x^2 \Sigma\, y^2}} \tag{2.1}$$

$$r_{xy} = \frac{\Sigma\, z_x z_y}{N} \tag{2.2}$$

$$r_{xy} = \frac{\Sigma\, xy}{N s_x s_y} \tag{2.3}$$

where x and y are deviations from the means of X and Y, or $x = X - \bar{X}$ and $y = Y - \bar{Y}$, where X and Y are raw scores and $\bar{X}$ and $\bar{Y}$ are the means of the X and Y subsets; z_x is a standard score of X, or $z_x = (X - \bar{X})/s_x = x/s_x$, where $s_x = $ standard deviation of X.

There are other ways to express relations. One of the best ways, too frequently neglected by researchers, is to graph the two sets of values. As we will see presently, graphing is important in regression analysis. A graph lays out a set of ordered pairs in a plot that can often tell the researcher not only about the direction and magnitude of the relation but also about the nature of the relation — for instance, linear or curvilinear — and something of deviant cases, cases that appear especially to "enhance" or "contradict" the relation. A graph is itself a relation because it is, in effect, a set of ordered pairs. Other ways to express relations are by crossbreaks or cross partitions of frequency counts, symbols, and diagrams. In short, the various ways to express relations are fundamentally the same: they are all ordered pairs.

To use and understand these ideas and others like them, we must now examine certain indispensable statistical terms and formulas: sums of squares and cross products and variances and covariances. We then return to correlation, and the interpretation of correlation coefficients. These ideas and tools will help us in the second half of the chapter when we formally begin the study of regression.

Sums of Squares and Cross Products

The *sum of squares* of any set of numbers is defined in two ways: by raw scores and by deviation scores. The raw score sum of squares, which we will often calculate, is ΣX_i^2, where $i = 1, 2, \ldots, N$, N being the number of cases. In the set of ordered pairs of Figure 2.1, $\Sigma X^2 = 0^2 + 1^2 + 2^2 + 3^2 = 14$, and $\Sigma Y^2 = 2^2 + 1^2 + 3^2 + 4^2 = 30$. The deviation sum of squares is defined:

$$\Sigma x^2 = \Sigma X^2 - \frac{(\Sigma X)^2}{N} \tag{2.4}$$

Henceforth we mean the deviation sum of squares when we say "sum of squares" unless there is possible ambiguity, in which case we will say "deviation sum of squares." The sums of squares of X and Y of Figure 2.1 are $\Sigma x^2 = 14 - 6^2/4 = 5$ and $\Sigma y^2 = 30 - 10^2/4 = 5$. Sums of squares will also be symbolized by ss, with appropriate subscripts.

The *sum of cross products* is the sum of the products of the ordered pairs of a set of ordered pairs. The raw score form is ΣXY, or $\Sigma X_i Y_i$, and the deviation score form is Σxy. The formula for the latter is

$$\Sigma xy = \Sigma XY - \frac{(\Sigma X)(\Sigma Y)}{N} \tag{2.5}$$

Again using the ordered pairs of Figure 2.1, we calculate $\Sigma XY = (0)(2) + (1)(1) + (2)(3) + (3)(4) = 19$, and $\Sigma xy = 19 - (6)(10)/4 = 4$. Henceforth, when we say "sum of cross products" we mean Σxy, or $\Sigma x_i x_j$.

Sums of squares and sums of cross products are the staples of regression analysis. The student must understand them thoroughly, be able to calculate them on a desk calculator routinely, and, hopefully, be able to program their calculation on a computer.[2] In other words, such calculations, by hand or by computer, must be second nature to the modern researcher. In addition, it is almost imperative that he know at least the basic elements of matrix algebra. The symbolism and manipulative power of matrix algebra make the often complex and laborious calculations of multivariate analysis easier and more comprehensible. The sums of squares and cross products, for example, can be expressed simply and compactly. If we let $\mathbf{X}$ be an N by k matrix of data, consisting of the measures on k variables of N individuals, then $\mathbf{X}_i \mathbf{X}_j$ expresses all the sums of squares and cross products. Or, for the deviation forms, one writes $\mathbf{x}_i \mathbf{x}_j$. A deviation sum of squares and cross products matrix, with $k = 3$, is given in Table 2.1. We will often encounter such matrices in this book.[3] For the most

[2] A computer program for doing multiple regression analysis, MULR, is given in Appendix C at the end of the book. The routines for calculating sums of squares and cross products are given in statements 300–305, 400–420, and 500–510. FORTRAN routines can be found in Cooley and Lohnes (1971, Chapters 1–2).

[3] The reader will profit from glancing at the first part of Appendix A on matrix algebra.

TABLE **2.1** DEVIATION SUMS OF SQUARES AND CROSS PRODUCTS
MATRIX, $k = 3$

$$\begin{pmatrix} \Sigma x_1{}^2 & \Sigma x_1 x_2 & \Sigma x_1 x_3 \\ \Sigma x_2 x_1 & \Sigma x_2{}^2 & \Sigma x_2 x_3 \\ \Sigma x_3 x_1 & \Sigma x_3 x_2 & \Sigma x_3{}^2 \end{pmatrix} = \mathbf{x}_i \mathbf{x}_j$$

part we will write statistical symbols, such as Σx^2, Σxy, and so on, rather than matrix symbols. Occasionally, however, matrix symbols will have to be used because writing out all the statistical symbols can become wearing and confusing. Moreover, certain conceptions and calculations are impossible without matrix algebra.

Variances and Covariances

A *variance* is the average of the squared deviations from the mean of a set of measures. Again using the numbers in Figure 2.1, the variance of the X subset is $V_x = \Sigma x^2 / N = 5/4 = 1.25$. The variance of the Y subset is $V_y = \Sigma y^2 / N = 5/4 = 1.25$.[4] (The *standard deviation* is of course the square root of the variance: $SD_x = \sqrt{\Sigma x^2 / N} = \sqrt{1.25} = 1.12$.) The magnitudes of variances and standard deviations, of course, express the variability of sets of scores.

The *covariance* is the variance of the intersection of the subsets, or $X \cap Y$. This is accomplished by calculating the arithmetic mean of the cross products, or $CoV_{xy} = \Sigma xy / N = 4/4 = 1$. [See the calculation done with equation (2.5), above.] The covariance expresses the relation between X and Y in another way. If we compare it to an average of the variance of X and Y, we get some notion of the "meaning" of the relation. The best way to do this amounts to another formula for the coefficient of correlation:

$$r_{xy} = \frac{CoV_{xy}}{\sqrt{V_x \cdot V_y}} \tag{2.6}$$

which is equivalent to formulas (2.1), (2.2), and (2.3).

Correlation and Common Variance

When the coefficient of correlation is squared, the resulting quantity is interpretable in a way that will later be useful. r_{xy}^2 expresses the variance shared in common by X and Y. It expresses, quantitatively, the intersection of the two subsets: $X \cap Y$.[5] In the above example, $r_{xy}^2 = .80^2 = .64$, which means, as we saw in Chapter 1, that 64 percent of the variance is common to both X and Y.

---

[4]For these simple demonstrations N is used in the denominators. Later, $N-1$ will be used. N is used to calculate *parameters* or population values. $N-1$ is used to calculate *statistics* or sample values because it yields an unbiased estimate of the parameter.

[5]"$\cap$" means "the intersection of." Thus, "$X \cap Y$" is read "X intersection Y," or "the intersection of X and Y." The intersection of sets X and Y consists of the elements common to both sets. The notion can be transferred to variance thinking: the variance common to both X and Y, or the covariance.

This can be conceptualized by altering formula (2.6) to

$$r^2_{xy} = \frac{(CoV_{xy})^2}{V_x \cdot V_y} \tag{2.7}$$

Substituting the values calculated earlier, we obtain: $r^2_{xy} = 1^2/(1.25 \cdot 1.25) = .64$. Analogous operations will occur again and again in subsequent discussions of simple and multiple regression and especially in discussions of multiple correlation.

Rather than using variances, which we have done for conceptual reasons, we will use deviation sums of squares because they are additive and permit us to state certain relations clearly and unambiguously. Moreover, their use will enable us to make a smooth transition to analysis of variance operations. The sums of squares formula analogous to formula (2.7) is

$$r^2_{xy} = \frac{(\Sigma\, xy)^2}{\Sigma\, x^2 \Sigma\, y^2} \tag{2.8}$$

Again substituting values calculated above, we obtain: $r^2_{xy} = 4^2/(5)(5) = .64$.

r^2 is called a *coefficient of determination* because it expresses the proportion of variance of Y "determined" by X, considering X the independent variable and Y the dependent variable. Since r is in effect a relation between standard scores the variance of Y is 1.00, and it is easy to see that the coefficient of determination is the proportion of the variance of Y determined by X. It is also easy to calculate the proportion of the variance of Y *not* accounted for by X: $1 - r^2_{xy} = 1 - .64 = .36$. This quantity is sometimes called the *coefficient of nondetermination* or *coefficient of alienation*. A better term, one used later, is *variance of the residuals*.

Let us clothe the statistical frame with research variables. Suppose X is authoritarianism and Y is ethnocentrism (prejudice). Then 64 percent of the variance of ethnocentrism (as measured) is determined by authoritarianism, and 36 percent is not. That is, in explaining ethnocentrism, we can say that 64 percent of its variance is shared by authoritarianism. Do not be misled. We do not say directly that X *causes* Y when we say "determined by." The meaning, more conservatively stated, is that the two variables share 64 percent of the total variance. And 36 percent of the variance of Y is not shared and is due, presumably, to other sources, other independent variables — and error. Despite these cautionary words, we will see later that it is necessary to come close to making causal inferences. This should not disturb us too much; much scientific work is cautious causal inference (see Blalock, 1964).

The notion of common variance is so important in correlation and regression theory and analysis that we explain the above example in a somewhat different way. What is done in any regression analysis, basically, is to explain the sources of variance of Y, the dependent variable. In the above case, two variables, X and Y, authoritarianism and ethnocentrism, are correlated .80. When this coefficient is squared, $.80^2 = .64$, an estimate of the proportion of Y determined by X is obtained. If, in addition, the total variance of Y is represen-

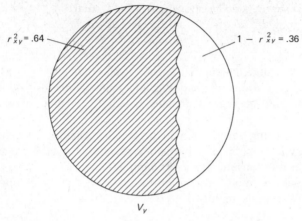

$r_{xy}^2 = .64$ $1 - r_{xy}^2 = .36$

V_y

FIGURE 2.2

ted by 1.00, then $1.00 - .64 = .36$ represents the proportion of variance of Y *not* accounted for by X. The situation is depicted in Figure 2.2. The whole circle represents the total variance of Y, V_y. The shaded portion represents that portion of the variance of Y that is determined by X. The unshaded portion is the remainder of the Y variance, the variance *not* determined by X, a residual variance, so to speak.

Spurious Correlation and Causation

Any student of elementary statistics has heard that one cannot infer causation from correlation. Many students in the behavioral sciences have had it drummed into them that they should not use or even think the word "cause" in scientific research. This is an extreme point of view. Scientists, after all, pursue causes and use causal reasoning. Even though, strictly speaking, the scientist cannot say that X causes Y, just as he cannot say that any scientific evidence "proves" anything, he assumes causal laws (Blalock, 1964, p. 12). Causal thinking is strictly theoretical, and empirical knowledge is probabilistic in nature. Nevertheless, we need not fear or eschew the word "cause." We must simply be careful with it, especially when working with and interpreting correlation.

Let us look at some examples of spurious correlation and lighten the discussion a bit. A famous example of a substantial spurious correlation is that supposedly calculated between the number of stork nests in areas in and around Stockholm, including rural areas, and the numbers of babies born in the areas. Suppose $r = .80$. Therefore storks bring babies! (Assuming that the correlation is actually .80, what might the explanation be?) Here is a more blatant example: The correlation between numbers of fire engines at fires and the damage caused is .90. Therefore the engines caused the damage!

An example about the authors of this book may be helpful. We were curious about the quality of our writing, but especially curious about its clarity.

We noticed that both of us smoked more when writing, one of us a pipe and the other cigarettes. [*Note*: Cigarette-smoking causes lung cancer; pipe-smoking causes lip and mouth cancer.] We had three judges rate for clarity on a five-point scale twenty samples of our writing, randomly selected from completed drafts of chapters of this book. We had also kept records of how much we smoked when writing the chapters. The correlation between rated clarity and amount of smoking was .74. Therefore . . .

These examples are fairly obvious. Let us take one or two examples of *possibly* spurious correlation from actual research data. Wilson (1967) found that Negro students from integrated schools did better than Negro students from segregated schools. This relation, however, was evidently spurious because when Wilson took account of the students' primary school cognitive development (measured by a mental maturity test in the first grade) and home influence (for example, number of objects in the home), the relation virtually disappeared. In other words, it was the mental maturity and the home environment that presumably affected achievement and not whether schools were integrated.

An even more difficult problem emerges from *Equality of Educational Opportunity* (Coleman et al., 1966, Section 9.10, pp. 91 and 119). Correlations were calculated between families owning an encyclopedia and the verbal ability of teachers. In a Northern Negro sample the correlation was .46, whereas in a Northern white sample it was −.05, virtually zero. If one were told that the correlation was zero, one would hardly be surprised. But with a Negro sample the correlation, especially for these kinds of variables, was substantial.[6] Clearly, one would hardly talk about a causal relation between parents owning encyclopedias and the verbal ability of teachers. But is there another variable, or more than one variable, influencing both of these variables? Evidently there must be. It may be social class. But if it were, why is the correlation near zero among white students?

The point of this example is twofold: to show that a correlation between two variables can be substantial when in fact there is probably no direct relation and certainly not a causal relation between the variables, and to show, too, that a correlation that is substantial in one sample may be zero in another sample — which of course may be a clue to the underlying relation.

Simple Linear Regression

The notion of regression is of course close to that of correlation. Indeed, the *r* used to indicate the coefficient of correlation really means regression. It is said that we study the regression of *Y* scores on *X* scores. How do *Y* scores "go back to," how do they "depend upon," the *X* scores? Galton, who was evidently

[6]In the Southern Negro sample, *r* = .59, and in the total Negro sample, *r* = .64, leaving little doubt as to the "reality" of the correlation. The Southern white *r* was .13, and the total white *r* was .03. Moreover, among Puerto Ricans and Mexican Americans, the *r*'s were .53 and .52.

the first to work out the idea of correlation, got the idea from the notion of "regression toward mediocrity," a phenomenon observed in studies of inheritance. Tall men tend to have shorter sons, and short men taller sons. The sons' heights tend to "regress," or "go back to," the mean of the population.

Statistically, if we want to predict Y from X and the correlation between X and Y is zero, then the best prediction is to the mean. For any given X, say X_5, we can only predict to the mean of Y. The higher the correlation, however, the better the prediction. If $r = 1.00$, then prediction is perfect. To the extent the correlation decreases from 1.00, to that extent predictions from X to Y are less than perfect and "regress" to the mean. We say, therefore, that if $r = 0$, the "best," and only, prediction is to the mean. If the X and Y values are plotted when $r = 1.00$, they will all lie on a straight line. The higher the correlation, whether positive or negative, the closer the plotted values will be to the regression line.

Two Examples: Correlation Present and Absent

To explain and illustrate statistical regression, we use two fictitious examples with simple numbers.[7] The examples are given in Table 2.2. Note immediately that the numbers of both examples are exactly the same; they are only arranged differently. In the example on the left, labeled A, the correlation between the X and Y values is .90, while in the example on the right, labeled B, the correlation is 0. Certain calculations necessary for regression analysis are also given in the table: the sums and means, the deviation sums of squares of X and Y ($\Sigma x^2 = \Sigma X^2 - (\Sigma X)^2/n$), the deviation cross products ($\Sigma xy = \Sigma XY - (\Sigma X)(\Sigma Y)/n$), and certain regression statistics to be explained shortly.

First, note the difference between the A and B sets of scores. They differ only in the order of the scores of the second or X columns. These different orders produce very different correlations between the X and Y scores. In the A set, $r = .90$, and in the B set, $r = .00$. Second, note the statistics at the bottom of the table. Σx^2 and Σy^2 are the same in both A and B, but Σxy is 9 in A and 0 in B.

The basic equation of simple linear regression is[8]

$$Y' = a + bX \qquad (2.9)$$

---

[7] These examples are taken from Kerlinger (1964, pp. 250–251), where they were used to illustrate the effect on sums of squares and the F test of correlation between experimental groups.

[8] Since most of the reasoning and analysis in this book is based on linear relations, a definition of "linear," or "linear relation," or "linear regression," is appropriate. *Linear*, of course, relates to straight lines. A linear equation is one in which the highest degree term in the variables is of the first degree. The equation $Y = a + 2X$ is of the first degree because X is to the first power. This expresses a linear relation. If different values of X and Y are plotted they will form a straight line. On the other hand, $Y = a + 2X^2$ is an equation of the second degree. It expresses a nonlinear relation. A linear regression simply means a regression whose equation is of the first degree. It is important to know if a relation is linear or nonlinear. Linear methods applied to nonlinear data will yield misleading results. Actually, we deal with nonlinear relations later, and, as we will see, the problem is not a difficult one — at least as far as we will treat it. For a thorough discussion of kinds of relations (or "functions") and equations, see Guilford (1954, Chapter 3).

TABLE **2.2**   REGRESSION ANALYSIS OF CORRELATED AND
UNCORRELATED SETS OF SCORES

(A)		$r = .90$			(B)		$r = .00$		
Y	X	XY	Y'	d	Y	X	XY	Y'	d
1	2	2	1.2	$-.2$	1	5	5	3	-2
2	4	8	3.0	-1.0	2	2	4	3	-1
3	3	9	2.1	.9	3	4	12	3	0
4	5	20	3.9	.1	4	6	24	3	1
5	6	30	4.8	.2	5	3	15	3	2
Σ: 15	20	69		0	15	20	60		0
M: 3	4				3	4			
Σ^2: 55	90			$\Sigma d^2 = 1.90$					$\Sigma d^2 = 10.00$

$$\Sigma y^2 = 55 - \frac{(15)^2}{5} = 10 \qquad\qquad \Sigma y^2 = 55 - \frac{(15)^2}{5} = 10$$

$$\Sigma x^2 = 90 - \frac{(20)^2}{5} = 10 \qquad\qquad \Sigma x^2 = 90 - \frac{(20)^2}{5} = 10$$

$$\Sigma xy = 69 - \frac{(15)(20)}{5} = 9 \qquad\qquad \Sigma xy = 60 - \frac{(15)(20)}{5} = 0$$

$$b = \frac{\Sigma xy}{\Sigma x^2} = \frac{9}{10} = .90 \qquad\qquad b = \frac{0}{10} = 0$$

$$a = \overline{Y} - b\overline{X} = 3 - (.90)(4) = -.60 \qquad a = 3 - (0)(4) = 3$$

$$Y' = a + bX = -.60 + .90X \qquad\qquad Y' = 3 + (0)X$$

where Y' = predicted scores of the dependent variable; X = scores of the independent variable; a = intercept constant; and b = regression coefficient. As we saw earlier, a regression equation is a prediction formula: Y values are predicted from X values. The correlation between the observed X and Y values determines how the prediction "works." The intercept constant, a, and the regression coefficient, b, will be explained shortly.

The two sets of X and Y values of Table 2.2 are plotted in Figure 2.3. The two plots are quite different. Lines have been drawn to "run through" the plotted points. If there was a way of placing these lines so that they would simultaneously be as close to all the points as possible, then the lines should express the relation between X and Y, the regression of Y on X.[9] The method for placing the lines is part of regression analysis. The line in the top plot, where $r = .90$, runs close to the plotted XY points. In the bottom plot, where $r = .00$, it

---

[9] A complete discussion of regression would include the study and meaning of the regression of X on Y, as well as the regression of Y on X. We will not go this far because our purposes do not require doing so. We will always be talking of Y as a dependent variable and there is little point to enlarging the discussion. The student, in his study of elementary statistics, will have learned that the two regressions are usually different. See Hays (1963, Chapter 15) for a complete discussion.

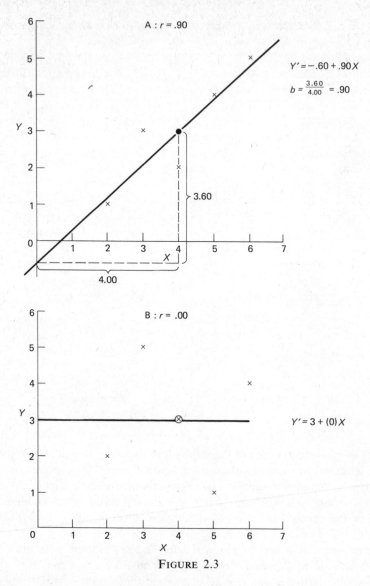

FIGURE 2.3

is not possible to run the line close to all the points or to most of them. Since $r =$ .00, the points, after all, are in effect scattered randomly.

    The *slope*, b, indicates the change in Y with a change of one unit of X. In Example A, we predict a change of .90 in Y with a change of 1 unit in X. The slope can be expressed trigonometrically: it is the length of the line opposite the angle made by the regression line with the X axis, divided by the length of the line adjacent to the angle. In Figure 2.3, if we drop a perpendicular from the little circle on the regression line, the point where the X and Y means intersect, to a line drawn horizontally from the point where the regression line intersects the Y axis, or at $Y = -.60$, then $3.60/4.00 = .90$. A change of 1 in X means a change of .90 in Y.

The plot of the X and Y values of Example B, bottom of Figure 2.3, is very different. In A, one can rather easily and visually draw a line through the points and achieve a fairly accurate approximation to the actual regression line. But in B this is hardly possible. The line can be drawn using other guidelines, which we discuss shortly. It is important to note, too, the scatter or dispersion of the plotted points around the two regression lines. In A, they cling closely to the regression line. If $r = 1.00$, they would all be on the line, as we said before. When $r = .00$, on the other hand, they scatter widely about the line. The lower the correlation, the greater the scatter.

To calculate the regression statistics of the two examples, we must calculate the deviation sums of squares and cross products. This has been done at the bottom of Table 2.2. The formula for the *slope*, or *regression coefficient*, b, is

$$b = \frac{\Sigma xy}{\Sigma x^2} \qquad (2.10)$$

The two b's are .90 and .00. In Figure 2.3, Example A, b has been calculated as the tangent of the angle: side opposite over side adjacent, as explained earlier. The *intercept constant*, a, is calculated with the formula

$$a = \bar{Y} - b\bar{X} \qquad (2.11)$$

The a's for the two examples are $-.60$ and 3; for example, for Example A, $a = 3 - (.90)(4) = -.60$. The intercept constant is the point where the regression line intercepts the Y axis. To draw the regression line, lay a ruler between the intercept constant on the Y axis and the point where the mean of Y and the mean of X intersect. (In Figure 2.3, these points are indicated with small circles.)

The final steps in the process, at least as far as it will be taken here, are to write regression equations and then, using the equations, to calculate the predicted values of Y, or Y', given the X values. The two equations are given in the last line of Table 2.2. First look at the regression equation for $r = .00$: $Y' = 3 + (0)X$. This means, of course, that all the predicted Y's are 3, the mean of Y. When $r = 0$, the best prediction is the mean, as indicated earlier. When $r = 1.00$, at the other extreme, the reader can see that one can predict exactly: one simply predicts the Y score corresponding to the X score. When $r = .90$, prediction is less than perfect and one predicts Y' values calculated with the regression equation. For example, to predict the first Y' score, we calculate

$$Y'_1 = -.60 + (.90)(2) = 1.20$$

The predicted scores of the A and B sets have been given in Table 2.2. (See columns labeled Y'.) Note an important point: If, for Example A, we plot the X and the predicted Y, or Y', values, the plotted points all lie on the regression line. That is, the regression line of the figure represents the set of predicted Y

values, given the X values and the correlation between the X and the observed Y values.

The higher the correlation the more accurate the prediction. The accuracy of the predictions of the two sets of scores can be clearly shown by calculating the differences between the original Y values and the predicted Y values, or $Y - Y' = d$, and then calculating the sums of squares of these differences. Such differences are called *residuals*. In Table 2.3, the two sets of residuals and their sums of squares have been calculated (see columns labeled d). The two values of Σd^2, 1.90 for A and 10.00 for B, are quite different, just as the plots in Figure 2.3 are quite different: that of the B, or $r = .00$, set is much greater than that of the A, or $r = .90$, set. That is, the higher the correlation, the smaller the deviations from prediction and thus the more accurate the prediction.

Simple Regression, Analysis of Variance, and Tests of Significance

Tests of significance for both simple and multiple regression are similar to those of analysis of variance. Therefore, in addition to describing such tests, we take the opportunity to lay a foundation for later developments. In Part II of the book, we will show the close relation between multiple regression and analysis of variance. Some of the basic conceptualization and statistics, how-ever, can be introduced now and extended in both Parts I and II.

In analysis of variance, the total variance of a set of dependent variable measures can be broken down into systematic variance and error variance. The simplest form of such a breakdown is: the variance between groups (experimental variance) and the variance within groups (error variance), which are parts of the total variance. Actually, statisticians work with sums of squares because they are additive. In regression analysis, we do virtually the same thing. The main difference is that the regression approach is more general: It fits and is applicable to most research problems with one dependent variable.

A basic equation of analysis of variance is

$$ss_t = ss_b + ss_w$$

where ss_t = total sum of squares; ss_b = between groups sum of squares; and ss_w = within groups sum of squares. The transition to regression analysis is direct. We write

$$ss_t = ss_{reg} + ss_{res} \qquad (2.12)$$

where ss_t = total sum of squares of Y; ss_{reg} = sum of squares of Y due to regression; and ss_{res} = sum of squares of the residuals, or the deviations from regression.

Before giving the regression formulas for the different sums of squares, we calculate these sums of squares using the data and statistics of Table 2.2. The total sums of squares, ss_t, are found simply by calculating the sums of squares of the Y columns of Table 2.2: $\Sigma y_t^2 = \Sigma Y^2 - (\Sigma Y)^2/N = (1^2 + 2^2 + 3^2 + 4^2 + 5^2) -$

$15^2/5 = 55 - 45 = 10$, for both A and B. The sums of squares for the Y' columns are $(1.2^2 + 3.0^2 + 2.1^2 + 3.9^2 + 4.8^2) - 15^2/5 = 53.10 - 45 = 8.10$, for A, and $(3^2 + 3^2 + 3^2 + 3^2 + 3^2) - 45 = 0$, for B. We now repeat the symbolic equation and follow it with the numerical values of A and B.

$$ss_t = ss_{reg} + ss_{res}$$

A: $10 = 8.10 + 1.90$

B: $10 = 0 + 10$

This is the foundation of most further developments. We have the total sum of squares of Y, the dependent variable measures, the sum of squares of Y due to regression, which is analogous to the between groups sum of squares, and the residual sum of squares, which is analogous to the within groups or error sum of squares. Actually, then, analysis of variance and multiple regression analysis are virtually the same. If this is so, then we should also be able to calculate and interpret F ratios and statistical significance for the regression as in the analysis of variance. The formula in one-way analysis of variance is

$$F = \frac{ss_b/df_1}{ss_w/df_2}$$

where df_1 = degrees of freedom associated with ss_b, and df_2 = degrees of freedom associated with ss_w. Similarly, the formula in regression analysis is

$$F = \frac{ss_{reg}/df_1}{ss_{res}/df_2} \tag{2.13}$$

The degrees of freedom are $df_1 = k$, where $k = 1$, and $df_2 = N - k - 1 = 5 - 1 - 1 = 3$. So

$$F = \frac{8.10/1}{1.90/3} = \frac{8.10}{.633} = 12.80$$

which is significant at the .05 level. (See Appendix D for a table of the F distribution.) Thus we can say that, in Example A, the regression of Y on X is statistically significant.

The test of significance can be done in two or three other ways. One, the correlation of .90 can be checked for significance in a table of r's significant at various levels. (See, for example, Fisher & Yates, 1963, p. 63, Table VII.) In the present case, $r = .90$ is significant at the .05 level. This approach, however, is not helpful when we want to do similar tests with more than one independent variable. The second way is to use the t test (Snedecor & Cochran, 1967, pp. 184–185). Such a test can be done in two ways: by testing the significance of the regression coefficient or by testing the significance of r directly. Actually, the two tests amount to the same thing.

In using the F test, above, the sums of squares were calculated from the values of Table 2.2. Another way, a more useful one as we will see later, is to calculate the total sum of squares of Y from the observed values of Y and then calculate the regression sum of squares with the following formula:

$$ss_{\text{reg}} = \frac{(\Sigma xy)^2}{\Sigma x^2} \qquad (2.14)$$

The formula requires the cross products of deviation scores, $x = X - \bar{X}$. The cross products of the X and Y scores are given in Table 2.2. Their sum, for the A data, is 69. Thus, $\Sigma xy = \Sigma XY - (\Sigma X)(\Sigma Y)/N = 69 - (15)(20)/5 = 9$. Now,

$$ss_{\text{reg}} = \frac{9^2}{10} = \frac{81}{10} = 8.10$$

The residual sum of squares is obtained by subtraction:

$$ss_{\text{res}} = ss_t - ss_{\text{reg}}$$
$$= 10 - 8.10 = 1.90$$

These values, of course, agree with those calculated earlier. Still another method, one that can be conveniently used with multiple regression, will be taken up later.

Standard Scores and Regression Weights

Although we will, in this book, emphasize the use of raw scores and the analysis of variance type of statistics — sums of squares, mean squares, F ratios, for example — we must also study and be quite aware of standard scores and their use in regression theory and analysis. Furthermore, we should also be aware of the difference between correlation and regression.

Standard Scores

A standard score, we recall, is a standard deviation score. If we divide deviations from the mean, $x = X - \bar{X}$, by the standard deviation of the set of scores, s, we obtain standard scores. Here is the formula:

$$z_x = \frac{X - \bar{X}}{s_x} = \frac{x}{s_x} \qquad (2.15)$$

where z_x = standard score, X = a raw score, $\bar{X}$ = mean of X scores, s_x = standard deviation of the set of X scores, and $x = X - \bar{X}$ (deviation scores).[10] Standard scores, as defined by formula (2.15), have a mean of 0 and a standard deviation of 1.

It is possible and theoretically satisfying to use standard scores in developing regression analysis. Hays (1963, Chapter 15) does so almost exclusively. Snedecor and Cochran (1967, Chapters 6 and 13) do not. In the present book, we use raw scores and deviation scores for the most part because most research uses of multiple regression do so. The researcher must be able to use both. The

---

[10]Standard scores are *not* normalized scores. That is, formula (2.15), above, does not change the shape of the distribution of the scores. Normalized scores are scores whose distribution has been made normal by a special operation. z scores are simply linear transformations of raw scores.

major purpose of this section is simply to introduce standard scores so that our next subject, regression weights, can be clarified. Later in the book the differences between the two kinds of scores will be further clarified.

Regression Weights: b and β

There are two, even three, kinds of regression weights. In theoretical treatments, β (beta) is the population regression weight which is unknown. The sample regression weight, b, is considered to be an estimate of β. Earlier we wrote the regression equation as $Y' = a + bX$. The population form of this equation is

$$Y' = \alpha + \beta X \tag{2.16}$$

where α (alpha) = mean of population corresponding to $X = 0$, and β = regression weight in the population, or the slope of the regression line. In short, β is the population regression coefficient, which is not known and must be estimated with fallible data. The estimate of β is b, which is calculated by formula (2.10). For the data of A in Table 2.3, $b = 9/10 = .90$. This use of b and β need not detain us; there is another usage to which we turn, if only briefly.

In this second usage, b is defined as in formula (2.10), and β is defined

$$\beta = b \frac{s_x}{s_y} \tag{2.17}$$

where s_y = standard deviation of the Y scores, and s_x = standard deviation of the X scores. We see, then, that another formula for b, when β is known, is

$$b = \beta \frac{s_y}{s_x} \tag{2.18}$$

With only one independent variable, formula (2.18) can be written

$$b = r_{xy} \frac{s_y}{s_x} \tag{2.19}$$

Thus $\beta = r_{xy}$ (when only X and Y are involved). It is not in general true, however, that b equals r. In the case of the data of Table 2.3, b is equal to r (.90) only because $s_y = s_x$.

Formulas (2.17), (2.18), and (2.19) tell us something more about the relation between b and β. β is the regression weight in standard score form. That is, if we first calculate standard scores and then use formula (2.10), suitably altered in symbols, $\Sigma z_x z_y / \Sigma z_x^2$, we will obtain β. All this can be shown much better with a simple example. In Table 2.3, we have taken the A data of Table 2.2 and altered the X scores slightly (lowered the first score by 1 and raised the fifth score by 1) so that the standard deviations of X and Y are different. The calculations for correlation and regression statistics are included in the table. The deviation scores, $y = Y - \bar{Y}$ and $x = X - \bar{X}$, are given beside the raw scores. The z scores, calculated with formula (2.15), are also given. The sums, means, sums of squares, and standard deviations are given immediately below the raw

TABLE **2.3**   FICTITIOUS DATA AND CORRELATION AND REGRESSION
CALCULATIONS TO SHOW RELATION BETWEEN r, b, AND β

Y	y	z_y	X	x	z_x
1	-2	-1.4142	1	-3	-1.5000
2	-1	$-.7071$	4	0	0
3	0	0	3	-1	$-.5000$
4	1	$.7071$	5	1	$.5000$
5	2	1.4142	7	3	1.5000

$$\Sigma Y = 15 \qquad\qquad \Sigma X = 20$$
$$\bar{Y} = 3 \qquad\qquad \bar{X} = 4 \qquad\qquad \Sigma XY = 73$$

$$\Sigma Y^2 = 55 \qquad\qquad \Sigma X^2 = 100 \qquad\qquad \Sigma xy = 73 - \frac{(15)(20)}{5} = 13$$

$$\Sigma y^2 = 10 \qquad\qquad \Sigma x^2 = 20 \qquad\qquad \Sigma z_x z_y = 4.5962$$
$$s_y = 1.5811 \qquad\qquad s_x = 2.2361$$

(a) $r_{xy} = \dfrac{\Sigma xy}{\sqrt{\Sigma x^2 \, \Sigma y^2}} = \dfrac{13}{\sqrt{(20)(10)}} = .9192$

(b) $r_{z_x z_y} = \dfrac{\Sigma z_x z_y}{N} = \dfrac{4.5962}{5} = .9192$

(c) $b = \dfrac{\Sigma xy}{\Sigma x^2} = \dfrac{13}{20} = .65$

(d) $\beta = b \dfrac{s_x}{s_y} = (.65)\left(\dfrac{2.2361}{1.5811}\right) = .9192$

(e) $\beta = \dfrac{\Sigma z_x z_y}{\Sigma z_x^2} = \dfrac{4.5962}{5.0000} = .9192$

scores and the deviation and z scores. The sums of the cross products of x and y and of z_x and z_y, $\Sigma z_x z_y$, are also given

The important calculations to show the relations of r, b, and β are given at the bottom of the table. In line (a), r_{xy} is calculated: .9192. r is also calculated with the standard score formula, which is

$$r_{xy} = \frac{\Sigma z_x z_y}{N} \tag{2.20}$$

Substituting $\Sigma z_x z_y = 4.5962$ and $N = 5$ in this formula yields, of course, .9192. The regression coefficient, b, is calculated in line (c) by formula (2.10), $b = .65$. In line (d), β is calculated with formula (2.17): it is .9192. Thus, we show that $r_{xy} = \beta$. Finally, in line (e), β is also calculated with the z scores, using a formula analogous to formula (2.10):

$$\beta = \frac{\Sigma z_x z_y}{\Sigma z_x^2} \tag{2.21}$$

This simple demonstration shows the relations between r, b, and β rather clearly. First, with one independent variable $r_{xy} = \beta$. Second, β is the regression coefficient that is used with standard scores. It is, in other words, in standard score form. The regression equation, in standard score form, is

$$z'_y = \beta z_x \qquad (2.22)$$

(Note that an intercept constant, a, is not needed since the mean of z scores is 0.) In later chapters, we will find that β weights become important both in the calculation of multiple regression problems and in the interpretation of data.

Study Suggestions

1. The student will do well to study simple regression from a standard text. The following two chapters are excellent — and quite different: Hays (1963, Chapter 15), and Snedecor and Cochran (1967, Chapter 6). While these two chapters are somewhat more difficult than certain other treatments, they are both worth the effort.
2. What is a relation? What is a correlation? How are the two terms alike? How are they different? Does a graph express a relation? How?
3. $r = .80$ between a measure of verbal achievement and an intelligence test. Explain what this means. Square r, $r^2 = (.80)^2 = .64$. Explain what this means. Why is r^2 called a coefficient of determination?
4. Explain the idea of residual variance. Give a research example (that is, make one up).
5. In the text, an example was given in which $r = .80$ between the numbers of stork nests in the Stockholm area and the numbers of babies born in the area. Explain how such a high and obviously spurious correlation can be obtained. Suppose in a study the correlation between racial integration in schools and achievement of black children was .60. Is it possible for this correlation to be spurious? How?
6. What does the word "regression" mean in common sense terms? What does it mean statistically? How are the concepts "regression" and "prediction" related?
7. What does the regression coefficient, b, mean? Use Figure 2.3 to illustrate your explanation. Clothe your explanation with the variables intelligence and verbal achievement. Do the same with the variables air pollution and respiratory disease.
8. Here are two regression equations. Assume that $X =$ intelligence and $Y =$ verbal achievement. Interpret both equations (ignore the intercept constant).

$$Y' = 4 + .3X$$
$$Y' = 4 + .9X$$

Suppose a student has $X = 100$. What is his predicted Y score, using the first equation? the second equation?
(*Answers:* 34 and 94.)
9. Here are a set of X and a set of Y scores (the second, third, and fourth pairs of columns are simply continuations of the first pair of columns):

X	Y	X	Y	X	Y	X	Y
2	2	4	4	4	3	9	9
2	1	5	7	3	3	10	6
1	1	5	6	6	6	9	6
1	1	7	7	6	6	4	9
3	5	6	8	8	10	4	10

Calculate the means, sums of squares and XY cross products, standard deviations, the correlation between X and Y, and the regression equation. Think of X as authoritarianism and Y as ethnocentrism (prejudice). Interpret the results including the square of r_{xy}.

(*Answers:* $\bar{X} = 4.95$; $\bar{Y} = 5.50$; $s_x = 2.6651$; $s_y = 2.9469$; $\Sigma x^2 = 134.95$; $\Sigma y^2 = 165.00$; $\Sigma xy = 100.50$; $r_{xy} = .6735$; $r_{xy}^2 = .4536$

$$Y' = a + bX$$
$$Y' = 1.8136 + .7447X)$$

Elements of Multiple Regression Theory and Analysis: Two Independent Variables

We are now ready to extend regression theory and analysis to more than one independent variable. In this chapter two independent variables are considered. In Chapter 4, the theory and method are extended to any number of independent variables. The advantage in separating the discussion of two and three or more variables lies in the relatively simple ideas and calculations with only two independent variables. With three or more variables, complexities of conceptualization and calculation arise that can impede understanding. The study of regression analysis with two variables should give us a firm foundation for understanding the k-variable case.

The Basic Ideas

In Chapter 2, a basic equation of simple linear regression was given by equation (2.9). The equation is repeated here with a new number. (For the convenience of the reader, we will follow this procedure of repeating equations but attaching new numbers to them.)

$$Y' = a + bX \tag{3.1}$$

where Y' = predicted Y (raw) scores, a = intercept constant, b = regression coefficient or weight, and X = raw scores of an independent variable.[1] The

---

[1]Throughout this text we use capital letters to designate variables in raw score form; for example, X and Y. Scores in deviation form, or $X - \bar{X}$, where $\bar{X}$ = the mean of a set of X scores, will be designated by small letters: x and y. If standard scores are meant, we will use z with appropriate subscripts. Subscript notation will be explained as we go along.

equation means that knowing the values of the constants a and b, we predict from X to Y.

The idea can be extended to any number of independent variables or X's:

$$Y' = a + b_1X_1 + b_2X_2 + \cdots + b_kX_k \tag{3.2}$$

where $b_1, b_2, \ldots, b_k$ are regression coefficients associated with the independent variables $X_1, X_2, \ldots, X_k$. Here we predict from the X's to Y using a and the b's.

The Principle of Least Squares

In simple regression, the calculation of a and b is easy. In all regression problems, simple and multiple, the *principle of least squares* is used. In any prediction of one variable from other variables there are errors of prediction. All scientific data are fallible. The data of the behavioral sciences are considerably more fallible than most data of the natural sciences. This really means that errors of prediction are larger and more conspicuous in analysis. In a word, error variance is larger. The principle of least squares tells us, in effect, to so analyze the data that the squared errors of prediction are minimized.

For any group of N individuals, there are N predictions, Y_i' (i runs from 1 through N). That is, we want to predict, from the X's we have, the X's observed, the Y scores of all the individuals. In doing so, we will be more or less in error. The principle of least squares tells us to calculate the predicted Y's so that the squared errors of prediction are a minimum. In other words, we want to minimize $\Sigma(Y_i - Y_i')^2$, where $i = 1, 2, \ldots, N$, $Y_i =$ the set of observed dependent variable scores, and $Y_i' =$ the set of predicted dependent variable scores. If the reader will turn back to Table 2.2 of Chapter 2, he will find that the fourth and fifth columns of examples A and B, the Y' and d columns, express the notions under discussion. The text of Chapter 2 explained the calculation of the sum of squares of the deviations from prediction, ss_{res}, or Σd^2. It is this sum of squares that is minimized.

It is not necessary to discuss the least squares principle mathematically and in detail. It is sufficient for the purpose at hand if we get a firm intuitive grasp of the principle.[2] The idea is to calculate a, the intercept constant, and b, the regression coefficient, to satisfy the principle. In Chapter 2, the following formula was used to calculate a:

$$a = \bar{Y} - b\bar{X} \tag{3.3}$$

The constant calculated with this formula helps to reduce errors of prediction. In multiple regression the formula for a is merely an extension of formula (3.3):

$$a = \bar{Y} - b_1\bar{X}_1 - \cdots - b_k\bar{X}_k \tag{3.4}$$

This formula will take on more meaning later when the data of a problem are analyzed.

---

[2]See Hays (1963, pp. 496–499) for a good discussion.

One of the main calculation problems of multiple regression is to solve equation (3.2) for the b's, the regression coefficients. With only two independent variables, the problem is not difficult. We show how it is done later in this chapter. With more than two X's, however, it is considerably more difficult. The method is discussed in Chapter 4. We will first take a fictitious problem with two independent variables, going through all the calculations. Second, we will interpret as much of the analysis as we are able to at this point.[3] It is important to point out before going further that the principles and interpretations discussed with two independent variables apply to problems with more than two independent variables.

An Example with Two Independent Variables

Suppose we have reading achievement, verbal aptitude, and achievement motivation scores on 20 eighth-grade pupils. (There will, of course, usually be many more than 20 subjects.) We want to calculate the regression of Y, reading achievement, on *both* verbal aptitude *and* achievement motivation. We already know that since the correlation between verbal aptitude and reading achievement is substantial we can predict reading achievement from verbal aptitude rather well. We wonder, however, whether we can substantially improve the prediction if we know something of the pupils' achievement motivation. Certain research (for example, McClelland et al., 1953) has indicated that achievement motivation may be useful in predicting school achievement. We decide to use both verbal aptitude and achievement motivation measures.

Calculation of Basic Statistics

Suppose the measures obtained for the 20 pupils are those given in Table 3.1.[4] In order to do a complete regression analysis, a number of statistics must be calculated. The sums, means, and sums of squares of raw scores on the three sets of scores are given in the three lines directly below the table. In addition, however, we will need other statistics: the deviation sums of squares of the three variables, their deviation cross products, and their standard deviations.

---

[3]In Chapter 4, alternative and more efficient methods of calculation than those used in this chapter will be described and illustrated. The calculations used in this chapter are a bit clumsy. They have the virtue, however, of helping to make the basic ideas of multiple regression rather clear.

[4]We "contrived" to have these and other fictitious scores in this chapter "come out" approximately as we wanted them to. We could simply have used correlation coefficients, an easier alternative. Doing so, however, would have deprived us of certain advantageous opportunities for learning. Although we have tried to make our examples as realistic as possible—that is, empirically and substantively plausible and close to results obtained in actual research—we have not always been completely successful. Moreover, as we indicated earlier, we wanted to use only simple numbers and very few of them. Such constraints sometimes make it difficult for examples to "come out right."

They are calculated below:

$$\Sigma y^2 = \Sigma Y^2 - \frac{(\Sigma Y)^2}{N} = 770 - \frac{(110)^2}{20} = 770 - 605 = 165.00$$

$$\Sigma x_1^2 = \Sigma X_1^2 - \frac{(\Sigma X_1)^2}{N} = 625 - \frac{(99)^2}{20} = 625 - 490.05 = 134.95$$

$$\Sigma x_2^2 = \Sigma X_2^2 - \frac{(\Sigma X_2)^2}{N} = 600 - \frac{(104)^2}{20} = 600 - 540.80 = 59.20$$

$$\Sigma x_1 y = \Sigma X_1 Y - \frac{(\Sigma X_1)(\Sigma Y)}{N} = 645 - \frac{(99)(110)}{20} = 645 - 544.50 = 100.50$$

$$\Sigma x_2 y = \Sigma X_2 Y - \frac{(\Sigma X_2)(\Sigma Y)}{N} = 611 - \frac{(104)(110)}{20} = 611 - 572 = 39.00$$

$$\Sigma x_1 x_2 = \Sigma X_1 X_2 - \frac{(\Sigma X_1)(\Sigma X_2)}{N} = 530 - \frac{(99)(104)}{20} = 538 - 514.80 = 23.20$$

$$s_y = \sqrt{\frac{\Sigma y^2}{N-1}} = \sqrt{\frac{165}{20-1}} = \sqrt{8.6842} = 2.9469$$

$$s_{x_1} = \sqrt{\frac{\Sigma x_1^2}{N-1}} = \sqrt{\frac{134.95}{20-1}} = \sqrt{7.1026} = 2.6651$$

$$s_{x_2} = \sqrt{\frac{\Sigma x_2^2}{N-1}} = \sqrt{\frac{59.20}{20-1}} = \sqrt{3.1158} = 1.7652$$

These statistics are staples of multivariate analysis and are almost always calculated by computer programs. We pull the results of the calculations together
for visual convenience in Table 3.2. Since the correlations between the variables will be needed later, we have inserted them below the principal diagonal
of the matrix (.6735, .3946, and .2596).

There is more than one way to calculate the essential statistics of multiple
regression analysis. Ultimately we will cover most of them. Now, however, we
concentrate on calculations that use sums of squares. Sums of squares have the
virtues of being additive and intuitively comprehensible. In addition, they
spring directly from the data. Their use also enables us to keep our discussion
closely related to analysis of variance procedures and calculations.

Reasons for the Calculations

Before proceeding with the calculations we need to review why we are doing
all this. First, we want to fill in the constants of the prediction equation, $Y' = a + b_1 X_1 + b_2 X_2$. That is, we have to calculate a and b_1 and b_2 so that we can, if
we wish, use the X's of individuals and predict the Y's. This means, in our
example, that if we have scores of individuals on verbal aptitude and achieve-

Y	X_1	X_2	Y'	$Y-Y'=d$
2	2	4	3.0305	-1.0305
1	2	4	3.0305	-2.0305
1	1	4	2.3534	-1.3534
1	1	3	1.9600	$-.9600$
5	3	6	4.4944	.5056
4	4	6	5.1715	-1.1715
7	5	3	4.6684	2.3316
6	5	4	5.0618	.9382
7	7	3	6.0226	.9774
8	6	3	5.3455	2.6545
3	4	5	4.7781	-1.7781
3	3	5	4.1010	-1.1010
6	6	9	7.7059	-1.7059
6	6	8	7.3125	-1.3125
10	8	6	7.8799	2.1201
9	9	7	8.9504	.0496
6	10	5	8.8407	-2.8407
6	9	5	8.1636	-2.1636
9	4	7	5.5649	3.4351
10	4	7	5.5649	4.4351

Σ: 110   99   104
M: 5.50   4.95   5.20
Σ^2: 770.   625.   600. $\Sigma\, d^2 = 81.6091$

	y	x_1	x_2
y	165.00	100.50	39.00
x_1	.6735	134.95	23.20
x_2	.3946	.2596	59.20
s	2.9469	2.6651	1.7652

[a]The tabled entries are as follows: the first line gives, successively, Σy^2, the deviation sum of squares of Y, the cross product of the deviations of X_1 and Y, or $\Sigma x_1 y$, and finally $\Sigma x_2 y$. The entries in the second and third lines, on the diagonal or above, are Σx_1^2, $\Sigma x_1 x_2$, and (in the lower right corner) Σx_2^2. The italicized entries *below* the diagonal are the correlation coefficients. The standard deviations are given in the last line.

ment motivation, we can easily insert them into the equation and obtain predicted Y's, predicted reading achievement scores.

Second, we want to know the proportion of the variance that the regression equation "accounts for." That is, we want to know how much of the total variance of Y, reading achievement, is due to the regression of Y on the X's, on verbal aptitude and achievement motivation, or the relation between a linear combination of the independent variables and the dependent variable. In Chapter 2, we saw that the sum of squares due to regression (and its accompanying mean square, or variance) expressed this relation. The multiple correlation coefficient squared, R^2, to be explained shortly, also expresses it.

Third, we need to know the relative importance of the different X's in making the predictions to Y. We need to know, in this case, the relative importance of X_1 and X_2, verbal aptitude and achievement motivation, in the prediction equation. The regression weights, b_1 and b_2, will answer this question in part, although we will see later that there are other more accurate and readily interpretable measures for this purpose. The question will also be answered by certain calculations of sums of squares and R^2's.

Finally, we want to be able to say whether the regression of Y on the X's, the relation between Y and the "best" linear combination of the X's, is statistically significant.

Calculation of Regression Statistics

The calculation of the b's of the regression equation is done rather mechanically with formulas for two X variables. They are

$$b_1 = \frac{(\Sigma x_2^2)(\Sigma x_1 y) - (\Sigma x_1 x_2)(\Sigma x_2 y)}{(\Sigma x_1^2)(\Sigma x_2^2) - (\Sigma x_1 x_2)^2}$$

$$b_2 = \frac{(\Sigma x_1^2)(\Sigma x_2 y) - (\Sigma x_1 x_2)(\Sigma x_1 y)}{(\Sigma x_1^2)(\Sigma x_2^2) - (\Sigma x_1 x_2)^2}$$

(3.5)

Taking the appropriate values from Table 3.1 and substituting them in the formulas, we calculate the b's:

$$b_1 = \frac{(59.20)(100.50) - (23.20)(39.00)}{(134.95)(59.20) - (23.20)^2} = \frac{5949.60 - 904.80}{7989.04 - 538.24} = \frac{5044.80}{7450.80} = .6771$$

$$b_2 = \frac{(134.95)(39.) - (23.20)(100.50)}{(134.95)(59.20) - (23.20)^2} = \frac{5263.05 - 2331.60}{7989.04 - 538.24} = \frac{2931.45}{7450.80} = .3934$$

Now, calculate a. The formula for two independent variables, a special case of formula (3.4), is

$$a = \bar{Y} - b_1 \bar{X}_1 - b_2 \bar{X}_2$$

Substituting the appropriate values yields

$$a = 5.50 - (.6771)(4.95) - (.3934)(5.20) = .1027$$

The whole regression equation can now be written with the calculated values of a and the b's:

$$Y' = .1027 + .6771X_1 + .3934X_2$$

As examples of the use of the equation in prediction, calculate the predicted Y's for the fifth and the twentieth subjects of Table 3.1:

$$Y'_5 = .1027 + (.6771)(3) + (.3934)(6) = 4.4944$$

$$Y'_{20} = .1027 + (.6771)(4) + (.3934)(7) = 5.5649$$

The *obtained* Y's are: $Y_5 = 5$ and $Y_{20} = 10$. The deviations of the predicted scores from the obtained scores, or $d = Y - Y'$, are

$$d_5 = 5 - 4.4944 = .5056$$

$$d_{20} = 10 - 5.5649 = 4.4351$$

One deviation is quite small and the other quite large. In fact, these are the smallest and largest deviations in the set of 20 deviations. The predicted Y's and the deviations or residuals, d, are given in the last two columns of Table 3.1. About half the d's are positive and about half negative, and most of them are relatively small. This is as it should be, of course. The a and the b's of the regression equation, recall, were calculated to satisfy the least squares principle, that is, to minimize the d's, or errors of prediction — rather, to minimize the squares of the errors of prediction. If we square each of the residuals or d's and add them, as we did in Chapter 2, we obtain $\Sigma d^2 = 81.6091$. (Note that $\Sigma d = 0$.) This can be symbolized Σy^2_{res} or ss_{res}, as we also showed earlier. In short, the deviation or residual sum of squares expresses that portion of the total Y sum of squares, Σy^2_i, that is *not* due to the regression. Actually, as we will soon see, there is no need to go through these involved calculations. The deviation or residual sum of squares can be calculated much more readily. We went through the lengthy calculations to show clearly what this sum of squares is.

The regression sum of squares is calculated with the following general formula:

$$ss_{\text{reg}} = b_1 \Sigma x_1 y + \cdots + b_k \Sigma x_k y \tag{3.6}$$

where $k =$ the number of X or independent variables. In the case of two X variables, $k = 2$, the formula reduces to[5]

$$ss_{\text{reg}} = b_1 \Sigma x_1 y + b_2 \Sigma x_2 y \tag{3.7}$$

Taking the b values calculated earlier and finding the deviation sums of cross products from Table 3.2, we substitute in (3.7):

$$ss_{\text{reg}} = (.6771)(100.50) + (.3934)(39.00) = 83.3912$$

This is that portion of the total sum of squares of Y, or Σy^2_i, that is due to the regression of Y on the two X's. Note that the total sum of squares of Y is 165.00

---

[5]The derivation of this equation is shown in Snedecor and Cochran (1967, pp. 388–389).

(from Table 3.2). If the regression sum of squares is added to the residual sum of squares, the sum equals the total sum of squares of Y, as equation (2.12) of Chapter 2 showed. We write this equation again with a new number and then substitute our calculated values:

$$ss_t = ss_{reg} + ss_{res} \tag{3.8}$$

$$ss_t = 83.3912 + 81.6091 = 165.0003$$

This is of course the same value given in Table 3.2, within errors of rounding.

The Coefficient of Multiple Correlation and Its Square

One of the most valuable statistics of multiple regression is the coefficient of multiple correlation, R. The square of this coefficient, R^2, is even more valuable for reasons to be given presently. The calculation of R^2 is simple. One formula that is particularly useful and easily interpreted is

$$R^2 = \frac{ss_{reg}}{ss_t} \tag{3.9}$$

The square root of R^2, of course, gives R. Substituting the sums of squares calculated above, we obtain

$$R^2 = \frac{83.3192}{165.0000} = .5054$$

$$R = \sqrt{.5054} = .7109$$

R is the product-moment correlation of the predicted Y's, Y'_i, which are of course linear combinations of the X's, and the obtained (or observed) Y's. This is shown by borrowing a formula from Snedecor and Cochran (1967, p. 402):

$$R^2 = \frac{(\Sigma yy')^2}{\Sigma y^2 \Sigma y'^2} \tag{3.10}$$

$$R = \frac{\Sigma yy'}{\sqrt{\Sigma y^2 \Sigma y'^2}} \tag{3.11}$$

The values of equation (3.10) can be calculated from the Y and Y' columns of Table 3.1. We already have $\Sigma y^2 = 165$. The comparable value of $\Sigma y'^2$ is calculated:

$$\Sigma y'^2 = \Sigma Y'^2 - \frac{(\Sigma Y')^2}{N} = 688.3969 - \frac{(110)^2}{20} = 83.3969$$

The sum of the deviation cross products is

$$\Sigma yy' = \Sigma YY' - \frac{(\Sigma Y)(\Sigma Y')}{N} = 688.3939 - \frac{(110)(110)}{20} = 83.3939$$

(The difference of .003 in the two sums of squares is due to errors of rounding. Actually, $\Sigma y'^2$ must equal $\Sigma yy'$. We will use calculated values as they are, however, since it makes no difference in the R^2 calculation.) Substituting in equation (3.10), R^2 is obtained:

$$R^2 = \frac{(83.3939)^2}{(83.3969)(165.)} = \frac{6954.5426}{13,760.4885} = .5054$$

and

$$R = \sqrt{.5054} = .7109$$

Finally, we calculate the F ratio, first repeating the formula given in Chapter 2:

$$F = \frac{R^2/k}{(1-R^2)/(N-k-1)} \tag{3.12}$$

where k = the number of independent variables.

$$F = \frac{.5054/2}{(1-.5054)/(20-2-1)} = \frac{.2527}{.0291} = 8.684$$

We can, of course, calculate F using the appropriate sums of squares. The formula is

$$F = \frac{ss_{reg}/df_{reg}}{ss_{res}/df_{res}} \tag{3.13}$$

The degrees of freedom associated with ss_{reg} is $k = 2$, the number of independent variables. The degrees of freedom associated with ss_{res} is $N-k-1 = 17$. Therefore,

$$F = \frac{83.3912/2}{81.6091/17} = \frac{41.6956}{4.8005} = 8.686$$

This agrees with the F calculated using R^2 (within rounding error). It is significant at the .01 level.[6]

Graphing the Regression

To help solidify understanding of the problem and its analysis, and especially to show graphically the nature of regression, the predicted and observed Y values of Table 3.1 are graphed in Figure 3.1. Y', the predicted Y values — which, remember, were really calculated from X_1 and X_2 — and Y, the observed Y values, are plotted. Y' is the abscissa and Y the ordinate. This is the same kind of plot we would use if we were plotting the graph of a simple two-variable correlation and regression problem. The difference is that the present independent variable, Y', is a regression composite of X_1 and X_2 instead of a single X.

---

[6]The curious student may wonder whether the rather elaborate procedure of least squares is really necessary. Why not simply add X_1 and X_2 and use this composite? If we do so, we obtain, in this case, $r = .70$, a value almost the same as R. The answer is that R calculated by the least squares procedure is, as said earlier, the maximum R possible, given the data. In some cases, the least squares R can be considerably higher than the R calculated in some other way. (Another problem, of course, is when the different independent variables have different metrics.)

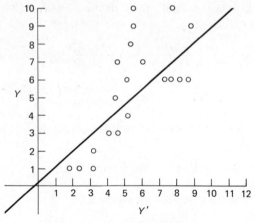

FIGURE 3.1

$R_{y.12} = .71$, a fairly substantial correlation. We would expect the plotted points of Y and Y' to lie fairly close to the regression line. They do. $R_{y.12} = .71$ expresses symbolically and quantitatively what the plot expresses graphically. To the extent that the points lie near the regression line (which, remember, is ruled in by drawing a straight line through a, the intercept constant, plotted on the Y axis, and the point where the two means, $\bar{Y}' = 5.50$ and $\bar{Y} = 5.50$, meet), the magnitude of R is high. If all the plotted points were on the regression line, then $R = 1.00$. If the points are scattered on the graph at random, then R will be close to zero. In other words, we can interpret the graph of Y' and Y much as we interpret an ordinary graph.

Interpretation of the Analysis

A great deal of space has been taken and detailed calculations made to try to clarify the basic business of multiple regression analysis. Even so, the calculations are incomplete; a good deal more analysis is still not only possible but desirable, especially if we are to be able to interpret certain highly significant studies published recently. We must pause, however, to interpret what has been done.

The F ratio tells us that the regression of Y on X_1 and X_2 is statistically significant. The probability of an F ratio this large occurring by chance is less than .01 (it is actually about .003). This means that the relation between Y and a linear least squares combination of X_1 and X_2 could probably not have occurred by chance. It tells us little or nothing about the magnitude of the relation. If the F ratio were not statistically significant, on the other hand, we would not ask about the magnitude of the relation.

The measures R and R^2, especially the latter, tell us explicitly about the magnitude of the relation. In this case, $R^2 = .51$, which means that approximately 51 percent of the variance of Y is accounted for by X_1 and X_2 in combination. (Recall that R^2 is called the *coefficient of determination*.) $R = .71$ can be interpreted much like an ordinary coefficient of correlation, except that values of R range from 0 to 1.00 unlike r, which ranges from -1.00 through 0 to $+1.00$. We will work mostly with R^2 in this book because its interpretation is unambiguous.[7]

Returning to the substance of the original research problem, recall that Y = reading achievement, X_1 = verbal aptitude, and X_2 = achievement motivation. $R^2 = .51$ — and $F = 8.686$, which says that R^2 is statistically significant — means that of the total variance of the reading achievement of the 20 children studied, 51 percent is accounted for by a linear *combination* of verbal aptitude and achievement motivation. In other words, a good deal of the children's reading achievement is explained by verbal aptitude and achievement motivation.

So far there is little difficulty. We now come to a more difficult problem: What are the relative contributions of X_1 and X_2, of verbal aptitude and achievement motivation, to Y, reading achievement? This question can be answered in two or three ways. Eventually we will look at all these ways. Now, however, we study b_1 and b_2, the regression coefficients. Unfortunately, b coefficients are not easy to interpret in multiple regression analysis. So as not to deflect the reader from the main thrust of our discussion, we will only interpret b's rather roughly. Later, we will give a more precise and correct analysis and interpretation.

In Chapter 2, it was said that a single b coefficient in the equation $Y' = a + bX$ indicated that as X changes one unit Y changes b units. The regression coefficient b is called the *slope*. We say that the slope of the regression line is at the rate of b units of Y for one unit of X. If the regression equation were $Y = 4. + .50X$, $b = .50$, and it would mean that as X changes one unit, Y changes half a unit. In multiple regression, on the other hand, the interpretation plot thickens because we have more than one b. In general, if the scales of values of X_1 and X_2 are the same or about the same, for example, all 20 values in each case range from 1 to 10, as ours do, then the b's are weights that show roughly the relative importance of the different variables in predicting Y. This is seen by merely studying the regression equation with the values of the b's calculated earlier:

$$Y' = .1026 + .6771X_1 + .3934X_2$$

X_1, verbal aptitude, is weighted more heavily than X_2, achievement motivation. But the situation is more complex than it seems. We have only given this explanation for *present* pedagogical purposes. The statement is *not* true in all cases.[8]

---

[7]The values of R and R^2 can be and often are inflated. This problem is discussed in Chapter 11.

[8]The relative importance of X_1 and X_2 is indeed different than the above b weights indicate. When X_1 and X_2 are in the order given, their actual contributions to R^2 are about .45 and .05. If, however, the independent variables are reversed, the contributions are about .16 for X_2 and .35 for X_3. In either case, it is evident that X_1 contributes considerably more than X_2. Later, the problem of the relative contributions of the independent variables will be studied in detail.

Alternative Calculation, Analysis, and Interpretation

In order to reinforce our understanding of multiple regression, let us look rather precisely at the various sums of squares. By doing so, we can see rather clearly what X_1 and X_2, separately and together, add to the regression. An important question we must ask is: Does adding X_2 to the regression equation add significantly to our prediction of Y? In the words of our example, how effective is X_2, achievement motivation, in increasing the accuracy of the prediction? A major purpose of adding independent variables is of course to increase accuracy of prediction. Put another way, a major purpose of adding independent variables is to reduce deviations from prediction or regression. In this case, did the addition of X_2 to X_1 materially reduce the residual sum of squares?

Remember that the total sum of squares we have to work with is the sum of squares of the Y scores, or $\Sigma y_t^2 = 165$. (See the earlier calculations that were done when we first introduced the present problem.) No matter how many or how few X variables we have, Σy_t^2 is always the same, 165. And remember that the regression sum of squares and the residual sum of squares always add up to the total sum of squares.

We now do the simple regression of Y on X_1 alone by using the values of Table 3.2 and calculating b, ss_{reg}, and ss_{res}:

$$b = \frac{\Sigma x_1 y}{\Sigma x_1^2} = \frac{100.50}{134.95} = .7447$$

$$ss_{\text{reg}} = \frac{(\Sigma x_1 y)^2}{\Sigma x_1^2} = \frac{(100.50)^2}{134.95} = 74.8444$$

$$ss_{\text{res}} = \Sigma y_t^2 - ss_{\text{reg}} = 165. - 74.8444 = 90.1556$$

In addition, we calculate R^2, R, and the F ratio, using formulas (3.9) and (3.12):

$$R_{y.1}^2 = \frac{ss_{\text{reg}}}{ss_t} = \frac{74.8444}{165.0000} = .4536$$

$$R_{y.1} = \sqrt{R_{y.1}^2} = \sqrt{.4536} = .6735$$

$$F = \frac{R_{y.1}^2/k}{(1 - R_{y.1}^2)/(N-k-1)} = \frac{.4536/1}{(1-.4536)/(20-1-1)} = \frac{.4536}{.0304}$$

$$= 14.9211$$

Or using formula (3.13) and $df_{\text{reg}} = k = 1$ and $df_{\text{res}} = N-k-1 = 20-1-1 = 18$,[9]

---

[9]The discrepancy between the values of the F ratios calculated by the two methods is due to rounding errors. The value of the F ratio as calculated on a large computer is 14.9432, the second of the two values, above. We will constantly be confronted with such small discrepancies. The reader should not be disturbed by them. Concentrate rather on understanding basic regression ideas. In actual use, of course, calculations will be done by a computer and most values will be accurate enough. For a discussion of rounding errors, see Draper and Smith (1966, pp. 52–53 and 143–144).

$$F = \frac{ss_{reg}/df_{reg}}{ss_{res}/df_{res}} = \frac{74.8444/1}{90.1556/18} = \frac{74.8444}{5.0086} = 14.9432$$

The F ratio is significant at the .01 level. Therefore, the regression of Y on X_1 alone is statistically significant. Since $R^2 = .45$, we can say that 45 percent of the variance of Y, reading achievement, is accounted for by X_1, verbal aptitude. Note that this is the multiple regression way of talking about ordinary correlation. The correlation between X_1 and Y, or r_{x_1y}, is .67. $r_{x_1y}^2$, therefore, is .45. In other words, we can regard ordinary two-variable correlation and regression as a special case of multiple correlation and multiple regression.

Now calculate the regression of Y on X_2 alone and R^2, R, and the F ratio:

$$b = \frac{\Sigma x_2 y}{\Sigma x_2^2} = \frac{39.00}{59.20} = .3946$$

$$ss_{reg} = \frac{(\Sigma x_2 y)^2}{\Sigma x_2^2} = \frac{(39.00)^2}{59.20} = 25.6926$$

$$ss_{res} = \Sigma y_t^2 - ss_{reg} = 165. - 25.6926 = 139.3074$$

$$R_{y.2}^2 = \frac{ss_{reg}}{ss_t} = \frac{25.6926}{165.0000} = .1557$$

$$R_{y.2} = \sqrt{.1557} = .3946$$

$$F = \frac{ss_{reg}/df_{reg}}{ss_{res}/df_{res}} = \frac{25.6926/1}{139.3074/(20-1-1)} = 3.320 \text{ (n.s.)}$$

While $R_{y.2}^2 = .16$ and $R_{y.2} = .39$, both appreciable quantities, the F ratio of 3.32 is not significant at the conventional level of .05. Since the probability is actually .08, a borderline case, we can pursue the matter further.[10] It is clear, however, that X_1 is a much better predictor of Y than X_2, considering them separately. If we had to choose, say, between X_1 and X_2, there would be no question which we would choose—provided our interest was in accuracy of prediction.

We are not quite ready to answer a more interesting question, although we can nibble at its edges: Does X_2 add significantly to prediction *when added to* X_1? The answer is that it adds to prediction: $R_{y.1}^2 = .45$, as we just saw, and $R_{y.12}^2 = .51$, as we saw earlier. The addition to the R^2 is .05. (The actual figures are $.5054 - .4536 = .0518$.) This additional contribution to the regression, however, is not statistically significant. Now note an interesting thing: when we calculated the R^2 from the regression of Y on X_2 alone, we obtained .16. When, however, we calculated the additional contribution of X_2 after X_1, we obtained .05. We return to this fundamental and important characteristic of multiple regression at the end of the chapter.

---

[10] It should be emphasized that ordinarily the separate regression of Y on X_2 would not be calculated. It is done here to make a point and to lay a foundation for a similar but more appropriate form of analysis later in the book.

Additional Regression Calculations

Even though X_2 does not add significantly to the prediction, we pull together our summary regression analyses in Tables 3.3, 3.4, and 3.5. In Table 3.3, we present the 20 obtained Y scores and the predicted Y and deviation scores, Y' and $d = Y - Y'$, for the two regressions, Y on X_1 and Y on X_1 and X_2. The obtained scores, Y, are given in the first column. The Y scores predicted from the regression of Y on X_1 alone, Y'_1, are given in the second column. The deviations from regression, d_1, are listed in the third column. In the fourth and fifth

TABLE **3.3**   DEPENDENT VARIABLE SCORES (Y) AND PREDICTED
SCORES (Y') FROM ONE AND TWO INDEPENDENT VARIABLES
(X_1, X_2), DATA OF TABLE 3.1

Y	Y'_1	d_1	Y'_{12}	d_{12}
2	3.3031	−1.3031	3.0305	−1.0305
1	3.3031	−2.3031	3.0305	−2.0305
1	2.5584	−1.5584	2.3534	−1.3534
1	2.5584	−1.5584	1.9600	−.9600
5	4.0478	.9522	4.4944	.5056
4	4.7925	−.7925	5.1715	−1.1715
7	5.5372	1.4628	4.6684	2.3316
6	5.5372	.4628	5.0618	.9382
7	7.0266	−.0266	6.0226	.9774
8	6.2819	1.7181	5.3455	2.6545
3	4.7925	−1.7925	4.7781	−1.7781
3	4.0478	−1.0478	4.1010	−1.1010
6	6.2819	−.2819	7.7059	−1.7059
6	6.2819	−.2819	7.3125	−1.3125
10	7.7713	2.2287	7.8799	2.1201
9	8.5160	.4840	8.9504	.0496
6	9.2607	−3.2607	8.8407	−2.8407
6	8.5160	−2.5160	8.1636	−2.1636
9	4.7925	4.2075	5.5649	3.4351
10	4.7925	5.2075	5.5649	4.4351
Σ: 110.	110.	0	110.	0
Σ²: 770.	679.8326		688.3969	
ss: 165.	74.8326	90.1556	83.3969	81.6091

columns, the Y scores predicted on the basis of X_1 and X_2, and Y'_{12}, and the accompanying deviation scores, d_{12}, are listed. The last three rows of the table give the sums (Σ) and the sums of squares (ss) of the five columns. The equations for the regressions of Y on X_1 alone and Y on X_1 and X_2 are

$$Y'_1 = 1.8137 + .7447 X_1 \tag{3.14}$$

$$Y'_{12} = .1027 + .6771 X_1 + .3934 X_2 \tag{3.15}$$

TABLE **3.4** SUMS OF SQUARES AND REDUCTION IN RESIDUAL SUMS OF
SQUARES OF REGRESSION ANALYSES, ADDITION OF X_2 TO X_1

X	ss_y	ss_{reg}	ss_{res}	Reduction
1	165.0000	74.8444	90.1556	
$1+2$	165.0000	83.3912	81.6091	8.5465

The values of equation (3.14) were calculated as follows (see Table 3.1
for the means and Table 3.2 for the sums of squares and cross products):

$$b_1 = \frac{\Sigma x_1 y}{\Sigma x_1^2} = \frac{100.50}{134.95} = .7447$$

$$a = \bar{Y} - b_1 \bar{X}_1 = 5.50 - (.7447)(4.95) = 1.8137$$

To calculate the 20 predicted Y scores, simply enter the X_1 values of Table 3.1
in equation (3.14). For example, Y_1' and Y_8' are calculated:

$$Y_1' = 1.8137 + (.7447)(2) = 3.3031$$

$$Y_8' = 1.8137 + (.7447)(5) = 5.5372$$

Now, calculate the d's that accompany these values:

$$d = Y - Y'$$

$$d_1 = 2 - 3.3031 = -1.3031$$

$$d_8 = 6 - 5.5372 = .4628$$

To calculate the predicted Y's of the regression of Y on X_1 and X_2 insert
the values of both X_1 and X_2 of Table 3.1 in equation (3.15). [The values of
equation (3.15) were calculated earlier.] For example, Y_1' and Y_8', the comparable values to those just calculated for the regression of Y on X_1 alone, are

$$Y_1' = .1027 + (.6771)(2) + (.3934)(4)$$

$$= .1027 + 1.3542 + 1.5736 = 3.0305$$

$$Y_8' = .1027 + (.6771)(5) + (.3934)(4)$$

$$= .1027 + 3.3855 + 1.5736 = 5.0618$$

TABLE **3.5** ANALYSES OF VARIANCE AND R^2'S OF THE TWO
REGRESSIONS OF Y ON X_1 AND X_2 AND Y ON X_1

Source	df	ss	ms	F	p	R^2
X_1, X_2	2	83.3909	41.6955	8.686	.003	.5054
Deviations	17	81.6091	4.8005			
X_1	1	74.8444	74.8444	14.943	.001	.4536
Deviations	18	90.1556	5.0086			
Total	19	165.0000				

The residual scores are

$$d_1 = 2 - 3.0305 = -1.0305$$
$$d_8 = 6 - 5.0618 = \quad .9382$$

We should probably pause at this point to emphasize just what we are try-ing to do. We are trying to show, in as straightforward a way as possible, what goes into multiple regression analysis. The nonmathematical student of educa-tion and the behavioral sciences is frequently confused when confronted with regression equations, regression weights, R^2's, and F ratios. We are trying to clear up at least some of this confusion by calculating many of the necessary quantities of multiple regression analysis rather directly. We would *not* use these methods in actual practice. They are too cumbersome. But they are good for pedagogical purposes because they approach multiple regression directly by working, as much as possible, with the original data and the sums of squares generated from the original data. Hopefully, one can "see" where the various quantities "come from."

Since part of the calculations of Table 3.3 were explained earlier in connec-tion with Table 3.1, we need only touch upon them here. Our main point is that the addition of X_2 to the regression reduced the residual or deviation sum of squares from 90.1556 to 81.6091 and increased the regression sum of squares from 74.8326 to 83.3969. (These last two figures, taken from the bottom line of Table 3.3, are slightly off due to rounding errors. The more accurate values are 74.8444 and 83.3909, as calculated by a computer.) The last line of the table, then, is the important one. It gives the sums of squares for Y, Y_1', d_1, Y_{12}', and d_{12}. Except for $\Sigma y_1'^2$, these sums of squares were calculated earlier.

The sums of squares of Table 3.3 are brought together for convenience in Table 3.4. In addition, the reduction in the sum of squares of the residuals or, conversely, the increase in the regression sum of squares — is given in the table: 8.55. In short, the table shows that the addition of X_2 to the regression reduces the deviations from regression (the residuals) by 8.55 — or, the addition of X_2 increases the regression sum of squares by 8.55. As we saw earlier, this is a decrease (or increase) of 5 percent: $8.55/165.00 = .05$.

Table 3.4 and its figures can show rather nicely just what the multiple correlation coefficient is — in sums of squares. Think of the extreme case. If we had a number of independent variables, $X_1, X_2, \ldots, X_k$, and they completely "explained" the variance of Y, the dependent variable, then $R^2_{y.12\ldots k} = 1.00$, and the sum of squares due to the regression would be the same as the total sum of squares, namely 165, and the residual sum of squares would be zero. But we do not "know" all these independent variables; we only "know" two of them, X_1 and X_2. The sum of squares for the regression of Y on X_1 alone is 74.84. The proportion of the dependent variable variance is $74.84/165.00 = .45$. The sum of squares of the regression of Y on X_1 *and* X_2 is 83.39. The proportion of the Y variance is $83.39/165.00 = .51$. The quantities .45 and .51, of course, are the R^2's.

In Table 3.5, the analyses of variance of the two regressions are summarized. In the upper part of the table, the analysis of variance of the regression of Y on X_1 and X_2 is given. The lower part of the table gives the analysis of variance for the regression of Y on X_1 alone. (The values given in Table 3.5 were taken from computer output. Some of them differ slightly from the values of Tables 3.3 and 3.4, again due to errors of rounding.)

It is evident from these analyses that X_2 does not add much to X_1. It increases our predictive power by only 5 percent. Returning to the substance of our variables and problem, verbal aptitude alone (X_1) accounts for about 45 percent of the variance of reading achievement (Y). If achievement motivation (X_2) is added to the regression equation, the amount of variance of reading achievement accounted for by verbal aptitude and achievement motivation together $(X_1$ and $X_2)$ is about 51 percent, an increase of about 5 to 6 percent.

A Set Theory Demonstration of Certain Multiple Regression Ideas

In many regression situations it may be found that the regression of Y on each of the independent variables, when added individually and in combination to the regression equation after the first independent variable has been entered, may add little to R^2. The reason is that the independent variables are themselves correlated. If the correlation between X_1 and X_2, for example, were zero, then the r^2 between X_2 and Y can be added to the r^2 between X_1 and Y to obtain $R^2_{y.12}$. That is, we can write the equation

$$R^2_{y.12} = r^2_{y.1} + r^2_{y.2} \qquad (\text{when } r_{12} = 0)$$

But this is almost never the case in the usual regression analysis situation. r_{12} will seldom be zero, at least with variables of the kind under discussion. In general, the larger r_{12} is, the less effective adding X_2 to the regression equation will be. If we add a third variable, X_3, and it is correlated with X_1 and X_2 — even though it may be substantially correlated with Y — it will add still less to the prediction. If the data of Table 3.1 were actually obtained in a study it would mean that verbal aptitude, X_1, alone predicts reading achievement, Y, almost as well as the combination of verbal aptitude, X_1, and achievement motivation, X_2. In fact, the addition of achievement motivation is not statistically significant, as shown earlier.

These ideas can perhaps be clarified by Figure 3.2 where each set of circles represents the sum of squares (or variance, if you will) of a Y variable and two X variables, X_1 and X_2. The set on the left, labeled (a), is a simple situation where $r_{y1} = .50$, $r_{y2} = .50$, and $r_{12} = 0$. If we square the correlation coefficients of X_1 and X_2 with Y and add them — $(.50)^2 + (.50)^2 = .25 + .25 = .50$ — we obtain the variance of Y accounted for by both X_1 and X_2, or $R^2_{y.12} = .50$.

But now study the situation in (b). We cannot add r^2_{y1} and r^2_{y2} because r_{12} is not equal to 0. (The degree of correlation between two variables is expressed

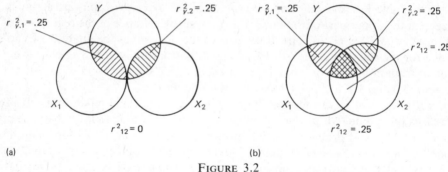

FIGURE 3.2

by the amount of overlap of the circles.[11]) The hatched areas of overlap represent the variances common to the pairs of depicted variables. The one doubly hatched area represents that part of the variance of Y that is common to the X_1 and X_2 variables. Or, it is part of r_{y1}^2; it is part of r_{y2}^2; and it is part of r_{12}^2. Therefore, to determine accurately that part of Y determined by X_1 and X_2, it is necessary to subtract this doubly hatched overlapping part so that it will not be counted twice.[12]

Careful study of Figure 3.2 and the relations it depicts should help the student grasp the principle stated earlier. Look at the right side of the figure. If we want to predict more Y, so to speak, we have to find other variables whose variance circles will intersect the Y circles and, at the same time, not intersect each other, or at least minimally intersect each other. In practice, this is not easy to do. It seems that much of the world is correlated — especially the world of the kind of variables we are talking about. Once we have found a variable or two that correlates substantially with school achievement, say, then it becomes increasingly difficult to find other variables that correlate substantially with school achievement and *not* with the other variables. For instance, one might think that verbal ability would correlate substantially with school achievement. It does. Then one would think that the need to achieve, or *n* achievement, as McClelland calls it, would correlate substantially with school achievement. Evidentally it does (McClelland et al., 1953). One might also reason that *n* achievement should have close to zero correlation with verbal ability, since verbal ability, while acquired to some extent, is in part a function of genetic endowment. It turns out, however, that *n* achievement is significantly related

[11]The above figure is used only for pedagogical purposes. It is not always possible to express all the complexities of possible relations among variables with such figures.

[12]Set theory helps to clarify such situations. Let $V(Y)$ = variance of Y, $V(X_1)$ = variance of X_1, and $V(X_2)$ = variance of X_2. $X_1 \cap Y$, $X_2 \cap Y$, and $X_1 \cap X_2$ represent the three separate intersections of the three variables. $X_1 \cap X_2 \cap Y$ represents the common intersection of all three variables. Then $V(X_1 \cap Y)$ represents the variance common to X_1 and Y, and so on, to $V(X_1 \cap X_2 \cap Y)$, which is the variance common to all three variables. We can now write the equation:

$$V_y = V(X_1 \cap Y) + V(X_2 \cap Y) - V(X_1 \cap X_2 \cap Y)$$

where V_y = the variance accounted for by X_1 and X_2. Actually, $V_y = R_{y.12}^2$.

to verbal ability (*ibid.*, pp. 234–238, especially Table 8.8, p. 235). This means that, while it may increase the R^2 when added to the regression equation, its predictive power in the regression is decreased by the presence of verbal ability because part of its predictive power is already possessed by verbal ability.

Assumptions

Like all statistical techniques, multiple regression analysis has several assumptions behind it that should be understood by researchers. Unfortunately, preoccupation with assumptions seems to frighten students, or worse, bore them. Many students, overburdened by zealous instructors with admonitions on when analysis of variance, say, can and cannot be used, come to the erroneous conclusion that it can only be rarely used. We do not like to see students turned off analysis and statistics by admonitions and cautionary precepts that usually do not matter that much. They sometimes do matter, however. Moreover, intelligent use of analytic methods requires knowledge of the rationale and thus the assumptions behind the methods. We therefore look briefly at certain assumptions behind regression analysis.[13]

First, in regression analysis it is assumed that the Y scores are normally distributed at each value of X. (There is no assumption of normality about the X's.) This assumption and others discussed below are necessary for the tests of statistical significance discussed in this book. The validity of an F test, for instance, depends on the assumption that the dependent variable scores are normally distributed in the population. The assumption is not needed to *calculate* correlation and regression measures (see McNemar, 1960). There is no need to assume anything to calculate r's, b's, and so on. (An exception is that, if the distributions of X and Y, or the combined X's and Y, are not similar, then the range of r may not be from -1 through 0 to $+1$.) It is only when we make inferences from a sample to a population that we must pause and think of assumptions.

A second assumption is that the Y scores have equal variances at each X point. The Y scores, then, are assumed to be normally distributed and to have equal variances at each X point.

Note the following equation:

$$Y = a + b_1 X_1 + \cdots + b_k X_k + e \qquad (3.16)$$

where e = error, or residual. These errors are assumed to be random and normally distributed, with equal variances at each X point. The latter point can be said: the distribution of the deviations from regression (the residuals) are the same at all X points. These assumptions about the e's are of course used in statistical estimation procedures.

It has convincingly been shown that the F and t tests are "strong" or "robust" statistics, which means that they resist violation of the assumptions

---

[13]The reader will find a good discussion of the assumptions, with a simple numerical example, in Snedecor and Cochran (1967, pp. 141–143).

(Anderson, 1961; Baker, Hardyck, & Petronovich, 1966; Boneau, 1960; Boneau, 1961; Games & Lucas, 1966; Lindquist, 1953, pp. 78–86). In general, it is safe to say that we can ordinarily go ahead with analysis of variance and multiple regression analysis without worrying too much about assumptions. Nevertheless, researchers must be aware that serious violations of the assumptions, and especially of combinations of them, can distort results. We advise students to examine data, especially by plotting, and, if the assumptions appear to be violated, to treat obtained results with even more caution than usual. The student should also bear in mind the possibility of transforming recalcitrant data, using one or more of the transformations that are available and that may make the data more amenable to analysis and inference (see Kirk, 1968, pp. 63–67; Mosteller & Bush, 1954).

Further Remarks on Multiple Regression and Scientific Research

The reader should now understand, to a limited extent at least, the use of multiple regression analysis in scientific research. In Chapters 1, 2, and 3, enough of the subject has been presented to enable us now to look at larger issues and more complex procedures. In other words, we are now able to generalize multiple regression to the k-variable case. A severe danger in studying a subject like multiple regression, however, is that we become so preoccupied with formulas, numbers, and number manipulations we lose sight of larger purposes. This is particularly true of complex analytic procedures like analysis of variance, factor analysis, and multiple regression analysis. We become so enwrapped by techniques and manipulations that we become the servants of methods rather than the masters. While we are forced to put ourselves and our readers through a good deal of number and symbol manipulation, we worry constantly about losing our way. These few paragraphs are to remind us why we are doing what we are doing.

In Chapter 1, we talked about the two large purposes of multiple regression analysis, prediction and explanation, and we said that prediction is really a special case of explanation. The student should now have somewhat more insight into this statement. To draw the lines clearly but oversimply, if we were interested only in prediction, we might be satisfied simply with R^2 and its statistical significance and magnitude. Success in high school or college as predicted by certain tests is a classic case. In much of the research on school success, the interest has been on prediction of the criterion. One need not probe too deeply into the whys of success in college; one wants mainly to be able to predict successfully. And this is of course no mean achievement to be lightly derogated.

In much behavioral research, however, prediction, successful or not, is not enough. We want to know why; we want to "explain" the criterion performance, the phenomenon under study. This is the main goal of science. To explain a phenomenon, moreover, we must know the relations between independent variables and a dependent variable and the relations among the independent variables. This means, of course, that R^2 and its statistical significance and magnitude are not enough; our interest focuses more on the whole regression equation and the regression coefficients.

Take a difficult and important psychological and educational phenomenon, problem solving. Educators have said that much teaching should focus on problem solving rather than only on substantive learning (Bloom, 1969; Dewey, 1916, especially Chapter XII, 1933; Gagné, 1970). It has been said, but by no means settled, that so-called discovery instruction will lead to better problem-solving ability. Here is an area of research that is enormously complex and that will not yield to oversimplified approaches. Nor will it yield to a "successful prediction" approach. Even if researchers are able to find independent variables that predict well to successful problem solving, it must become possible to state with reasonable specificity and accuracy what independent variables lead to what sort of problem-solving behavior. Moreover, the interrelations and interactions of such independent variables in their influence on problem solving must be understood (see Chapter 10 and Berliner & Cahen, 1973; Cronbach & Snow, 1969).

Here, for example, are some variables that may help to explain problem-solving behavior: teaching principles or discovering them (Kersh & Wittrock, 1962), intelligence, convergent and divergent thinking (Guilford, 1967, Chapters 6 and 7), anxiety (Sarason et al., 1960), and concreteness-abstractness (Harvey, Hunt, & Schroder, 1961). That these variables interact in complex ways need hardly be said. That the study of their influence on problem-solving ability and behavior needs to reflect this complexity does need to be said. It should be obvious that prediction to successful problem solving is not enough. It will be necessary to drive toward explanation of problem solving using these and other variables in differing combinations.

All this means is that researchers must focus on explanation rather than on prediction, at least in research focused on complex phenomena such as problem solving, achievement, creativity, authoritarianism, prejudice, organizational behavior, and so on. In a word, theory, which implies both explanation and prediction, is necessary for scientific development. And multiple regression, while well-suited to predictive analysis, is more fundamentally oriented to explanatory analysis. We do not simply throw variables into regression equations; we enter them, wherever possible, at the dictates of theory and reasonable interpretation of empirical research findings.

What we are trying to do here is set the stage for our study of the analytic and technical problems of Chapters 4 and 5 by focusing on the relation between predictive and explanatory scientific research and multiple regression analysis. It is our faith, in short, that analytic problems are solved and mastered more by understanding their purpose than by simply studying their technical aspects. The study and mastery of the technical aspects of research are necessary but not sufficient conditions for the solution of research problems. Understanding of the purposes of the techniques is also a necessary condition.

In Chapter 4, our study is extended to k independent variables and the general solution of the regression equation. Both predictive and explanatory uses of regression analysis will be examined in greater complexity and depth. In Chapter 5, we explore some of the basic meaning and intricacies of explanatory analysis by trying to deepen our knowledge of statistical control in the multiple regression framework.

Study Suggestions

1. Suppose that an educational psychologist has two correlation matrices, A and B, calculated from the data of two different samples:

$$
\begin{array}{c}
\begin{array}{ccc}
X_1 & X_2 & Y
\end{array} \\
\begin{array}{c}
X_1 \\
X_2 \\
Y
\end{array}
\begin{pmatrix}
1.00 & 0 & .70 \\
0 & 1.00 & .60 \\
.70 & .60 & 1.00
\end{pmatrix} \\
A
\end{array}
\qquad
\begin{array}{c}
\begin{array}{ccc}
X_1 & X_2 & Y
\end{array} \\
\begin{array}{c}
X_1 \\
X_2 \\
Y
\end{array}
\begin{pmatrix}
1.00 & .40 & .70 \\
.40 & 1.00 & .60 \\
.70 & .60 & 1.00
\end{pmatrix} \\
B
\end{array}
$$

    (a) Which matrix will yield the higher R^2? Why?
    (b) Calculate the R^2 of matrix A.
    (*Answers*: (a) Matrix A; (b) $R^2 = .85$.)
2. b and β are called regression coefficients or *weights*. Why are they called "weights"? Give two examples with two independent variables each.
3. Suppose that two regression equations that express the relation between age, X, and intelligence (mental age), Y, of middle-class and working-class children are:

    Middle class: $Y' = .30 + .98X$

    Working class: $Y' = .30 + .90X$

    The regression weights, b, are therefore .98 and .90.
    (a) Accepting these weights at face value, what do they imply about the relation between chronological age and mental age in the two samples?
    (b) Calculate Y' values with both equations at ages 8, 10, 15, and 20. What do the Y''s tell us?
    (c) Can the analysis between slopes (for example, the differences between slopes) sometimes be important? Why?
4. Do this exercise with the data of Table 3.1. Calculate the sums (or means) of each X_1 and X_2 pair. Correlate these sums (or means) with the Y scores. Compare the square of this correlation with $R^2_{y.12} = .51$. ($r^2 = .70^2 = .49$.) Since the values of .51 and .49 are quite close, why should we not simply use the sums or means of the independent variables and not bother with the complexity of multiple regression analysis?
5. In regression analysis with one independent variable, $r_{xy} = \beta$. Suppose that a sociologist has calculated the correlation between a composite environmental variable and intelligence and $r_{xy} = .40$. Suppose, further, that a 3-year program was initiated to enrich deprived children's environments. Assuming that the program is successful (after 3 years),
    (a) What should happen to the correlation between the environmental variables and intelligence, other things being equal?
    (b) What should happen to the β weight?
6. Review the regression calculations of the data of Table 3.1. Now study equations (3.12) and (3.13). Note that the denominator of (3.12) is $(1 - R^2)/(N - k - 1)$ and the denominator of (3.13) is ss_{res}/df_{res}.

(a) What does $1 - R^2$ appear to be?

(b) Does equation (3.13) look like an analysis of variance formula for the F ratio? If so, to what analysis of variance statistical term is ss_{res} related?

7. Take the results of the regression calculations associated with the data of Table 3.1. Suppose the three variables are: $X_1 =$ intelligence; $X_2 =$ social class; $Y =$ verbal achievement. Write down the main regression statistics and interpret them. What can you say about the probable relative contributions of the independent variables to the dependent variable?

8. Examine the comparative situations of (a) and (b) in Figure 3.2.

(a) In which of these situations can we clearly and unambiguously talk about the relative contributions of the independent variables? Why?

(b) In (b), what makes the interpretation of the relative contributions of the independent variables difficult? As best you can at this stage explain why. Can you draw a principle of interpretation from this example?

9. Why is multiple regression analysis with two independent variables in general "better" than such analysis with one independent variable? Under what circumstances will this tend not to be true?

10. Fictitious data for three variables, X_1, X_2, and Y, are given below. The X_1 and Y scores are the same as those of Table 3.1; the X_2 scores, however, are different.

X_1	X_2	Y	(X_1	X_2	Y)
2	5	2	4	3	3
2	4	1	3	6	3
1	5	1	6	9	6
1	3	1	6	8	6
3	6	5	8	9	10
4	4	4	9	6	9
5	6	7	10	4	6
5	4	6	9	5	6
7	3	7	4	8	9
6	3	8	4	9	10

(The second set of three columns is merely a continuation of the first set.)

(a) Calculate the following statistics, using the methods described in the text: sums, means, standard deviations, sums of squares and cross products, and the three r's.

[*Note:* Some of these were of course calculated in the chapter. We suggest calculating everything and then checking the text. Set up a table like Table 3.2 to keep things orderly.]

(b) Calculate b's, a, ss_{reg}, ss_{res}, R^2, R, F.

(c) Write the regression equation.

(d) Calculate $R_{y.12}^2 - R_{y.1}^2$. What does this amount indicate? Compare it to the similar calculation of the example of Tables 3.1 and 3.2.

(e) Interpret the above statistics. Use the variables given in the text in your interpretation.

(*Answers*: (a) Means: 4.95, 5.50, 5.50; standard deviations: 2.6651, 2.1151, 2.9469; sums of squares: 134.95, 85.00, 165.00; sums of cross products: $\Sigma x_1 x_2 = 15.50$, $\Sigma x_1 y = 100.50$, $\Sigma x_2 y = 63.00$; $r_{12} = .1447$, $r_{y1} = .6735$, $r_{y2} = .5320$. (b) $b_1 = .6737$, $b_2 = .6183$; $a = -1.2356$; $ss_{reg} = 106.6614$; $ss_{res} = 58.3386$; $R_{y.12}^2 = .6464$; $R = .8040$; $F = 15.5407$ ($df =$

2, 17). (c) $Y' = -1.2356 + .6737X_1 + .6183X_2$. (d) $R^2_{y.12} - R^2_{y.1} = .6464 - .4536 = .1928$.)

11. Using the regression statistics calculated for No. 10, above, calculate the predicted Y's, Y', and the d's $(d = Y - Y')$.
    (a) Calculate Σd^2. What is this sum of squares?
    (b) Calculate the correlation between Y' and Y. What should this equal; that is, what statistic calculated in No. 10, above, should be the same (within rounding error)? Why?
    (*Answers*: (a) $\Sigma d^2 = 58.3386$; (b) $r_{yy'} = .8040$, same as $R_{y.12}$.)

General Method of Multiple Regression Analysis

The main purpose of Chapter 3 was to give the reader an intuitive understanding of multiple regression and its rationale. Some of the formulas and methods used were general; they apply to any multiple regression problem. There are several significant and necessary forms of analysis, however, that were not discussed because their exposition and explication might have obscured understanding.

This chapter has four related purposes. One, a general method of multiple regression analysis that can be applied to all multiple regression problems will be expounded. This method, as we will see, can handle any number of independent variables. Two, the differences between types of regression coefficients will be clarified. Three, the method of testing the statistical significance of adding variables to the regression equation introduced in Chapter 3 will be elaborated. Four, a dichotomous or dummy variable using 1's and 0's will be introduced. We will see that there is no essential difference in method between analyses with continuous variables and those with categorical, or coded, variables.

A Special Note to the Student

Before proceeding, we must say something of the peculiar difficulties of this chapter. Chapters 3 and 4 are the conceptual and technical foundation of the book. In them we present most of the substance of multiple regression theory and analysis. The explanations of the material of Chapter 3 were fairly straightforward; there were no great difficulties that could not be surmounted with reasonable application. Chapter 4, however, is somewhat different. The solution of a set of regression equations with more than two independent variables requires a higher level of conceptualization that is not as easy to grasp and work

53

with — and to explain. Some knowledge and familiarity with matrix algebra and matrix thinking are required. In addition, the student has to know and understand the distinction between kinds of regression coefficients and their use and interpretation in actual research and one or two other aspects of the subject the lack of understanding of which will impede our study.

Such knowledge and understanding are not easy to attain. The trouble is that there is a gap between the levels of analysis of Chapter 3 and the present chapter. Although we have tried to bridge the gap by working through problems in some detail, there are technical points that cannot be grasped without special methods. The main key to mastery of these points is matrix algebra. Matrix algebra, at least as much of it as we need, is not difficult. But it does require special study. No knowledge of advanced algebra is necessary. It is sufficient if the student knows elementary algebra. In short, the student and researcher who expect to understand and use multiple regression have to know the basic elements of matrix algebra.[1]

At this point, then, we urge the student to divert his study to Appendix A, which is a systematic elementary treatment of certain fundamental aspects of matrix algebra. The treatment of the subject is specifically geared to the multiple regression needs of this book. It is not a comprehensive treatment the technical details of which might distract us from our goals. Nevertheless, it should be sufficient to enable the student to follow much of the literature on multivariate analysis.

We particularly urge study to the point of being easily familiar with the following aspects of matrix algebra: matrix manipulations analogous to simple algebraic manipulations; matrix operations, but especially the multiplication of matrices; the notion of the inverse of a matrix; and the solution of the so-called normal equations using matrix algebra. Although not essential, it will be helpful if the student also knows and understands, if only to a limited extent, determinants of matrices. Appendix A outlines and discusses these topics.

Matrices and Subscripts

The uninitiated person can easily be confused, even overwhelmed, by notation problems. Moreover, different references use different systems of symbols, which of course confuse the student even more. In this book, we have adopted a relatively simple system that works in most situations.

The first source of difficulty is the use of the subscripts i and j for different purposes. The student must learn the different uses. For a matrix of raw data, we write X_{ij}, which is shorthand for the matrix given in Table 4.1. The subscript on the left, accompanying each X in the table, stands for the case number (the subject number) and the cases are the rows. The subscript on the right stands for the variable number; thus the columns of the matrix are variables. i = rows, and j = columns. i runs from 1 through N, the number of cases or subjects, and j runs from 1 through k, the number of variables.

---

[1]For those students who are unfamiliar with matrix algebra, we suggest, in addition to Appendix A, study of Kemeny, Snell, and Thompson (1966, Chapter V).

TABLE **4.1**   A MATRIX, X_{ij}, OF RAW DATA

Cases (i)	Variables (j)						
	1	2	3	·	·	·	k
1	X_{11}	X_{12}	X_{13}	·	·	·	X_{1k}
2	X_{21}	X_{22}	X_{23}	·	·	·	X_{2k}
·	·	·	·	·	·		·
·	·	·	·	·	·		·
·	·	·	·	·	·		·
N	X_{N1}	X_{N2}	X_{N3}	·	·	·	X_{Nk}

The difficulty and possible source of confusion occur when i and j are used somewhat differently. In designating a correlation matrix, or a variance-covariance matrix, we again use i for rows and j for columns. R_{ij}, for example, means a correlation matrix with i rows and j columns. In this case, since correlation matrices are symmetric, $i = j$. But there are matrices that are not symmetric. X_{ij}, the matrix of Table 4.1, is of course not symmetric. Actually, there is no real difficulty. The only difficulty might arise because i and j, in a correlation matrix, for example, both stand for variable numbers, whereas in Table 4.1 they stand for cases and variables, respectively.

Vectors – 1 by N or 1 by k matrices – usually have single subscripts, for example, X_i or X_j, or β_i and β_j. It makes little difference whether i or j is used as long as we know what we mean. We will usually use j. We can write β_j or b_j. If $j = 1, 2, 3, 4$, then $\beta_j = (\beta_1, \beta_2, \beta_3, \beta_4)$. Vectors will sometimes need two subscripts, but one of them always remains fixed. We can write R_{yj}, where y remains fixed – there is, after all, only one Y variable – and j varies, say, from 1 through 3. This notation is used in equations (4.3), below, where $R_{yj} = (r_{y1}, r_{y2}, r_{y3})$. The only difference is that here we write it as a row vector but in (4.3) as a column vector.

One other usage sometimes causes trouble. Matrices can be and are written in two ways, in matrix boldface letters, $\mathbf{A}$, $\mathbf{X}$, and the like, and in italicized letters with subscripts. Whenever there is no ambiguity, it is easier to write and use the boldface letters. Subscripts can be used with boldface letters, for example, when a summation may be indicated. We will ordinarily not do so in this book. Instead, we will use the italicized letters, such as X_{ij}, or β_j. Remember, however, that X_{ij} can be symbolized $\mathbf{X}$ when it is clear what the subscripts are.

General Analysis with Three Independent Variables

When there are one or two independent variables, it is possible, as we saw in Chapter 3, to use special formulas to obtain the unknown quantities of the regression equation. For example, to obtain the b coefficients in a problem with two independent variables we used equations (3.5). When there are more than two independent variables it is not practically feasible to calculate the b coefficients by using special formulas. Instead of such formulas, one can use the more elegant and conceptually simpler methods of matrix algebra. Fortunately, contemporary users of multiple regression do not have to worry too much about calculations. Most computing centers have multiple regression programs that accomplish the calculations expeditiously and effectively.[2]

The general regression equation for any number of variables was given earlier [equation (3.2)]. As usual, we repeat it here with a new number:

$$Y' = a + b_1X_1 + b_2X_2 + \cdots + b_kX_k \tag{4.1}$$

It is necessary to determine a and the b's of this equation. The objective of the determination is to find those values of a and the b's that will minimize the sum of squares of the deviations or residuals. The calculus provides the method of differentiation for doing this. If the calculus is used, we arrive at a set of simultaneous linear equations called *normal* equations (no relation to the normal distribution). These equations contain the coefficients of correlation among all the independent variables and between the independent variables and the dependent variable and a set of so-called beta weights, β_j.

General Solution of Regression Equations

When it was said above that the regression equation (4.1) is solved for a and the b's, the real meaning is that the *set* of k equations is solved simultaneously. It is possible to solve the implied set of equations of (4.1) using the X's, but the details are more cumbersome than the method to be described, which uses the correlations among all the variables. Coefficients of correlation are in standard score form, as was shown in Chapter 2 from the equation:

$$r_{xy} = \frac{\sum z_x z_y}{N}$$

where z_x = standard scores of X, z_y = standard scores of Y, and N = number of cases. If all the obtained scores in a set of data are transformed to z scores, the normal equations obtained from the calculus, for three independent variables, are

$$\begin{aligned}
\beta_1 + r_{12}\beta_2 + r_{13}\beta_3 &= r_{y1} \\
r_{21}\beta_2 + \beta_2 + r_{23}\beta_3 &= r_{y2} \\
r_{31}\beta_3 + r_{32}\beta_2 + \beta_3 &= r_{y3}
\end{aligned} \tag{4.2}$$

where β_j = the beta weights; r_{ij} = the correlations among the independent variables; r_{yj} = the correlations between the independent variables and the dependent variable, Y. (Note that $r_{12} = r_{21}$, $r_{13} = r_{31}$, and so on. Note, too, that β_1, β_2, and β_3 on the diagonal can be understood to be $r_{11}\beta_1$, $r_{22}\beta_2$, and $r_{33}\beta_3$ since $r_{11} = r_{22} = r_{33} = 1.00$.)

The correlations among all the variables are of course calculated from the data. In Appendix A, the matrix conceptualization and calculation of the raw

---

[2]Computer programs to accomplish multiple regression are discussed in Appendices B and C. In addition, we give the FORTRAN listing for one such program (Appendix C).

score sums of squares and cross products are shown. The matrix symbols, $\mathbf{X'X}$, mean $\Sigma X_i X_j$, and the symbols $\mathbf{x'x}$ mean $\Sigma x_i x_j$, the deviation sums of squares and cross products. Because of the importance in our work of the latter matrix, we write it out for three variables:

$$\mathbf{x'x} = \Sigma x_i x_j = \begin{pmatrix} \Sigma x_1^2 & \Sigma x_1 x_2 & \Sigma x_1 x_3 \\ \Sigma x_2 x_1 & \Sigma x_2^2 & \Sigma x_2 x_3 \\ \Sigma x_3 x_1 & \Sigma x_3 x_2 & \Sigma x_3^2 \end{pmatrix}$$

where $i = 1, 2, \ldots, k$ variables and j similarly, and $\Sigma x_1^2 = \Sigma X_1^2 - (\Sigma X_1)^2/N$, $\Sigma x_1 x_2 = \Sigma X_1 X_2 - (\Sigma X_1)(\Sigma X_2)/N$ and so on. (If all the terms in the matrix are divided by N, the number of cases, a matrix of variances and covariances is obtained.) From the above matrix, the correlation matrix, $\mathbf{R}$, can be obtained using a formula given in Chapter 2:

$$\mathbf{R} = R_{ij} = r_{x_i x_j} = \frac{\Sigma x_i x_j}{\sqrt{\Sigma x_i^2 \Sigma x_j^2}}$$

The $\mathbf{R}$ matrix given on the left of (4.3), below, is such a matrix. Note that the correlation matrix can be written $\mathbf{R}$ or R_{ij}. Both are matrix notations, as indicated earlier.

We have repeated these computational and statistical details for three reasons. One, almost all computer programs for multiple regression analysis calculate the above matrices, although they may not print them out. Two, they are the basic calculations for virtually all the analyses of this book. And three, we wanted to be sure that the student knows clearly the ingredients of subsequent calculations. The student should also bear in mind that the calculations outlined above for three variables are easily extended to any number of variables and that one of the variables can be the Y or dependent variable.[3]

Since the correlations are calculated from the data, the only unknowns of the set of equations (4.2) are the beta weights, β_j. Therefore the equations must somehow be solved for the β's. This can be done in several ways. One can solve the set as we learned to solve simultaneous linear equations in high school or college (see Kemeny, Snell, & Thompson, 1966, pp. 251ff.). Such a method is cumbersome and difficult, however, especially for a large number of independent variables. One can also solve the equations by using various methods, for example, the Doolittle method (see most statistics texts), devised to cut down the labor involved. Perhaps the "best" and most elegant method is to use matrix algebra. We outline this method not only because it is elegant but also because it is, for the most part, the method used in computer programs.[4]

Equations (4.2) can be conceived as matrices and broken down into matrices. There is the matrix of intercorrelations of the independent variables, the

[3]The student may be puzzled by the absence of a Y variable in the matrices $\mathbf{X'X}$ and $\mathbf{x'x}$, above. Usually, all the data are read into the computer as one raw score data matrix, or, to save computer storage, one row at a time may be read in and the calculations done seriatim. Such a matrix may be called $\mathbf{X}$. Then, at a later stage, the programmer somehow indicates that one of the k variables read in, perhaps the first variable or the last one, is the Y or dependent variable.

[4]We repeat here some of the discussion of Appendix A, but we will shortly clothe the matrix algebraic frame with a numerical example.

matrix or vector of β coefficients, and the matrix or vector of the correlations between the independent variables and the dependent variable. We write these matrices as follows:

$$
\begin{pmatrix} r_{11} & r_{12} & r_{13} \\ r_{21} & r_{22} & r_{23} \\ r_{31} & r_{32} & r_{33} \end{pmatrix} \begin{pmatrix} \beta_1 \\ \beta_2 \\ \beta_3 \end{pmatrix} = \begin{pmatrix} r_{y1} \\ r_{y2} \\ r_{y3} \end{pmatrix} \tag{4.3}
$$
$$
\quad\quad R_{ij} \quad\quad\quad\quad \beta_j \quad\quad R_{yj}
$$

To multiply two matrices the elements of the rows of the first matrix, R_{ij}, are multiplied by the elements of the columns of the second matrix, β_j, and then the products are added. If this is done to equations (4.3), equations (4.2) are obtained: $r_{11}\beta_1 + r_{12}\beta_2 + r_{13}\beta_3 = r_{y1}$; $r_{21}\beta_1 + r_{22}\beta_2 + r_{23}\beta_3 = r_{y2}$; and so on.

Let the matrix of independent variable r's in (4.3) be called R_{ij}, the vector of β's β_j, and the vector of r's between the independent variables and the dependent variable R_{yj}. Then (4.3) can be written

$$
R_{ij}\beta_j = R_{yj} \tag{4.4}
$$

As indicated earlier, the matrix equation must be solved for the β_j. If the symbols of equation (4.4) represented single quantities and not matrices, the procedure would be algebraically simple: just divide R_{ij} into R_{yj}. Although matrices can be added, subtracted, and multiplied, they cannot be divided, at least in the usual sense of the word. There is, in matrix algebra, however, an operation analogous to division: calculate the *inverse* of the matrix—in this case R_{ij}, the matrix of independent variable correlations—and multiply it by R_{yj}. The matrix equations are

$$
R_{ij}\beta_j = R_{yj}
$$
$$
\beta_j = R_{ij}^{-1}R_{yj} \tag{4.5}
$$

where R_{ij}^{-1} is the inverse of R_{ij}. The problem then becomes one of calculating the inverse.

The inverse of matrix **A** is that matrix which, when multiplied by **A**, produces an identity matrix, **I**. An identity matrix is a matrix with 1's in the principal diagonal (from upper left to lower right) and 0's in the off-diagonal cells. That is,

$$
\mathbf{A}^{-1}\mathbf{A} = \mathbf{I}
$$

If we did not have computers, this sort of solution of (4.3) and (4.4) would be difficult because the calculation of the inverse of a matrix is a weary and error-prone job. Fortunately, computer routines for matrix inversion are common and easily used.[5] We now illustrate, though briefly, these points.

---

[5]There is one case when the solution $\beta_j = R_{ij}^{-1}R_{yj}$ will not work. If a matrix is singular—for example, if one or more of the rows of the matrix are dependent upon one or more of the other rows of the matrix—then it has no inverse. Fortunately, most correlation matrices have inverses, but not all. It is wise, when using matrix inversion routines, to insert into the program a check routine that will multiply the inverse by the original matrix to produce the identity matrix: $\mathbf{A}^{-1}\mathbf{A} = \mathbf{I}$. If a data matrix has no inverse, then the computer will of course produce nonsense, or it will abort the program.

An Example with Three Independent Variables

Note the fictitious data of Table 4.2, which are like the data of Table 3.1 in Chapter 3. The Y and X_1 values are the same, in fact. But the X_2 values have been altered so that X_2 contributes more to the prediction of Y, and X_3, a new variable which does not contribute significantly to the prediction of Y, has been added.

TABLE **4.2**   FICTITIOUS EXAMPLE: ATTITUDES TOWARD OUTGROUPS (Y), AUTHORITARIANISM (X_1), DOGMATISM (X_2), AND RELIGIOSITY (X_3) MEASURES; MULTIPLE REGRESSION

Y	X_1	X_2	X_3	Y'	$Y-Y' = d$
2	2	5	1	2.5396	−.5396
1	2	4	2	2.1029	−1.1029
1	1	5	4	2.4831	−1.4831
1	1	3	4	1.2351	−.2351
5	3	6	5	4.5312	.4688
4	4	4	6	4.0889	−.0889
7	5	6	3	5.3934	1.6066
6	5	4	3	4.1454	1.8546
7	7	3	7	5.5074	1.4926
8	6	3	7	4.8890	3.1110
3	4	3	8	3.8395	−.8395
3	3	6	9	5.2804	−2.2804
6	6	9	5	8.2584	−2.2584
6	6	8	4	7.4471	−1.4471
10	8	9	5	9.4952	.5048
9	9	6	5	8.2416	.7584
6	10	4	7	7.9866	−1.9866
6	9	5	8	8.1795	−2.1795
9	4	8	8	6.9595	2.0405
10	4	9	7	7.3962	2.6038

Σ: 110	99	110	108	110	$\Sigma d = 0$
M: 5.50	4.95	5.50	5.40		$\Sigma d^2 = 55.4866$
Σ^2: 770	625	690	676	714.5168	

$$\Sigma y'^2 = 109.5168$$

To make interpretation interesting, suppose that a social psychological researcher is interested in determinants or correlates of attitudes toward outgroups, and that he believes that authoritarianism, dogmatism, and religiosity (degree of religious commitment) each contribute to such attitudes.[6] Suppose that the data of Table 4.2 are the result of his first attempt to study the relations among these variables. Y, then, is attitudes toward outgroups, X_1 is authoritarianism, X_2 is dogmatism, and X_3 is religiosity. We want to know how well the three independent variables predict to the dependent variable.

[6]It is assumed that an appropriate scale has been used to measure each of the variables. We are not concerned here with measurement details.

TABLE **4.3**   DEVIATION SUMS OF SQUARES AND CROSS PRODUCTS,
CORRELATION COEFFICIENTS, AND STANDARD DEVIATIONS OF DATA
OF TABLE 4.1[a]

	y	x_1	x_2	x_3
y	165.00	100.50	63.00	43.00
x_1	.6735	134.95	15.50	39.40
x_2	.5320	.1447	85.00	2.00
x_3	.3475	.3521	.0225	92.80
s	2.9469	2.6651	2.1151	2.2100

[a]The table entries are as follows. The first line gives, successively, Σy^2, the deviation sum of squares of Y, the cross products of the deviations of X_1 and Y, $\Sigma x_1 y$, X_2 and Y, $\Sigma x_2 y$, and X_3 and Y, $\Sigma x_3 y$. The entries in the second, third, and fourth lines, on the diagonal and above, are: Σx_1^2, $\Sigma x_1 x_2$, and $\Sigma x_1 x_3$; Σx_2^2, $\Sigma x_2 x_3$; and Σx_3^2. The italicized entries *below* the diagonal are the correlation coefficients. The standard deviations are given in the last line.

The tabled data are in the same form as those of Table 3.1 of the previous chapter: the sums, means, and sums of squares are given at the bottom of the table. The predicted Y's, or Y', and the deviations from prediction are given in the last two columns of the table. These latter columns, of course, were calculated from the complete regression equation to be discussed presently. In Table 4.3 the deviation sums of squares, Σx_j^2 and Σy^2, are given in the diagonal cells, and the cross products, $\Sigma x_1 x_2$, $\Sigma x_1 x_3$, $\Sigma x_2 x_3$, and $\Sigma x_j y$, are given in the upper half of the table. The intercorrelations are given below the diagonal. Finally, the standard deviations of the variables are given in the last line of the table. We do not discuss details of these calculations since such details were given in Chapter 3. We wish, rather, to concentrate on the more important regression calculations.[7]

Before anything else, we must solve the following set of equations for the β's [see equations (4.2), *supra*]:

$$\beta_1 + .1447\beta_2 + .3521\beta_3 = .6735$$
$$.1447\beta_1 + \beta_2 + .0225\beta_3 = .5320$$
$$.3521\beta_1 + .0225\beta_2 + \beta_3 = .3475$$

[The correlation coefficients of Table 4.3 have simply been substituted for the r symbols of equations (4.2).] To solve for the β's, the matrix equations (4.4) and (4.5) are used. Substituting the correlation coefficients in the extended matrix form of (4.3), we obtain

$$\begin{pmatrix} 1.0000 & .1447 & .3521 \\ .1447 & 1.0000 & .0225 \\ .3521 & .0225 & 1.0000 \end{pmatrix} \begin{pmatrix} \beta_1 \\ \beta_2 \\ \beta_3 \end{pmatrix} = \begin{pmatrix} .6735 \\ .5320 \\ .3475 \end{pmatrix}$$

[7]We have done all the calculations in this chapter with a desk calculator and by computer. The figures reported in the tables and elsewhere are those calculated with a desk calculator. The student, who is advised to do all these calculations with a desk calculator (with one possible exception to be mentioned later), will find certain relatively minor discrepancies between his own calculations and the calculations given here. These are due to the inevitable errors of rounding that occur in complex calculations. See footnote 9, Chapter 3, for earlier remarks on such errors.

and

$$\begin{pmatrix} \beta_1 \\ \beta_2 \\ \beta_3 \end{pmatrix} = \begin{pmatrix} 1.0000 & .1447 & .3521 \\ .1447 & 1.0000 & .0225 \\ .3521 & .0225 & 1.0000 \end{pmatrix}^{-1} \begin{pmatrix} .6735 \\ .5320 \\ .3475 \end{pmatrix}$$

$$\beta_j = \qquad R_{ij}^{-1} \qquad R_{yj}.$$

It is apparent that the major problem is to obtain the inverse of the matrix of the intercorrelations of the independent variables, R_{ij}^{-1}. Although obtaining matrix inverses is nowadays invariably done with computer routines, as we said earlier, it is possible to obtain them and the necessary β values by using certain desk calculator routines. Effective routines can be found in Dwyer (1951, Chapter 13) and Walker and Lev (1953, pp. 332–336). The inverse of R_{ij} in the present problem was obtained using Walker and Lev's outline of the Fisher-Doolittle method. It is given below. We give it here rather than the more accurate computer solution to familiarize the student with the effects of rounding errors and to encourage him to try, at least once or twice, to perform such calculations. In any case, the ultimate results obtained with different methods should not differ too much from each other.

The inverse of R_{ij} is inserted in the above extended layout, the necessary matrix multiplication performed, and the β_j obtained, as follows:

$$\begin{pmatrix} \beta_1 \\ \beta_2 \\ \beta_3 \end{pmatrix} = \begin{pmatrix} 1.1665 & -.1595 & -.4071 \\ -.1595 & 1.0223 & .0331 \\ -.4071 & .0331 & 1.1426 \end{pmatrix} \begin{pmatrix} .6735 \\ .5320 \\ .3475 \end{pmatrix} = \begin{pmatrix} .5593 \\ .4479 \\ .1405 \end{pmatrix}$$

$$\beta_j = \qquad R_{ij}^{-1} \qquad R_{yj} = \beta_j \;.$$

The three β_j, then, are $\beta_1 = .5593$, $\beta_2 = .4479$, and $\beta_3 = .1405$.

To obtain the b_j, the following equation, which was introduced in Chapter 2 [equation (2.18)], is used:

$$b_j = \beta_j \frac{s_y}{s_j} \tag{4.6}$$

where b_j regression or b weights, and $j = 1, 2, 3$; s_y = standard deviation of the dependent variable, Y; s_j = the standard deviations of the independent variables. Taking the standard deviations given in the last line of Table 4.3 and the above β_j, we find

$$b_1 = \beta_1 \frac{s_y}{s_1} = (.5593)\left(\frac{2.9469}{2.6651}\right) = .6184$$

$$b_2 = \beta_2 \frac{s_y}{s_2} = (.4479)\left(\frac{2.9469}{2.1151}\right) = .6240$$

$$b_3 = \beta_3 \frac{s_y}{s_3} = (.1405)\left(\frac{2.9469}{2.2100}\right) = .1873$$

To calculate the value a of the regression equation, equation (3.4) of Chapter 3 is adapted to the present problem:

$$a = \bar{Y} - b_1\bar{X}_1 - b_2\bar{X}_2 - b_3\bar{X}_3$$

$$a = 5.50 - (.6184)(4.95) - (.6240)(5.50) - (.1873)(5.40)$$

$$= -2.0045$$

The full regression equation is now written in two forms, one with the b's and one with the β's. Take first the one with the b's, the one with which we are familiar:

$$Y' = -2.0045 + .6184X_1 + .6240X_2 + .1873X_3$$

This equation is used to calculate the predicted Y's. The calculated values are given in the Y' column of Table 4.2. The d's are also calculated: $d = Y - Y'$. They are entered in the last column of the table. The summation of the Y''s and the d's yields 110 and 0, as it should.

Next, calculate the sums of squares due to regression, $\Sigma y'^2$, or ss_{reg}, and to the deviations from regression (the residual sum of squares), Σd^2 or ss_{res}. $\Sigma y'^2 = 109.5168$ and $\Sigma d^2 = 55.4866$. Their sum is actually 165.0034, the difference being due to errors of rounding. (The values obtained from the computer are $\Sigma y'^2 = 109.5134$ and $\Sigma d^2 = 55.4866$.)

The second form of the regression equation uses the β's and the standard scores, z_j. Standard scores and the use of β with standard scores were discussed in Chapter 2. Recall that the β's are standard partial regression coefficients.

$$z'_y = \beta_1 z_1 + \beta_2 z_2 + \beta_3 z_3$$

where $z = (X - \bar{X})/s$. Although we will not use this form in our work, we will have more to say about the β's and z scores later in the book.

R and R^2, as shown in Chapter 3, can be calculated in several ways. We illustrate three of them, again borrowing from formulas of Chapter 3.

$$R = \frac{\Sigma yy'}{\sqrt{\Sigma y^2 \Sigma y'^2}} = \frac{109.5151}{\sqrt{(165)(109.5168)}} = .8147$$

As shown in Chapter 3, this is the most intuitively obvious formula because it expresses multiple correlation as the product-moment correlation between the observed and the predicted Y scores. It is *not* recommended, however, for the routine calculation of R. R^2, of course, is $(.8147)^2 = .6637$. The other two formulas, with the calculations of the present problem, are

$$R^2 = \frac{ss_{reg}}{ss_t} \tag{4.7}$$

$$= \frac{109.5168}{165.0000} = .6637$$

$$R^2 = 1 - \frac{ss_{res}}{ss_t} \tag{4.8}$$

$$= 1 - \frac{55.4066}{165.0000} = .6637$$

The first of the two formulas is familiar: it is equation (3.9) of Chapter 3. The second formula is new. Its relation to the first formula should be obvious.

The statistical significance of R^2, and thus the regression, is as usual determined by the F ratio, using equation (3.12), and substituting the R^2 just calculated and the degrees of freedom:

$$F = \frac{R^2/k}{(1-R^2)/(N-k-1)} \qquad (4.9)$$

$$= \frac{.6637/3}{(1-.6637)/(20-3-1)} = 10.526$$

(k = the number of independent variables.) At 3 and 16 degrees of freedom, this is significant at the .01 level.

The results of the analysis mean that the three independent variables contribute significantly to the variance of the dependent variable. In fact, the three variables authoritarianism, dogmatism, and religiosity, as a group, account for about 66 percent of the variance of attitudes toward outgroups. To say just how much of this variance is contributed by each of the three variables is not clearly possible without further analysis. The relative sizes of the b and β weights seem to indicate that authoritarianism (X_1) and dogmatism (X_2) contribute about equally and that religiosity (X_3) contributes little, but such interpretations are shaky and dangerous. We must now take up part of this problem. Before doing so, however, we again discuss b and β weights.

Regression Weights and Scales of Measurement

The "purpose" of regression weights, as explained briefly earlier, is to so weight the individual X's, the measures of the independent variables, that the best prediction is possible under the conditions of the relations among the X's themselves and between the X's and Y, the dependent variable. The criterion for achieving this "best" prediction is that the sum of squares of the deviations from regression (or the residual sum of squares) be a minimum. That is, the weights must be so chosen that Σd^2 is the minimum quantity possible, given the relations among the independent variables themselves and the relations between the independent variables and the dependent variable.

One can use other methods of weighting. For example, one can simply take the means of each individual's X scores and use these means to predict the Y scores. (To do this all the independent variables must have the same scale of measurement.) Or one might devise weights on the basis of intuitive conjectures as to the relative importance of the different variables in predicting Y. Teachers do this when they give scores or weights to different items in a test. They are in effect predicting pupil performance on the basis of a homemade regression equation that predicts pupil grades on the basis of the weighted items. Much intuitive and experimental prediction follows some such procedure. Such systems may yield useful predictions. The system of using means, for instance, yields an r^2 of .61 between the original Y and the predicted Y

values of the problem in the preceding section. The system of least squares, however, has the virtue, among other virtues, of yielding the maximum correlation possible between a linear combination of the independent variables and the dependent variable. And this is accomplished by the regression weights.

In Chapter 2, the regression weights, b and β, were defined and the relation between them discussed. Equations (2.17) and (2.18) of that discussion are valid here, as is most of the discussion. [Equation (2.19) is not valid with more than one independent variable.] The importance of regression weights, however, requires further discussion, especially in the context of more than one independent variable. In this section, therefore, we again tackle the problem of what regression weights are, the relation between b's and β's, and their interpretation.

The two kinds of regression coefficients are sometimes confused and usage is not always consistent. We try here to adopt a relatively straightforward usage that agrees with most references and with the mathematics of multiple regression presented above. One distinction between b and β, brought out briefly in Chapter 2, is that β's are population values or parameters and b's are sample values used to estimate β's, the population values. This notion of the b's as sample estimates of the β's is theoretically useful.[8] Both the b's and the β's are slopes and mean the change in Y with each unit change in X. That is, β_1 is the average change in Y when X_1 is changed by one unit, other X variables in the regression held constant.

Another usage is possible. In this usage the b's have the same function and interpretation mentioned above. The β's, however, are conceived as the regression coefficients to be used with standard scores (see McNemar, 1962, pp. 171ff.). We use this conception of the β's because it fits nicely with the solution of the normal equations when correlation coefficients are used in the equations and because it makes interpretation of research data somewhat easier. In short, the β's are in standard score form in this usage, they are first cousins of correlation coefficients, and they lend themselves well to the interpretation of research data because, as standard scores, all the independent variables, the X's, have the same scale of measurement.[9]

Recall that the β's were obtained from the solution of the normal equations:

$$R_{ij}\beta_j = R_{yj}$$

$$\beta_j = R_{ij}^{-1}R_{yj}$$

The β's have a precise meaning here. To repeat part of an earlier discussion, they are *standard partial regression coefficients*. "Standard" means they are used when all the variables are in standard score form. "Partial" means that the effects of variables other than the one to which the weight applies are held

---

[8]For a complete presentation of this correct theoretical usage, see Snedecor and Cochran (1967, Chapter 13). Draper and Smith's (1966, Chapter 1) discussion is also good.

[9]Some writers use different symbols for the β's as we use them. Presumably to avoid confusion, they use, for example, b^* or $\hat{\beta}$.

constant. We can write β_1, in a three-variable problem, as $\beta_{y1.23}$, β_2 as $\beta_{y2.13}$, and β_3 as $\beta_{y3.12}$. $\beta_{y1.23}$, for example, means the standard partial regression weight of variable 1 with variables 2 and 3 held constant.

Recall, too, that correlation coefficients are in standard form: $r_{ij} = \Sigma z_i z_j / N$. This means that when we calculate R_{ij}^{-1}, the inverse of R_{ij}, in the above equation the resulting β_j will be in standard form. The b weights, on the other hand, although they are partial regression coefficients, are not in standard form. As we saw in Chapter 3 [see formula (3.5)], they can be calculated with deviation sums of squares and cross products. They can also be calculated from the β's using formula (4.6). Formula (4.6) showed that the β's are multiplied by s_y/s_j. This multiplication converts the β's into weights that can be used with the original units of measurement of the variables.[10]

In general, then, b coefficients are partial regression weights that can be used with the original X measures, or with the deviations of X's from their means, or x, in the regression equation to calculate predicted Y's or y's. β weights are standardized partial regression coefficients that can only be used in the regression equation if all the X measures have been converted to standard or z scores. Moreover, the β's are produced by the matrix equation $\beta_j = R_{ij}^{-1} R_{yj}$ in the solution of the simultaneous equations discussed earlier. This means that they are on the same level of discourse as the correlation coefficients from which they are calculated. It can be shown, too, that they have other interesting and important properties, some of which will be discussed later.[11]

Testing Statistical Significance

To this point two of the four purposes stated at the beginning of the chapter have been accomplished. The general method of multiple regression analysis has been explained and illustrated, and the distinctions between the two kinds of regression coefficients, b and β, have been made. Now we must explain how to test the statistical significance of the regression weights or the individual variables in the regression equation and the statistical significance of adding

---

[10]The standard deviations of the raw scores that are used in formula (4.6) reflect the original units of measurement. The multiplication of the β's, which are in standard score form with a mean of 0 and a standard deviation of 1, by s_y/s_j, both components of which were calculated from the raw scores, has the effect of transforming the β's to the b's, weights that are in raw score form.

[11]The β's and R^2 can be calculated in another way that is mathematically simpler and more elegant than the method used in this chapter. If the correlations among the independent variables and the correlations between the independent variables and the dependent variable are all included in one matrix, the inverse of this matrix — call it R^{-1} — can be used to calculate the R^2's and β's. Let r^{yy} indicate the diagonal value of the Y cell of R^{-1}, and r^{yj} the off-diagonal cells. The following formulas then apply:

$$R^2 = 1 - \frac{1}{r^{yy}}$$

$$\beta_j = \frac{-r^{yj}}{r^{yy}}$$

A full discussion can be found in Guttman (1954, pp. 291–293). The above notation is slightly different from Guttman's. He discusses a different problem.

variables to the equation. The first of these methods uses t ratios (or F ratios) and the second R^2's and F ratios. The first has not been considered until now; the second was taken up in Chapter 3 in some detail. In applying the t ratio to test the significance of the regression weights, we ask the important question: Is the regression of the dependent variable on an independent variable statistically significant after taking the effect of the other independent variables into account? Take the problem of authoritarianism, dogmatism, religiosity, and attitudes toward outgroups. We may wish to ask if the regression of attitudes toward outgroups, Y, on dogmatism, X_2, is statistically significant after controlling for the effects of authoritarianism, X_1, and religiosity, X_3. In effect, we are asking whether b_2 (or $b_{y2.13}$) is statistically significant.

Standard Errors, t Ratios, and the Statistical Significance of Regression Coefficients

In multiple regression work, there are several kinds of standard errors, namely standard errors of estimate, of β weights, and of b weights. [We omit consideration of the standard error of β weights since the standard error of b weights accomplishes our purposes. The test of significance of b applies to β. That is, if b is statistically significant, so is β. See Anderson (1966, p. 164).] Standard errors are of course standard deviations. They are measures of the variability of error; thus the name "standard error."

The *standard error of estimate* is simply the standard deviation of the residuals, d_i. Its raw score formula is

$$SE_{est} = \sqrt{\frac{ss_{res}}{N-k-1}} \qquad (4.10)$$

For the data of Table 4.2 (last column),

$$SE_{est} = \sqrt{\frac{55.4866}{20-3-1}} = \sqrt{3.4679} = 1.8623$$

The square of SE_{est} is the *variance of estimate* or the mean square of the residuals. It is the error term used in the F test [see formula (3.13), Chapter 3].

The standard error of estimate is an index of the variation or dispersion of the predicted Y measures about the regression. If SE_{est} is relatively large when compared to the standard deviation the estimate of Y on the basis of the X's is poor. The smaller SE_{est} is the better the prediction. SE_{est} for the data of Table 4.2 is 1.86. Compare this to the standard deviation of the Y scores (from Table 4.3): 2.95. If R^2 approaches 0, the standard error of estimate approaches — and can exceed — the standard deviation of Y. With zero correlation between Y and the X's the best prediction is to the mean of Y. If so, then $d = Y - \bar{Y} = y$, and ss_{res} approaches Σy_t^2, or ss_t. In the present case, the standard error of estimate is considerably smaller than the standard deviation of Y. Thus the prediction seems to be successful.

Although the standard error of estimate is useful, we are really more interested in it because it enters the formulas for the standard errors of the regression weights. The standard error of a regression weight is like any other standard error: its purposes are to indicate the variability of errors and to provide a measure with which to compare the statistics whose significance is being tested. Many tests of statistical significance are fractions of the form

$$\frac{\text{statistic}}{\text{standard error of the statistic}}$$

like the t ratio and the F ratio.

The standard errors of b coefficients can be calculated in several ways. The method to be used here has the virtue of being closely related to the approach and methods of calculation used in this book. The formula is

$$SE_{b_j} = \sqrt{\frac{SE_{\text{est}}^2}{ss_{x_j}(1 - R_j^2)}} \qquad (4.11)$$

where SE_{b_j} = the standard error of the jth b weight; SE_{est}^2 = the squared standard error of estimate, or the variance of estimate; ss_{x_j} = sum of squares of variable j; and R_j^2 = the squared multiple correlation between variable j, used as a dependent variable, and the remaining independent variables. Adapting this formula to the first independent variable of the data of Table 4.2, gives[12]

$$SE_{b_1} = \sqrt{\frac{SE_{\text{est}}^2}{ss_{x_1}(1 - R_{1.23}^2)}} \qquad (4.12)$$

We have all the values required except that for $R_{1.23}^2$, which, of course, means the R^2 of the regression of variable 1 on variables 2 and 3. This value can be calculated from the inverse of the correlation matrix of independent variables, a matrix given earlier. What we do now illustrates the usefulness of R^{-1}, the inverse of the correlation matrix. The R^2's between any one of the independent variables and the remaining independent variables can be readily calculated with the following formula taken from Anderson (1966, p. 164) or Guttman (1954, p. 293):

$$R_j^2 = 1 - \frac{1}{r^{jj}} \qquad (4.13)$$

where R_j^2 = the squared multiple correlation between the jth independent variable and the remaining independent variables, and r^{jj} = the diagonal values of the inverse of the matrix of correlations among the independent variables.

---

[12]Most texts give different formulas. One is: $SE_{b_j} = s\sqrt{c_{jj}}$, where s = standard error of estimate, as calculated by equation (4.10), and c_{jj} = the values in the diagonal of the inverse of the variance-covariance matrix of the independent variables. Since, in this book, the correlation matrix and its inverse are used almost entirely, this method is not used. For a good presentation of the method, see Snedecor and Cochran (1967, pp. 389–392). For other formulas, see Anderson (1966, p. 164) and Ezekiel and Fox (1959, p. 283). Formula (4.11) is merely another version of the formulas given in these two references.

Using the diagonal values of R^{-1} given earlier, we calculate the three R^2's:

$$R_1^2 = R_{1.23}^2 = 1 - \frac{1}{1.1665} = .1427$$

$$R_2^2 = R_{2.13}^2 = 1 - \frac{1}{1.0223} = .0218$$

$$R_3^2 = R_{3.12}^2 = 1 - \frac{1}{1.1426} = .1248$$

Substituting the value of the standard error of estimate calculated above, 1.8622, the sums of squares of variables 1, 2, and 3 (Table 4.3), and the above values of R_j^2 in equation (4.13), we obtain the standard errors of the three b weights. These calculations have been done in Table 4.4. The t ratio is, of course, $t_j = b_j/SE_{b_j}$, with $df = N - k - 1 = 20 - 3 - 1 = 16$. The three t ratios have also been calculated in Table 4.4. (The t ratio values calculated by computer are given in parentheses beside the t ratios calculated by desk calculator.

TABLE **4.4**   CALCULATION OF STANDARD ERRORS AND t RATIOS OF
b WEIGHTS, DATA OF TABLES 4.2 AND 4.3

$$SE_{b_1} = \sqrt{\frac{(1.8622)^2}{(134.95)(1 - .1427)}} = \sqrt{\frac{3.4678}{115.6926}} = \sqrt{.0300} = .1732$$

$$t_1 = \frac{.6184}{.1732} = 3.5704 \qquad (3.5715)$$

$$SE_{b_2} = \sqrt{\frac{(1.8622)^2}{(85.)(1 - .0218)}} = \sqrt{\frac{3.4678}{83.1470}} = \sqrt{.0417} = .2042$$

$$t_2 = \frac{.6240}{.2042} = 3.0558 \qquad (3.0554)$$

$$SE_{b_3} = \sqrt{\frac{(1.8622)^2}{(92.80)(1 - .1248)}} = \sqrt{\frac{3.4678}{81.2186}} = \sqrt{.0427} = .2066$$

$$t_3 = \frac{.1874}{.2066} = .9071 \qquad (.9069)$$

The small differences are as usual due to errors of rounding.) At 16 degrees of freedom, b_1 and b_2 are statistically significant ($p < .01$), but b_3 is not.

The interpretation of these t ratios is a bit complex. The first t, $t_1 = 3.57$, significant at the .01 level, indicates that b_1 is significantly different from 0 and that X_1 contributes significantly to the regression after X_2 and X_3 are taken into account. The second t, $t_2 = 3.06$, also significant at the .01 level, indicates that X_2 contributes significantly to the regression after X_1 and X_3 are taken into account. And the third t, $t_3 = .91$, which is not significant, indicates that, after X_1 and X_2 are taken into account, X_3 does not contribute significantly to the regression, and that b_3 does not differ significantly from 0. These t tests, in other words, are tests of the b's, which are *partial* regression coefficients. They indicate the slope of the regression of Y on X_j, after controlling the effect of the other independent variables. In this case, X_1 and X_2, or authoritarianism and dogmatism, contribute significantly to the prediction of Y, attitudes toward outgroups, but X_3, religiosity, does not. In the next chapter, such partial relations are studied rather thoroughly.

An F test approach, with identical results, can also be used (Snedecor & Cochran, 1967, pp. 386–388). It has the virtue of making the situation a bit clearer. Rather than verbalize the method in detail, we take part of the above problem to show the reader what the t test of the regression coefficient means.

The sum of squares due to the regression of Y on all three independent variables has already been calculated: $ss_{reg} = 109.5168$. Now calculate the sum of squares due to the regression of Y on X_2 and X_3. This can be done as follows: Calculate b_2 and b_3 with equations (3.5), Chapter 3, or calculate β_2 and β_3 with the method of this chapter [invert the R matrix that includes only r_{23}, and then use equation (4.5)]. Next, use equation (3.6), Chapter 3, to calculate the sum of squares due to the regression of Y on X_2 and X_3. The sum of squares of Y on X_2 and X_3, or $ss_{y.23}$, is 44.2422 (the two b's are $b_2 = .7306$ and $b_3 = .4476$). If $ss_{y.23}$ is subtracted from $ss_{y.123}$, the remainder should be the sum of squares due to the regression of Y on X_1, *after* removing the effect of the regression of Y on X_2 and X_3.[13] All this is set up in Table 4.5. In the table we also do an analysis of variance, which simply requires the additional calculation of the mean squares (divide the sums of squares by the appropriate degrees of freedom). The resulting $F = 44.2422/3.4679 = 12.7576$, at 1 and 16 degrees of freedom, is significant at the .01 level.

TABLE **4.5**   ANALYSIS OF VARIANCE OF THE REGRESSION OF Y ON X_1, AFTER REMOVING THE REGRESSION OF Y ON X_2 AND X_3, DATA OF TABLES 4.2 AND 4.3

Source	df	ss	ms	F
$Y.123$	3	109.5168		
$Y.23$	2	65.2746		
$Y.1 = Y.123 - Y.23$	1	44.2422	44.2422	12.7576
Residual	16	55.4866	3.4679	
Total	19			

$$t_1 = \sqrt{12.7576} = 3.5718$$

The analysis of variance shows that the regression of Y on X_1, after excising the effect of X_2 and X_3, is statistically significant. Substantively, this means that authoritarianism contributes significantly to the prediction of attitudes

---

[13]These calculations are left as an exercise for the student. The student's results should be close to those given above, but there will probably be relatively minor errors of rounding. If the results agree, however, to two decimal places, they are essentially "correct." [*Suggestion*: Use five or six decimal places in the calculations.]

toward outgroups, after taking dogmatism and religiosity into account. It also means that b_1 is statistically significant. At the bottom of Table 4.5, the square root of the F ratio has been calculated: $\sqrt{12.7576} = 3.5718$. But this is the t ratio calculated with b_1 earlier. So the t ratio and the analysis of variance of Table 4.5 amount to the same thing. The analysis of variance approach has been labored because it makes the nature of the t test of b weights quite clear. Most computer programs report the t ratios. But a program could just as well calculate and report F ratios. The important thing is that the student know what the t and F ratios mean.

The Statistical Significance of Variables Added to the Regression Equation

The problem now to be discussed, the statistical significance of variables added to the regression equation, was introduced in some detail in Chapter 3. Because of its importance and the need to generalize to any number of independent variables, it is taken up again, but in more detail.

Suppose that the data of Table 4.2 consisted of the first and second independent variables, X_1 and X_2, and the dependent variable, Y, and that a multiple regression analysis has been done. The necessary basic statistics are of course given in Tables 4.2 and 4.3. As usual, the sums of squares of regression and deviations from regression, ss_{reg} and ss_{res}, the a and b coefficients, and R^2 and F are calculated. The results are as follows:

$$b_1 = .6737, \qquad b_2 = .6183$$
$$Y' = -1.2356 + .6737X_1 + .6183X_2$$
$$ss_{reg} = 106.6614, \qquad ss_{res} = 58.3386$$
$$R^2_{y.12} = .6464, \qquad F = 15.541$$

Since this F ratio, at 2 and 17 degrees of freedom, is significant at the .01 level, the regression is statistically significant.

Now, compare the R^2's and F ratios to those calculated for the four-variable problem:

$$R^2_{y.123} = .6637, \qquad F = 10.526 \quad (p < .01)$$

We must answer the question: Does the addition of X_3 add significantly to the regression or the prediction? To answer this question, another F ratio must be calculated, the F ratio of the difference between the two R^2's, $R^2_{y.123}$ and $R^2_{y.12}$. The formula for the F ratio is, in this case,[14]

$$F = \frac{(R^2_{y.123} - R^2_{y.12})/(k_1 - k_2)}{(1 - R^2_{y.123})/(N - k_1 - 1)} \tag{4.14}$$

where N = total number of cases; k_1 = number of independent variables of the larger R^2, in this case 3; k_2 = number of independent variables of the smaller

---

[14]See Cohen (1968, p. 435) and Guilford (1956, p. 400).

R^2, in this case 2. Substituting the values just calculated gives

$$F = \frac{(.6637 - .6464)/(3-2)}{(1 - .6637)/(20-3-1)} = \frac{.0173/1}{.3363/16} = \frac{.0173}{.0210}$$
$$= .824 \text{ (not significant)}$$

The F ratio is less than 1 and, of course, is not significant. Variable X_3, religiosity, does not significantly add to the prediction of Y.[15]

It is important to note that this testing technique can be generalized to other comparisons. To do so we write the equation in general form:

$$F = \frac{(R^2_{y.12...k_1} - R^2_{y.12...k_2})/(k_1 - k_2)}{(1 - R^2_{y.12...k_1})/(N - k_1 - 1)} \tag{4.15}$$

where $R^2_{y.12...k_1}$ = the squared multiple correlation coefficient for the regression of Y on k_1 variables (the larger coefficient); and $R^2_{y.12...k_2}$ = the squared multiple correlation coefficient for the regression of Y on the k_2 variables, where k_2 = the number of independent variables of the smaller R^2.

Now test the addition of variable 2 to variable 1. That is, use $R^2_{y.12}$ and and $R^2_{y.1}$ in the equation

$$F = \frac{(R^2_{y.12} - R^2_{y.1})/(k_1 - k_2)}{(1 - R^2_{y.12})/(N - k_1 - 1)}$$

$R^2_{y.1}$ is simply the square of the correlation between X_1 and Y: $(.6735)^2 = .4536$. $R^2_{y.12}$, as calculated above, is .6464. Therefore,

$$F = \frac{(.6464 - .4536)/(2-1)}{(1 - .6464)/(20-2-1)} = \frac{.1928}{.3536/17} = 9.269$$

which, at 1 and 17 degrees of freedom, is significant at the .01 level. Variable 2, dogmatism, adds significantly to the regression.

In sum, variable 1, authoritarianism, is a significant predictor of Y, attitudes toward outgroups. The addition of variable 2, dogmatism, significantly increases the accuracy of the prediction. But variable 3, religiosity, adds little to the prediction even though $R^2_{y.123}$ is greater than $R^2_{y.12}$.

Calculating R^2's in this manner and using the F test to evaluate the statistical significance of increments to prediction, as it were, is a powerful method of analysis. It enables us to determine the relative efficacies of different variables in the regresssion equation, at least as far as statistical significance is concerned. It should be borne in mind, however, that the relative efficacies of the variables are affected by the order of the variables in the equation. It is quite possible for a variable to be by itself a significant predictor of a dependent variable, but, when added to another variable, which is itself a significant predictor of the dependent variable, not to add anything to the prediction. In the present example, X_3 is not by itself a significant predictor of Y because the

---

[15]A test of the significance of the regression of Y on X_3 alone also shows the regression to be not significant. The R^2 to use in equation (3.12) is of course the square of the correlation between X_3 and Y, or, from Table 4.3, $(.3475)^2 = .1208$.

correlation between X_3 and Y is .35 which, at 18 degrees of freedom, is not statistically significant at the .05 level. But in an example to be presented in the next section we will find that a third variable, which by itself *is* a significant predictor of Y, does not add significantly to the prediction.

The order in which variables are entered in a regression equation, then, is highly important. A variable entered as X_1 may act quite differently when entered as X_2 or X_3. The higher the correlation between X_1 and X_2, the more pronounced will be the difference. In the present example, X_2, dogmatism, would have accounted for more of the variance of Y, attitudes toward out-groups, if it had been entered into the regression equation as X_1 rather than X_2. (It is important to note that the order of the variables does not affect the regression coefficients.)

Before leaving this subject, another note of caution is needed. We must clearly distinguish, as we always should, between statistical significance and the magnitude and importance of the relations among variables. An F or t ratio can be statistically significant when the magnitude of a relation is actually trivial. For instance, suppose a multiple regression has been calculated with two independent variables and 150 cases and $R_{y.12} = .22$. The F ratio is 3.87, significant at the .05 level. But $R^2_{y.12}$ is only .05, hardly a relation of much importance. Similarly, the addition of a variable to a multiple regression may increase R^2 significantly, but the increase may be, and often is, quite small. The student will profit from reading Hays' (1963, pp. 323–327) fine discussion of statistical significance and magnitude of relations.

An Example with a Coded Variable

To reinforce our learning of the points made in Chapter 3 and in this chapter, and to introduce another important idea in the analysis of behavioral science data, we add another example. We take the data given in Tables 4.2 and 4.3, drop variable 3, or X_3, and replace it with a coded or dummy variable. Dummy variables will not be explained now. Explanation is saved for a later more thorough discussion. For present purposes, let us say that 1's and 0's are assigned to individuals or cases depending on their status on attributes like sex, social class, political preference, marital status, and the like. Variable 3 in the new problem is social class. If an individual is working class we assign him a 1. If, however, he is middle class we assign him a 0.[16] We believe, say, that knowledge of the social class membership or status of our subjects should perhaps enhance the prediction of prejudice toward outgroups. Suppose there is research evidence to support such a notion. Let us also suppose that there is little or no knowledge of the relations between social class and authoritarianism and dogmatism. And there is no evidence at all on how the three variables together may be related to prejudice toward outgroups. In any case, the new data

---

[16]The social class variable is capable of more precise measurement. A five-point scale, for instance, is feasible. A better example might have been a true dichotomous variable like sex. But we wanted our correlation and regression outcomes to be as realistic as possible.

are given in Tables 4.6 and 4.7, which are like Tables 4.2 and 4.3. Note that the various sums of squares and correlations are calculated, as before. The calculation with X_3 and its 1's and 0's proceeds exactly as though the 1's and 0's were continuous measures.

Note first, in Table 4.7, the substantial correlation of variable 3 with Y, .5571. Note, too, that X_1 and X_3 are substantially correlated, .4812, but that X_2 and X_3 have a low correlation, .0970. The remaining correlations are the same as those in Table 4.3. One would expect that the correlation of .5571 between

TABLE **4.6** FOUR-VARIABLE EXAMPLE WITH CODED VARIABLE: ATTITUDES TOWARD OUTGROUPS (Y), AUTHORITARIANISM (X_1), DOGMATISM (X_2), AND SOCIAL CLASS (X_3) MEASURES

Y	X_1	X_2	X_3
2	2	5	1
1	2	4	0
1	1	5	0
1	1	3	0
5	3	6	0
4	4	4	0
7	5	6	1
6	5	4	1
7	7	3	1
8	6	3	1
3	4	3	0
3	3	6	0
6	6	9	0
6	6	8	1
10	8	9	1
9	9	6	1
6	10	4	1
6	9	5	0
9	4	8	0
10	4	9	1
Σ: 110	99	110	10
M: 5.50	4.95	5.50	.50
Σ^2: 770	625	690	10

X_3 and Y would definitely enhance the prediction. And the correlation of .0970 between X_2 and X_3 supports the notion of an enhanced prediction. But how about the substantial correlation between X_1 and X_3? Remember the ideal multiple regression situation: high correlations between the predictors and the criterion and low correlations among the predictors. In this case we have a nice mixture of correlations and an intriguing question. This is a good example of a situation where we cannot be at all sure of the answer to the question until we have done the analysis.

TABLE **4.7**   DEVIATION SUMS OF SQUARES AND CROSS PRODUCTS,
CORRELATION COEFFICIENTS, AND STANDARD DEVIATIONS OF DATA OF
TABLE 4.6[a]

	y	x_1	x_2	x_3
y	165.00	100.50	63.00	16.00
x_1	.6735	134.95	15.50	12.50
x_2	.5320	.1447	85.00	2.00
x_3	.5571	.4812	.0970	5.00
s	2.9469	2.6651	2.1151	.5130

[a]See footnote a, Table 4.3, for an explanation of the entries.

As usual, the β's and b's must be calculated. We again use equation (4.5): $\beta_j = R_{ij}^{-1} R_{yj}$. After calculating the inverse of R_{ij}, R_{ij}^{-1}, it is multiplied by R_{yj} to obtain β_j.

$$\begin{pmatrix} 1.3180 & -.1305 & -.6215 \\ -.1305 & 1.0223 & -.0365 \\ -.6215 & -.0365 & 1.3028 \end{pmatrix} \begin{pmatrix} .6735 \\ .5320 \\ .5570 \end{pmatrix} = \begin{pmatrix} .4720 \\ .4356 \\ .2878 \end{pmatrix}$$

$$\qquad\qquad R_{ij}^{-1} \qquad\qquad\qquad R_{yj} \qquad\quad \beta_j$$

Now, calculate b_j:

$$b_1 = \beta_1 \frac{s_y}{s_1} = (.4720)\left(\frac{2.9469}{2.6651}\right) = .5219$$

$$b_2 = \beta_2 \frac{s_y}{s_2} = (.4356)\left(\frac{2.9469}{2.1151}\right) = .6069$$

$$b_3 = \beta_3 \frac{s_y}{s_3} = (.2878)\left(\frac{2.9469}{.5130}\right) = 1.6533$$

To write the regression equation, calculate a:

$$a = \bar{Y} - b_1 \bar{X}_1 - b_2 \bar{X}_2 - b_3 \bar{X}_3$$

$$= 5.50 - (.5219)(4.95) - (.6069)(5.50) - (1.6533)(.50)$$

$$= -1.2480$$

The full regression equation, therefore, is

$$Y' = -1.2480 + .5219 X_1 + .6069 X_2 + 1.6533 X_3$$

To calculate the regression sum of squares, formula (3.6) of Chapter 3 can be used. It is given here with a new number:

$$ss_{reg} = b_1 \Sigma x_1 y + b_2 \Sigma x_2 y + b_3 \Sigma x_3 y \qquad (4.16)$$

Use the b's just calculated, take the appropriate sums of cross products from Table 4.7, and substitute in (4.16):

$$ss_{reg} = (.5219)(100.50) + (.6069)(63.00) + (1.6532)(16.00)$$
$$= 117.1385$$

Then

$$R^2_{y.123} = \frac{ss_{reg}}{ss_t} = \frac{117.1385}{165.0000} = .7099$$

R^2 is substantial. Taking the calculated R^2 at face value, 71 percent of the variance of Y is accounted for by the linear least squares combination of the three independent variables.

It is useful at this point to show another method of calculating R^2 and the regression sum of squares. The formulas are

$$R^2_{y.12...k} = \beta_1 r_{y1} + \beta_2 r_{y2} + \cdots + \beta_k r_{yk} \qquad (4.17)$$

$$ss_{reg} = R^2 ss_t \qquad (4.18)$$

where the symbols are defined as before. In studying formula (4.18), recall that $R^2 = ss_{reg}/ss_t$, and that formula (4.18) is merely obtained through algebraic manipulation. Substituting in formula (4.17) the β's (calculated earlier) and the correlations between the three independent variables and the dependent variable (see Table 4.7), we obtain

$$R^2_{y.123} = (.4720)(.6735) + (.4356)(.5320) + (.2878)(.5571)$$
$$= .7100$$

This is the same, within slight error of rounding, as the value of .7099 calculated with the formula $R^2 = ss_{reg}/ss_t$.

Using formula (4.18) to calculate the regression sum of squares yields

$$ss_{reg} = (.7099)(165.) = 117.1335$$

which is close to the earlier value of 117.1385. (The value, as calculated by computer, is 117.1405.) Formula (4.18) shows that the regression sum of squares is that part of the total sum of squares of Y due to the multiple correlation of the three variables with the dependent variable. If $R^2 = 1.00$, then $ss_{reg} = 165$: all of the total sum of squares of Y is due to the regression. At the other extreme, if $R^2 = 0$, then $ss_{reg} = 0$: none of the total sum of squares of Y is due to the regression.

From earlier calculations we know that the R^2 obtained from the regression of Y on X_1 and X_2, or $R^2_{y.12}$ is .6464. Is the difference, $.7099 - .6464 = .0635$, statistically significant? The answer is found by calculating the F ratio using formula (4.15):

$$F = \frac{(R^2_{y.123} - R^2_{y.12})/(k_1 - k_2)}{(1 - R^2_{y.123})/(N - k_1 - 1)} = \frac{(.7099 - .6464)/(3-2)}{(1 - .7099)/(20 - 3 - 1)}$$

$$= \frac{.0635}{.0181} = 3.508$$

which is statistically not significant. The addition of the coded social class variable has not enhanced the prediction.

If the results of this multiple regression analysis came from real data and other things were equal, then we could come to the following tentative conclusions.[17] A large part of the variance of prejudice toward outgroups is accounted for by authoritarianism, dogmatism, and social class. If these results are to be trusted, in other words, knowledge of individuals' degrees of authoritarianism and dogmatism and their social class membership will enable us to predict their attitudes toward outgroups quite well. Although we have some idea of the relative contributions of the three independent variables to the dependent variable—from the correlation coefficients—we really do not know yet which variable is most important in the prediction and which is the next most important, though we do know which is the least important. Nor do we know anything of the interactions among the independent variables in their influence on the dependent variable. We will take up these important matters in subsequent chapters.

Some Problems

The general method presented in this chapter is applicable to any number of independent variables. The general equation (4.1) is used just as it was used with the four-variable (three independent variables) problem of the chapter. Additional terms are merely added to equations (4.2) and (4.3) to obtain the β coefficients, and additional equations used to find the b coefficients. The R^2 and F formulas are the same. The calculations, of course, become more complex.

In addition, the F ratio for testing the statistical significance of a certain variable or certain variables in a regression can be applied, as equation (4.15) indicates, to any combinations of variables and not just to the difference between the R^2 produced by all the variables and the R^2 produced by all the variables less one variable. That is, not only can we test $R^2_{y.123} - R^2_{y.12}$; we can also test, say, $R^2_{y.1234} - R^2_{y.12}$.

There are difficulties, of course, when we add variables. One of these is computational. To do a multiple regression with more than three or four independent variables on a desk calculator is difficult, time-consuming, and highly vulnerable to computational error. Fortunately, the electronic computer has obviated the necessity of desk calculator calculations. A second difficulty is errors. It is quite easy to make mistakes when doing problems by hand. More important are errors of rounding. We have already seen how these can accumulate. With a large problem they can be troublesome indeed. The neophyte would do well to study computer output carefully. Most good regression programs will be tolerably accurate. But if the numbers of a problem are large and if there are many cases, output can be inaccurate.

---

[17]The phrase "other things equal" is important. It means that the sample is representative, that a duplication of the study will produce very similar correlations among the variables and very similar β and b weights—a doubtful assumption—and so on. These points and others like them will be discussed later in the book, especially after study of actual research examples of the use of multiple regression analysis.

Another difficulty is the instability of regression coefficients. When a variable is added to a regression equation, all the regression coefficients change. In addition, regression coefficients may change from sample to sample as a result of sampling fluctuations, especially when the independent variables are highly correlated (Darlington, 1968). All this means, of course, that substantive interpretation of regression coefficients is difficult and dangerous, and it becomes more difficult and dangerous as predictors are more highly correlated with each other.

Another problem in multiple regression analysis should be clear by now because we have met it two or three times before. In fact, we explained it in some detail in Chapter 3. Its importance warrants repetition in the context of this chapter, however. The problem is that the addition of variables to the regression equation results in decreasing prediction payoff. If all the independent variables in regression analyses were correlated zero, this principle would not be valid. The only condition for prediction for any variable or variables would be for the independent variables to be substantially correlated with the dependent variable. Under such a condition, regression analysis, but particularly the interpretation of data obtained by regression analysis, would be greatly simplified. Unfortunately, the reality is that, with the exceptions that we will discuss in later chapters, independent variables are usually correlated. Consequently, interpretation of regression analysis data is often complex, difficult, even misleading.

Concluding Remarks

We have presumably accomplished the objectives set at the beginning of the chapter. In so doing we have tried to lay the foundations for a basic understanding of the rationale of multiple regression, for how multiple regression works, for performing most of the calculations involved, and for interpreting the results of the analysis. In our desire to keep the discussion fairly straightforward and near a level that will permit intuitive grasp of the notions involved, we have excluded certain ideas and techniques that are essential parts of the complete multiple regression fabric. Some of these neglected ideas and techniques we will take up as we go along.

An important fundamental point made earlier should now be made again. Regression is perhaps more closely, clearly, and tightly related to science and scientific investigation than most other analytic techniques of the behavioral sciences. The core of science and scientific investigation is to understand and explain natural phenomena through controlled and empirical inquiry. To explain natural phenomena means to explain dependent variables. In the two contrived examples we have used, the phenomena to be explained, the dependent variables, were reading achievement and prejudice toward outgroups. But what does "explain" mean? A shorthand and powerful mode of explanation is implied by the mathematical function of the form $y = f(x)$, which is a rule of correspondence. It is a rule—in this case the rule is f—that assigns to each member of a set of objects some *one* member of another set. Functions are a special form of relations or sets of ordered pairs.[18] For example, when we set up the social class independent variable in the example of this chapter, we in effect "created" a function of the form $y = f(x)$, where y = measures of prejudice toward outgroups, x = measures of social class membership, or $\{1,0\}$, and f = the rule for setting up the function, or, in this case, the rule of measurement.

A simple linear regression equation is in essence a function: $y = a + bx$. In function form it can be written $y = f(x)$, and f, the rule, says or implies how to find the constants a and b. The regression analysis method, in other words, tells us how to set up the function so that the deviations from regression are a minimum. That is, the regression techniques we are learning are rules for establishing new least squares functions from the sets of ordered pairs of empirical data. From the empirical set of ordered pairs, $\{y, x\}$ — the data — we calculate a and b and then, using the regression equation, calculate another set of ordered pairs, $\{y, y'\}$, the set of pairs of the obtained dependent variable, y, and the predicted dependent variable, y', calculated from the independent variable.

Multiple regression is in principle no different. The function can still be conceived as $y = f(x)$, but x stands for more than one independent variable and f is a more complicated rule. In addition, we have several sets of functional ordered pairs, $\{y, x_1\}$, $\{y, x_2\}$, and so on, and a final set of functional ordered pairs, $\{y, y'\}$ having been calculated from the many x's rather than a single x.

It should be obvious why we said that regression is directly and tightly tied to science and scientific investigation. It deals directly and explicitly with functions and relations and thus seeks directly and explicitly to explain natural phenomena. Of course it can be shown (see Kerlinger, 1964, pp. 84ff.) that most scientific investigation deals essentially with relations and functions. We are simply saying that regression analysis, by its very nature and method, is more closely, directly, and explicitly related to one of the fundamental aims of science than most other analytic methods.

These points have been emphasized because we are anxious *not* to have our readers become so method oriented and method conscious that they forget what analysis is all about. This can easily happen and does happen. Working with multiple regression, analysis of variance, and factor analysis can be so fascinating that one can lose sight of the scientific reason for using these powerful techniques. The reason is to study complex relations and functions through appropriate analysis of data in order to understand and explain natural phenomena.

Study Suggestions

1. X_{ij} stands for the X of the ith row and the jth column.
    (a) Identify the following elements, and repeat the identification until it

---

[18]For an elementary mathematical discussion of functions and relations see Kemeny et al. (1958, pp. 70ff.). Discussion more oriented toward the behavioral sciences can be found in Kerlinger (1964, Chapter 6) and McGinnies (1965, Chapters 3 and 4).

becomes easy:

$$X_{14}, X_{27}, X_{54}, X_{26}, X_{71}, X_{43}, X_{11}, X_{33}, X_{35}, X_{53}, X_{62}, X_{55};$$
$$X_{i3}, X_{4j}, X_{6j}, X_{i7}, X_{ij}$$

(b) Write out a 5×5 matrix, X_{ij}, labeling each X with two numerical subscripts.

(c) What do the following symbols mean:

$$X_{1k}, X_{4k}, X_{Nk}, X_{N3}, X_{N5}, b_j, r_{yj}, r_{y2}, r_{y4}, \beta_4, X_i, X_j, \Sigma X_i, \Sigma X_j$$

(d) Study the explanation of statistical symbols in Hays (1963, Appendix A).

2. Explain the meaning of equations (4.2) in the text. (Just say what each line of the set of equations means.)

3. Write out the full matrix, with numerical subscripts, of the expression $\mathbf{x}'\mathbf{x} = \Sigma x_i x_j$, for four variables. What are the diagonals? What are the off-diagonal elements?

4. Multiply the left side of equations (4.3), using the matrix multiplication rule of matrix algebra. (See Appendix A.) Compare the results with equations (4.2).

5. Explain the meaning of the following matrix equation:

$$\beta_j = R_{ij}^{-1} R_{yj}$$

6. If an investigator has the intercorrelations among four independent variables and, in addition, the correlations between each independent variable and the dependent variable, can he do a multiple regression analysis? That is, can he do such an analysis without having the raw data matrix, X_{ij}? [*Hint*: See equation (4.5) and the discussion that follows it with the correlations of the data of Table 4.2.]

   If we know nothing but the correlations and do a multiple regression analysis, what regression statistics do we forego? Can we calculate R^2 knowing only the correlations? [See equation (4.13).]

7. Add the X_1, X_2, and X_3 scores of Table 4.2 to form a new composite vector of scores; for example, $2+5+1 = 8$; $2+4+2 = 8$; $1+5+4 = 10$; and so on. Correlate this vector of sums or composite scores with the Y vector of scores.

   (a) Is this like a multiple regression analysis?

   (b) Is the result close to that obtained by multiple regression? If so, will it always be close?

   (*Answers*: (a) Yes; (b) Yes: $r = .78$, compared to $R = .82$; No.)

8. In Study Suggestion 10 of Chapter 3, a three-variable (two independent variables) problem was given. Using the statistics already calculated, now estimate the statistical significance of X_2 after accounting for X_1. Interpret, using the variables $X_1 =$ intelligence, $X_2 =$ social class, $Y =$ verbal achievement. [*Hint*: See formula (4.15) and the equation immediately below it in the text.]

   (*Answer*: $R_{y.12}^2 - R_{y.1}^2 = .6464 - .4536 = .1928$; $F = 9.271$, $df = 1, 17$, significant at the .01 level.)

9. Although the calculation and interpretation of the t ratios associated with regression weights are not easy, the student should go through the procedure to understand it. Use the statistics given in Study Suggestion 10 of Chapter 3 (see *Answers*) and calculate and interpret the t ratios for b_1 and b_2. You will need another statistic, R_j^2, or $R_{1.2}^2$ and $R_{2.1}^2$ (see text). With

two variables these are easy to obtain: $r_{12} = .1447$, and $R^2_{1.2} = (.1447)^2 = .0209$, and $R^2_{2.1} = (.1447)^2 = .0209$. (With three or more independent variables, the R^2's have to be obtained from R^{-1}, the inverse matrix of R, as described in the text.) Now calculate the standard error of estimate using formula (4.10) and the standard errors of the b's using formula (4.11). The t ratio is: $t_j = b_j/SE_{b_j}$. The degrees of freedom are the same as those for the residual: $N - k - 1 = 20 - 2 - 1 = 17$.
(*Answers*: $t_1 = 4.180$ and $t_2 = 3.045$, both significant at the .001 level.)
[*Note*: Computer programs usually give these t ratios routinely.]

10.   Does the order in which the independent variables are entered in the regression equation make any difference in evaluating the effect, say, of the first variable entered? The last variable? If so, why is this? Does the order of entry of variables affect the regression coefficients? Does the order of entry affect R^2? If the order of entry makes a difference, what difficulties does this raise in the interpretation of the data of actual research problems?

11.   Read two or three of the following studies and note how the authors used multiple regression analysis. As best you can at your present stage of learning, criticize the authors' usage.

Cutright (1963). A sociological study of considerable interest.
Knief and Stroud (1959). A relatively simple study.
Lave and Seskin (1970). Impressive article and interesting use of multiple regression.
Layton and Swanson (1958).
Scannell (1960).
Worell (1959). Dramatic increase in R demonstrated by addition of variables.

12.   Here are eight variables. Select from these variables and write three research hypotheses (using the variables appropriately as independent and dependent variables) that can be tested using multiple regression analysis:

verbal achievement, self-concept, intelligence, social class, level of aspiration, race, reading achievement, achievement motivation

13.   Suppose you had a research problem that required a scientific explanation of prejudice, and further, suppose that you know that six independent variables correlate with prejudice, for example, authoritarianism, religious conviction, education, conservatism, social class, and age. These independent variables are also known to be intercorrelated to different degrees.
(a) Under what correlation conditions will you achieve the best prediction to prejudice?
(b) Is it likely that after entering any four of these variables that the addition of the fifth and sixth variables will add substantially to the prediction? Why?

Statistical Control: Partial and Semipartial Correlation

Preceding chapters have been aimed at helping the student understand and use standard multiple regression analysis. Discussions were focused for the most part on the multiple regression equation and attendant statistics and calculations: the multiple correlation coefficient and its square, R and R^2, the F ratio associated with R^2, and the calculations of the constants of the multiple regression equation, a, b, and β. These are basic and necessary staples of multiple regression analysis. But there is much more to the subject. In this chapter we attack two or three other facets of correlation and multiple regression: statistical control through partial correlation and the calculation and interpretation of the individual and joint contributions of independent variables to the variance of the dependent variable. Of course, we have already studied the individual and joint contributions of the independent variables to the dependent variable, but now we probe deeper into the subject.

Statistical Control of Variables

Studying the relations among variables is not easy. The most severe problem is expressed in the question: Is this relation I am studying really the relation I think it is? This can be called the problem of the validity of relations. Science is basically preoccupied with formulating and verifying statements of the form, if p, then q—if dogmatism, then ethnocentrism, to borrow an example from Chapter 4. The problem of validity of relations boils down essentially to the question of whether it is *this p* that is related to q, or, in other words, whether the discovered relation between *this* independent variable and the dependent variable is truly the relation we think it is. In order to have some confidence in the validity of any particular if p, then q statement, we have to have some confidence that it is really p that is related to q and not r or s or t. To attain such confidence scientists invoke techniques of control.

Reflecting the complexity and difficulty of studying relations, control is itself a complex subject. Although fundamentally important, we cannot discuss it in detail.[1] Nevertheless, we believe that the technical analytic notions to be discussed in this chapter are best understood if they are approached as part of the subject of control. Therefore we are forced to discuss control, to some extent at least.

In scientific research, control means control of variance. There are a number of ways to control variance. The best-known is to set up an experiment, whose most elementary form is an experimental group and a so-called control group. The scientist tries to increase the difference, the variance, between the two groups by his experimental manipulation. To set up a research design is itself a form of control. One designs a study, in part, to maximize systematic variance, minimize error variance, and control extraneous variance. Another well-known form of variance control is matching subjects. One also controls variance by subject selection. If one wants to control the variable sex, for example, one can select as subjects only males or only females. This of course reduces sex variability to zero.

Potentially the most powerful form of control in research is to assign subjects randomly to experimental groups. Other things being equal, if random assignment has been used, one can assume that one's groups are equal in all possible characteristics. In a word, all variables except the one that forms the basis for the groups – different methods of changing attitudes, say – are controlled. Control here means that the variations among the subjects due to anything that makes them different are scattered, by definition at random, throughout the several groups. Unfortunately, in much behavioral research random assignment is not possible because such research is ex post facto in nature. Ex post facto research is that research in which the independent variable or variables have already "occurred," so to speak, and the investigator cannot control them directly by manipulation. The hypothetical study of the relations among authoritarianism, dogmatism, social class, and attitudes toward outgroups of Chapter 4, for instance, would be labeled ex post facto research. The independent variables are beyond the manipulative control of the researcher.

Testing alternative hypotheses to the hypothesis under study is a form of control, although different in kind from those discussed above and below. The point of this whole discussion is that different forms of control are similar in function. They are different expressions of the one principle: control is control of variance. And so it is with the statistical form of control to be discussed in this chapter. Statistical control means that one uses statistical methods to identify, isolate, or nullify variance in a dependent variable that is presumably "caused" by one or more independent variables that are extraneous to the

---

[1] For detailed discussions, see Kerlinger (1964, pp. 280–286, 360–361, 369–371; 1969).

particular relation or relations under study. Statistical control is particularly important when one is interested in the joint or mutual effects of more than one independent variable on a dependent variable because one has to be able to sort out and control the effects of some variables while studying the effects of other variables. Multiple regression and related forms of analysis provide ways to achieve such control.

Two Examples

First, take a rather artificial example. Suppose one is studying the relation between the size of the right-hand palm and verbal ability. There is bound to be a substantial correlation between these two variables since age underlies both of them. Palms get larger as children get older, and verbal ability increases with age. In order to ascertain the "real" relation between the two variables age must be controlled. It can easily be controlled by studying children of one age or within a narrow age range. In so doing, the variance of the variable age is reduced to zero or near zero.

Extending the example to numbers, suppose the correlation between size of palm and verbal ability, or r_{12}, is .50, the correlation between size of palm and age, or r_{13}, is .70, and that between verbal ability and age, or r_{23}, is also .70. If we now "control" the age variable and calculate the correlation between size of palm and verbal ability we find it to be .04. The method used to calculate this reduced correlation is called *partialing*, or *partial correlation*. It is an important form of statistical control.

Now, a more realistic example. Suppose we are studying how well a college selection test predicts grade-point average. Let us say that the correlation between these two variables, r_{12}, in a well-chosen sample is .50. But we know that intelligence is an important factor determining both performance on the selection test and grade-point average. We are interested only in the relation between the test and grade-point average uninfluenced by intelligence (if that is possible). We want, in other words, to "partial out" intelligence from the correlation; we want to hold it constant. The correlation between the selection test and intelligence, r_{13}, is .70, and that between grade-point average and intelligence, r_{23}, is .50. If we calculate the correlation between the selection test and grade-point average and control intelligence by partialing its effect out, the actual correlation is .24, a sharp drop indeed from the original .50.

The Nature of Control by Partialing

The formulas for calculating partial correlation coefficients are comparatively simple. What they accomplish, however, is not so simple. To help achieve the understanding we need in order to interpret multiple regression analysis results adequately, we first present the partial correlation formulas with examples and then present a detailed analysis of what is behind the statistical operations. The student is encouraged to work with us through the calculations and the reasoning presented.

The symbol for expressing the correlation between two variables with a third variable partialed out is $r_{12.3}$, which means the correlation between variables 1 and 2, partialing out variable 3. There is another way to express the latter idea: $r_{12.3}$ is the correlation that would be obtained between variables 1 and 2 in a group in which variable 3 is constant. If, somehow, all the members of the group were the same on variable 3 and the correlation between 1 and 2 was calculated, this correlation would be tantamount to $r_{12.3}$. Take a simple though unrealistic example. If, in a study, we use subjects all with IQ's of 100, we have controlled intelligence, or held it constant. Consequently, the correlation between variables 1 and 2 will be unaffected by intelligence. Partial correlations accomplish the same thing statistically.

In the selection test, grade-point average, and intelligence example just considered, the main interest was in the "controlled correlation" between the selection test, variable 1, and grade-point average, variable 2, partialing out the effect of intelligence, variable 3. The formula for the coefficient of partial correlation is

$$r_{12.3} = \frac{r_{12} - r_{13}r_{23}}{\sqrt{1 - r_{13}^2}\ \sqrt{1 - r_{23}^2}} \tag{5.1}$$

The calculation of the example in which the correlation dropped from .50 to .24, where $r_{12} = .50$, $r_{13} = .50$, and $r_{23} = .70$, is

$$r_{12.3} = \frac{.50 - (.50)\,(.70)}{\sqrt{1 - (.50)^2}\ \sqrt{1 - (.70)^2}} = \frac{.50 - .35}{.6185} = .24$$

There is not much point in trying to attain intuitive understanding of this formula. We therefore take another tack and set up a rather lengthy demonstration of what the partial correlation formula accomplishes – in the context of regression analysis.

Partial Correlation and Multiple Regression

Partial correlation coefficients can be calculated using regression analysis. To illustrate this, we again set up a simple fictitious example that the student can easily work through with us. The example is given in Table 5.1. We also append below the raw data the deviation sums of squares, variances, and standard deviations of each of the three groups.[2] The immediate problem is to calculate $r_{12.3}$, the correlation between variables 1 and 2, partialing out variable 3. First, we calculate $r_{12.3}$ using formula (5.1). To do so, however, we need to know the correlations among the three variables. These are easily calculated from the deviation sums of squares, Σx^2, and the deviation cross products, $\Sigma x_i x_j$, which

---

[2]The student should not be misled by the simplicity and the uniformity of the numbers and variables of this example. For instance, in almost no problem would the sums of squares and the standard deviations be the same. We chose these very simple numbers so that our points could be followed easily.

TABLE **5.1** FICTITIOUS DATA AND
STATISTICS FOR REGRESSION AND
PARTIAL CORRELATION ANALYSIS,
THREE-VARIABLE PROBLEM

X_1	X_2	X_3
1	3	3
2	1	2
3	2	1
4	4	4
5	5	5
Σx^2: 10	10	10
V: 2.5	2.5	2.5
s: 1.5811	1.5811	1.5811

TABLE **5.2** DEVIATION SUMS
OF SQUARES AND CROSS
PRODUCTS AND CORRELATIONS,
DATA OF TABLE 5.1[a]

	X_1	X_2	X_3
X_1	10.	7.	6.
X_2	*.70*	10.	9.
X_3	*.60*	*.90*	10.

[a]The sums of squares are in the diagonal, the cross products above the diagonal, and the correlations below the diagonal. The latter are italicized.

are given in Table 5.2, the former in the diagonal and the latter above the diagonal. The calculated correlation coefficients are given below the diagonal and are italicized.

Using formula (5.1), we obtain

$$r_{12.3} = \frac{.70 - (.60)(.90)}{\sqrt{1 - (.60)^2}\sqrt{1 - (.90)^2}} = \frac{.1600}{.3487} = .4588 = .46$$

The partialing of variable 3 from the correlation between variables 1 and 2 has sharply reduced the correlation: from .70 to .46.[3]

TABLE **5.3** REGRESSION OF X_1 ON
X_3 WITH CALCULATIONS OF
PREDICTED X'S AND DEVIATIONS
FROM REGRESSION, $d = X - X'$

X_1	X_3	$1.2 + .6X_3 = X_1'$	d_1
1	3	$1.2 + 1.8 = 3.0$	-2.0
2	2	$1.2 + 1.2 = 2.4$	$-.4$
3	1	$1.2 + .6 = 1.8$	1.2
4	4	$1.2 + 2.4 = 3.6$	.4
5	5	$1.2 + 3.0 = 4.2$	.8

It will be useful to pursue two relations that involve the residuals obtained in regression analyses. Suppose we calculate the residuals, d, obtained by the regression of X_1 on X_3, one of the predictor variables of Table 5.1. This has been done in Table 5.3. (The actual calculations of the regression statistics will be done later.) These two sets of values have been entered in Table 5.4, but, to clarify matters, the designations of the variables of Table 5.3 have been changed: X_3 becomes X, X_1' becomes Y', but d remains the same.

First calculate the correlation between the independent variable, X, and

[3]Note, however, that large reductions like this are unusual. In most cases, the reduction is more modest (Nunnally, 1967, p. 154).

TABLE **5.4**   VALUES OF
THE PREDICTOR,
$X = X_3$, THE PREDICTED,
$Y' = X_1'$, AND THE
RESIDUAL, d, VARIABLES,
DATA OF TABLE 5.3

X	Y'	d
3	3.0	-2.0
2	2.4	$-.4$
1	1.8	1.2
4	3.6	$.4$
5	4.2	$.8$

the predicted variable, Y'. The sum of squares of the cross products is $\Sigma xy' = 51 - (15)^2/5 = 10$, and $\Sigma y'^2 = 48.60 - (15)^2/5 = 3.60$. The correlation, then, is

$$r_{xy'} = \frac{\Sigma xy'}{\sqrt{\Sigma x^2 \Sigma y'^2}} = \frac{6}{\sqrt{(10)(3.6)}} = 1.00$$

This is an illuminating result. Is the correlation between a predictor variable and the predicted variable always 1? It is.

The reason is that the regression equation result is really a coding of X. A coding operation on a variable – for example, adding 2 to each value or multiplying each variable by .68 – does not affect the correlation of the variable with another variable because the correlation coefficient is actually calculated with standard scores. That is, the basic formula for r is

$$r_{xy} = \frac{\Sigma z_x z_y}{N} \tag{5.2}$$

or the mean of the cross products of the z, or standard, scores, $(X - \bar{X})/s$, where s = standard deviation. And, of course, coded scores and the original scores will yield the same standard scores. It can now easily be seen that the regression equation, $Y' = a + bX$, is a coding operation on the X scores: the X's are multiplied by a constant and another constant is added. In short, the predicted Y scores, Y', are merely coded X scores.

If we now correlate the X scores and the d scores, on the other hand, we obtain

$$r_{xd} = \frac{\Sigma xd}{\sqrt{\Sigma x^2 \Sigma d^2}} = \frac{0}{\sqrt{(10)(6.4)}} = 0$$

And this is always true because the d scores, the residuals, are the deviations from prediction, the errors made in predicting Y from X.

Another way to understand these outcomes is to note that the sum of squares due to regression is $\Sigma y'^2 = 3.6$, and this is entirely accounted for by X. On the other hand, the sum of squares due to the deviations from regression is $\Sigma d^2 = 6.4$, which is not at all accounted for by X. (Note, again, that $\Sigma y'^2 +$

$\Sigma d^2 = \Sigma y^2$, or $3.6 + 6.4 = 10.0$.) When a vector of residuals, $d = Y - Y'$, is created by using X as a predictor, it is said that Y is *residualized* on X. We now return to partial correlation.

A partial correlation is actually the correlation between two sets of residuals obtained as follows: Suppose we calculate the regression of X_1 on X_3 and of X_2 on X_3. The two regression equations are

$$X_1' = a + bX_3$$
$$X_2' = a + bX_3$$

After solving each of the equations separately and calculating the predicted dependent variable measures, X_1' and X_2', we calculate the two sets of residuals $d_1 = X_1 - X_1'$ and $d_2 = X_2 - X_2'$. The partial correlation, $r_{12.3}$, is the correlation between the residuals, d_1 and d_2. We now do these rather roundabout calculations to clearly understand the reasoning.

First, calculate the regression of X_1 on X_3, using the data of Table 5.1 and the regression equation, above: $X_1' = a + bX_3$. The value of the regression coefficient, b, is obtained from the formula

$$b = r_{xy}\frac{s_y}{s_x} \tag{5.3}$$

or the correlation coefficient times the ratio of the standard deviation of Y to the standard deviation of X. In the present problem this formula becomes

$$b = r_{13}\frac{s_1}{s_3}$$

but since $s_1 = s_3$, $s_1/s_3 = 1$ and b (in this case) is the correlation between variables 1 and 3. Thus $b = r_{13} = .60$. (We could of course have calculated b using an earlier formula: $b = \Sigma x_1 x_3 / \Sigma x_3^2 = 6/10 = .60$.)

The intercept constant, a, is obtained, as usual, from the formula

$$a = \bar{X}_1 - b\bar{X}_3 = 3.0 - (.60)(3.0) = 1.2$$

(Previously this formula was $a = \bar{Y} - b\bar{X}$.) The regression equation, then, is

$$X_1' = a + bX_3$$
$$X_1' = 1.2 + .6X_3$$

In Table 5.3, the original X_1 and X_3 values, the calculations of the five predicted X_1' values, and the deviations from the predicted values, $d_1 = X - X'$, were laid out. (The subscript 1 is used with d, the residuals, simply to distinguish the present operation from the second analysis using X_2 and X_3.) The main purpose of these calculations was to obtain the d_1 values. As we learned in previous chapters, they are the deviations from regression. They represent the factors other than X_3 that contribute to the variance of X_1.

If we treat the values of X_2 and X_3 similarly, we should find a similar vector of d's, which represent the factors other than X_3 that contribute to the variance of X_2. The regression equation is

$$X_2' = a + bX_3$$
$$X_2' = .3 + .9X_3$$

(The reader is left to calculate a and b.) The calculation of the predicted X_2''s and the d_2 values are shown in Table 5.5.

The vector of d's in Table 5.3 represents the errors in predicting X_1 from X_3. The vector of d's in Table 5.5 represents the errors in predicting X_2 from X_3. As said above, the d values of Table 5.3 represent factors other than X_3 that contribute to X_1. Similarly, the d values of Table 5.5 represent factors other than X_3 that contribute to X_2. Remember that we want to know the correlation between variables 1 and 2 after partialing out the influence of variable 3.

TABLE **5.5**   REGRESSION OF X_2 ON X_3 WITH CALCULATIONS OF
PREDICTED X'S AND DEVIATIONS FROM REGRESSION,
$d = X - X'$

X_2	X_3	$.3 + .9X_3 = X_2'$	d_2
3	3	$.3 + 2.7 = 3.0$	0
1	2	$.3 + 1.8 = 2.1$	-1.1
2	1	$.3 + .9 = 1.2$	.8
4	4	$.3 + 3.6 = 3.9$	.1
5	5	$.3 + 4.5 = 4.8$	.2

If the d vector of the regression of variable 1 on variable 3 represents influences on variable 1 other than variable 3, and if the d vector of the regression of variable 2 on variable 3 represents influences on variable 2 other than variable 3, then the correlation between the two d vectors should be the correlation between variables 1 and 2 uninfluenced by variable 3. If the reader will take the trouble to calculate the correlation between the two d vectors, he will find the correlation to be the same as that calculated by the partial correlation formula, $r_{12.3} = .46$. In short, the correlation of two variables with the influence of a third variable held constant is the correlation between the residuals obtained from the regressions of each of the variables on the third variable.

Partial correlation, then, is a technique of control in which each of the two variables in a relation from which we want to remove the influence of a third variable is residualized on the third variable. We partialed out of the correlation between variables 1 and 2 the effect of variable 3 by creating two d vectors: the d vector from the regression of X_1 on X_3 and the d vector from the regression of X_2 on X_3. In our new terminology, we residualized X_1 on X_3 and X_2 on X_3. We then correlated the residuals to obtain the partial correlation. From the above demonstration that $r_{xd} = 0$, we can see that in the earlier example, $r_{x_3d_1} = 0$ and $r_{x_3d_2} = 0$. That is, the correlation between X_3 and the residual vector d_1 is zero, and the correlation between X_3 and the residual vector d_2 is zero. Therefore the two d vectors are free of the influence of X_3. Thus, if we correlate them they must represent the correlation between X_1 and X_2 free of the influence of X_3.

Other Partial Correlations

The partial correlations considered to this point may be called first-order partial r's: we remove the effect of one variable from the correlation between two other variables. With three variables, we can calculate three first-order partial r's: $r_{12.3}$, $r_{13.2}$, $r_{23.1}$. The formulas are of course the same as formula (5.1) except for the subscripts:

$$r_{13.2} = \frac{r_{13} - r_{12}r_{23}}{\sqrt{1 - r_{12}^2}\,\sqrt{1 - r_{23}^2}} \qquad (5.4)$$

$$r_{23.1} = \frac{r_{23} - r_{12}r_{13}}{\sqrt{1 - r_{12}^2}\,\sqrt{1 - r_{13}^2}} \qquad (5.5)$$

For the data of Table 5.1 the coefficients are $r_{13.2} = -.10$ and $r_{23.1} = .84$.

The student can profit from study of these correlations. If variable 1 is grade-point average in college, variable 2 scores on a college entrance examination, and variable 3 intelligence test scores, then the three partial r's will have more meaning. The main problem, solved by $r_{12.3}$, is that the correlation of basic interest, $r_{12} = .70$, indicates substantial predictive power of the entrance examination to grade-point average. But we must ask: Does some other variable inflate this r, for example, intelligence? Since $r_{12.3} = .46$, it is clear that intelligence does play an important role in the prediction and must be controlled if we want to know the "real" predictive efficacy of the entrance examination.

Now take the interpretation of $r_{13.2} = -.10$. This dramatic effect of the partial r equation means that the correlation between intelligence and grade-point average, which appeared to be .60, is actually close to zero after controlling for whatever the entrance examination measures. It is not likely, of course, that we would try to interpret such a correlation. It is quite unrealistic due to the contrived nature of the example. The third partial correlation, $r_{23.1} = .84$, means that the correlation between the entrance test and intelligence, which was originally .90, was reduced somewhat by controlling for grade-point average. Again, this partial r is not too useful. This set of r's illustrates the important truism that the use of statistics has to be governed by research problems and hypotheses and not be indiscriminately applied to all situations and variables. The only really sensible partial correlation, in this case, is $r_{12.3} = .46$. It means that the more accurate estimate of the correlation between the entrance examination and grade-point average, the correlation presumably unaffected by intelligence, is .46, a reduction in the predictive power of about 28 percent ($.70^2 - .46^2 = .49 - .21 = .28$).

Partial correlation is not limited to three variables. So-called higher-order partial correlations can be calculated. The order of partial r's is determined by the number of variables being partialed. $r_{12.3}$ is a first-order partial, while $r_{12.34}$ is a second-order partial, since variables 3 and 4 are partialed from variables 1 and 2.

The reasoning and procedure outlined above apply to higher-order partial r's, which can be calculated by using successive partialing. For example,

$$r_{12.34} = \frac{r_{12.3} - r_{14.3}r_{24.3}}{\sqrt{1 - r_{14.3}^2}\ \sqrt{1 - r_{24.3}^2}}. \tag{5.6}$$

Note that the formula uses first-order partials. For third-order partials the formulas and calculations are cumbersome, but the pattern is the same.[4]

Another Method of Viewing and Calculating Partial Correlations

The above method of calculating partial correlations is the traditional one presented in most texts. Unfortunately, while the formulas work, they are, as said earlier, rather cumbersome. More important, they hardly give the student an intuitive feeling for what is behind partial correlations, nor do they reflect the relation between multiple regression analysis and partial correlation. We now present a more general and simpler method—at least conceptually—suggested by Ezekiel and Fox (1959, pp. 193–194) and Quenouille (1950, pp. 124–125). We do not necessarily recommend the method for actual calculations because it has a drawback. To calculate some partial correlations, it requires terms not ordinarily calculated in multiple regression analysis. Like the previous demonstration that a partial correlation is the correlation between two residuals, the purpose of the following presentation is mainly pedagogical.

Partial correlation can be viewed as a relation between residual variances in a somewhat different way than described earlier. $R_{y.123}^2$ expresses the variance in Y accounted for by X_1, X_2, and X_3. $R_{y.12}^2$ expresses the variance in Y accounted for by X_1 and X_2. We also learned earlier that $1 - R_{y.123}^2$ expresses the variance in Y *not* accounted for by the regression of Y on X_1, X_2, and X_3. Similarly, $1 - R_{y.12}^2$ expresses the variance *not* accounted for by the regression of Y on X_1 and X_2. Let us borrow some calculations done on the main data of Table 4.2, Chapter 4, in which there were three independent variables and a dependent variable. $R_{y.123}^2$ was equal to .6637. $R_{y.12}^2$ was also calculated in the latter part of Chapter 4; it was $R_{y.12}^2 = .6464$. With only these two R^2's, one of the three partial correlations can be calculated, $r_{y3.12}$, the correlation between Y and X_3, partialing out X_1 and X_2.

The formula for the *square* of $r_{y3.12}$ is (Ezekiel and Fox, 1959, p. 193)

$$r_{y3.12}^2 = \frac{(1 - R_{y.12}^2) - (1 - R_{y.123}^2)}{1 - R_{y.12}^2}$$

Substituting the values given above, we obtain

$$r_{y3.12}^2 = \frac{(1 - .6464) - (1 - .6637)}{1 - .6464} = \frac{.0173}{.3536} = .0489$$

The square root is $\sqrt{.0489} = .2211$, which is the partial correlation between Y

---

[4]Certain multiple regression computer programs, which we shall discuss at the end of the book, provide partial correlations as part of their output.

and X_3, partialing out X_1 and X_2. (It should be noted that the numerator of the equation is the squared semipartial correlation to be discussed later.)

In like manner, $r_{y1.23}$ and $r_{y2.13}$ can be calculated, provided that the appropriate R^2's have been calculated. In a problem with only two independent variables, $r_{y1.2}$ and $r_{y2.1}$ can be calculated similarly. The formula for $r^2_{y1.2}$, for example, is

$$r^2_{y1.2} = \frac{(1 - R^2_{y.2}) - (1 - R^2_{y.12})}{1 - R^2_{y.2}}$$

The only flaw in the method is that all the R^2's needed are not usually calculated, even with computer programs. (There is no compelling reason, however, why they could not be routinely calculated by the computer.)

In the formula for $r^2_{y3.12}$, above, $1 - R^2_{y.123}$ indicates the variance of Y not accounted for by X_1, X_2, and X_3. It is the variance of the deviations from regression, the residual variance. The expression $1 - R^2_{y.12}$ is the variance in Y not accounted for by X_1 and X_2. We have, then, two residual variances, one from the regression of Y on X_1, X_2, and X_3 and one from the regression of Y on X_1 and X_2. The latter is larger than the former because, in general, the more variables put into the regression equation the greater R^2 and the smaller $1 - R^2$. What can be called the *partial variance* is the ratio of the difference between the larger and smaller residual variances to the larger residual variance. The partial correlation is then the square root of this partial variance.

The nature of partial variance and thus partial correlation can perhaps be clarified by study of Figure 5.1. The figure has been drawn to represent the situation in calculating $r^2_{y3.12}$. The area of the whole square represents the total variance of Y: it equals 1. The horizontally hatched area represents $1 - R^2_{y.12} = 1 - .6464 = .3536$. The vertically hatched area (it is also doubly hatched due to

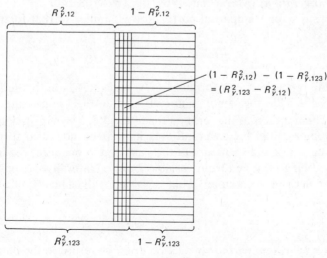

FIGURE 5.1

the overlap with the horizontally hatched area) represents $R^2_{y.123} - R^2_{y.12} =$ .6637 − .6464 = .0173. It also represents $(1 - R^2_{y.12}) - (1 - R^2_{y.123}) = (1 -$ .6464) − (1 − .6637) = .3536 − .3363 = .0173. (The areas $R^2_{y.12}$ and $R^2_{y.123}$ are labeled in the figure.) The partial variance is simply the ratio of the doubly hatched area to the horizontally hatched area, or (.3536 − .3363)/.3536 = .0173/ .3536 = .0489. Or, it can be interpreted as the squared correlation between Y and X_3, after excising the effect of X_1 and X_2. The shared variance, expressed by the doubly hatched area of Figure 5.1, is the basis of this interpretation.

Semipartial Correlation

In a sense, the preceding discussion of partial correlation was preparatory to discussion and understanding of semipartial correlation. Semipartial correlation is important and pertinent in multiple regression analysis and particularly important in the interpretation of multiple regression data.

The partial correlation procedure discussed above partials the unwanted variance from *both* variables under study. In the example used, the effect of intelligence was partialed out of both the entrance test and the grade-point averages. This means that the partial r expresses the relation between the two variables with intelligence entirely controlled. Suppose, however, that the researcher does not want intelligence completely ruled out. Suppose he wants intelligence partialed out of the entrance examination scores and *not* out of the grade-point averages. He may believe for some reason that whatever intelligence is part of the grade-point average variable should remain in the variable and not be partialed out of it. Intelligence can then be partialed only from the entrance test and not from the grade-point averages. Such correlations are called *semipartial correlations* (Nunnally, 1967, p. 155). They have also been called *part correlations* (McNemar, 1962, pp. 167–168).

The formula for semipartial correlation is similar to that for partial correlation:

$$r_{1(2.3)} = \frac{r_{12} - r_{13}r_{23}}{\sqrt{1 - r^2_{23}}} \tag{5.7}$$

The only difference between formulas (5.1) and (5.7) is in the denominators. The term $\sqrt{1 - r^2_{13}}$ in the denominator of formula (5.1) is missing in formula (5.7). Note the notation of the term on the left, $r_{1(2.3)}$. The expression used for a partial correlation was $r_{1.23}$. Parentheses have been added in formula (5.7). They indicate the selective nature of the partialing procedure: the influence of variable 3 is being removed from variable 2 only. Naturally, we might want to remove the influence of variable 3 from variable 1 only. The formula is

$$r_{2(1.3)} = \frac{r_{12} - r_{23}r_{13}}{\sqrt{1 - r^2_{13}}} \tag{5.8}$$

To show what semipartial correlation does we return to the data of Tables 5.1 and 5.5. In Table 5.5 we calculated the predicted values, X'_2, and the deviations from prediction, $d_2 = X_2 - X_2'$, in the regression of X_2 on X_3. Recall that d_2 represented influences other than X_3 in the relation between X_2 and X_3. If we now calculate the correlation between d_2 and X_1 (of Table 5.1), this correlation will express the relation between X_1 and X_2 with X_2 purged of the influence of X_3. For the reader's convenience, the X_1 and d_2 vectors are given in Table 5.6. If we now calculate the correlation between the two vectors we obtain .37. Using formula (5.7) and substituting the original correlations between the three variables (see Table 5.2), we obtain the same value:

$$r_{1(2.3)} = \frac{.70 - (.60)(.90)}{\sqrt{1 - (.90)^2}} = \frac{.1600}{.4358} = .37$$

TABLE **5.6**   VECTORS X_1 AND d_2 FROM TABLES 5.1 AND 5.5

X_1	d_2
1	0
2	− 1.1
3	.8
4	.1
5	.2

We can calculate semipartial correlations of any order, as we do with partial correlations. Indeed, such higher-order semipartials become important in further study of multiple regression. For example, $r_{1(2.34)}$ is a second-order and $r_{1(2.345)}$ a third-order semipartial correlation. $r_{1(2.34)}$ expresses the correlation between variables 1 and 2, with variables 3 and 4 partialed out of variable 2 only. Variable 2, in other words, is residualized on variables 3 and 4. To make it quite clear, we express the meaning of $r_{1(2.34)}$ a little differently: it is the correlation between 1 and 2 after having subtracted from 2 whatever it shares with 3 and 4.

The main reason so much space has been devoted to partial correlation and semipartial correlation is that they are important in a deeper understanding of multiple regression and correlation, but especially important in the substantive interpretation of multiple regression results. We now turn, then, to the relation between multiple regression and semipartial correlation.

Multiple Regression and Semipartial Correlation

The conceptual and computational complexity and difficulty of multiple correlation arise from the intercorrelations of the independent variables. This complexity was discussed to some extent in Chapter 4, but we had to defer discussion in depth until this chapter.

If the correlations among the independent variables are all zero, the situation is simple. The multiple correlation squared is simply the sum of each of the squared correlations with the dependent variable:

$$R^2_{y.12...k} = r^2_{y1} + r^2_{y2} + \cdots + r^2_{yk} \qquad (5.9)$$

Furthermore, it is possible to state unambiguously the proportion of variance in the dependent variable accounted for by each of the independent variables. For each independent variable it is the square of its correlation with the dependent variable. The simplicity of the case in which the correlations among the independent variables are zero is easily explained: each variable offers unique information not shared with any of the other independent variables.

In most behavioral research, however, the picture is not so clear and simple. The independent variables are usually correlated, sometimes substantially. Cutright (1969), for instance, studied the presumed effects of communication, urbanization, education, and agriculture on the political development of 77 nations. The intercorrelations of his independent variables were very high: absolute values from .69 to .88. Cutright in a footnote (*ibid.*, p. 376) points out the difficulty of interpreting the results because of the high intercorrelations. We will return to this study in a later chapter. The ubiquity of smaller and larger intercorrelations of independent variables and the difficulty of unambiguous interpretation of data are especially well illustrated in the large and important study, *Equality of Educational Opportunity* (Coleman et al., 1966, Appendix). The intercorrelations of the independent variables ranged from negative to positive and from low to high. We will also return to this study in a later chapter.[5]

There is a way out of the difficulty, even though it does not solve the problem completely. We can adjust correlated variables so that their correlations are zero. In fact, the main point of our lengthy discussions of partial and semipartial correlation is just this. When correlated variables are "uncorrelated" — or the correlations are made zero — it is said that they are orthogonalized. ("Orthogonal" means right angled. Two axes, which can represent variables, are orthogonal if the angle between them is 90 degrees.) Formula (5.9) for calculating R^2 can be altered so that it is applicable to correlated variables. The altered formula, for four independent variables, is

$$R^2_{y.1234} = r^2_{y1} + r^2_{y(2.1)} + r^2_{y(3.12)} + r^2_{y(4.123)} \qquad (5.10)$$

(We use only four independent variables rather than a general formula for simplicity. Once the idea is grasped, the formula can be extended to more variables.)

Formula (5.9) is a special case of (5.10) (except for the number of variables). If the correlations among the independent variables are all zero, then (5.10) reduces to (5.9). Note what the formula says. The first independent variable, 1, since it is the first to enter the formula, expresses the variance shared by variable y and 1. Subsequent expressions will have to express the variance

---

[5]We cannot overemphasize the difficulty of the subject we are now entering. It is difficult not only because of the difficulty of interpretation of multiple regression data but also because of the great complexity of the world of behavioral science data and because we seek, as scientists, to use multiple regression to mirror some of the complexity of this world.

of added variables without duplicating or overlapping this first variance contribution. The second expression, $r^2_{y(2.1)}$, is the semipartial correlation (squared) between variables y and 2, partialing out variance shared by variables 1 and 2.

The third term, $r^2_{y(3.12)}$, is the next higher semipartial correlation. When variable 3 is introduced, we want to take out of it whatever it shares with variables 1 and 2 so that the variance it contributes to the prediction of the dependent variable is not redundant to that already contributed by 1 and 2. It expresses the variance common to variables y and 3, partialing out the variances of variables 1 and 2. The last term, $r^2_{y(4.123)}$, is the variance common to variables y and 4, partialing out the influence of variables 1, 2, and 3. In short, the formula spells out a procedure which residualizes each successive independent variable on the independent variables that preceded it. It is tantamount to orthogonalizing the independent variables. Since this is so, each term indicates the proportion of variance in the dependent variable that that variable contributes to R^2, which itself indicates the proportion of the total variance of the dependent variable that all the independent variables in the regression account for.

As far as the calculation of R^2 is concerned, it makes no difference in what order the independent variables enter the equation and the calculations. That is, $R^2_{y.123} = R^2_{y.213} = R^2_{y.312}$. But the order in which the independent variables are entered into the equation makes a great deal of difference in the amount of variance accounted for by each variable. A variable, if entered first, almost invariably will account for a much larger proportion of the variance than if it is entered second or third. In general, when the independent variables are correlated, the more they are correlated and the later they are entered in the regression equation, the less the variance accounted for.

To illustrate the calculations associated with a formula like (5.10), and to consolidate our understanding of semipartial correlation and its relation to multiple regression, we write the formula for three independent variables and then calculate the value of $R^2_{y.123}$ for the three-variable problem of Chapter 4 (data of Tables 4.2 and 4.3). The formula for three independent variables is

$$R^2_{y.123} = r^2_{y1} + r^2_{y(2.1)} + r^2_{y(3.12)} \tag{5.11}$$

The intercorrelations of the three independent variables and between the independent variables and the dependent variable are reproduced in Table 5.7.

TABLE **5.7**   CORRELATIONS AMONG THE INDEPENDENT VARIABLES
AND BETWEEN THE INDEPENDENT VARIABLES AND THE
DEPENDENT VARIABLE, DATA OF TABLE 4.3

	1	2	3	y
1	1.0000	.1447	.3521	.6735
2		1.0000	.0225	.5320
3			1.0000	.3475
y				1.0000

These correlations have merely been taken from Table 4.3 and reproduced here for the reader's convenience.

The first term on the right of formula (5.11), r_{y1}^2, is merely the squared correlation between variable 1 and the dependent variable $(.6735)^2 = .4536$. The next term is $r_{y(2.1)}^2$. The formula, adapted from formula (5.8), is

$$r_{y(2.1)} = \frac{r_{y2} - r_{y1}r_{12}}{\sqrt{1 - r_{12}^2}} \qquad (5.12)$$

Using the values of Table 5.7, we obtain

$$r_{y(2.1)} = \frac{.5320 - (.6735)(.1447)}{\sqrt{1 - (.1447)^2}} = \frac{.43454}{.98949}$$

$$= .43915$$

We need now to calculate the last term of formula (5.11), $r_{y(3.12)}^2$. The formula is

$$r_{y(3.12)} = \frac{r_{y(3.1)} - r_{y(2.1)}r_{3(2.1)}}{\sqrt{1 - r_{3(2.1)}^2}} \qquad (5.13)$$

It contains terms that we have not calculated: $r_{y(3.1)}$ and $r_{3(2.1)}$. The formulas and the calculations for the present problem are

$$r_{y(3.1)} = \frac{r_{y3} - r_{y1}r_{13}}{\sqrt{1 - r_{13}^2}} \qquad (5.14)$$

$$= \frac{.3475 - (.6735)(.3521)}{\sqrt{1 - (.3521)^2}} = \frac{.11036}{.93595} = .11791$$

$$r_{3(2.1)} = \frac{r_{23} - r_{13}r_{12}}{\sqrt{1 - r_{12}^2}} \qquad (5.15)$$

$$= \frac{.0225 - (.3521)(.1447)}{\sqrt{1 - (.1447)^2}} = \frac{-.02845}{.98949} = -.02875$$

Substituting in formula (5.13), we obtain

$$r_{y(3.12)} = \frac{.11791 - (.43915)(-.02875)}{\sqrt{1 - (-.02875)^2}} = \frac{.13054}{.99960} = .13059$$

Finally, we can calculate $R_{y.123}^2$ using formula (5.11):

$$R_{y.123}^2 = (.6735)^2 + (.43915)^2 + (.13059)^2$$

$$= .4536 + .19285 + .01705 = .6635$$

This value is the same, within rounding error, as the value of .6637 calculated by other methods in Chapter 4. In this particular problem with this particular order of the independent variables, variables 1, 2, and 3 contribute, respectively, 45 percent, 19 percent, and 2 percent to the variance of Y.[6]

---

[6]The above calculations have been done for pedagogical purposes only. The student can profit by doing them for himself. In general, however, we advise dependence on the computer.

Before leaving these manipulations, let us revert to the method developed in Chapter 4 of estimating the statistical significance of variables added to the regression equation. Recall that we subtracted one R^2 from another R^2 and tested the significance of the difference between them. For example, we can test the difference, $R^2_{y.123}$ and $R^2_{y.12}$, and if the difference is statistically significant we can say that variable 3 contributes significantly to the prediction. Since we have learned the use of squared semipartial correlations, we can be more explicit about what is actually happening when we perform such subtractions and statistical tests.

Take the above example. $R^2_{y.123} = .6637$ and $R^2_{y.12} = .6464$. The difference, $.6637 - .6464 = .0173$, is not statistically significant (see Chapter 4). This difference is really $r^2_{y(3.12)}$, the third term of Equation (5.11). (The difference between .0173 and the earlier value, .0171, is due to errors of rounding.) In general, such differences between R^2's are squared semipartial correlations. The F test applied to the differences between R^2's, then, is really a test of the statistical significance of semipartial correlations.

We have devoted a good deal of space to semipartial correlations and their calculation because of their difficulty and, more important, their usefulness in the ultimate interpretation of data obtained from multiple regression analysis. The assessment of the relative contributions of independent variables is a shaky and undependable business, as we indicated earlier. If a researcher has a reasonably sound basis for the particular order of entry of independent variables in the regression equation, however—for instance, on the basis of theory—then squared semipartial correlations provide an adequate and comparatively dependable way to estimate the relative contributions of the independent variables to the variance of the dependent variable. Although their interpretation has nothing absolute about it, it is probably the best method of estimating the relative contributions, especially when used in conjunction with other methods. In our later chapters we will use squared semipartial correlations and regression weights together to help us interpret the data of published studies.[7]

Control, Explication, and Interpretation

The main points of this chapter have been the control and explication of variables and the use and calculation of partial and semipartial correlation to help achieve control and explication of variables. From a research view, the main

---

[7]Until recently the literature on this problem has been sparse. Certain methods of analysis and interpretation, for example, using the squares of the beta weights as indices of the magnitudes of the contributions of the independent variables, are inadequate. For further discussion, see Chapter 11. See also Darlington (1968), Gordon (1968), and Pugh (1968). The Darlington article is authoritative and is required reading for any serious student of multiple regression. For a method of calculating semipartial correlations that avoids the complexities of the formulas given above, see Nunnally (1967, pp. 154–155, 165–171).

point of the chapter was the interpretation of the data yielded by multiple regression analysis. In a multiple regression analysis we not only want to know how well a combination of independent variables predicts a dependent variable; we also want to know how much each variable contributes to the prediction. R^2 indicates the portion of the total variance of the dependent variable that the independent variables account for. In the problem used in the last section, $R^2 = .66$: the combination of the three independent variables accounted for 66 percent of the total variance of the dependent variable. In substantive terms, the combination of authoritarianism (X_1), dogmatism (X_2), and religiosity (X_3) accounted for about two-thirds of the variance of attitudes toward outgroups (Y). It makes no difference, as we said before, how we arrange the independent variables: we will always obtain $R^2 = .66$.

We also said, however, that as far as the amount of variance accounted for by the individual variables is concerned, the order of entry of the variables into the regression equation does make a difference. Take the example of the last section. When entered second in the equation variable 2 accounted for 19 percent of the total regression variance. If it had been entered first in the equation it would have accounted for 28 percent of the variance. Similarly, variable 3's contribution, if it had been entered first instead of third, would have jumped from 2 percent to 12 percent. In other words, religiosity adds almost nothing to the prediction when entered third, but if entered first it contributes considerably more to the prediction of attitudes toward outgroups.[8]

All this means that the researcher has to plan his analysis on the basis of his research problem and hypotheses and on the basis of previous knowledge, if any. After a research study is completed one should not try all possibilities of variable order and then pick the one that suits him most. With four variables one has 24 possibilities of variable order! Which one is the "correct" one? Only the researcher can say, and his choice has to come from some sort of theoretical reasoning. Otherwise he is confronted with a bewildering number of possibilities and a chaos of interpretation. There is a consoling aspect to these complexities of interpretation. If the researcher is interested only in the overall prediction success of his set of variables, then the order of entering variables does not matter. As we have already said, he will always obtain the same R^2 and the same predicted Y's no matter what the order of the variables.

Partial r's and semipartial r's have different though related purposes and uses. For the most part, partial r's are used for control purposes. One wants to know the relation between variables when other variables and sources of influence on the dependent variable are controlled or held constant. In the main example of this chapter, we wanted to know the correlation between the selection test and grade-point average uninfluenced by intelligence. With partial correlation the unwanted influence is removed from both variables of the correlation. The effect of intelligence is removed from both the selection test and

[8]Because the order in which variables enter the regression equation affects their contribution to the variance of the dependent variable, we discuss different approaches to the ordering of variables in a later chapter.

grade-point averages. Partial correlations are used primarily for the control purposes mentioned.

Semipartial correlations, on the other hand, are more central to the multiple regression analysis picture. They represent the correlation between two variables with the influence of another variable or other variables removed from one of the variables being correlated. But more important from our point of view, the squares of semipartial correlation coefficients, as calculated and used earlier, tell us the amount of variance contributed by the separate independent variables of the regression equation. They tell us the variances contributed, however, *only for that particular order of the independent variables*. More accurately, they tell us the contribution to the variance of the dependent variable that each independent variable adds *after* the variance contribution of preceding variables. With three independent variables, for instance, equation (5.11) tells us that variable 1 contributes r_{y1}^2, the square of the correlation between Y and variable 1, variable 2 contributes or adds $r_{y(2.1)}^2$, or the variance represented by the square of the correlation between Y and variable 2, with the influence of variable 1 partialed from variable 2, and variable 3 contributes $r_{y(3.12)}^2$, or the variance indicated by the square of the correlation between Y and variable 3, with the influence of variables 1 and 2 partialed from variable 3.

This sort of analysis is part of what we mean by "explanation." We "explain" the variance of the dependent variable by indicating the relative contributions of the independent variables to the prediction of the dependent variable. And, of course, our ultimate explanation is of the substance of the variables and the relations among the variables. We are not interested merely in the statistics. We are interested in the explanation of the phenomenon represented by the dependent variable. In our example, we want to "explain" grade-point averages, which are of course indices of academic achievement, and we want to explain achievement in substantive language. The statistical language helps us arrive at this ultimate goal of interpreting and explaining the substantive relations among variables. And multiple regression analysis is one of the most powerful parts of statistical language. As we will see later, multiple regression analysis can be a potent tool in the development and testing of theory. It should be fairly clear by now how multiple regression analysis fits into the larger scientific picture of testing the propositions derived from theory as well as into practical prediction research.[9]

Study Suggestions

1. The student will profit from studying some of the problems of this chapter as discussed by Snedecor and Cochran (1967, Chapter 13). See, especially, their treatment of partial correlation, pp. 400–403. (As indicated earlier, this chapter is generally valuable for multiple regression analysis.)

[9] A relatively new development in the analysis of behavioral research data, which leans heavily on partial correlation, is so-called causal analysis, one aspect of which is called path analysis. We will examine path analysis and causal analysis in Chapter 11.

2. (a) An educational psychologist has found the correlation between a college admissions test and grade-point average, partialing out intelligence, to be .38. The zero-order correlation between the test and grade-point average was .54. Interpret the partial correlation.

   (b) Suppose in a study of children that the correlation between strength and height is .70 and that between strength and weight is .80. The correlation between height and weight is .86. What is the best estimate of the "true" correlation between strength and height? [Use formula (5.1).]

   (c) The correlation between level of aspiration and school achievement was found by a researcher to be .51, and the correlation between social class and school achievement was .40. The correlation between level of aspiration and social class was .30. What is the correlation between level of aspiration and school achievement with social class partialed out?   (*Answers*: (b) $r_{12.3} = .04$; (c) $r_{y1.2} = .45$.)

3. Suppose the correlation between size of palm and verbal ability is .55, between size of palm and age .70, and between age and verbal ability .80. The correlation between size of palm and verbal ability after partialing out age is −.02. Accept these correlations at face value and explain them.

4. It has been said in the text that control is control of variance. What does this mean? What does it mean, for example, to say that sex is controlled, or that intelligence is controlled? How does partial correlation enter the picture?

5. Here are some correlations among three variables. In each case calculate partial correlations, using formula (5.1). Control variable 2. Interpret the correlations using the following variables: Y = group cohesiveness (members want to stay in group); X_1 = participation in decision-making; X_2 = group atmosphere (open-closed).

	r_{y1}	r_{y2}	r_{12}
(a)	.60	.40	.50
(b)	.60	.40	.90
(c)	.90	.70	.80
(d)	.70	.90	.80

   (*Answers*: (a) .50; (b) .60; (c) .79; (d) −.08.)

6. Explain the meaning of partial correlation by using residuals. [*Suggestion*: Work carefully through Tables 5.1, 5.2, 5.3, 5.4, and 5.5 and the accompanying text.]

7. Go back to Study Suggestion 10, Chapter 3. Here are the r's: $r_{y1} = .6735$; $r_{y2} = .5320$; $r_{12} = .1447$. The comparable r's for the data of Table 3.2, Chapter 3, are $r_{y1} = .6735$; $r_{y2} = 3946$; $r_{12} = .2596$. Calculate, in each case, $r_{y1.2}$. Note that one partial r is almost the same and the other is even higher. Why is this, do you suppose?   (*Answers*: $r_{y1.2} = .71$; $r_{y1.2} = .64$.)

8. How does a semipartial correlation coefficient differ from a partial correlation coefficient? How is the squared semipartial correlation coefficient related to the study of the addition of independent variables in multiple regression analysis?

9. Explain each term of formula (5.10).

10. Why is it important for researchers to plan their studies, including the relative importance, order, or priority of independent variables? On what are such considerations as relative importance, order, or priority based?

PART 2

The Regression Analysis of Experimental and Nonexperimental Data

Categorical Variables, Dummy Variables, and One-Way Analysis of Variance

Whenever one formulates a hypothesis about a relation between one or more independent variables and a dependent variable, one is in effect saying that a certain proportion of the variability of the dependent variable is accounted for by the independent variables. In preceding chapters methods of using information from continuous independent variables for the prediction of the dependent variable were presented. It should be obvious, however, that information from what may be called categorical variables may also be useful in accounting for the variance of the dependent variable.

A *categorical variable* is one in which subjects differ in type or kind. Each subject is assigned to one of a set of mutually exclusive categories that are not ranked. Although one may use numerals for the purpose of identifying the various categories, such numerals do not denote quantities. In contrast, a *continuous variable* is one in which subjects differ in amount or degree. A continuous variable can take on numerical values that form an ordinal, interval, or ratio scale. In short, a continuous variable expresses gradations, whereas a categorical variable does not.[1]

Examples of categorical variables are: sex, political party affiliation, race, marital status, and experimental treatments.[2] Some examples of continuous

[1]Strictly speaking, a continuous variable has infinite gradations. When measuring height, for example, one may resort to ever finer gradations. Any choice of gradations on such a scale depends on the degree of accuracy called for in the given situation. When, on the other hand, one is measuring the number of children in a set of families, one is dealing with a variable that has only discrete values. This type of variable, too, is referred to as a continuous variable in this book. Some writers use the terms *qualitative* and *quantitative* for what we call *categorical* and *continuous*.

[2]It is possible to have experimental treatments that can be ordered on a scale, for example different dosages of a drug. We return to this point in a later chapter. In the present context experimental treatments refer to distinct treatments, like different methods of teaching, different kinds of reinforcement, and so on.

variables are: intelligence, achievement, dosages of a drug, intensity of a shock, and frequency of reinforcement. When, for example, one wishes to study whether males and females differ in their attitudes toward American intervention in Southeast Asia, one is studying the relation between a categorical independent variable (sex) and a continuous dependent variable (attitudes). When one wishes to assess the effects of three different methods of teaching reading on the achievement of fifth graders, one is using a variable with three nonordered categories (teaching methods) to account for variability in reading achievement. Basically, one seeks to determine whether using knowledge of group membership will significantly reduce errors of prediction as compared to errors made when this information is not used.

Using Information of Group Membership

The basic linear equation referred to several times earlier is $Y' = a + bX$. It can be easily demonstrated that this formula is algebraically identical to the formula $Y' = \bar{Y} + bx$. Substituting $\bar{Y} - b\bar{X}$ for a in the first formula, we obtain

$$\begin{aligned} Y' &= \bar{Y} - b\bar{X} + bX \\ &= \bar{Y} + b(X - \bar{X}) \\ &= \bar{Y} + bx \end{aligned} \tag{6.1}$$

Formula (6.1) indicates that the application of the regression equation leads to the prediction of a score composed of the grand mean of the dependent variable plus the product of the regression coefficient and an individual's standing on the independent variable; that is, his deviation from the mean of the independent variable. Note that when $b = 0$, that is, when the correlation between X and Y is zero, formula (6.1) will lead to a prediction of a score equal to the mean of Y for each individual in the group. When no information other than group membership is available, or when available information is irrelevant, the predicted score for all subjects is the mean of the dependent variable.

A Numerical Example

As an illustration let us assume that we have the scores of attitudes toward divorce laws for a group of 15 individuals. These fictitious scores are listed in Table 6.1 under Y. The procedures and interpretations are, of course, generalizable to any set of scores with any number of subjects. If we were asked to guess at, or predict, each individual's score, our best prediction would be the mean of the group, which is 6.00. Since we also know the actual score each individual obtained, it is possible to calculate the errors in the predictions. These errors are listed as discrepancies of the actual scores from the predicted scores, under column 2 in Table 6.1. They sum to zero, as do all deviations from a mean. The squares of the deviations are listed in column 3, and their sum is 120. The sum of the squares of errors when predicting that each individual has a score equal to the mean is 120. It can be shown[3] that choosing a

---

[3]See, for example, Edwards (1964, pp. 5–6).

TABLE **6.1** FICTITIOUS DATA FOR FIFTEEN SUBJECTS

Group	Ss	1	2	3	4	5	6	7	8	9
		Y	$Y - \bar{Y}$	$(Y - \bar{Y})^2$	$Y - \bar{Y}_1$	$(Y - \bar{Y}_1)^2$	$Y - \bar{Y}_2$	$(Y - \bar{Y}_2)^2$	$Y - \bar{Y}_3$	$(Y - \bar{Y}_3)^2$
A_1	1	4	-2	4	-2	4	—	—	—	—
	2	5	-1	1	-1	1	—	—	—	—
	3	6	0	0	0	0	—	—	—	—
	4	7	1	1	1	1	—	—	—	—
	5	8	2	4	2	4	—	—	—	—
A_2	6	7	1	1	—	—	-2	4	—	—
	7	8	2	4	—	—	-1	1	—	—
	8	9	3	9	—	—	0	0	—	—
	9	10	4	16	—	—	1	1	—	—
	10	11	5	25	—	—	2	4	—	—
A_3	11	1	-5	25	—	—	—	—	-2	4
	12	2	-4	16	—	—	—	—	-1	1
	13	3	-3	9	—	—	—	—	0	0
	14	4	-2	4	—	—	—	—	1	1
	15	5	-1	1	—	—	—	—	2	4
Σ:		90	0	120	0	10	0	10	0	10

$\bar{Y} = 6$ $\bar{Y}_1(\text{Ss } 1\text{–}5) = 6$ $\bar{Y}_2(\text{Ss } 6\text{–}10) = 9$ $\bar{Y}_3(\text{Ss } 11\text{–}15) = 3$

constant other than the mean as the predicted score for all subjects will yield a sum of squared discrepancies, or deviations, larger than 120. In sum, our best prediction, under the circumstances, is that each individual's score is equal to the mean of the group.

Using Information about Membership in Distinct Groups

Until now we have dealt with the 15 subjects as if they belonged to one group. Suppose, however, that we are now told that they belong to three distinct groups. Suppose that subjects 1–5 are married males, subjects 6–10 are divorced males, and subjects 11–15 are single males.[4] Can we improve our prediction by using the additional information about membership in the three groups? In other words, how much can we reduce the sum of squares of errors of prediction by using the additional information? In order to answer this question we set up a new method of prediction, that is, *for each individual we predict a score equal to the mean of the group to which he belongs.* This has been done in columns 4–9 of Table 6.1. Column 4, for example, gives the deviations of individuals' scores from the mean of group A_1. The sum of squares of the discrepancies for each group is 10, so that for the three groups combined this sum of squares is 30. Note that by using the knowledge about membership in distinct groups the sum of squares of errors of prediction has decreased from 120 to 30, a considerable reduction. The magnitude of the error is now one-quarter of what it was when the 15 individuals were considered members of a single group. Stated in another way, 75 percent of what was previously error is now explainable by using information about group membership. No further reduction of error can be effected without additional information about the individuals. The reduction in error of prediction can be tested for significance, but we do not do so in this context since the main purpose of the presentation was to introduce the logic of using information about group membership for reducing errors of prediction.

Dummy Variables

One can show membership in a given category of the variable by the use of a dummy variable. A *dummy variable* is a vector in which members of a given category are assigned an arbitrary number, while all others — that is, subjects not belonging to the given category — are assigned another arbitrary number. For example, if the variable is sex, one can assign 1's to males and 0's to females. The resulting vector of 1's and 0's is a dummy variable. Dummy variables can be very useful in analysis of research data when the independent variables are categorical. Furthermore, as we shall see later, they have several other useful purposes. They can be used, for example, in combination with continuous independent variables or as a dependent variable.

---

[4]The three groups can, of course, represent three other kinds of categories, for example, three different experimental treatments, three different countries of origin, three religions, three professions, three political parties, and so on.

Probably the simplest method of creating a dummy variable is to assign 1's to subjects of a group one wishes to identify and 0's to all other subjects. This is basically a coding system. One can, of course, use other systems. For example, one can assign a 91 to all subjects in the group under consideration and a −3 to all others. There are, however, certain advantages in using some coding systems in preference to others.[5] For the present, we shall use the system of 1's and 0's to create dummy vectors for the three groups under consideration.

In Table 6.2 we repeat the Y vector of Table 6.1. In addition, we create two vectors, labeled X_1 and X_2. In vector X_1, subjects in group A_1 are assigned 1's, while the subjects not belonging to A_1 are assigned 0's. In vector X_2, subjects in group A_2 are assigned 1's, while subjects not belonging to group A_2 are assigned 0's. We can also create another vector in which subjects of group A_3 will be assigned 1's, and those not belonging to this group will be assigned 0's. Note, however, that such a vector is not necessary since information about group membership is exhausted by the two vectors that were created. A third vector will not add any information to the information contained in the first two vectors. Stated another way, knowing an individual's status in reference to the first two vectors, that is, knowing whether he is a member of either group A_1 or

[5]Coding systems are dealt with extensively in Chapter 7.

TABLE **6.2**   FICTITIOUS DATA FOR FIFTEEN SUBJECTS

Group	Y	X_1	X_2
	4	1	0
	5	1	0
A_1	6	1	0
	7	1	0
	8	1	0
	7	0	1
	8	0	1
A_2	9	0	1
	10	0	1
	11	0	1
	1	0	0
	2	0	0
A_3	3	0	0
	4	0	0
	5	0	0
Σ:	90	5	5
M:	6	.33333	.33333
s:	2.92770	.48795	.48795

$$r_{y1} = .00000 \qquad r_{y2} = .75000 \qquad r_{12} = -.50000$$

A_2, is sufficient information about his group membership. If he is not a member of either A_1 or A_2, he must be a member of A_3.[6] The necessary and sufficient number of vectors to code group membership is equal to the number of groups, or categories, minus one. For k number of groups we must create $k-1$ vectors, each of which will have 1's for members of a given group and 0's for subjects not belonging to the group.[7]

Multiple Regression with Dummy Variables

The principles and methods of multiple regression analysis presented in the preceding chapters apply equally to continuous and to categorical variables. When dealing with a categorical independent variable one can express it appropriately with dummy variables and do a regression analysis with the dummy variables as the independent variables.

We now do a multiple regression analysis of the data of Table 6.2, in which Y is the dependent variable and vectors X_1 and X_2 are the independent variables. The results of the basic calculations are also reported in Table 6.2. These were obtained by the methods presented in Chapter 2 and are not repeated here.

We repeat formula (3.5) with a new number:

$$b_1 = \frac{(\Sigma x_2^2)\,(\Sigma x_{1y}) - (\Sigma x_1 x_2)\,(\Sigma x_{2y})}{(\Sigma x_1^2)\,(\Sigma x_2^2) - (\Sigma x_1 x_2)^2}$$

$$b_2 = \frac{(\Sigma x_1^2)\,(\Sigma x_2 y) - (\Sigma x_1 x_2)\,(\Sigma x_1 y)}{(\Sigma x_1^2)\,(\Sigma x_2^2) - (\Sigma x_1 x_2)^2} \qquad (6.2)$$

$$a = \bar{Y} - b_1 \bar{X}_1 - b_2 \bar{X}_2$$

For the data of Table 6.2 we obtain

$$b_1 = \frac{(3.33333)\,(.00000) - (-1.66667)\,(15.00000)}{(3.33333)\,(3.33333) - (-1.66667)^2}$$

$$= \frac{25.00005}{8.33330} = 3.00002$$

$$b_2 = \frac{(3.33333)\,(15.00000) - (-1.66667)\,(.00000)}{(3.33333)\,(3.33333) - (-1.66667)^2}$$

$$= \frac{49.99995}{8.33330} = 6.00002$$

$$a = (6.00000) - (3.00002)\,(.33333) - (6.00002)\,(.33333)$$

$$= 6.00000 - 1.00000 - 1.99999 = 3.00001$$

[6]We can, of course, use two other vectors in which, for example, the 1's will be assigned to groups A_2 and A_3, while members of group A_1 will be assigned 0's throughout. We return to this point later. Note, however, that regardless of which groups are assigned 1's, the number of vectors necessary and sufficient for information about group membership is two.

[7]$k-1$ is actually the number of degrees of freedom associated with groups or categories. In the present problem, we have three groups, and therefore two degrees of freedom for groups, thus requiring two vectors.

The regression equation to two decimal places is

$$Y' = 3.00 + 3.00X_1 + 6.00X_2$$

Note that a (the Y intercept) is equal to the mean of group A_3, the group whose members were assigned 0's throughout. This will always be the case when using 1's and 0's as the coding system. Furthermore, each b is equal to the difference between the mean of a given group assigned 1's and the group assigned 0's throughout. In our problem, the mean of group A_1 is 6.00. The b associated with vector X_1 (b_1) is 3.00, which is the difference between the mean of A_1 and the mean of A_3; that is, $6.00 - 3.00 = 3.00$. Similarly, the mean of group A_2 is 9.00, and b_2 is equal to 6.00 ($9.00 - 3.00 = 6.00$).

Applying the regression equation to predict Y for a given X will, in each case, yield the mean of the group to which the X belongs. For example, for the first subject in groups A_1, A_2, and A_3, respectively (that is, subjects 1, 6, and 11 of Table 6.2):

$$Y'_1 = 3.00 + 3.00(1) + 6.00(0) = 6.00$$
$$Y'_6 = 3.00 + 3.00(0) + 6.00(1) = 9.00$$
$$Y'_{11} = 3.00 + 3.00(0) + 6.00(0) = 3.00$$

These are the means of the three groups. The regression equation leads to the same predictions made in the beginning of the chapter: for each subject a score equal to the mean of his group.

Continuing with the analysis, we calculate the regression sum of squares. We repeat formula (3.7):

$$ss_{reg} = b_1 \Sigma x_1 y + b_2 \Sigma x_2 y \qquad (6.3)$$

For the data of Table 6.2:

$$ss_{reg} = 3.00(.00) + (6.00)(15.00) = 90.00$$

The total sum of squares is 120.00 (see Table 6.2). Therefore,

$$ss_{res} = 120.00 - 90.00 = 30.00$$

It should be obvious that the results are identical to those obtained from the previous analysis. It is now possible to calculate R^2.

$$R^2_{y.12} = \frac{90}{120} = .75$$

Seventy-five percent of the sum of squares of Y is explained by group membership. This value can, of course, be tested for significance using formula (3.12), which is repeated here for convenience:

$$F = \frac{R^2/k}{(1 - R^2)/(N - k - 1)} \qquad (6.4)$$

$$F = \frac{.75/2}{(1 - .75)/(15 - 2 - 1)} = \frac{.375}{.25/12}$$

$$= \frac{(.375)(12)}{.25} = 18.00$$

An F ratio of 18, with 2 and 12 degrees of freedom, is significant at the .01 level. This indicates that the relation between group membership and attitudes toward divorce laws is significant. Stated another way, the information about group membership enhances the prediction of subjects' attitudes toward divorce laws.

To recapitulate, when dealing with a continuous dependent variable and a categorical independent variable, one creates k dummy vectors (k = number of groups or categories of the categorical variable minus one). In each vector membership in a given group is indicated by assigning 1's to the members of the group and 0's to all others who are not members of the group. $R^2_{y.12...k}$ is then computed. The R^2 indicates the proportion of variance in the dependent variable accounted for by the categorical independent variable. The F ratio associated with the R^2 indicates whether the proportion of variance accounted for is statistically significant at the level chosen by the investigator.

Alternative Approaches to the Calculation of R^2

In this section we illustrate the calculation of R^2 with two alternative methods. This presentation can be simplified since we are dealing with two independent variables only (two vectors for group membership). In the case of two independent variables the formula for R^2 can be expressed as follows:

$$R^2_{y.12} = \frac{r^2_{y1} + r^2_{y2} - 2r_{y1}r_{y2}r_{12}}{1 - r^2_{12}} \tag{6.5}$$

where $R^2_{y.12}$ = the squared multiple correlation of Y (the dependent variable) with 1 and 2 (two independent variables); r_{y1} and r_{y2} = the correlations of the dependent variable with variables 1 and 2, respectively. To apply the formula, all that is needed are the r's between the variables. These r's were reported in Table 6.2. The calculation of r between variables coded 1's and 0's can be simplified by using the following formula (see, for example, Cohen, 1968):

$$r_{ij} = -\sqrt{\frac{n_i n_j}{(n - n_i)(n - n_j)}} \tag{6.6}$$

where n_i = sample size in group i; n_j = sample size in group j; and n = total sample in the g groups. When the groups are of equal size (in our case, $n_1 = n_2 = n_3 = 5$), the formula reduces to

$$r_{ij} = -\frac{1}{g - 1}$$

where g = number of groups. In our problem $g = 3$ (three groups). Thus,

$$r_{ij} = -\frac{1}{3 - 1} = -\frac{1}{2} = -.5$$

Using now the figures from Table 6.2 we calculate

$$R^2_{y.12} = \frac{(.00)^2 + (.75)^2 - 2(.00)(.75)(-.50)}{1 - (-.50)^2} = \frac{.5625}{1 - .25} = \frac{.5625}{.75} = .75$$

The same value of R^2 was obtained with the previous calculations. Furthermore, it was shown earlier that R^2 indicates the proportion of variance in the dependent variable accounted for by the independent variables. Stated differently, R^2 indicates the proportion of the sum of squares of the dependent variable accounted for by the independent variables, or the proportion of the total sum of squares due to regression. Multiplying R^2 by the sum of squares of the dependent variable will therefore indicate the sum of squares due to regression. From the above it follows that multiplying the sum of squares of the dependent variable by $(1 - R^2)$ will indicate the sum of squares of the residuals, or error. For our problem, $\Sigma y^2 = 120$ (the total sum of squares we wish to explain). The sum of squares due to regression is

$$ss_{reg} = (R^2_{y.12})(\Sigma_y{}^2) = (.75)(120) = 90.00$$

This is the same value obtained before. The sum of squares due to error is

$$(1 - R^2_{y.12})(\Sigma_y{}^2) = (1 - .75)(120) = 30.00$$

Again, this is the same value obtained before.

We can now calculate the regression equation by first calculating the β's and then the b's. With two independent variables the formulas for the β's are

$$\beta_1 = \frac{r_{y1} - r_{y2}r_{12}}{1 - r^2_{12}} \qquad (6.7)$$

$$\beta_2 = \frac{r_{y2} - r_{y1}r_{12}}{1 - r^2_{12}} \qquad (6.8)$$

The transformation of β's to b's was discussed in Chapter 4 [formula (4.6)]. The formula is repeated with a new number:

$$b_j = \beta_j \frac{s_y}{s_j} \qquad (6.9)$$

where b_j = regression coefficients, and $j = 1, 2$; s_y = standard deviation of the dependent variable, Y; s_j = the standard deviations of the independent variables. Applying formulas (6.7)–(6.9) to the data of Table 6.2, we obtain

$$\beta_1 = \frac{.00 - (.75)(-.50)}{1 - (-.50)^2} = \frac{.375}{.75} = .50$$

$$\beta_2 = \frac{.75 - (.00)(-.50)}{1 - (-.50)^2} = \frac{.75}{.75} = 1.00$$

$$b_1 = .50\frac{2.92770}{.48795} = 3.00$$

$$b_2 = 1.00\frac{2.92770}{.48795} = 6.00$$

One can also calculate a and thus complete the regression equation, which will, of course, be identical to the one obtained earlier.

In Chapter 5 it was shown how one can calculate R^2 using zero-order and semipartial correlations. Formula (5.10) can be restated for the case of two independent variables as follows:

$$R^2_{y.12} = r^2_{y1} + r^2_{y(2.1)} \qquad (6.10)$$

where $R^2_{y.12}$ = squared multiple correlation of Y (the dependent variable) with 1 and 2 (the independent variables); r_{y1} = the zero-order correlation of Y with 1; $r_{y(2.1)}$ = the semipartial correlation of Y with 2, partialing 1 from 2.

As an alternative method of coding, let us change the assignment of the dummy variable. Vector X_1 will consist of 1's for members of group A_2, 0's for all others. Vector X_2 will consist of 1's for members of group A_3, 0's for all others. As a result, members of group A_1 will be assigned 0's throughout. Rather than displaying this in a table, which will look like Table 6.2, we report here the zero-order correlations and proceed with the calculations:

$$r_{y1} = .75 \qquad r_{y2} = -.75 \qquad r_{12} = -.50$$

$$r_{y(2.1)} = \frac{(-.75) - (.75)(-.50)}{\sqrt{1 - (-.50)^2}}$$

$$= \frac{-.75 - (-.375)}{\sqrt{1 - .25}} = \frac{-.37500}{\sqrt{.75}} = \frac{-.37500}{.86603} = -.43301$$

$$R^2_{y.12} = (.75000)^2 + (-.43301)^2 = .56250 + .18750 = .75000$$

Again, the same R^2 is obtained. If one were to calculate the regression equation by applying the formulas used earlier, one would find that a is equal to 6, the mean of the group assigned 0's throughout, while b_1 will be equal to 3.00 ($\bar{Y}_{A_2} - \bar{Y}_{A_1}$), and b_2 will be equal to -3.00 ($\bar{Y}_{A_3} - \bar{Y}_{A_1}$). This is left as an exercise for the student.

Another useful exercise is to analyze the same data by applying the following formula:

$$R^2_{y.12} = \beta_1 r_{y1} + \beta_2 r_{y2} \qquad (6.11)$$

This time vector X_1 may consist of 1's for members of group A_1, 0's for all others. Vector X_2 may consist of 1's for members of group A_3, 0's for all others. As a result, members of group A_2 will be assigned 0's throughout. If one does this, then

$$r_{y1} = .00 \qquad r_{y2} = -.75 \qquad r_{12} = -.50$$

You may now complete the analysis.

Regardless of the method of coding and the method of calculation, the R^2 will be the same. The reasons for choosing one system in preference to another are discussed in Chapter 7. It should also be noted that although the example presented had equal n's in the groups, the analysis with unequal n's is done in exactly the same manner.

One-Way Analysis of Variance

While following the above presentation you may have wondered whether it was not possible to subject the data to a one-way analysis of variance. Assuming you are familiar with the analysis of variance you may have questioned whether there is anything to be gained by learning what may seem to be a more complicated analysis. It is true that the data may be analyzed with analysis of variance, and, in order to demonstrate the identity of the two analyses, we now present a conventional analysis of variance. After this presentation the important question whether one method of analysis is preferable to the other will be attacked.

The data for the three groups, as well as the calculations of the analysis of variance, are presented in Table 6.3. The F ratio of 18 with 2 and 12 degrees of freedom is identical to the one obtained by the regression analysis. To estimate the magnitude of the relation between the independent and dependent variable, we calculate the so-called correlation ratio, or E (often called η, eta):

$$E = \sqrt{\frac{ss_b}{ss_t}} = \sqrt{\frac{90}{120}} = \sqrt{.75} = .866$$

where ss_b = between groups sum of squares, and ss_t = total sum of squares. The magnitude of the relation is substantial. If we square E, we obtain $E^2 = (.866)^2 = .75$, which is the proportion of variance of the dependent variable, attitudes, accounted by the independent variable, marital status. E^2 is identical

TABLE **6.3**   FICTITIOUS ATTITUDE DATA AND ANALYSIS OF
VARIANCE CALCULATIONS

	A_1	A_2	A_3	
	4	7	1	
	5	8	2	
	6	9	3	
	7	10	4	
	8	11	5	
ΣX:	30	45	15	$\Sigma X_t = 90$
$\bar{X}$:	6	9	3	$(\Sigma X_t)^2 = 8100$
				$\Sigma X_t^2 = 660$

$$C = \frac{8100}{15} = 540$$

$$\text{Total} = 660 - 540 = 120$$

$$\text{Between} = \frac{30^2 + 45^2 + 15^2}{5} - 540 = 90$$

Source	df	ss	ms	F
Between	2	90	45.00	18.00
Within	12	30	2.50	
Total	14	120		

to R^2 obtained earlier. Note further the identity of the different sums of squares obtained in the two analyses: ss_t is 120 in both analyses; $ss_b = ss_{reg} = 90$; $ss_w = ss_{res} = 30$.

We can write two analogous equations for the decomposition of the total sum of squares. For the analysis of variance

$$ss_t = ss_b + ss_w \tag{6.12}$$

where $ss_t =$ total sum of squares, $ss_b =$ between groups sum of squares, and $ss_w =$ within groups sum of squares. For the regression analysis,

$$\Sigma y^2 = \Sigma y_{reg}^2 + \Sigma y_{res}^2 \tag{6.13}$$

where $\Sigma y^2 =$ total sum of squares of the Y scores, $\Sigma y_{reg}^2 =$ sum of squares due to regression, and $\Sigma y_{res}^2 =$ residual sum of squares. We can also write, of course,

$$ss_t = ss_{reg} + ss_{res} \tag{6.14}$$

which is the same as equation (6.13) but with different symbols. Although we will use equation (6.14) for its simplicity and clarity, we give (6.13) to show clearly that it is the sum of squares of Y that is under discussion.

It is obvious that the two methods of analysis are interchangeable. They amount to the same thing so far as numerical outcome is concerned. They can differ, however, in interpretation. In the analysis of variance, the significant F ratio means that the difference between the means is statistically significant with the usual connotations of departure from chance. The null hypothesis is H_o: $\mu_1 = \mu_2 = \mu_3$. (Greek letters are used to indicate population values. In this case, μ, mu, is a population mean.) In the regression analysis, on the other hand, the significant F ratio means that R^2 is statistically significant. It expresses the statistical significance of the relation between X and Y, X being group membership, and is, in a sense, more fundamental since it addresses itself "directly" to the point of main interest, the relation between X and Y rather than to the differences between the Y means. The null hypothesis is H_o: $R^2 = 0$. The statistically significant F ratio of 18 leads us to reject this hypothesis: R and R^2 differ significantly from zero. If this were the only difference between the two systems one might be justified in doubting whether one system is really preferable to the other. After all, it was demonstrated that E^2 is equal to R^2, and the former is easily obtainable from the ratio of the between groups sum of squares to the total sum of squares. There are, however, a number of advantages to the use of multiple regression analysis, which will become evident in subsequent chapters. For the present, therefore, we restrict ourselves to a listing of some of these advantages.

Some Advantages of the Multiple Regression Approach

Although analysis of variance and multiple regression analysis are interchangeable in the case of categorical independent variables, multiple regression analysis is superior or the only appropriate method of analysis in the following cases: (1) when the independent variable is continuous, that is experimental treatments with varying degrees of the same variable; (2) when the independent variables are both continuous and categorical, as in analysis of covariance, or treatments by levels designs; (3) when cell frequencies in a factorial design are unequal and disproportionate; (4) when studying trends in data: linear, quadratic, and so on. Moreover, multiple regression analysis provides a more direct method of calculating and interpreting certain statistics — for example, orthogonal comparisons between means — and it obviates the need for several computer programs for the various analysis of variance designs. One multiple regression program is sufficient for most designs.[8]

Summary

It was shown that when dealing with a categorical independent variable, one creates k vectors (k = number of groups, or categories, minus one), which serve to identify group membership. A multiple regression analysis is then done, using the k vectors as the independent variables. The resulting R^2 indicates the proportion of variance in the dependent variable accounted for by the independent variable. R^2 is tested for significance with an F ratio with k degrees of freedom for the numerator and $N - k - 1$ degrees of freedom for the denominator. When using 1's and 0's for coding group membership, the resulting regression equation has the following properties: a (the intercept) is equal to the mean of the group assigned 0's throughout, and each b (regression coefficient) is equal to the mean of the group assigned 1's in a given vector minus the mean of the group assigned 0's throughout.

The virtual identity of multiple regression analysis and the analysis of variance was demonstrated for the case of one categorical independent variable. The same holds true for the case of more than one categorical independent variable, as will be shown in subsequent chapters.

Although both the analysis of variance and multiple regression may be used with categorical independent variables, the latter is more versatile in that it is applicable to situations in which the former is not. A partial listing of such situations was given above. In conclusion, then, the multiple regression approach is more general and useful.

Study Suggestions

1. Distinguish between categorical and continuous variables. Give examples of each.
2. The relation between religious affiliation and moral judgment is studied in a sample of Catholic, Jewish, and Protestant children. What kind of a variable is religious affiliation? Explain.
3. What is a dummy variable? What is its use?

---

[8]This is not an exhaustive, but rather an illustrative, listing.

4. In a study with six different groups, how many dummy vectors are needed to exhaust the information of group membership? Explain. (*Answer*: 5.)
5. In a research study with four treatments, A_1, A_2, A_3, and A_4, dummy vectors were constructed as follows: vector X_1 consisted of 1's for subjects in treatment A_1, 0's for all others; vector X_2 consisted of 1's for subjects in treatment A_2, 0's for all others; vector X_3 consisted of 1's for subjects in treatment A_3, 0's for all others. A multiple regression analysis was done in which the dependent variable measure was regressed on the three dummy vectors. The regression equation obtained in the analysis was

$$Y' = 8.00 + 6.00\,X_1 + 5.00\,X_2 - 2.00\,X_3$$

On the basis of this equation, what are the means of the four treatment groups on the dependent variable measure?
(*Answers*: $\bar{Y}_{A_1} = 14.00$, $\bar{Y}_{A_2} = 13.00$, $\bar{Y}_{A_3} = 6.00$, $\bar{Y}_{A_4} = 8.00$.)
6. In a study of problem solving, subjects were randomly assigned to three different treatments. At the conclusion of the experiment, the subjects were given a set of problems to solve. The problem-solving scores for the three treatments were:

A_1	A_2	A_3
2	3	7
3	3	6
2	4	4
5	4	7
3	2	8
5	2	4

Use dummy vectors to code the treatments. Do a multiple regression analysis by regressing the problem-solving scores on the dummy vectors. Calculate: (a) R^2, (b) regression sum of squares, (c) residual sum of squares, (d) the regression equation, (e) F ratio. Interpret the results.
(*Answers*: (a) $R^2 = .5428$; (b) $ss_{reg} = 32.4444$; (c) $ss_{res} = 27.3333$; (d) $Y' = 6.00 - 2.67\,X_1 - 3.00\,X_2$; (e) $F = 8.90$, with 2 and 15 df.)

Dummy, Effect, and Orthogonal Coding of Categorical Variables

In Chapter 6 the coding of a categorical independent variable was introduced. In effect, the categories of the independent variables, or what corresponds to treatments in an analysis of variance design, were transformed into vectors, which were used as independent variables in a regression equation. The system of 1's and 0's, so-called dummy variables, was used, 1 meaning membership in a given category, or treatment group, and 0 no membership in that category or group. Vectors of 1's and 0's were treated like vectors of continuous measures and used as independent variables in regression equations and calculations.

In this chapter the idea of coding is formalized, and the simple idea of dummy variables with 1's and 0's is expanded to include other forms of coding. Although dummy variable coding is simple and effective, we will find that other systems of coding are preferable for certain purposes. We will also find that tests of statistical significance are related to the coding of the categorical variables and that, with appropriate coding, certain statistics of analysis of variance can easily be recovered when using regression calculations.

Coding and Methods of Coding

A *code* is a set of symbols to which meanings can be assigned. For example, the set of symbols {A,B,C} can be assigned to three different groups of people, such as Protestants, Catholics, and Jews. Or the set {1,0} can be assigned to males and females and the resulting vector of 1's and 0's used in numerical analysis. In our use of coding, symbols such as $0, 1, 0, -1$, and so on, are assigned to objects of mutually exclusive subsets of a defined universe to indicate subset or group membership.

The assignment of symbols follows a rule or a set of rules determined by independent means. For some variables the rule may be obvious and need little or no explanation, as in the assignment of 1's and 0's to males and females. There are, however, variables that require elaborate explication of rules, about which there may not be agreement among all or most observers. For example, the assignment of symbols indicating membership in different social classes may involve a complex set of rules about which there may not be universal agreement. An example of even greater complexity is the explication of rules for the assignment of symbols to extraverts and introverts.

Whatever the rules and whatever the coding, subjects given the same symbol are treated as equal to each other on a variable. If one were to define rules of membership in political parties, we might assign 1's to Democrats and 0's to Republicans. All subjects assigned 1, then, are considered equally as Democrats, no matter how different they may be on other variables and no matter how they may differ in their devotion, activity, and commitment to the Democratic party.

Note that the numbers assigned as symbols do not mean rank ordering of the categories. Any set of numbers can be used: $\{1,0\}$, $\{I,II,III\}$, $\{1,0,-1\}$, and so on. Some coding systems, however, have properties that make them more useful than other coding systems. This is especially so when the symbols are used in statistical analysis.

In multiple regression analysis, for example, some characteristic or aspect of the members of a population or sample is objectively defined, and it is then possible to create a set of ordered pairs, the first members of which constitute the dependent variable, Y, and the second members numerical indicators of group membership. Of the many numerical indicators one may adopt, there are three systems that seem to be most useful: 1's and 0's, where 1 indicates membership in a given group and 0 indicates no such membership. This method is called *dummy coding*. Another method is the assignment of 1's, 0's, and -1's, where 1's are assigned to members of a given group, 0's to members of all other groups but one, and the members of this one group are assigned -1's. We call this method *effect coding*. A third method uses orthogonal coefficients and is thus called *orthogonal coding*. (Orthogonal coding will be explained later.) It should be noted that the overall analysis and results are identical regardless of which of the three coding methods is used in the multiple regression analysis. Some of the intermediate results and the statistical tests of significance associated with the three methods are different. We now turn to a detailed treatment of each of the three systems of coding variables.

Dummy Coding

The simplest system of coding variables is dummy coding. In this system one generates a number of vectors such that in any given vector membership in a given group or category is assigned 1, while nonmembership in the category is assigned 0. The number of vectors necessary to exhaust the information about group membership is equal to the number of groups, or categories, minus one. The use of dummy coding was illustrated in Chapter 6 in connection with the analysis of fictitious data of attitudes toward divorce laws among married, divorced, and single males. We reproduce in Table 7.1 the data and some of the results of the analysis of Chapter 6.

It was pointed out in Chapter 6, that when calculating the equation for the regression of the dependent variable, Y, on the dummy vectors, a (the intercept) is equal to the mean of the group or category assigned 0's throughout. In the present example $a = 3.00$, which is equal to the mean of group A_3. Furthermore, each b (regression coefficient) is equal to the difference between the mean of the group assigned 1's in a given vector and the mean of the group assigned 0's throughout. Thus, the difference between the means of group A_1 and A_3 is 3.00, which is the same as the value of b_1, the weight associated with vector X_1 in which members of group A_1 were assigned 1's. Similarly for b_2 (6.00), which is equal to the difference between the means of group A_2 and A_3. Accordingly, the use of dummy coding results in treating the group assigned 0's as a control group. This becomes even more evident as we turn our attention to the tests of significance of the regression coefficients.

TABLE 7.1   DUMMY CODING AND RESULTS FOR ATTITUDES
TOWARD DIVORCE LAWS[a]

Group	Y	X_1	X_2
A_1	4	1	0
	5	1	0
	6	1	0
	7	1	0
	8	1	0
A_2	7	0	1
	8	0	1
	9	0	1
	10	0	1
	11	0	1
A_3	1	0	0
	2	0	0
	3	0	0
	4	0	0
	5	0	0
ss:	120	3.33333	3.33333
s:	2.92770	.48795	.48795

$r_{y1} = .00000$ $r_{y2} = .75000$ $r_{12} = -.50000$
$Y' = 3.00 + 3.00 X_1 + 6.00 X_2$
$ss_{reg} = 90$ $ss_{res} = 30$

[a]These data were given in Table 6.2. The calculations of the results may be found in Chapter 6, following Table 6.2.

Tests of Significance

In Chapter 4 the test of significance of regression coefficients was discussed and illustrated. It was shown that dividing b by its standard error results in a t ratio, which, with the appropriate degrees of freedom, is assessed for significance at a prespecified level. We repeat formulas (4.10) and (4.12) with new numbers and a different notation:[1]

$$s^2_{y.12...k} = \frac{ss_{res}}{N-k-1} \tag{7.1}$$

where $s^2_{y.12...k}$ = the variance of estimate of Y, the dependent variable, when regressed on variables 1 to k; ss_{res} = residual sum of squares; N = number of subjects; k = number of independent variables, or coded vectors.

$$s_{b_{y1.2...k}} = \sqrt{\frac{s^2_{y.12...k}}{\sum x^2_1(1-R^2_{1.2...k})}} \tag{7.2}$$

where $s_{b_{y1.2...k}}$ = standard error of b_1; $s^2_{y.12...k}$ = variance of estimate; $\sum x^2_1$ = sum of squares of variable 1; $R^2_{1.2...k}$ = the squared multiple correlation between variable 1, used as a dependent variable, and variables 2 to k as the independent variables.

In the present example $ss_{res} = 30$, $N = 15$, and $k = 2$. Therefore,

$$s^2_{y.12} = \frac{30}{15-2-1} = 2.50$$

Note that this value is the same as the mean square error obtained in the analysis of variance of these data (see Table 6.3). The degrees of freedom associated with this term are, of course, also the same: $N-k-1 = 12$.

Adapting formula (7.2) to the b weight for vector X_1 of the data of Table 7.1 gives

$$s_{b_1} = \sqrt{\frac{s^2_{y.12}}{\sum x^2_1(1-R^2_{1.2})}} \tag{7.3}$$

$$s_{b_1} = \sqrt{\frac{2.50000}{(3.33333)[1-(-.50000)^2]}} = \sqrt{\frac{2.50000}{(3.33333)(.75000)}}$$

$$= \sqrt{\frac{2.50000}{2.50000}} = \sqrt{1.00000} = 1.00$$

and the t ratio for b_1:

$$t_{b_1} = \frac{b_1}{s_{b_1}} = \frac{3.00}{1.00} = 3.00$$

The degrees of freedom for this t ratio are the same as those associated with $s^2_{y.12}$, the variance of estimate. We thus have for b_1 a t ratio of 3.00 with 12

---

[1]Earlier we used the symbol SE^2_{est} for the variance of estimate, and SE_{bj} for the standard error of the jth b coefficient. This was done for the sake of simplicity in introducing these terms. Henceforth, the symbols introduced here will be used. When there is no danger of ambiguity, however, this notation is abbreviated. For example, we write b_1 for $b_{y1.23...k}$ and s_{b_1} for $s_{b_{y1.23...k}}$.

degrees of freedom. s_{b_2} (the standard error of b_2) is also equal to 1.00, since $\Sigma x_2^2 = \Sigma x_1^2$ when 1's are used for group membership and the groups have equal n's, and $R_{1.2}^2 = R_{2.1}^2 = r_{12}^2$. Therefore,

$$t_{b_2} = \frac{6.00}{1.00} = 6.00, \quad \text{with 12 } df$$

We now demonstrate that the t ratios obtained above are identical to the t ratios obtained when following Dunnett (1955): one calculates multiple t ratios between each treatment mean and a control group mean. The formula for a t ratio subsequent to the analysis of variance is

$$t = \frac{\bar{X}_1 - \bar{X}_2}{\sqrt{MSE\left(\frac{1}{n_1} + \frac{1}{n_2}\right)}} \tag{7.4}$$

where $\bar{X}_1, \bar{X}_2 =$ means of the first and second groups respectively; $MSE =$ mean square error from the analysis of variance; $n_1, n_2 =$ number of subjects in groups 1 and 2, respectively. When $n_1 = n_2$, formula (7.4) can be stated as follows:

$$t = \frac{\bar{X}_1 - \bar{X}_2}{\sqrt{\frac{2MSE}{n}}} \tag{7.5}$$

where $n =$ number of subjects in one group; all other terms are as defined above. In the problem presented in Chapter 6 and further analyzed above:

$$\bar{Y}_{A_1} = 6.00 \qquad \bar{Y}_{A_2} = 9.00 \qquad \bar{Y}_{A_3} = 3.00$$

$$n_1 = n_2 = n_3 = 5$$

$$MSE = 2.50$$

Comparing group A_1 to A_3, the group that serves as a control because its members are assigned 0's in Table 7.1 is

$$t_1 = \frac{6.00 - 3.00}{\sqrt{\frac{2(2.50)}{5}}} = \frac{3.00}{\sqrt{\frac{5}{5}}} = \frac{3.00}{\sqrt{1}} = \frac{3.00}{1} = 3.00$$

Comparing group A_2 to A_3

$$t_2 = \frac{9.00 - 3.00}{\sqrt{\frac{2(2.50)}{5}}} = \frac{6.00}{1} = 6.00$$

The two t ratios are identical to the ones obtained for the two b weights associated with the dummy vectors of Table 7.1. In order to determine whether a given t ratio for the comparison of a treatment with a control group is significant at a prespecified level, one may check a special table prepared by Dunnett. This table is reproduced in various statistics books, for example, Edwards

(1968), Winer (1971). For the present case, where the analysis was performed as if there were two treatments and a control group, the tabled values for a one-tailed t with 12 df are 2.11 (.05 level), 3.01 (.01 level), and for a two-tailed test: 2.50 (.05 level), 3.39 (.01 level).

To recapitulate, the F ratio associated with the R^2 of the dependent variable with the dummy vectors indicates whether there are significant differences among some or all the means of the groups. This is, of course, equivalent to the overall F ratio of the analysis of variance. The t ratio for each b weight is equivalent to the t ratio for the difference between the mean of the group assigned 1's in the vector with which a given b is associated and the mean of the group assigned 0's throughout. In other words, the t ratios for the b's in effect serve to compare each treatment mean with the control group mean.[2]

Dummy coding is not restricted to situations in which there are several treatment groups and a control group. In fact, the data analyzed above came from three groups, none of which was a control group. It was only for the purpose of demonstrating the properties of dummy coding that group A_3 was treated as a control group. Dummy coding can be used whenever one is dealing with several groups, or several categories of a variable, such as sex, political preference, religious affiliation, and the like. In such usage, however, one will not be interested in interpreting the t ratios associated with the b weights. Instead, the overall F ratio for the R^2 is interpreted to determine whether some or all of the means of the groups differ significantly. In order to determine specifically which of the group means are significantly different, it is necessary to apply one of the methods for multiple comparisons between means. This topic is taken up in a subsequent section.

Effect Coding

Effect coding is so named because, as shown below, the regression coefficients yielded by its use reflect the effects of the treatments of the analysis. The code numbers used are 1's, 0's, and −1's. Effect coding is thus similar to dummy coding. The difference is that in dummy coding one group or category is assigned 0's in all the vectors, whereas in effect coding the same group is assigned −1's in all the vectors. (See −1's assigned to A_3 in Table 7.2.) Although it makes no difference which group is assigned −1's, it is convenient to assign them to members of the last group. One generates k vectors (k = number of groups minus one). In each vector, members of one group are assigned 1's, all other subjects are assigned 0's, except for members of the last group who are assigned −1's.

The application of effect coding to the data earlier analyzed by dummy coding is illustrated in Table 7.2. Note that in vector 1 of Table 7.2 members of group A_1 are assigned 1's, members of group A_2 are assigned 0's, and members

---

[2]Certain computer programs for regression analysis have as part of their output the b's and the t's associated with them. For a discussion of the use and interpretation of the output of such programs, see Chapter 8 and Appendix C.

TABLE **7.2** EFFECT CODING OF ATTITUDES TOWARD DIVORCE DATA, AND CALCULATIONS NECESSARY FOR MULTIPLE REGRESSION ANALYSIS[a]

Group	Ss	Y	1	2
	1	4	1	0
	2	5	1	0
A_1	3	6	1	0
	4	7	1	0
	5	8	1	0
	6	7	0	1
	7	8	0	1
A_2	8	9	0	1
	9	10	0	1
	10	11	0	1
	11	1	−1	−1
	12	2	−1	−1
A_3	13	3	−1	−1
	14	4	−1	−1
	15	5	−1	−1
Σ:		90	0	0
Σ²:		660	10	10
M:		6	0	0
s:		2.92770	.84515	.84515

$$\Sigma YX_1 = 15 \quad \Sigma YX_2 = 30 \quad \Sigma X_1 X_2 = 5$$
$$r_{y1} = .43301 \quad r_{y2} = .86603 \quad r_{12} = .50000$$

[a]Vector Y is repeated from Table 7.1.

of group A_3 are assigned −1's. In vector 2, members of A_1 are assigned 0's, those of A_2 are assigned 1's and those of A_3 are assigned −1's.

When the sample sizes in all groups are equal, the calculations of the zero-order correlations necessary for the multiple regression analysis are greatly simplified. First, it is not necessary to calculate the zero-order correlations between coded vectors, since the correlation between any two coded vectors, X_i and X_j, is .50. This is so because for any coded vector: $\Sigma X_i = \Sigma X_j = 0$; $\Sigma X_i^2 = \Sigma X_j^2 = 2n$; $\Sigma X_i X_j = n$, where ΣX = sum of raw scores; ΣX^2 = sum of squared raw scores; $\Sigma X_i X_j$ = sum of the cross products of raw scores; n = number of subjects in any one of the groups involved in the multiple regression analysis. In view of the above, the raw score formula for the coefficient of correlation,

$$r_{x_i x_j} = \frac{N \Sigma X_i X_j - (\Sigma X_i)(\Sigma X_j)}{\sqrt{N \Sigma X_i^2 - (\Sigma X_i)^2} \sqrt{N \Sigma X_j^2 - (\Sigma X_j)^2}}$$

reduces to

$$\frac{Nn}{\sqrt{N2n}\sqrt{N2n}} = \frac{Nn}{N2n} = .50 \qquad (N = \text{Total number of subjects})$$

Second, the zero-order correlation between Y and any one of the coded vectors, X, is also simplified. Considering the properties of the coded vectors, discussed above, the formula

$$r_{xy} = \frac{N \Sigma XY - (\Sigma X)(\Sigma Y)}{\sqrt{N \Sigma X^2 - (\Sigma X)^2} \sqrt{N \Sigma Y^2 - (\Sigma Y)^2}}$$

reduces to

$$r_{xy} = \frac{N \Sigma XY}{\sqrt{N2n} \sqrt{N \Sigma Y^2 - (\Sigma Y)^2}}$$

The denominator of the reduced formula is constant for the zero-order correlations between any coded vector and the dependent variable. Consequently, having calculated the denominator, all that is necessary to obtain the zero-order correlations is to divide each $N\Sigma XY$ by the denominator. For example, for the data of Table 7.2, the denominator of the zero-order correlation between any coded vector and Y is

$$\sqrt{N2n} \sqrt{N \Sigma Y^2 - (\Sigma Y)^2} = \sqrt{(15)(10)} \sqrt{(15)(660) - (90)^2} = 519.61525$$

From Table 7.2, $\Sigma YX_1 = 15$; $\Sigma YX_2 = 30$; $N = 15$. Therefore

$$r_{y1} = \frac{(15)(15)}{519.61525} = .43301 \qquad r_{y2} = \frac{(15)(30)}{519.61525} = .86603$$

Applying the formula for R^2 [formula (6.5)] to the data of Table 7.2, we obtain

$$R^2_{y.12} = \frac{r^2_{y1} + r^2_{y2} - 2r_{y1}r_{y2}r_{12}}{1 - r^2_{12}}$$

$$R^2_{y.12} = \frac{(.43301)^2 + (.86603)^2 - 2(.50000)(.43301)(.86603)}{1 - (.50000)^2}$$

$$= .75$$

As might have been expected, $R^2_{y.12}$ is identical to the value obtained with the dummy coding. The F ratio associated with $R^2_{y.12}$ is, of course, also the same as obtained earlier, namely 18.00, with 2 and 12 degrees of freedom.

We illustrate now that doing the calculations with the proportions of variance or with the sum of squares leads to the same results. Having calculated R^2 it is simple to obtain the regression sum of squares:

$$ss_{reg} = (R^2_{y.12...k})(\Sigma y^2) \tag{7.6}$$

where ss_{reg} = regression sum of squares; $R^2_{y.12...k}$ = squared multiple correlation of Y with k coded vectors, or the proportion of variance of Y accounted for by k coded vectors; Σy^2 = sum of squares of the dependent variable, Y.

For the present problem, $R^2_{y.12} = .75$ and $\Sigma y^2 = 120.00$.

$$ss_{reg} = (.75)(120.00) = 90.00$$

This is equivalent to the between groups sum of squares obtained in the analysis of variance of these data (see Table 6.3).

The residual, or error, sum of squares may be obtained by subtraction:

$$ss_{res} = \Sigma\, y^2 - ss_{reg} \qquad (7.7)$$

or by the following formula:

$$ss_{res} = (1 - R^2_{y.12...k})(\Sigma\, y^2) \qquad (7.8)$$

where ss_{res} = residual sum of squares; $(1 - R^2_{y.12...k})$ = proportion of variance not accounted for by the coded vectors. For the present problem

$$ss_{res} = (1 - .75)(120.00) = (.25)(120.00) = 30.00$$

This value is equivalent to the within groups sum of squares of the analysis of variance (see Table 6.3). The analysis is now summarized in Table 7.3, where for purposes of comparison part I is expressed in proportions and part II in sums of squares. In each part of Table 7.3 the symbolic expressions and the numerical results are presented.

Thus far, there is no difference between the present analysis and the one done with dummy coding. The two methods of coding do differ, however, in their regression equations and their properties.

TABLE **7.3** SUMMARY OF REGRESSION ANALYSIS FOR
DATA OF TABLE 7.2

I: *Proportions of Variance*

Source	df	Prop. of Variance	ms	F
Regression	k	R^2	R^2/k	$\dfrac{R^2/k}{(1-R^2)/(N-k-1)} = \dfrac{.75/2}{.25/12}$
	2	.75	.75/2	$= 18.00$
Residual	$N-k-1$	$1-R^2$	$(1-R^2)/(N-k-1)$	
	$15-2-1$	$1-.75$	$(1-.75)/12$	
Total	14	1.00		

II: *Sums of Squares*

Source	df	ss	ms	F
Regression	k	$R^2\,\Sigma\, y^2$	$(R^2\,\Sigma\, y^2)/k$	ms_{reg}/ms_{res}
	2	$(.75)(120.00)$	90.00/2	$\dfrac{45.00}{2.50} = 18.00$
Residual	$N-k-1$	$(1-R^2)\,\Sigma\, y^2$	$(1-R^2)(\Sigma\, y^2)/(N-k-1)$	
	$15-2-1$	$(1-.75)(120.00)$	30.00/12	
Total	14	120.00		

The Regression Equation

In order to calculate the regression equation for the data of Table 7.2, we apply formula (6.2):

$$b_1 = \frac{(10)(15) - (5)(30)}{(10)(10) - (5)^2} = \frac{150 - 150}{100 - 25} = 0$$

$$b_2 = \frac{(10)(30) - (5)(15)}{(10)(10) - (5)^2} = \frac{300 - 75}{75} = \frac{225}{75} = 3$$

$$a = 6 - (0)(0) - (3)(0) = 6$$

$$Y' = 6 + 0X_1 + 3X_2$$

Note that a (the intercept) is equal to the grand mean of the dependent variable, $\bar{Y}$, while each b is equal to the deviation of the mean of the group assigned 1's in the vector with which it is associated from the grand mean. Thus, b_1 is equal to the deviation of mean A_1 from the grand mean ($\bar{Y}$); that is, $6.00 - 6.00 = 0$. b_2 is equal to $\bar{Y}_{A_2} - \bar{Y} = 9.00 - 6.00 = 3.00$. It is evident, then, that *each b reflects a treatment effect*; b_1 reflects the effect of treatment A_1, while b_2 reflects the effect of treatment A_2. This method of coding thus generates a regression equation whose b coefficients reflect the effects of the treatments. In order to appreciate the properties of the regression equation that results from the use of effect coding, it is necessary to digress for a brief presentation of the linear model. After this presentation the discussion of the regression equation is resumed.

The Fixed Effects Linear Model

The fixed effects one-way analysis of variance is presented by some authors (for example, Graybill, 1961; Scheffé, 1959; Searle, 1971) in the form of the linear model:

$$Y_{ij} = \mu + \beta_j + \epsilon_{ij} \tag{7.9}$$

where Y_{ij} = the score of individual i in group or treatment j; μ = the population mean; β_j = the effect of treatment j; ϵ_{ij} = error associated with the score of individual i in group or treatment j. "Linear model" means that an individual's score is conceived as a linear composite of several components. In the present case [formula (7.9)] it is a composite of three parts: the grand mean, a treatment effect, and an error term. The error is the part of Y_{ij} not explained by the grand mean and the treatment effect. This can be seen from a restatement of formula (7.9):

$$\epsilon_{ij} = Y_{ij} - \mu - \beta_j \tag{7.10}$$

The method of least squares is used to minimize the sum of squared errors ($\Sigma\epsilon_{ij}^2$). In other words, an attempt is being made to explain as much of Y_{ij} as possible by the grand mean and a treatment effect. In order to obtain a unique solution to the problem, the restraint that $\Sigma\beta_g = 0$ is used (g = number of groups). The meaning of this condition is simply that the sum of the treatment effects is zero. It is shown below that such a restraint results in expressing each treatment effect as the deviation from the grand mean of the mean of the treatment whose effect is studied.

Although formula (7.9) is expressed in parameters, or population values, in actual analyses statistics are used as estimates of these parameters:

$$Y_{ij} = \bar{Y} + b_j + e_{ij} \qquad (7.11)$$

where $\bar{Y}$ = the grand mean; b_j = effect of treatment j; e_{ij} = error associated with individual i under treatment j.

The sum of squares, $\Sigma(Y - \bar{Y})^2$, can be expressed in the context of the regression equation. It will be recalled [formula (6.1)] that $Y' = \bar{Y} + bx$. Therefore,

$$Y = \bar{Y} + bx + e$$

A deviation of a score from the mean of the dependent variable can be expressed thus:

$$Y - \bar{Y} = \bar{Y} + bx + e - \bar{Y}$$

Substituting $Y - \bar{Y} - bx$ for e in the above formula, we obtain

$$Y - \bar{Y} = \bar{Y} + bx + Y - \bar{Y} - bx - \bar{Y}$$

Now, $\bar{Y} + bx = Y'$ and $Y - \bar{Y} - bx = Y - Y'$ By substitution,

$$Y - \bar{Y} = Y' + Y - Y' - \bar{Y}$$

Rearranging the terms on the right,

$$Y - \bar{Y} = (Y' - \bar{Y}) + (Y - Y') \qquad (7.12)$$

Since we are interested in explaining the sum of squares,

$$\Sigma y^2 = \Sigma [(Y' - \bar{Y}) + (Y - Y')]^2$$
$$= \Sigma (Y' - \bar{Y})^2 + \Sigma (Y - Y')^2 + 2\Sigma (Y' - \bar{Y})(Y - Y')$$

It can be demonstrated that the last term on the right equals zero. Therefore,

$$\Sigma y^2 = \Sigma (Y' - \bar{Y})^2 + \Sigma (Y - Y')^2 \qquad (7.13)$$

The first term on the right, $\Sigma(Y' - \bar{Y})^2$, is the sum of squares due to regression. It is analogous to the between groups sum of squares of the analysis of variance. $\Sigma(Y - Y')^2$ is the residual sum of squares, or what is termed the within groups sum of squares in the analysis of variance. $\Sigma(Y' - \bar{Y})^2 = 0$ means that Σy^2 is all due to residuals, and we thus have explained nothing by knowledge of X. If, on the other hand, $\Sigma(Y - Y')^2 = 0$, all the variability is explained by regression, or by the information X provides. We return now to the regression equation that resulted from the analysis with effect coding.

The Meaning of the Regression Equation.

From the foregoing discussion it can be seen that the use of effect coding results in a regression equation that reflects the linear model. This is illustrated by applying the regression equation obtained above ($Y' = 6 + 0X_1 + 3X_2$) to some of the subjects of Table 7.2. For subject number 1 we obtain

$$Y_1' = 6 + 0(1) + 3(0) = 6$$

This is, of course, the mean of the group to which this subject belongs, namely the mean of A_1. The residual for subject 1 is

$$e_1 = Y_1 - Y_1' = 4 - 6 = -2$$

Expressing the score of subject 1 in components of the linear model:

$$Y_1 = a + b_1 + e_1$$
$$4 = 6 + 0 + (-2)$$

Since a is equal to the grand mean ($\bar{Y}$), and for each group (except for the one assigned -1's) there is only one vector in which it is assigned 1's, the predicted score for each subject is a composite of a and the b for the vector in which the subject is assigned 1. In other words, *a predicted score is a composite of the grand mean and the treatment effect of the group to which the subject belongs.* Thus, for subjects in group A_1 the application of the regression equation results in $Y' = 6 + 0(1)$, because subjects in this group are assigned 1's in the first vector only, and 0's in all others, regardless of the number of groups involved in the analysis.

For subjects of group A_2 the regression equation is, in effect, $Y' = 6 + 3(1)$, 6 being the a (intercept), and 3 the b associated with the vector in which subjects of group A_2 are assigned 1's. Since the predicted score for any subject is the mean of his group expressed as a composite of $a + b$, and since a is the grand mean, it follows that b is the deviation of the group mean from the grand mean. As stated earlier, b is equal to the treatment effect for the group with which it is associated. For group A_1 the treatment effect is $b_1 = 0$, and for group A_2 the treatment effect is $b_2 = 3$.

Applying now the regression equation to subject number 6 (the first subject of group A_2) we obtain

$$Y_6' = 6 + (0)(0) + (3)(1) = 9$$
$$e_6 = Y_6 - Y_6' = 7 - 9 = -2$$

Expressing the score of subject 6 in components of the linear model:

$$Y_6 = a + b_2 + e_6$$
$$7 = 6 + 3 + (-2)$$

The treatment effect for the group assigned -1 is easily obtained when considering the constraint $\Sigma b_g = 0$. In the present problem this means

$$b_1 + b_2 + b_3 = 0$$

Substituting the values for b_1 and b_2 obtained above,

$$0 + 3 + b_3 = 0$$
$$b_3 = -3$$

In general, the treatment effect for the group assigned -1's is $-\Sigma b_k$ ($k =$ number of dummy vectors, or $g - 1$, number of groups minus one). For the present problem,

$$b_3 = -\Sigma (b_k) = -\Sigma (0 + 3) = -3$$

The mean of A_3 is 3, and its deviation from the grand mean ($\bar{Y} = 6.00$) is -3, which is the value of b_3, the treatment effect for A_3.

Applying the regression equation to subject 11 (the first subject in group A_3),

$$Y'_{11} = 6 + 0(-1) + 3(-1)$$
$$= 6 - 3 = 3$$

This is the mean of A_3. All other subjects in A_3 will have the same predicted Y.

$$e_{11} = Y_{11} - Y'_{11} = 1 - 3 = -2$$
$$Y_{11} = a + b_3 + e_{11}$$
$$1 = 6 + (-3) + (-2)$$

The R^2 obtained with effect coding is the same as that obtained with dummy coding. And the meaning of the regression equation is that a is equal to the mean of the dependent variable, $\bar{Y}$, and each b coefficient is equal to the treatment effect for the group with which it is associated.

Multiple Comparisons between Means

A significant F ratio for R^2 leads to the rejection of the null hypothesis that there is no relation between group membership, or treatments, and performance on the dependent variable. With a categorical independent variable the significant R^2 in effect means that the null hypothesis $\bar{Y}_1 = \bar{Y}_2 = \cdots = \bar{Y}_g$ ($g =$ number of groups, or categories) is rejected. Rejection of the null hypothesis, however, does not necessarily mean that all the means are significantly different from each other. In order to determine which of the means differ significantly from each other, one of the methods of multiple comparisons of means must be used.

There are two types of comparisons of means: planned comparisons and post hoc comparisons. *Planned comparisons* are hypothesized by the researcher prior to the overall analysis. Consequently, they are also referred to as a priori comparisons. *Post hoc*, or a posteriori, comparisons are performed following the rejection of the overall null hypothesis. Only when the F ratio associated with R^2 is significant may one proceed with post hoc comparisons between means in order to detect where significant differences exist.

The topic of post hoc comparisons is complex.[3] There are various methods available, but no universal agreement about their appropriateness. For a presentation and a discussion of the various methods the reader is referred to

---

[3]Planned comparisons are discussed later in this chapter, in the section devoted to orthogonal coding.

Kirk (1968), Miller (1966), Winer (1971), and Games (1971). The presentation here is limited to a method developed by Scheffé (1959). The Scheffé, or the S method, is the most general method of multiple comparisons. It enables one to make all possible comparisons between individual means as well as between combinations of means. In addition, it is applicable to equal as well as unequal frequencies in the groups or the categories of the variable. It is also the most conservative test. That is, it is less likely than other tests to show differences as significant. A *comparison*, a *contrast*, or a *difference*, is a linear combination of the form

$$D = C_1\bar{Y}_1 + C_2\bar{Y}_2 + \cdots + C_j\bar{Y}_j \tag{7.15}$$

where D = difference or contrast; C = coefficient by which a given mean, $\bar{Y}$, is multiplied; j = number of means involved in the comparison. For the kind of comparisons dealt with in this section it is required that $\Sigma C_j = 0$. That is, the sum of the coefficients in a given contrast must equal zero. Thus, contrasting $\bar{Y}_1$ with $\bar{Y}_2$ one can set $C_1 = 1$ and $C_2 = -1$. Accordingly,

$$D = (1)(\bar{Y}_1) + (-1)(\bar{Y}_2) = \bar{Y}_1 - \bar{Y}_2$$

A contrast is not limited to individual means. One may, for example, contrast the average of $\bar{Y}_1$ and $\bar{Y}_2$ with that of $\bar{Y}_3$. This can be accomplished by setting $C_1 = 1/2; C_2 = 1/2; C_3 = -1$. Accordingly,

$$D = (1/2)(\bar{Y}_1) + (1/2)(\bar{Y}_2) + (-1)(\bar{Y}_3)$$

$$= \frac{\bar{Y}_1 + \bar{Y}_2}{2} - \bar{Y}_3$$

A comparison is considered statistically significant if $|D|$ (the absolute value of D) exceeds a value S, which is defined as follows:

$$S = \sqrt{kF_\alpha; k, N-k-1} \sqrt{MSR\left[\Sigma \frac{(C_j)^2}{n_j}\right]} \tag{7.16}$$

where k = number of coded vectors, or number of groups minus one; F_α; $k, N-k-1$ = tabled value of F with k and $N-k-1$ degrees of freedom at a prespecified α level; MSR = mean square residuals or, equivalently, the mean square error from the analysis of variance; C_j = coefficient by which the mean of group j is multiplied; n_j = number of subjects in group j.

The method is now illustrated for some comparisons between the means of the example of Table 7.2. For this example $\bar{Y}_{A_1} = 6.00$; $\bar{Y}_{A_2} = 9.00$; $\bar{Y}_{A_3} = 3.00$; $MSR = 2.50$; $k = 2$; $N-k-1 = 12$ (see Tables 6.3 and 7.3). The tabled F ratio for 2 and 12 df for the .05 level is 3.89. Contrasting $\bar{Y}_{A_1}$ with $\bar{Y}_{A_2}$,

$$D = (1)(\bar{Y}_{A_1}) + (-1)(\bar{Y}_{A_2}) = 6.00 - 9.00 = -3.00$$

$$S = \sqrt{(2)(3.89)} \sqrt{2.50\left[\frac{(1)^2}{5} + \frac{(-1)^2}{5}\right]} = \sqrt{7.78} \sqrt{2.50\left(\frac{2}{5}\right)}$$

$$= \sqrt{7.78} \sqrt{\frac{5}{5}} = (2.79)(1.00) = 2.79$$

Since $|D|$ exceeds S it is concluded that $\bar{Y}_{A_1}$ is significantly different from $\bar{Y}_{A_2}$ at the .05 level. Because $n_1 = n_2 = n_3$, S is the same for any comparison between two means. It can therefore be concluded that the difference between $\bar{Y}_{A_1}$ and $\bar{Y}_{A_3}$ (6.00 − 3.00), and that between $\bar{Y}_{A_2}$ and $\bar{Y}_{A_3}$ (9.00 − 3.00) are also significant. In the present example all the possible comparisons between individual means are significant.

Suppose that one also wished to compare the average of the means for groups A_1 and A_3 with the mean of group A_2. This can be done in the following manner:

$$D = (1/2)(\bar{Y}_{A_1}) + (1/2)(\bar{Y}_{A_3}) + (-1)(\bar{Y}_{A_2})$$

$$D = (1/2)(6.00) + (1/2)(3.00) + (-1)(9.00) = -4.50$$

$$S = \sqrt{(2)(3.89)}\ \sqrt{(2.50)\left[\frac{(.5)^2}{5} + \frac{(.5)^2}{5} + \frac{(-1)^2}{5}\right]}$$

$$= \sqrt{7.78}\ \sqrt{(2.50)\frac{(1.50)}{5}} = \sqrt{7.78}\ \sqrt{\frac{3.75}{5}} = \sqrt{7.78}\ \sqrt{.75}$$

$$= (2.79)(.87) = 2.43$$

$|D|$ (4.50) is larger than S (2.43) and it is concluded that there is a significant difference between $\bar{Y}_{A_2}$ and $(\bar{Y}_{A_1} + \bar{Y}_{A3})/2$.

In order to avoid working with fractions one may multiply the coefficients by a constant. For the above comparison, for example, the coefficients may be multiplied by 2, thereby setting $C_1 = 1$, $C_2 = 1$, $C_3 = -2$.

$$D = (1)(6.00) + (1)(3.00) + (-2)(9.00) = -9.00$$

$$S = \sqrt{(2)(3.89)}\ \sqrt{(2.50)\left[\frac{(1)^2}{5} + \frac{(1)^2}{5} + \frac{(-2)^2}{5}\right]}$$

$$= 2.79\ \sqrt{(2.50)\left(\frac{6}{5}\right)} = 2.79\ \sqrt{3.00} = (2.79)(1.73) = 4.83$$

The ratios of D to S in both instances are the same, within rounding errors. The second D is twice as large as the first D, and the second S is twice as large as the first S. The conclusion from either the first or the second calculation is therefore the same.

Any number of means and any combination of means can be compared similarly. The only constraint is that the sum of the coefficients of each comparison must be zero.

When comparing means one is in effect contrasting treatment effects, or b coefficients when effect coding is used. Recall that the mean of a given group is a composite of the grand mean and the treatment effect for the group. For effect coding this was expressed as $\bar{Y}_j = a + b_j$, where $\bar{Y}_j$ = mean of group j; a = intercept, or the grand mean, $\bar{Y}$; and b_j = the effect of treatment j, or $\bar{Y}_j - \bar{Y}$.

Accordingly, when contrasting, for example, $\bar{Y}_{A_1}$ with $\bar{Y}_{A_2}$,

$$D = (1)(\bar{Y}_{A_1}) + (-1)(\bar{Y}_{A_2}) = (1)(a+b_1) + (-1)(a+b_2)$$
$$= a + b_1 - a - b_2 = b_1 - b_2$$

It should be clear that the comparison of the two group means, $\bar{Y}_{A_1}$ and $\bar{Y}_{A_2}$, is the same as the comparison between b_1 and b_2, or $\bar{Y}_{A_1} - \bar{Y}_{A_2} = 6 - 9 = -3$, and $b_1 - b_2 = 0 - 3 = -3$.

Orthogonal Coding

In the preceding section post hoc comparisons between means were illustrated using the Scheffé method. It was pointed out that such comparisons are performed subsequent to a significant R^2 in order to determine which means, or treatment effects, differ significantly from each other. Instead of a post hoc approach, however, it is possible to take an a priori approach in which differences between means, or treatment effects, are hypothesized prior to the analysis of the data. The tests of significance for a priori, or planned, comparisons are more powerful than those for post hoc comparisons. In other words, it is possible for a specific comparison to be not significant when tested by post hoc methods but significant when tested by a priori methods. This advantage of the planned comparisons stems from the demands on the researcher: he must hypothesize the differences prior to the analysis, and he is limited to those comparisons about which hypotheses were formulated.

Post hoc comparisons, on the other hand, enable the researcher to engage in so-called data snooping by performing any or all of the conceivable comparisons between means. Tests of significance using this approach are thus more conservative than those for the planned comparisons approach, as they should be. The choice between the two approaches depends on the state of the knowledge in the area under study and the goals of the researcher. The greater the knowledge, the less one has to rely on omnibus tests and data snooping, and the more one is in a position to formulate planned comparisons.

There are two types of planned comparisons: orthogonal and nonorthogonal. The presentation here is limited to orthogonal comparisons.[4] We try to show how one can use orthogonal coding to obtain the results for orthogonal comparisons. Before doing this, however, a brief discussion of orthogonal comparisons is necessary.

Orthogonal Comparisons

Two comparisons are orthogonal when the sum of the products of the coefficients for their respective elements is zero. Consider the following two comparisons:

$$D_1 = (1)(\bar{Y}_1) + (-1)(\bar{Y}_2) + (0)(\bar{Y}_3)$$
$$D_2 = (-1/2)(\bar{Y}_1) + (-1/2)(\bar{Y}_2) + (1)(\bar{Y}_3)$$

[4]For a good discussion of planned nonorthogonal comparisons, see Kirk (1968).

In the first comparison, D_1, $\bar{Y}_1$ is contrasted with $\bar{Y}_2$. In comparison D_2, the average of $\bar{Y}_1 + \bar{Y}_2$, is contrasted with $\bar{Y}_3$. To determine whether these two comparisons are orthogonal we multiply the coefficients for each element in the two comparisons and sum. Accordingly:

$$1. \quad (1) + (-1) + (0)$$

$$2. \quad (-1/2) + (-1/2) + (1)$$

$$1 \times 2: \quad (1)(-1/2) + (-1)(-1/2) + (0)(1) = 0$$

D_1 and D_2 are obviously orthogonal.

Consider now the following two comparisons:

$$D_3 = (1)(\bar{Y}_1) + (-1)(\bar{Y}_2) + (0)(\bar{Y}_3)$$

$$D_4 = (-1)(\bar{Y}_1) + (0)(\bar{Y}_2) + (1)(\bar{Y}_3)$$

The sum of the products of the coefficients of these comparisons is

$$(1)(-1) + (-1)(0) + (0)(1) = -1$$

Comparisons D_3 and D_4 are not orthogonal.

The number of orthogonal comparisons one can perform within a given analysis is equal to the number of groups minus one, or the number of coded vectors necessary to describe group membership. For three groups, for example, one can perform two orthogonal comparisons. Several possible comparisons for three groups are listed in Table 7.4. Comparison 1, for example, contrasts the mean of A_1 with the mean of A_2, while comparison 2 contrasts the mean of A_3 with the average of the means A_1 and A_2. It was shown above that comparisons 1 and 2 are orthogonal. Comparisons 1 and 3, on the other hand, are not orthogonal.[5]

Table 7.4 contains three alternative sets of two orthogonal comparisons, namely 1 and 2, 3 and 4, 5 and 6. The specific set of orthogonal comparisons one chooses is dictated by the theory from which the hypotheses are derived. If, for example, A_1 and A_2 are two experimental treatments while A_3 is a control

[5]For a further treatment of this topic see Appendix A.

TABLE 7.4   SOME POSSIBLE COMPARISONS BETWEEN MEANS
OF THREE GROUPS

Comparison	Groups		
	A_1	A_2	A_3
1	1	-1	0
2	$-1/2$	$-1/2$	1
3	1	$-1/2$	$-1/2$
4	0	1	-1
5	1	0	-1
6	$-1/2$	1	$-1/2$

group, one may wish, on the one hand, to contrast means A_1 and A_2, and, on the other hand, to contrast the average of means A_1 and A_2 with the mean of A_3. (Comparisons 1 and 2 of Table 7.4 will accomplish this.) Referring to the example of attitudes toward divorce laws analyzed earlier, comparison 1 will contrast the mean of married males with the mean of divorced males. Comparison 2 will contrast the mean of married and divorced males with the mean of single males.

Regression Analysis with Orthogonal Coding

When hypothesizing orthogonal comparisons it is possible to use the coefficients of the hypothesized contrasts as the numbers in the coded vectors in regression analysis. The application of this method, referred to here as orthogonal coding, yields results that are directly interpretable. In addition, it is shown below that the use of orthogonal coding simplifies the calculations of the regression analysis.

Orthogonal coding is now applied to an analysis of the data earlier analyzed with dummy coding and effect coding. Using the three methods of coding with the same fictitious data enables one to compare the overall results as well as to study the unique properties of each method.

In Table 7.5 we repeat the Y vector reported earlier in Tables 7.1 and 7.2. It will be recalled that this vector represents attitudes toward divorce laws of groups A_1, A_2, and A_3 — married males, divorced males, and single males, respectively. Vector 1 in Table 7.5 represents a contrast between mean A_1 and mean A_2. Vector 2 represents a contrast between the average of means A_1 and of A_2, and the mean of A_3. It was shown earlier that these two comparisons are orthogonal. While some of the coefficients of comparison 2 in Table 7.4 were expressed as fractions, the same comparison is accomplished here with integers. (Multiplying the coefficients of comparison 2 in Table 7.4 by a constant of 2 yields the coefficients $-1 -1\ 2$.)

We can now use multiple regression analysis, with Y as the dependent variable and vectors 1 and 2 as the independent variables. The advantage of the computational ease in using orthogonal coding should be evident from the zero-order correlations reported in Table 7.5. Note that $r_{12} = .00$, as it should, since the meaning of orthogonality is vectors at right angles (90 degrees), which means zero correlation. Whenever orthogonal coding is used, all the correlations between the coded vectors are zero. It was shown earlier [formula (5.9)] that when the correlations between all the independent variables are zero, the formula for R^2 is

$$R^2_{y.12...k} = r^2_{y1} + r^2_{y2} + \cdots + r^2_{yk} \qquad (7.17)$$

For the present problem,

$$R^2_{y.12} = r^2_{y1} + r^2_{y2}$$

$$= (-.43301)^2 + (-.75000)^2 = .18750 + .56250 = .75000$$

The same value for $R^2_{y.12}$ was obtained when the two other coding methods

TABLE **7.5**   ORTHOGONAL CODING OF ATTITUDES TOWARD DIVORCE
DATA, AND CALCULATIONS NECESSARY FOR MULTIPLE REGRESSION
ANALYSIS[a]

Group	Y	1	2
	4	1	−1
	5	1	−1
A_1	6	1	−1
	7	1	−1
	8	1	−1
	7	−1	−1
	8	−1	−1
A_2	9	−1	−1
	10	−1	−1
	11	−1	−1
	1	0	2
	2	0	2
A_3	3	0	2
	4	0	2
	5	0	2
Σ:	90	0	0
M:	6	0	0
ss:	120	10	30
s:	2.92770	.84515	1.46385
$r_{y1} = -.43301$		$r_{y2} = -.75000$	$r_{12} = .00000$

[a]Vector Y is repeated from Table 7.1.

were used. Furthermore, because there is no correlation among the coded vectors, the square of the zero-order correlation of each coded vector with the dependent variable indicates the proportion of variance of the dependent variable accounted for by each vector. To obtain the sum of squares accounted for by a coded vector one may apply the following formula:

$$ss_j = (r_{yj}^2)(\Sigma y^2) \tag{7.18}$$

where r_{yj} = correlation between the dependent variable, Y, and a coded vector j; Σy^2 = sum of squares of the dependent variable. The sum of squares accounted for by the first coded vector of Table 7.5 is

$$ss_1 = (.18750)(120.00) = 22.50$$

and for the second vector.

$$ss_2 = (.56250)(120.00) = 67.50$$

Note that $22.50 + 67.50 = 90.00$, which is equal to the regression sum of squares obtained in the earlier analysis of these data. The residual sum of squares is also the same as the one obtained earlier, namely 30.00, and so are the degrees of freedom: 2 for the regression sum of squares (k), and 12 for the residual sum of squares ($N-k-1$). It is possible to divide each sum of squares by its degrees of freedom to obtain mean squares and then calculate the F ratio, which will be the same as obtained earlier ($F = 18.00$, with 2 and 12 df).

When working with orthogonal comparisons, however, it is not the overall F ratio that is of interest. The researcher is not interested in determining whether there are significant differences somewhere in his data, but rather in knowing whether the a priori hypothesized differences are significant. The testing of such hypotheses can be accomplished by using tests with individual degrees of freedom, each related to a specific hypothesis. It was shown above that the regression sum of squares was broken into two orthogonal components, each with one degree of freedom. To calculate the F ratios for the individual degrees of freedom, first calculate the mean square of the residuals (MSR), or $s_{y.12}^2$ (variance of estimate). Although we know the value of MSR from previous calculations, it is again calculated for completeness of presentation. By formula (7.8),

$$ss_{res} = (1-R_{y.12}^2)(\Sigma\, y^2) = (1-.75)(120.00) = 30.00$$

and by formula (7.1), the mean square residuals is

$$MSR = \frac{ss_{res}}{N-k-1} = \frac{30.00}{15-2-1} = \frac{30.00}{12} = 2.50$$

Since in the case under consideration each sum of squares due to regression has one degree of freedom, it follows that the mean square for the numerator of each F ratio for individual degrees of freedom is equal to the regression sum of squares due to a given comparison. The denominator of each F ratio is the MSR obtained above. Accordingly,

$$F_1 = \frac{ss_{reg(1)}}{MSR} = \frac{22.50}{2.50} = 9.00$$

with 1 and 12 degrees of freedom, $p < .05$. This F ratio indicates that the difference between the means of groups A_1 and A_2 (that is, $6.00 - 9.00 = -3.00$) is significant at the .05 level.

For the second comparison,

$$F_2 = \frac{ss_{reg(2)}}{MSR} = \frac{67.50}{2.50} = 27.00$$

with 1 and 12 degrees of freedom, $p < .005$. The average of the means of married and divorced males (groups A_1 and A_2), 7.50, is significantly different from the average of single males (group A_3), 3.00, at the .005 level of significance.[6]

---

[6]Although the sums of squares of each comparison are independent, the F ratios associated with them are not, because the same mean square error is used for all the comparisons. When the number of degrees of freedom for the mean square error is large, the comparisons can be viewed as independent. For a discussion of this point the reader is referred to Hays (1963) and Kirk (1968).

Note the interesting relation between the F ratios for the individual degrees of freedom and the overall F ratio. The latter is an average of all the F ratios obtained from the orthogonal comparisons. In the present case, $(9.00 + 27.00)/2 = 18.00$, which is the value of the overall F ratio. This demonstrates the advantage of orthogonal comparisons. Unless the treatment effects are equal, some orthogonal comparisons will have F ratios larger than the overall F ratio. Even when the overall F ratio is not significant, some of the orthogonal comparisons may have significant F ratios. Furthermore, while a significant overall F ratio is a necessary condition for the application of post hoc comparisons between means, the calculation of the overall F ratio is not necessary for orthogonal comparison analysis. The interest in such analysis is in the F ratios for the individual degrees of freedom corresponding to the specific differences hypothesized prior to the analysis.

The foregoing analysis is summarized in Table 7.6, where it is possible to see clearly how the total sum of squares is broken into the various components. When doing an analysis with orthogonal comparisons, one can calculate a t ratio for each comparison rather than an F ratio. The calculation of t ratios enables one to set confidence intervals around the differences between means. With t ratios, moreover, one can use one-tailed tests of significance for individual comparisons. Since $t^2 = F$ when the numerator of the F ratio has one degree of freedom, as is the case with the type of comparisons under consideration, t may be obtained by taking the square root of F. For the present example, the t ratios are 3.00 and 5.20 ($\sqrt{9}$ and $\sqrt{27}$). The degrees of freedom for each t ratio are equal to the degrees of freedom associated with the residual sum of squares: $N - k - 1$. In the present example, each t ratio has 12 degrees of freedom.

The Regression Equation

The calculation of the regression equation for the case of orthogonal coding is probably simplest when, from the various expressions for the R^2, one considers the following:

$$R^2_{y.12...k} = \beta_1 r_{y1} + \beta_2 r_{y2} + \cdots + \beta_k r_{yk} \tag{7.19}$$

TABLE **7.6** SUMMARY OF THE ANALYSIS WITH ORTHOGONAL CODING, ATTITUDES TOWARD DIVORCE DATA

Source	df	ss	ms	F
Total Regression	2	90.00	45.00	18.00
Regression due to Vector 1	1	22.50	22.50	9.00
Regression due to Vector 2	1	67.50	67.50	27.00
Residual	12	30.00	2.50	
Total	14	120.00		

where $R^2_{y.12...k}$ = squared multiple correlation of Y with k independent variables; $\beta_1, \beta_2 \ldots \beta_k$ = standardized regression coefficients for the k independent variables; $r_{y1}, r_{y2} \ldots r_{yk}$ = zero-order correlations between the dependent variable, Y, and each of the independent variables. From formulas (7.17) and (7.19) it follows that when the independent variables are not correlated with each other, each β is equal to the zero-order correlation with which it is associated. For the present problem in which the coded vectors are orthogonal, β_1, the standardized coefficient associated with vector 1 of Table 7.5, is $-.43301$ (r_{y1}), and $\beta_2 = -.75000$ (r_{y2}). To calculate the b's we apply formula (6.9), which is repeated here with a new number:

$$b_j = \beta_j \frac{s_y}{s_j} \qquad (7.20)$$

where b_j = regression coefficient for the jth independent variable; β_j = standardized regression coefficient for the jth independent variable; s_y = standard deviation of the dependent variable, Y; s_j = standard deviation of the jth independent variable. From Table 7.5: $s_y = 2.92770$; $s_1 = .84515$; $s_2 = 1.46385$. Therefore,

$$b_1 = -.43301 \frac{2.92770}{.84515} = -1.50$$

$$b_2 = -.75000 \frac{2.92770}{1.46385} = -1.50$$

$$a = \bar{Y} - b_1\bar{X}_1 - b_2\bar{X}_2$$
$$= 6.00 - (-1.50)(0) - (-1.50)(0) = 6.00$$

Since the mean of any comparison is zero, it follows that a is always equal to the grand mean of the dependent variable, $\bar{Y}$.

The regression equation is

$$Y' = 6.00 - 1.50\,X_1 - 1.50\,X_2$$

Applying this regression equation to the scores of any subject will, of course, yield a predicted score equal to the mean of the group to which the subject belongs. As noted above, a is equal to the grand mean. What is the meaning of each of the b's? Look at vector 1 of Table 7.5 and note that the scores in group A_1 are associated with a $+1$, while scores in group A_2 are associated with a -1. Since $b_1 = -1.50$, then $(-1.50)(1) - (-1.50)(-1) = -3.00$, which is equal to the difference between $\bar{Y}_{A_1}$ and $\bar{Y}_{A_2}$ (that is, $6.00 - 9.00 = -3.00$). Applying b to the codes in the vector with which it is associated results in the value for the hypothesized comparison. Note, too, that multiplying each coefficient in the coded vector by the b of that vector, squaring and summing, one obtains

$$[(-1)(-1.50)]^2(5) + [(1)(-1.50)]^2(5) = (2.25)(5) + (2.25)(5) = 22.50$$

This is the sum of squares obtained above for the contrast of $\bar{Y}_{A_1}$ and $\bar{Y}_{A_2}$. In other words, out of the regression sum of squares, 90.00, 22.50 is explained by the first contrast: (1) $(\bar{Y}_{A_1}) + (-1)(\bar{Y}_{A_2})$. As noted earlier, the sum of squares for a comparison is independent of the sum of squares for another comparison to which it is orthogonal.

Now look at b_2, which is associated with vector 2 of Table 7.5. Vector 2 contrasts the mean of A_3 with the average of the means of A_1 and A_2. That is $3.00 - (6.00 + 9.00)/2 = 3.00 - 7.50 = -4.50$. This is the same as applying b_2, -1.50, to the codes of vector 2: $(-1.5)(2) - (-1.5)(-1) = -4.50$. Calculate the sum of squares associated with vector 2:

$$[(-1.50)(2)]^2(5) + [(-1.50)(-1)]^2(10) = 45.00 + 22.50 = 67.50$$

This is the same value as that obtained earlier when mean A_3 was contrasted with the average of means A_1 and A_2.

Tests of Significance

The standard error of each b can be obtained by the application of formula (7.2), which for the special case when all the independent variables are not correlated, reduces to

$$s_{bj} = \sqrt{\frac{s^2_{y.12...k}}{\Sigma x_j^2}} \qquad (7.21)$$

where s_{bj} = standard error of b for the jth variable; $s^2_{y.12...k}$ = variance of estimate; Σx_j^2 = sum of squares for the jth variable. *Formula (7.21) applies only when the independent variables are orthogonal.* When the independent variables are correlated, formula (7.2) must be used. In the present problem $s^2_{y.12} = 2.50$; $\Sigma x_1^2 = 10$; $\Sigma x_2^2 = 30$ (see Table 7.5).

$$s_{b_1} = \sqrt{\frac{2.50}{10}} = \sqrt{.25} = .50$$

and the t ratio for b_1 is

$$t_{b_1} = \frac{b_1}{s_{b_1}} = \frac{-1.50}{.50} = -3.00$$

$$s_{b_2} = \sqrt{\frac{2.50}{30}} = \sqrt{.08333} = .28867$$

$$t_{b_2} = \frac{-1.50}{.28867} = -5.19624$$

Recall that the square of each t ratio is equal to the F ratio. Thus, $t^2_{b_1} = (-3.00)^2 = 9.00$, the F ratio associated with vector 1; $t^2_{b_2} = (-5.19624)^2 = 27.00$, the F ratio for vector 2. Consequently, when orthogonal coding is used, the t ratio for each b is in effect the test of significance for the contrast with which the given b is associated. To further demonstrate this point we apply the formula for the t ratio for orthogonal comparisons used subsequent to the application of the analysis of variance (see, for example, Kirk, 1968,

p. 74):

$$t = \frac{C_1 \bar{Y}_1 + C_2 \bar{Y}_2 + \cdots + C_j \bar{Y}_j}{\sqrt{MSE\left[\sum \frac{(C_j)^2}{n_j}\right]}} \tag{7.22}$$

where $t = t$ ratio for a comparison; $C =$ coefficient by which a given mean, $\bar{Y}$, is multiplied; $j =$ number of means involved in the comparison; $MSE =$ mean square error, which is equal to the mean square residuals in the context of regression analysis. Applying formula (7.22) to the contrast reflected in vector 1 of Table 7.5, that is, to the contrast of $\bar{Y}_{A_1}$ and $\bar{Y}_{A_2}$:

$$t = \frac{(1)(6.00) + (-1)(9.00)}{\sqrt{2.50\left[\frac{(1)^2}{5} + \frac{(-1)^2}{5}\right]}} = \frac{6.00 - 9.00}{\sqrt{2.50\left(\frac{2}{5}\right)}} = \frac{-3.00}{1} = -3.00$$

The same t ratio was obtained above for b_1. And for the second contrast,

$$t = \frac{(-1)(6.00) + (-1)(9.00) + (2)(3.00)}{\sqrt{2.50\left[\frac{(-1)^2}{5} + \frac{(-1)^2}{5} + \frac{(2)^2}{5}\right]}} = \frac{-9.00}{\sqrt{2.50\left(\frac{6}{5}\right)}} = \frac{-9.00}{\sqrt{3}}$$

$$= \frac{-9.00}{1.73205} = -5.19615$$

Again, this is the same value obtained as the t ratio for b_2 (within rounding error).

When using orthogonal coding, then, the tests of significance for the b coefficients are in effect the tests of significance for the a priori comparisons. Most computer programs for regression analysis report the t ratio for each b, in addition to the overall results. Accordingly, one obtains the tests of significance for each comparison directly when orthogonal coding is used.

Orthogonal Coding and Ease of Calculations

It is obvious from the above presentation that regression calculations are greatly simplified when orthogonal coding is used.[7] When a computer program is available, of course, this is not much of an advantage. When one does not have access to a computer, however, the use of orthogonal coding may be useful in reducing and simplifying the calculations necessary for the regression analysis. One first calculates zero-order correlations between each coded

[7]This is even more evident when, unlike the present example, the analysis involves more than three groups. In the present example there are two coded vectors and it is therefore possible to apply the relatively simple formula for R^2 with two independent variables [formula (6.5)]. With more than two independent variables, however, the most efficient method of analysis is with matrix algebra, as described in Chapter 4 and Appendix A. The virtue of orthogonal coding is that in the basic formula $b = (\mathbf{X'X})^{-1}\mathbf{X'y}$, $\mathbf{X'X}$ is a diagonal matrix. The inverse of such a matrix is a diagonal matrix whose elements are the reciprocals of $\mathbf{X'X}$. It thus becomes easy to solve for the vector of the b's, and to obtain the standard errors of the b's. The reader who has a facility with matrix algebra may wish to repeat the analysis with this method.

vector and the dependent variable.[8] Then formula (7.17), which is simply the sum of the squared zero-order correlations, will yield R^2. When the orthogonal coding is used for the purpose of facilitating calculations, it should not be difficult to develop a system of comparison for any number of groups. One such system is to contrast the first group with the second, then the first two groups with the third, then the first three groups with the fourth, and so forth, successively until all the possible orthogonal comparisons are exhausted. This will perhaps become clear with an example. Suppose there are five groups. Four possible orthogonal comparisons as suggested above are indicated in Table 7.7. There is, of course, a large number of other possible orthogonal comparisons, which may serve as well. When orthogonal coding is used for facility of calculations, one does not interpret the individual comparisons, but simply adds their contributions to obtain R^2 or the regression sum of squares. When the F ratio associated with R^2 is significant, one may proceed with post hoc comparisons between means, as discussed earlier in the section on multiple comparisons.

Unequal Sample Sizes in Groups

It is desirable that sample sizes of groups be equal. There are two major reasons: the statistical tests presented in this chapter are more sensitive when they are based on equal n's, and equal n's minimize distortions that may occur when there are departures from certain assumptions underlying these statistical tests.[9] Moreover, calculations with equal n's are simpler than with unequal n's. It

[8]The calculation of the zero-order correlations is also simplified, since ΣX in any coded vector is zero as a result of the requirement that the sum of the coefficients of a comparison equal zero. Consequently, the raw score formula for

$$r = \frac{N \Sigma XY - (\Sigma X)(\Sigma Y)}{\sqrt{N \Sigma X^2 - (\Sigma X)^2} \sqrt{N \Sigma Y^2 - (\Sigma Y)^2}}$$

reduces to

$$r = \frac{N \Sigma XY}{\sqrt{N \Sigma X^2} \sqrt{N \Sigma Y^2 - (\Sigma Y)^2}}.$$

Furthermore, the second term in the denominator of the reduced formula is constant for all the zero-order correlations between the coded vectors and the dependent variable. The other terms in the reduced formula are easily obtainable.

[9]For further discussion of the advantages of equal sample sizes, see Li (1964, I, 147–148; 197–198).

TABLE 7.7   FOUR ORTHOGONAL COMPARISONS FOR FIVE GROUPS

	Groups				
Comparisons	A_1	A_2	A_3	A_4	A_5
1	1	-1	0	0	0
2	-1	-1	2	0	0
3	-1	-1	-1	3	0
4	-1	-1	-1	-1	4

frequently happens, however, that a researcher must deal with unequal n's. Even though he started with equal n's, he may end up with unequal n's because of errors in the recording of scores, breakdown of apparatus used in an experiment, subject attrition, and the like. The present section is devoted to the analysis of data with unequal n's. We present first analyses with dummy coding and effect coding. Then we explain unequal $=n$ analysis with orthogonal coding.

Dummy and Effect Coding for Unequal n's

The analysis of data for unequal n's with dummy or effect coding proceeds in the same manner as that for equal n's. This is illustrated with part of the fictitious data of the earlier part of the chapter. The example analyzed with the three coding methods consisted of three groups, each composed of 5 subjects. For the present analysis we delete the scores of the fourth and the fifth subjects from group A_1, and the score of the fifth subject from group A_2. Group A_3 remains intact. Accordingly, there are 3, 4, and 5 subjects in groups A_1, A_2, and A_3, respectively. The scores for these groups, along with dummy and effect coding are reported in Table 7.8. Note that the methods used are identical to those used with equal n's (see Tables 7.1 and 7.2). Vectors 1 and 2 of Table 7.8 are dummy coding, while vectors 3 and 4 are effect coding. The calculations of

TABLE 7.8   BASIC REGRESSION CALCULATIONS, UNEQUAL n's
DUMMY AND EFFECT CODING[a]

Group	Y	1	2	3	4
A_1	4	1	0	1	0
	5	1	0	1	0
	6	1	0	1	0
A_2	7	0	1	0	1
	8	0	1	0	1
	9	0	1	0	1
	10	0	1	0	1
A_3	1	0	0	-1	-1
	2	0	0	-1	-1
	3	0	0	-1	-1
	4	0	0	-1	-1
	5	0	0	-1	-1
Σ:	64	3	4	-2	-1
M:	5.33333	.25000	.33333	$-.16667$	$-.08333$
ss:	84.66667	2.25000	2.66667	7.66667	8.91667
s:	2.77434	.45227	.49237	.83485	.90034

$\Sigma y1 = -1.00000 \quad \Sigma y2 = 12.66667 \qquad \Sigma y3 = 10.66667 \quad \Sigma y4 = 24.33333$
$r_{y1} = -.07245 \quad r_{y2} = .84298 \qquad\qquad r_{y3} = .41867 \qquad r_{y4} = .88562$
$r_{12} = -.40825 \qquad\qquad\qquad\qquad\qquad r_{34} = .58459$

[a]Vectors 1 and 2 are dummy coding; vectors 3 and 4 are effect coding.

the multiple regression analysis are done in the same way as with equal n's. We repeat formula (6.5) with a new number:

$$R^2_{y.12} = \frac{r^2_{y1} + r^2_{y2} - 2r_{y1}r_{y2}r_{12}}{1 - r^2_{12}} \tag{7.23}$$

Applying formula (7.23) to the dummy coding:

$$R^2_{y.12} = \frac{(-.07245)^2 + (.84298)^2 - 2(-.07245)(.84298)(-.40825)}{1 - (-.40825)^2}$$

$$= \frac{.71586 - .04987}{1 - .16667} = \frac{.66599}{.83333} = .79919$$

and for effect coding,

$$R^2_{y.34} = \frac{(.41867)^2 + (.88562)^2 - 2(.41867)(.88562)(.58459)}{1 - (.58459)^2}$$

$$= \frac{.95961 - .43351}{1 - .34175} = \frac{.52610}{.65825} = .79924$$

The same R^2, within rounding error, is obtained in both analyses. To calculate the F ratio for this R^2 we observe that in each analysis $k = 2$ and $N - k - 1 = 12 - 2 - 1 = 9$. Accordingly,

$$F = \frac{.7992/2}{(1 - .7992)/9} = \frac{.3996}{.0223} = 17.92$$

with 2 and 9 degrees of freedom, $p < .01$. We turn now to the regression equation for each analysis.

The Regression Equation for Dummy Coding

We repeat formulas (6.7)–(6.9) with new numbers:

$$\beta_1 = \frac{r_{y1} - r_{y2}r_{12}}{1 - r^2_{12}} \tag{7.24}$$

$$\beta_2 = \frac{r_{y2} - r_{y1}r_{12}}{1 - r^2_{12}} \tag{7.25}$$

$$b_j = \beta_j \frac{s_y}{s_j} \tag{7.26}$$

Applying formulas (7.24)–(7.26) to the data of Table 7.8, we obtain for the dummy coding (vectors Y, 1, and 2):

$$\beta_1 = \frac{(-.07245) - (.84298)(-.40825)}{1 - (-.40825)^2} = \frac{-.07245 + .34415}{1 - .16667}$$

$$= \frac{.27170}{.83333} = .32604$$

$$b_1 = .32604 \frac{2.77434}{.45227} = 2.00001$$

$$\beta_2 = \frac{(.84298) - (-.07245)(-.40825)}{1 - (-.40825)^2} = \frac{.84298 - .02958}{1 - .16667}$$

$$= \frac{.81340}{.83333} = .97608$$

$$b_2 = .97608 \frac{2.77434}{.49237} = 5.49988$$

$$a = 5.3333 - (2.00001)(.25000) - (5.49988)(.33333)$$

$$= 5.33333 - .50000 - 1.83328 = 3.00005$$

The regression equation, to two decimals, is

$$Y' = 3.00 + 2.00\,X_1 + 5.50\,X_2$$

Note that the properties of this equation are the same as those with equal n's. That is, a is equal to the mean of the group assigned 0's throughout ($\bar{Y}_{A_3} = 3.00$; see Table 7.8), b_1 is equal to the difference between $\bar{Y}_{A_1}$ and $\bar{Y}_{A_3}$ ($5.00 - 3.00 = 2.00$), and b_2 is equal to the difference between $\bar{Y}_{A_2}$ and $\bar{Y}_{A_3}$ ($8.50 - 3.00 = 5.50$). The application of the regression equation to any subject of Table 7.8 will yield the value of the mean of the group to which the subject belongs, as it did with equal n's.

It was stated earlier that with dummy coding the group assigned 0's throughout acts as a control group, and that testing each b for significance amounts to testing the difference between the mean of the group with which the given b is associated and the mean of the control group. The same holds true for unequal n's and is demonstrated for b_1.
By formula (7.8):

$$ss_{\text{res}} = (1 - .7992)(84.66667) = 17.00107$$

By formula (7.1):

$$s_{y.12}^2 = \frac{17.00107}{9} = 1.88901$$

By formula (7.2):

$$s_{b_1} = \sqrt{\frac{1.88901}{(2.25000)[(1 - (-.40825)^2]}} = \sqrt{\frac{1.88901}{(2.25000)(.83333)}}$$

$$= \sqrt{\frac{1.88901}{1.87499}} = \sqrt{1.00748} = 1.00373$$

$$t_{b_1} = \frac{2.00001}{1.00373} = 1.99$$

By formula (7.4):

$$t = \frac{5.00 - 3.00}{\sqrt{1.88901\left(\frac{1}{3} + \frac{1}{5}\right)}} = \frac{2.00}{\sqrt{1.00747}} = \frac{2.00}{1.00373} = 1.99$$

The same t ratio was obtained when b_1 was tested for significance and when the conventional formula (7.4) for testing the significance of the difference between two means was applied. The degrees of freedom associated with this t ratio are 9 (the degrees of freedom associated with the residual sum of squares: $N-k-1$). The t ratio for b_2 is 5.97. Its calculation is left as an exercise for the student.

The Regression Equation for Effect Coding

Applying formulas (7.24)–(7.26) to the data of Table 7.8, for the effect coding (vectors Y, 3, and 4):

$$\beta_3 = \frac{(.41867) - (.88562)(.58459)}{1 - (.58459)^2} = \frac{.41867 - .51772}{1 - .34175}$$

$$= \frac{-.09905}{.65825} = -.15047$$

$$b_3 = -.15047 \frac{2.77434}{.83485} = -.50003$$

$$\beta_4 = \frac{(.88562) - (.41867)(.58459)}{1 - (.58459)^2} = \frac{.88562 - .24475}{1 - .34175}$$

$$= \frac{.64087}{.65825} = .97360$$

$$b_4 = .97360 \frac{2.77434}{.90034} = 3.00008$$

$$a = 5.33333 - (-.50003)(-.16667) - (3.00008)(-.08333)$$

$$= 5.33333 - .08334 + .25000 = 5.49999$$

The regression equation, to two decimals, is

$$Y' = 5.50 - .50\,X_3 + 3.00\,X_4$$

While this regression equation has the same properties as the one obtained from effect coding with equal n's, note that a is not equal to the grand mean of the dependent variable ($\bar{Y} = 5.33333$, see Table 7.8). In the case of unequal n's, a is equal to the unweighted mean of the group means. In the present example: $\bar{Y}_{A_1} = 5.00$, $\bar{Y}_{A_2} = 8.50$, $\bar{Y}_{A_3} = 3.00$. $a = (5.00 + 8.50 + 3.00)/3 = 16.50/3 = 5.50$. To obtain a weighted mean for unequal n's, each group has to be weighted by the number of subjects on which it is based. In the present example,

$$\bar{Y} = \frac{(3)(5.00) + (4)(8.50) + (5)(3.00)}{3 + 4 + 5} = \frac{64}{12} = 5.33$$

which is the same as the value obtained in Table 7.8. When the sample sizes are equal, the mean of the means is the same as the grand mean, since all the means are weighted by a constant (the sample size).

As shown previously, each b weight is the effect of the treatment with which it is associated, or the deviation of the group mean with which it is associated from the overall (unweighted) mean. $b_3 = -.50$, which is equal to the deviation of $\bar{Y}_{A_1}$ from the overall mean ($5.00 - 5.50$); $b_4 = 3.00$ is the deviation of $\bar{Y}_{A_2}$ from the overall mean ($8.50 - 5.50$). The effect for A_3 is $-\Sigma(-.50 + 3.00) =$

-2.50. Again, this is equal to the deviation of $\bar{Y}_{A_3}$ from the overall mean $(3.00 - 5.50)$. The application of the regression equation to any subject of Table 7.8 will, of course, yield the mean of the group to which the subject belongs.

R^2 for this analysis is significant, as shown above. Consequently, it is possible to proceed with multiple comparisons between means using the Scheffé method. This is illustrated for the difference between $\bar{Y}_{A_1}$ and $\bar{Y}_{A_2}$. The test involves the application of formulas (7.15) and (7.16). The following information is necessary: $\bar{Y}_{A_1} = 5.00$; $\bar{Y}_{A_2} = 8.50$; $k = 2$. The tabled F at .05 level with 2 and 9 degrees of freedom $= 4.26$; $MSR = s^2_{y.12} = 1.88901$.

By formula (7.15):

$$D = (1)(5.00) + (-1)(8.50) = -3.50$$

By formula (7.16):

$$S = \sqrt{(2)(4.26)} \sqrt{1.88901 \left[\frac{(1)^2}{3} + \frac{(-1)^2}{4} \right]}$$

$$= \sqrt{8.52} \sqrt{1.10192} = \sqrt{9.38836} = 3.06$$

Since $|D|$ is greater than S it is concluded that $\bar{Y}_{A_1}$ is significantly different from $\bar{Y}_{A_2}$ (.05 level). It is of course possible to test other comparisons between means or combinations of means.

Orthogonal Coding with Unequal n's

For samples with unequal n's, a comparison is defined as

$$D = n_1 C_1 + n_2 C_2 + \cdots + n_j C_j = 0 \qquad (7.27)$$

where $D = $ difference, or comparison; $n_1, n_2 \ldots n_j = $ number of subjects in groups $1, 2 \ldots j$, respectively; $C = $ coefficient. Note that with equal n's formula (7.27) reduces to the requirement stated earlier in this chapter, namely that $\Sigma C_j = 0$. For the example with unequal n's, analyzed above with dummy and effect coding, the number of subjects is 3, 4, and 5 in groups A_1, A_2, and A_3, respectively. Suppose we want to compare $\bar{Y}_{A_1}$ with $\bar{Y}_{A_2}$ and assign 1's to members of group A_1, -1's to members of group A_2, and 0's to members of group A_3. By formula (7.27):

$$D = (3)(1) + (4)(-1) + (5)(0) = -1$$

It will be noted that these coefficients are not appropriate, since $D \neq 0$. It is necessary to find a set of coefficients that will satisfy the requirement that D equal zero. The simplest way to satisfy equation (7.27) is to use n_2 (4) as the coefficient for group A_1, and $-n_1$ (-3) as the coefficient for group A_2. Accordingly the comparison between groups A_1 and A_2 is

$$D = (3)(4) + (4)(-3) + (5)(0) = 0$$

Suppose we now wish to contrast groups A_1 and A_2 with group A_3. For this comparison we use $-n_3$ (-5) as the coefficients for groups A_1 and A_2, and $n_1 + n_2 = 7$ as the coefficient for group A_3. This comparison, too, satisfies the requirement of formula (7.27):

$$D = (3)(-5) + (4)(-5) + (5)(7) = 0$$

Are these two comparisons orthogonal? With unequal n's two comparisons are orthogonal if

$$n_1 C_{11} C_{21} + n_2 C_{12} C_{22} + n_3 C_{13} C_{23} = 0 \tag{7.28}$$

where the first subscript for each C refers to the number of the comparison, and the second subscript refers to the number of the group. For example, C_{11} means the coefficient of the first comparison for group 1, and C_{21} is the coefficient of the second comparison for group 1, and similarly for the other coefficients. For the two comparisons under consideration:

$$D_1 = (3)(4) + (4)(-3) + (5)(0)$$

$$D_2 = (3)(-5) + (4)(-5) + (5)(7)$$

$$(3)(4)(-5) + (4)(-3)(-5) + (5)(0)(7) = (3)(-20) + (4)(15) + 0 = 0$$

The two comparisons are orthogonal. Using the coefficients of the two comparisons, two coded vectors are generated, each reflecting one of the orthogonal comparisons. It is now possible to do a multiple regression analysis where Y is the dependent variable and the two coded vectors are treated as independent variables. The data for the three groups, along with the orthogonal vectors, are reported in Table 7.9, where vector 1 reflects the comparison between $\bar{Y}_{A_1}$ and $\bar{Y}_{A_2}$, vector 2 reflects the comparison between the weighted average of $\bar{Y}_{A_1}$ and $\bar{Y}_{A_2}$ and the mean of group A_3, $\bar{Y}_{A_3}$. By formula (7.17):

$$R^2_{y.12} = r^2_{y1} + r^2_{y2} = (-.49803)^2 + (-.74242)^2$$
$$= .24803 + .55119 = .79922$$

The same R^2 was obtained with dummy and effect coding for these data. Obviously, the F ratio for this R^2 is also the same as that obtained in the earlier analyses: 17.92, with 2 and 9 degrees of freedom.

When the orthogonal comparisons are hypothesized prior to the analysis, however, it is more meaningful to calculate F ratios for each comparison than to calculate the overall F ratio. We proceed on the assumption that the two comparisons under consideration were hypothesized a priori and calculate the F ratios for the individual degrees of freedom. First, the regression sum of squares due to each coded vector is calculated. The total sum of squares of the dependent variable, Σy^2, is 84.66667 (see Table 7.9). The regression sum of squares due to vector 1 is

$$ss_{\text{reg}(1)} = (r^2_{y1})(\Sigma y^2) = (.24803)(84.66667) = 20.99987$$

where $ss_{\text{reg}(1)} =$ regression sum of squares due to vector 1 of Table 7.8. And

TABLE **7.9**   ORTHOGONAL CODING FOR UNEQUAL n's AND CALCULATIONS
NECESSARY FOR MULTIPLE REGRESSION ANALYSIS[a]

Group	Y	1	2
	4	4	-5
A_1	5	4	-5
	6	4	-5
	7	-3	-5
	8	-3	-5
A_2	9	-3	-5
	10	-3	-5
	1	0	7
	2	0	7
A_3	3	0	7
	4	0	7
	5	0	7
Σ:	64	0	0
M:	5.33333	0	0
ss:	84.66667	84.00000	420.00000
s:	2.77434	2.76340	6.17914
	$r_{y1} = -.49803$	$r_{y2} = -.74242$	$r_{12} = \quad .00000$

[a]Vector 1 reflects the comparison between $\bar{Y}_{A_1}$ and $\bar{Y}_{A_2}$. Vector 2 reflects the comparison
between the weighted average of $\bar{Y}_{A_1}$ and $\bar{Y}_{A_2}$ with $\bar{Y}_{A_3}$.

for vector 2,

$$ss_{\text{reg}(2)} = (r_{y2}^2)(\Sigma\, y^2) = (.55119)(84.66667) = 46.66742$$

where $ss_{\text{reg}(2)}$ = regression sum of squares due to vector 2 of Table 7.9. These
two sums of squares are independent, and their sum is equal to the total regression sum of squares. That this is so is demonstrated by calculating the total
regression sum of squares using formula (7.6):

$$ss_{\text{reg}} = (R_{y.12}^2)(\Sigma\, y^2) = (.79922)(84.66667) = 67.66730$$

which is equal to the sum of the two components obtained above.

To calculate the F ratios for the individual degrees of freedom it is necessary to calculate the mean square residuals (MSR), or $s_{y.12}^2$ (variance of estimate). By formula (7.8):

$$ss_{\text{res}} = (1 - R_{y.12}^2)(\Sigma\, y^2) = (1 - .79922)(84.66667) = 16.99937$$

And by formula (7.1), the mean square residuals is

$$MSR = \frac{ss_{\text{res}}}{N - k - 1} = \frac{16.99937}{12 - 2 - 1} = \frac{16.99937}{9} = 1.88882$$

This value is, within rounding error, equal to MSR obtained in the analyses of these data with dummy and effect coding. Since in the case under consideration each sum of squares due to regression has one degree of freedom, it follows that the mean square for the numerator of each F ratio for individual degrees of freedom is equal to the regression sum of squares due to a given comparison. The denominator of each F ratio is the MSR obtained above. Accordingly,

$$F_1 = \frac{ss_{\text{reg(1)}}}{MSR} = \frac{20.99987}{1.88882} = 11.12$$

with 1 and 9 degrees of freedom, $p < .01$.

$$F_2 = \frac{ss_{\text{reg(2)}}}{MSR} = \frac{46.66742}{1.88882} = 24.71$$

with 1 and 9 degrees of freedom, $p < .01$. Since both F ratios are significant beyond the .01 level, it is concluded that the two a priori hypotheses are supported. That is, the comparison between $\bar{Y}_{A_1}$ and $\bar{Y}_{A_2}$ (vector 1 of Table 7.9) is significant beyond the .01 level, as is the comparison between the weighted mean of groups A_1 and A_2 with the mean of group A_3 (vector 2 of Table 7.9).

The overall F ratio is equal to the average of the F ratios for the individual degrees of freedom, just as it was with equal n's. In the present example, the overall F ratio is 17.92 (2 and 9 df), which is equal to

$$\frac{F_1 + F_2}{2} = \frac{11.12 + 24.71}{2} = 17.92$$

The foregoing analysis is summarized in Table 7.10, where one can see clearly how the total sum of squares is broken into the various components.

TABLE 7.10   SUMMARY OF ANALYSIS WITH ORTHOGONAL CODING FOR UNEQUAL n's

Source	df	ss	ms	F
Total Regression	2	67.66730	33.83365	17.91
Regression due to Vector 1	1	20.99987	20.99987	11.12
Regression due to Vector 2	1	46.66742	46.66742	24.71
Residual	9	16.99937	1.88882	
Total	11	84.66667		

The Regression Equation for Orthogonal Coding

Recall that with orthogonal coding each β is equal the zero-order correlation between the dependent variable and the coded vector with which it is associated. Accordingly (from Table 7.9): $\beta_1 = r_{y1} = -.49803$; $\beta_2 = r_{y2} = -.74242$.

And by formula (7.26),

$$b_1 = \beta_1 \frac{s_y}{s_1} = -.49803 \frac{2.77434}{2.76340} = -.50000$$

$$b_2 = \beta_2 \frac{s_y}{s_2} = -.74242 \frac{2.77434}{6.17914} = -.33333$$

$$a = \bar{Y} - b_1\bar{X}_1 - b_2\bar{X}_2$$
$$= 5.33333 - (-.50000)(0) - (-.33333)(0) = 5.33333$$

The regression equation, to two decimals, is

$$Y' = 5.33 - .50X_1 - .33X_2$$

When orthogonal coding is used with unequal n's, a is equal to the weighted mean of the group means. Stated differently, a is equal to the grand mean of the dependent variable, $\bar{Y}$ (see Table 7.9). One can obtain the regression sum of squares due to a given vector by multiplying each coefficient in the vector by its b, then squaring and summing for all the elements in the vector, just as one does with orthogonal coding and equal n's. For example, b_1 (the b for vector 1 of Table 7.9) is $-.50000$. Vector 1 consists of three coefficients each with a value of 4, four coefficients with a value of -3, and five coefficients with a value of 0. Accordingly, the regression sum of squares due to vector 1 is

$$ss_{reg(1)} = [(4)(-.50000)]^2(3) + [(-3)(-.50000)]^2(4) = 21.00000$$

The same value, within rounding errors, was obtained earlier when the squared zero-order correlation of the dependent variable and vector 1 was multiplied by the total sum of squares: $[(r_{y1}^2)(\Sigma y^2)]$. In the manner shown above one may also calculate the regression sum of squares due to vector 2.

Tests of Significance

The tests of significance of the b weights for orthogonal coding with unequal n's have the same properties as do the tests of significance of the b weights for orthogonal coding with equal n's. The t ratio for each b weight is equal to the square root of the F ratio for the individual degree of freedom to which the b corresponds. In other words, testing the significance of a b weight amounts to testing the significance of the difference between the means involved in the comparison associated with the given b. This is illustrated for the b weights obtained above: $b_1 = -.50000$, and $b_2 = -.33333$. Using $s_{y.12}^2 = 1.88882$, which was calculated above, and the appropriate sum of squares from Table 7.9, the standard error of b_1 [formula (7.21)] is

$$s_{b_1} = \sqrt{\frac{s_{y.12}^2}{\Sigma x_1^2}} = \sqrt{\frac{1.88882}{84.00000}} = \sqrt{.02249} = .14998$$

$$t_1 = \frac{b_1}{s_{b_1}} = \frac{-.50000}{.14998} = -3.33378, \quad \text{with } 9 \; df$$

$t_1{}^2 = (-3.33378)^2 = 11.11$, which is equal to F_1 obtained above (see Table 7.10). The standard error of b_2 is

$$s_{b_2} = \sqrt{\frac{s_{y.12}^2}{\Sigma x_2^2}} = \sqrt{\frac{1.88882}{420.00000}} = \sqrt{.00450} = .06708$$

$$t_2 = \frac{b_2}{s_{b_2}} = \frac{-.33333}{.06708} = -4.96914, \quad \text{with 9 } df$$

$t_2^2 = (-4.96914)^2 = 24.69$, which, within rounding errors, is equal to F_2 (see Table 7.10). The conclusions based on the F ratios for the individual degrees of freedom and the t ratios for the b weights are of course the same. With t ratios, however, it is also possible to calculate confidence intervals as well as to do one-tailed tests of significance.

Summary

Three methods of coding categorical variables were presented in this chapter. They were called: dummy coding, effect coding, and orthogonal coding. Regardless of the coding method used, the results of the overall analysis are the same. When a regression analysis is done with Y as the dependent variable and k coded vectors (k = number of groups minus one) reflecting group membership as the independent variables, the overall R^2, regression sum of squares, residual sum of squares, and the F ratio are the same with any coding method. The predictions based on the regression equations resulting from the different coding methods are also identical. In each case the predicted score is equal to the mean of the group to which the subject belongs. The coding methods do differ in the properties of their regression equations. A brief summary of the major properties of each method follows.

With *dummy coding,* k coded vectors consisting of 1's and 0's are generated. In each vector, in turn, subjects of one of the groups are assigned 1's and all others are assigned 0's. Since k is equal to the number of groups minus one, it follows that members of one of the groups are assigned 0's in all the vectors. This group is treated as a control group in the analysis. In the regression equation, the intercept, a, is equal to the mean of the control group. Each b coefficient is equal to the difference between the mean of the group assigned 1's in the vector associated with the b, and the mean of the control group. The test of significance of a given b is a test of significance between the mean of the group associated with the b and the mean of the control group. Although dummy coding is particularly useful when one does in fact have several experimental groups and one control group, it may also be used in situations in which no particular group serves as a control for all others. In the latter case, multiple comparisons between means may be performed subsequent to a significant R^2. One of the methods of multiple comparisons, the Scheffé test, was presented in the chapter. The properties of dummy coding are the same for equal or unequal sample sizes.

Effect coding is similar to dummy coding. The difference is that in dummy coding one of the groups is assigned 0's in all the coded vectors, while in effect coding one of the groups is assigned -1's in all the vectors. As a result, the regression equation reflects the linear model. In other words, the intercept, a, is equal to the grand mean of the dependent variable, $\bar{Y}$, and each b is equal to the treatment effect for the group with which it is associated, or the deviation of the mean of the group from the grand mean, $\bar{Y}$. When effect coding is used with unequal sample sizes, the intercept of the regression equation is equal to the unweighted mean of the group means. Each b is equal to the deviation of the mean of the group with which it is associated from the unweighted mean. Subsequent to obtaining a significant R^2, for equal or unequal n's, one does multiple comparisons between means, as described in the chapter.

Orthogonal coding consists of k coded vectors of orthogonal coefficients. The selection of orthogonal coefficients for equal and unequal sample sizes was discussed and illustrated. In the regression equation, a is equal to the grand mean, $\bar{Y}$, for equal as well as unequal sample sizes. Each b reflects the specific comparison with which it is related. Testing a given b for significance amounts to testing the specific hypothesis that the comparison reflects. The t ratio for each b is the same as the t ratio obtained from orthogonal comparisons subsequent to a conventional analysis of variance. With orthogonal comparisons one is not concerned with the overall F ratio for the R^2, but rather with the testing of the hypotheses for the specific comparisons formulated prior to the analysis.

Which method of coding one chooses depends on one's purpose and interest. When, for example, one wishes to compare several treatment groups with a control group, dummy coding is the appropriate method. Needless to say, for planned orthogonal comparisons, orthogonal coding reflecting the specific hypotheses is the most appropriate. It was pointed out, however, that even when one does not hypothesize orthogonal comparisons, orthogonal coding may still be used for the purpose of simplifying calculations, especially when a computer is not available. Effect coding also simplifies calculations. Its main virtue, however, is that the resulting regression equation reflects the linear model.

Study Suggestions

1. Under what conditions is dummy coding particularly useful?
2. What are the properties of the regression equation obtained from an analysis with dummy coding?
3. A regression equation is obtained from a regression analysis with dummy coding. The t ratio for the first b coefficient is 2.15. What in effect is being tested by this t ratio? How is the t ratio interpreted?
4. What are the properties of the regression equation obtained from an analysis with effect coding?
5. What is meant by the linear model? With one independent categorical variable, what does the linear model consist of?
6. The following regression equation was obtained from an analysis with

effect coding for four groups with equal n's:

$$Y' = 102.5 + 2.5\,X_1 - 2.5\,X_2 - 4.5\,X_3$$

(a) What is the grand mean, $\bar{Y}$?

(b) What are the means of the four groups?

(*Answers*: (a) $\bar{Y} = 102.5$; (b) $\bar{Y}_1 = 105$, $\bar{Y}_2 = 100$, $\bar{Y}_3 = 98$, $\bar{Y}_4 = 107$)

7.   What is accomplished by the Scheffé test? Under what conditions may it be used?

8.   In a study consisting of four groups, each with ten subjects, the following results were obtained:

$$\bar{Y}_1 = 16.5 \qquad \bar{Y}_2 = 12.0 \qquad \bar{Y}_3 = 16.0 \qquad \bar{Y}_4 = 11.5 \qquad MSR = 7.15$$

(a) Write the regression equation that will be obtained if effect coding is used. Assume that subjects in the fourth group are assigned -1's.

(b) Write the regression equation that will be obtained if dummy coding is used. Assume that subjects in the fourth group are assigned 0's.

(c) Do a Scheffé test for the following comparisons, at the .05 level: (1) between $\bar{Y}_1$ and $\bar{Y}_2$; (2) between the mean of $\bar{Y}_1$ and $\bar{Y}_2$, and $\bar{Y}_3$; (3) between the mean of $\bar{Y}_1$, $\bar{Y}_2$, $\bar{Y}_4$, and $\bar{Y}_3$.

(*Answers*: (a) $Y' = 14.0 + 2.5\,X_1 - 2.0\,X_2 + 2.0\,X_3$. (b) $Y' = 11.5 + 5.0\,X_1 + .5\,X_2 + 4.5\,X_3$. (c) (1) $|D| = 4.5$; $S = 3.5$; significant; (2) $|D| = 3.5$; $S = 6.1$; not significant; (3) $|D| = 8.0$; $S = 8.6$; not significant.)

9.   A researcher studied the relationship between political party affiliation and attitudes toward school busing. He administered an attitude scale to samples of Conservatives, Republicans, Liberals, and Democrats, and obtained the following scores. (The scores are fictitious.)

Conservatives	Republicans	Liberals	Democrats
2	3	5	4
3	3	6	5
4	4	6	5
4	4	7	7
6	5	7	7
6	6	9	7
7	8	10	9
7	8	10	9
8	9	11	10
8	10	12	10

(a) Using dummy coding, do a regression analysis of the data. Calculate the following: (1) R^2; (2) regression equation; (3) F ratio.

(b) Using effect coding, do a regression analysis of the above data. What is the regression equation?

(c) Do Scheffé tests for the differences between all possible pairs of means? Which differences are significant at the .05 level?

(d) Assume that the researcher had the following a priori hypotheses: that Republicans have more favorable attitudes toward school busing than do Conservatives; that Liberals are more favorable than Democrats; that Liberals and Democrats are more favorable toward school busing than are Conservatives and Republicans.

Use orthogonal coding to express these hypotheses and do a regression analysis. Calculate the following: (1) R^2; (2) regression equation; (3) t ratios for each of the b coefficients; (4) sum of squares due to each hypothesis; (5) residual sum of squares; (6) F ratios for each hypothesis. Interpret the results obtained under (a)–(d) above.

(*Answers*: (a) (1) $R^2 = .1987$; (2) $Y' = 7.3 - 1.8 X_1 - 1.3 X_2 + 1 X_3$; (3) $F = 2.98$, with 3 and 36 df. (b) $Y' = 6.775 - 1.275 X_1 - .775 X_2 + 1.525 X_3$. (c) S for a comparison between any two groups is 3.05. The largest D (that between Liberals and Conservatives) is 2.8. Therefore, none of the comparisons is significant. (d) (1) $R^2 = .1987$; (2) $Y' = 6.775 + .250 X_1 + .500 X_2 + 1.025 X_3$; (3) $t_1 = .48$; $t_2 = .96$; $t_3 = 2.79$. Each t ratio has 36 df; (4) $ss_{reg(1)} = 1.250$; $ss_{reg(2)} = 5.000$; $ss_{reg(3)} = 42.025$; (5) $ss_{res} = 194.700$; (6) $F_1 = .23$; $F_2 = .92$; $F_3 = 7.77$. Each F ratio has 1 and 36 df.)

Multiple Categorical Variables and Factorial Designs

It should be clear by now that many, perhaps most, studies in the behavioral sciences are multivariate in nature. The complex phenomena studied by behavioral scientists can rarely be explained adequately with one independent variable. In order to explain a substantial proportion of the variance of the dependent variable, it is almost always necessary to study the independent and combined effects of several independent variables.

In earlier chapters (see, particularly, Chapters 3 and 4) it was shown how multiple continuous independent variables are used to explain variance of the dependent variable. The use of coding in the analysis of data with one categorical independent variable was explained in Chapters 6 and 7. The present chapter is devoted to a treatment of designs with multiple categorical independent variables. We will try to show how the same methods of coding categorical variables presented in Chapter 7 may be used with multiple categorical variables.

In the context of the analysis of variance, independent variables are also referred to as *factors*. A factor is a variable; for example, methods of teaching, sex, levels of motivation. The two or more subdivisions or categories of a factor are, in set theory language, *partitions* (Kemeny, Snell, & Thompson, 1966, Chapter 3). The subdivisions in a partition are subsets and are called cells. If a sample is divided into male and female, there are two cells, A_1 and A_2, with males in one cell and females in the other. In a factorial design, two or more partitions are combined to form a *cross partition*, which consists of all subsets formed by the intersections of the original partitions. For instance, the intersection of two partitions or sets, $A_i \cap B_j$, is a cross partition. (The cells must be disjoint and they must exhaust all the cases.) It is possible to have 2×2, 2×3, 3×3, 4×5, and, in fact, $p \times q$ factorial designs. Three or more factors with two or more subsets per factor are also possible: $2 \times 2 \times 2$, $2 \times 3 \times 3$, $3 \times 3 \times 5$, $2 \times 2 \times 3 \times 3$, $2 \times 3 \times 3 \times 4$, and so on.

A factorial design is customarily displayed as in Figure 8.1. There are two independent variables, A and B, with two subsets of A: A_1 and A_2, and three subsets of B: B_1, B_2, and B_3. The cells obtained by the cross partitioning are indicated by A_1B_1, A_1B_2, and so on.

	B_1	B_2	B_3
A_1	A_1B_1	A_1B_2	A_1B_3
A_2	A_2B_1	A_2B_2	A_2B_3

FIGURE 8.1

Advantages of Factorial Designs

There are several advantages to studying the effects on a dependent variable of several independent variables. The first and perhaps most important advantage is that it is possible to determine whether the independent variables interact in their effect on the dependent variable. An independent variable can "explain" a relatively small proportion of variance of a dependent variable, while its inter-action with other independent variables may explain a relatively large propor-tion of the variance. Studying the effects of independent variables in isolation cannot reveal the interaction between them.

Second, factorial designs afford the researcher greater control, and, con-sequently, more sensitive statistical tests compared to the statistical tests used in analyses with single variables. When a single independent variable is used, the variance not explained by it is relegated to the error term. Needless to say, the larger the error term the less sensitive is the statistical test in which it is used. One method of reducing the magnitude of the error term is to identify as many sources of systematic variance of the dependent variable as is possible, feasible, and meaningful under a given set of circumstances. For example, suppose one is studying the effect of different styles of leadership on group productivity. If no other variable is included in the design, all the variance not explained by leadership styles becomes part of the error term. Suppose, however, that each group has an equal number of males and females, and that there is a correlation between sex and the type of productivity under study. In other words, some of the variance of productivity is due to sex. Under such circumstances, the introduction of sex as another independent variable will result in a reduction in the error estimate by reclaiming that part of the de-pendent variable variance due to sex. Note that the proportion of variance due to leadership styles will remain unchanged. But since the error term will be decreased the test of significance for the effect of leadership styles will be more sensitive. The same reasoning of course applies to testing the effect of sex. In addition, as noted above, an interaction between the two factors may be detected. For example, one style of leadership may lead to greater productivity among males, while another style may lead to greater productivity among females.

Third, factorial designs are efficient. One can test the separate and combined effects of several variables using the same number of subjects one would have to use for separate experiments.

Fourth, in factorial experiments the effect of a treatment is studied across different conditions of other treatments. Consequently, generalizations from factorial experiments are broader than generalizations from single-variable experiments. Factorial designs are examples of efficiency, power, and elegance. They also expeditiously accomplish scientific experimental purposes.

Analysis of a Three-by-Three Design

Analysis with multiple categorical independent variables is illustrated for the case of two independent variables, each having three categories. The same procedure, however, applies to any number of independent variables with any number of categories. A set of fictitious data for two factors (A and B), each with three categories, is given in Table 8.1. Assume that A_1, A_2, and A_3 represent surburban, urban, and rural schools, respectively. Further, assume that B_1 and B_2 represent two "experimental" methods of instruction, and that B_3 represents a "traditional" method. The dependent variable, Y is a measure of, say, verbal learning. It is also assumed that the researcher was guided by valid principles of research design,[1] and that he is interested in making inferences only about the categories included in this design. In other words, he is concerned with a fixed effects model.

[1] For a treatment of principles of research design, see Campbell and Stanley (1963).

TABLE 8.1   FICTITIOUS DATA FROM AN EXPERIMENT WITH THREE TEACHING METHODS IN THREE RESIDENTIAL REGIONS[a]

| Regions | Teaching Methods | | | |
	B_1	B_2	B_3	$\bar{Y}_A$
Suburban	16	20	10	
A_1	14	16	14	15
Urban	12	17	7	
A_2	10	13	7	11
Rural	7	10	6	
A_3	7	8	4	7
$\bar{Y}_B$:	11	14	8	$\bar{Y} = 11$

[a] $\bar{Y}_A$ = means for the three categories of A; $\bar{Y}_B$ = means for the three categories of B; $\bar{Y}$ = grand mean.

Orthogonal Coding for a Three-by-Three Design

We first analyze the data as if no planned comparisons were contemplated. Accordingly, the data are treated in the most general form of a factorial design. Although any of the coding methods presented in Chapter 7 may be used, we begin with orthogonal coding, since it has the advantages of convenience and simplicity of calculation. Regardless of the method of coding, however, one codes each factor or categorical variable separately as if it were the only independent variable in the design. In other words, while one variable is being coded all other variables are ignored. The Y vector is the dependent variable. For each categorical independent variable, coded vectors are generated, the number of vectors being equal to the number of categories minus one, or the number of degrees of freedom associated with a given variable. Thus, each set of coded vectors identifies one independent variable, for example, group membership or treatment effects. Any other designation required by the specific design is handled similarly. In the present example it is necessary to generate two coded vectors for each of the categorical variables.

In Table 8.2 we repeat the scores of Table 8.1, this time in the form of a single vector, Y. The coded vectors (1 through 8) necessary for the analysis of this 3×3 design are given beside the Y vector. Vectors 1 and 2 represent orthogonal coding for factor A. In vector 1 subjects of category A_1 are assigned 1's, those of category A_2 are assigned -1's, and those of category A_3 are assigned 0's. Accordingly, vector 1 contrasts category A_1 with category A_2. [*Remember*: when coding one factor, ignore the other factor.] Thus, for example, subjects assigned 1's in vector 1 all belong to category A_1 but to different categories of B.

In vector 2 of the table, subjects in categories A_1 and A_2 are assigned 1's, while subjects in category A_3 are assigned -2's. Accordingly, vector 2 contrasts categories A_1 and A_2 with category A_3.

Vectors 1 and 2 are orthogonal. Since these vectors are used solely for convenience of calculation and not to represent planned comparisons, another set of orthogonal vectors could have been used for the same purpose. In Chapter 7, when orthogonal coding was used for convenience of calculation, the simplest method for generating the coded vectors was shown to be as follows. The first vector is generated so that the first category of the variable being coded is contrasted with the second category of the variable. A second vector is then generated in which the first two categories of the variable are contrasted with the third category. In a third vector the first three categories are contrasted with the fourth category. One proceeds and generates vectors in this manner until their number equals the number of degrees of freedom associated with the variable being coded. Factor A of Table 8.2 has 2 degrees of freedom, and therefore two orthogonal vectors were generated to represent it. Similarly, vectors 3 and 4 were generated to represent factor B.

To repeat, vectors 1 and 2 of Table 8.2 represent factor A, and vectors 3 and 4 represent factor B. These four vectors represent the main effects of factors A and B. It is necessary now to represent the interaction between A and

TABLE **8.2** ORTHOGONAL CODING FOR A 3×3 DESIGN FOR DATA OF TABLE 8.1[a]

Cell	Y	1	2	3	4	5 (1×3)	6 (1×4)	7 (2×3)	8 (2×4)
A_1B_1	16	1	1	1	1	1	1	1	1
	14	1	1	1	1	1	1	1	1
A_2B_1	12	−1	1	1	1	−1	−1	1	1
	10	−1	1	1	1	−1	−1	1	1
A_3B_1	7	0	−2	1	1	0	0	−2	−2
	7	0	−2	1	1	0	0	−2	−2
A_1B_2	20	1	1	−1	1	−1	1	−1	1
	16	1	1	−1	1	−1	1	−1	1
A_2B_2	17	−1	1	−1	1	1	−1	−1	1
	13	−1	1	−1	1	1	−1	−1	1
A_3B_2	10	0	−2	−1	1	0	0	2	−2
	8	0	−2	−1	1	0	0	2	−2
A_1B_3	10	1	1	0	−2	0	−2	0	−2
	14	1	1	0	−2	0	−2	0	−2
A_2B_3	7	−1	1	0	−2	0	2	0	−2
	7	−1	1	0	−2	0	2	0	−2
A_3B_3	4	0	−2	0	−2	0	0	0	4
	6	0	−2	0	−2	0	0	0	4
ss:	340	12	36	12	36	8	24	24	72
M:	11	0	0	0	0	0	0	0	0
s:	4.47214	.84017	1.45521	.84017	1.45521	.68599	1.18818	1.18818	2.05798
r:		.37574	.65079	−.28180	.48810	.03834	−.06642	−.06642	.11504

[a] r = the correlation between the coded vector under which the value appears and the dependent variable Y. Thus, for example, the correlation between vector 1 and Y is .37574, the value listed under vector 1. Similarly for all other vectors. The correlation between any two coded vectors is, of course, zero.

B. The degrees of freedom for an interaction between variables equal the product of the degrees of freedom associated with each of the variables whose interaction is being considered. Accordingly, in the present example there are 4 degrees of freedom for the interaction between factors *A* and *B* (the product of the degrees of freedom associated with these factors, that is, 2×2). The four vectors for the interaction are generated by cross multiplying, in succession, each of the vectors of one factor with each of the vectors of the other factor. Vectors 5 through 8 in Table 8.2 are obtained in this manner. That is, the product of vectors 1 and 3 yields vector 5; 1×4 yields vector 6; 2×3 yields vector 7; and 2×4 yields vector 8. Note that the sum of any of these vectors, or the sum of the products of any two of them is zero. All the coded vectors of Table 8.2 are orthogonal.

Calculation of R^2 and the Overall F Ratio

Having generated all the necessary coded vectors, it is now possible to do a multiple regression analysis in which *Y* is the dependent variable and the coded vectors are treated as the independent variables. Since all the coded vectors are orthogonal, the calculation of R^2 is simple. All that is necessary is to sum the squared zero-order correlations of each coded vector with the dependent variable.[2] Applying formula (7.17) to the *r*'s of Table 8.2, we obtain

$$R^2_{y.12345678} = r^2_{y1} + r^2_{y2} + r^2_{y3} + r^2_{y4} + r^2_{y5} + r^2_{y6} + r^2_{y7} + r^2_{y8}$$

$$= (.37574)^2 + (.65079)^2 + (-.28180)^2 + (.48810)^2 + (.03834)^2$$
$$+ (-.06642)^2 + (-.06642)^2 + (.11504)^2$$

$$= .14118 + .42353 + .07941 + .23824 + .00147$$
$$+ .00441 + .00441 + .01323$$

$$= .90588$$

Thus the proportion of variance in the dependent variable accounted for by the eight coded vectors is .90588. Now test R^2 for significance. Repeating formula (6.4) with a new number:

$$F = \frac{R^2/k}{(1 - R^2)/(N - k - 1)} \tag{8.1}$$

where R^2 = squared multiple correlation; k = number of independent variables, or coded vectors; N = total number of subjects. For the present example,

$$F = \frac{.90588/8}{(1 - .90588)/(18 - 8 - 1)} = \frac{.11324}{.01046} = 10.83$$

with 8 and 9 degrees of freedom, $p < .01$.

What does this *F* ratio refer to? Actually it refers to the significance of

---

[2] The calculation of the zero-order correlations is also simplified, considering that the ΣX for each column is zero. The raw score formula for the correlation between a coded vector, X, and the dependent variable, Y, reduces to $r_{xy} = (N \Sigma XY)/(\sqrt{N \Sigma X^2} \sqrt{N \Sigma Y^2 - (\Sigma Y)^2})$. The second term in the denominator is, of course, constant for all the *r*'s in a given analysis.

$R^2_{y.12345678}$. Had this been a conventional regression analysis with eight independent variables, we would have concluded on the basis of the F ratio that R^2 was
significant beyond the .01 level. What is the meaning of the squared multiple
correlation in the present context? With two independent variables, each having three categories, there are nine distinct combinations that can be treated as
nine separate groups. There is, for example, a group under conditions A_1B_1, a
suburban group (A_1) taught by one experimental method (B_1). Another group,
A_1B_2, is a suburban group (A_1) taught by another experimental method (B_2).
And so forth for the rest of the combinations. If one were to perform a multiple
regression analysis of Y with the nine distinct categories (or a one-way analysis
of variance for nine groups) one would obtain the same R^2 as that above. The
F ratio associated with $R^2_{y.12345678}$ is the overall F ratio that would be obtained if
one were to perform a regression analysis in which each cell was treated as a
separate group. In other words, the overall R^2 indicates what proportion of the
variance is explained by all the available information.

Instead of working with proportions of variance it is possible to work with
sums of squares (as in Chapters 6 and 7). The total sum of squares for the
dependent variable (Σy^2) is 340 (see Table 8.2). By formula (7.6), the regression sum of squares is

$$ss_{reg} = (R^2_{y.12...8})(\Sigma y^2) = (.90588)(340.00000) = 307.99920$$

and the residual sum of squares is [formula (7.7)]

$$ss_{res} = \Sigma y^2 - ss_{reg} = 340.00000 - 307.99920 = 32.00080$$

The degrees of freedom for regression are 8 (k). The mean square regression is

$$ms_{reg} = \frac{ss_{reg}}{k} = \frac{307.99920}{8} = 38.49990$$

The degrees of freedom for the residual sum of squares are 9 $(N-k-1)$. The
mean squares residuals is

$$ms_{res} = \frac{ss_{res}}{N-k-1} = \frac{32.00080}{9} = 3.55564$$

Dividing the mean square regression by the mean square residuals one obtains
the F ratio:

$$F = \frac{ms_{reg}}{ms_{res}} = \frac{38.49990}{3.55564} = 10.83$$

with 8 and 9 degrees of freedom. The same F ratio was obtained for the proportion of variance accounted for by the coded vectors (R^2).

In factorial designs, however, it is not sufficient to know that the overall
sum of squares due to regression is significant. The main interest is in the
proportions of variance, or the sums of squares, accounted for by each of the
factors and by the interactions of the factors. Toward this end we partition the
regression sum of squares into its respective components, which in the present
problem are A, B, and $A \times B$.

Partitioning the Regression Sum of Squares

The partitioning of the regression sum of squares can be readily accomplished considering that the coded vectors representing the main effects (A and B) and their interaction ($A \times B$) are orthogonal. Under such circumstances, the squared zero-order correlation of each coded vector with the dependent variable, Y, indicates the proportion of variance, or sum of squares, explained by the vector. To obtain the proportion of variance due to a given factor, simply sum the proportions of variance accounted for by the vectors representing the factor. Thus the proportion of variance accounted for by factor A of Table 8.2 is equal to $r_{y1}^2 + r_{y2}^2$, since vectors 1 and 2 represent this factor. Multiplying the proportion of variance accounted for by a given factor by the sum of squares of the dependent variable (Σy^2) one obtains the regression sum of squares due to the factor under consideration. The degrees of freedom value for a sum of squares thus obtained equals the number of coded vectors from which the sum of squares was obtained.

Apply the procedure outlined above to the data of Table 8.2. The sum of squares regression due to factor A is

$$ss_{\text{reg}(A)} = (r_{y1}^2 + r_{y2}^2)(\Sigma y^2)$$
$$= (.14118 + .42353)(340.00000) = 192.00140$$

The sum of squares due to factor B is

$$ss_{\text{reg}(B)} = (r_{y3}^2 + r_{y4}^2)(\Sigma y^2)$$
$$= (.07941 + .23824)(340.00000) = 108.00100$$

and the sum of squares due to $A \times B$ is

$$ss_{\text{reg}(A \times B)} = (r_{y5}^2 + r_{y6}^2 + r_{y7}^2 + r_{y8}^2)(\Sigma y^2)$$
$$= (.00147 + .00441 + .00441 + .01323)(340.00000) = 7.99680$$

The residual sum of squares is, as calculated above, 32.00080.

In the present example, one obtains three F ratios: for factors A and B and for their interaction, $A \times B$. The foregoing analysis, along with the three F ratios, is summarized in Table 8.3. The F ratios for the main effects (A and B) are significant beyond the .01 level, while the F ratio for the interaction is less than one. The fact that there is no significant interaction indicates that the effects of the two factors on verbal learning are independent. Consequently, it is possible to make statements about the effects of one of the factors without having to refer to the other. It will be noted that factor A accounts for about 56 percent of the variance, factor B accounts for about 32 percent, while the interaction between A and B accounts for about 2 percent (see Table 8.3).

On the basis of the statistical tests it is concluded that there are significant differences between the teaching methods (factor B), as well as between the residential regions (factor A). Assuming that the researcher had not formulated planned comparisons, it is now possible to make post hoc comparisons between

TABLE **8.3**   SUMMARY OF MULTIPLE REGRESSION ANALYSIS FOR DATA
OF TABLE 8.2

Source	df	Prop. of Variance	ss	ms	F
Total Regression	8	.90588	307.99920		
Due to A	2	.56471	192.00140	96.00070	27.00
Due to B	2	.31765	108.00100	54.00050	15.19
Due to $A \times B$	4	.02352	7.99680	1.99920	< 1
Residual	9	.09412	32.00080	3.55564	
Total	17	1.00000	340.00000		

the means of the categories of the main effects to determine which of the categories are significantly different from each other.

Post Hoc Comparisons

The Scheffé method for multiple comparisons between means was presented and illustrated in Chapter 7 for the case of one categorical independent variable. The procedure is basically the same for comparisons between means in a factorial design. It will be noted, however, that in factorial designs when multiple comparisons between category means of a factor are made, other factors are ignored. It is, of course, possible to compare, in turn, the category means of each factor. Such comparisons are meaningful only when the main effects are significant, and the interaction effects are not significant. When the interactions are significant, it is more meaningful to perform comparisons between means of specific cells, or factor combinations.

As in Chapter 7 [formula (7.15)], a comparison is defined as follows:

$$D = C_1 \bar{Y}_1 + C_2 \bar{Y}_2 + \cdots + C_j \bar{Y}_j \qquad (8.2)$$

where D = difference or comparison; C = coefficient by which a given mean, $\bar{Y}$, is multiplied; j = number of means involved in the comparisons. For any one comparison it is required that $\Sigma C_j = 0$. That is, the sum of the coefficients in a given comparison has to equal zero. In order for a given comparison to be considered significant, $|D|$ (the absolute value of D) has to exceed a value of S. The calculation of S depends on the particular comparison as well as the factor from which it is obtained. This is illustrated for a pq design; that is, a factorial design with p categories for factor A and q categories for factor B.

For comparisons between category means of factor A:

$$S = \sqrt{(p-1)F_\alpha; p-1, pq(n-1)} \sqrt{MSR \left[\sum \frac{(C_j)^2}{n_j} \right]} \qquad (8.3)$$

where p = number of categories in the factor; F_α; $p-1$, $pq(n-1)$ = tabled value of F with $p-1$ and $pq(n-1)$ degrees of freedom at a prespecified α

level; n = number of subjects in any cell; MSR = mean square residuals or, equivalently, the mean square error from the analysis of variance; C_j = coefficient by which the mean of category j is multiplied; n_j = number of subjects in category j.

For comparisons between category means of factor B:

$$S = \sqrt{(q-1)F_\alpha; q-1, pq(n-1)} \sqrt{MSR\left[\sum \frac{(C_j)^2}{n_j}\right]} \qquad (8.4)$$

where q = number of categories in the factor. All other terms as defined above under formula (8.3). For comparisons between means of any two cells:

$$S = \sqrt{(pq-1)F_\alpha; pq-1, pq(n-1)} \sqrt{MSR\left[\sum \frac{(C_j)^2}{n_j}\right]} \qquad (8.5)$$

where all the terms are as defined under formula (8.3).

Note that in each of the formulas for S [formulas (8.3)–(8.5)], the degrees of freedom for the numerator of the tabled F ratio are the degrees of freedom associated with the factor from which the specific comparison is made. Thus, for example, the degrees of freedom for the numerator of the tabled F ratio for factor A above are $p-1$ (p = number of categories in factor A). The degrees of freedom for the denominator of the tabled F ratio in formulas (8.3)–(8.5) always equal the degrees of freedom associated with the residual, or error term. Note, too, that when comparing any two cell means, the comparison is treated as if it came from an analysis with one factor consisting of pq categories. Consequently, the degrees of freedom for the numerator of the tabled F ratio in the formula for S [formula (8.5)] equal $pq-1$.

The Scheffé method is now applied to the data presented in Table 8.1. It will be recalled that both main effects for this analysis (A and B) are significant, while their interaction is not significant. For the purpose of illustration, comparisons between means of factor B (teaching methods) are made. From Table 8.1, $\bar{Y}_{B_1} = 11$; $\bar{Y}_{B_2} = 14$; $\bar{Y}_{B_3} = 8$. The present design consists of $p = 3$ categories for factor A and $q = 3$ categories for factor B. $n = 2$, the number of subjects in each cell, or factor combination. $MSR = 3.56$ (see Table 8.3). Applying formula (8.2) to the comparison between $\bar{Y}_{B_1}$ and $\bar{Y}_{B_2}$,

$$D = C_1\bar{Y}_{B_1} + C_2\bar{Y}_{B_2}$$
$$= (1)(11) + (-1)(14) = -3$$

In order to obtain S [formula (8.4)] it is necessary to find the tabled F ratio with $q-1$ and $pq(n-1)$ degrees of freedom at a specified level of significance. For the present example, $q-1 = 3-1 = 2$, and $pq(n-1) = (3 \times 3)$ $\times (2-1) = 9$. Assuming the researcher selected the .05 level of significance, the tabled value of F with 2 and 9 degrees of freedom at this level of significance is 4.26. n_j, the number of subjects in each category being compared, is 6 (there are two subjects per cell and each category consists of three cells).

Applying formula (8.4),

$$S = \sqrt{(q-1)F_{\alpha}; q-1, pq(n-1)} \sqrt{MSR\left[\sum \frac{(C_j)^2}{n_j}\right]}$$

$$= \sqrt{(2)(4.26)} \sqrt{3.56\left[\frac{(1)^2}{6} + \frac{(-1)^2}{6}\right]}$$

$$= \sqrt{8.52}\,\sqrt{1.19} = (2.92)(1.09) = 3.18$$

Since $|D| = 3$ is smaller than S, it is concluded that $\bar{Y}_{B_1}$ is not significantly different from $\bar{Y}_{B_2}$. The S obtained above (3.18) is, of course, the same for any comparison between two means of factor B. Consequently, the comparison between $\bar{Y}_{B_1}$ and $\bar{Y}_{B_3}$ is

$$D = (1)(11) + (-1)(8) = 3$$

is not significant ($D < S$; 3 and 3.18, respectively). And the comparison between $\bar{Y}_{B_2}$ and $\bar{Y}_{B_3}$

$$D = (1)(14) + (-1)(8) = 6$$

is significant.

The two experimental teaching methods, B_1 and B_2, do not differ significantly from each other. Moreover, teaching method B_1 does not differ significantly from teaching method B_3, while teaching methods B_2 and B_3 do differ significantly. Suppose, for the sake of illustration, that it is desired to compare the mean of categories B_1 and B_2 with the mean of B_3. This comparison is

$$D = (1/2)(\bar{Y}_{B_1}) + (1/2)(\bar{Y}_{B_2}) + (-1)(\bar{Y}_{B_3})$$

$$= (1/2)(11) + (1/2)(14) + (-1)(8) = 4.5$$

and S for this comparison is

$$S = \sqrt{(q-1)F_{\alpha}, q-1, pq(n-1)} \sqrt{MSR\left[\sum \frac{(C_j)^2}{n_j}\right]}$$

$$= \sqrt{(2)(4.26)} \sqrt{3.56\left[\frac{(.5)^2}{6} + \frac{(.5)^2}{6} + \frac{(-1)^2}{6}\right]}$$

$$= \sqrt{8.52}\,\sqrt{.89} = (2.92)(.94) = 2.74$$

Since $|D| = 4.5$ is larger than $S = 2.74$, it is concluded that the mean of the two experimental methods (B_1 and B_2) is significantly different from the mean of the traditional method (B_3).

In the manner illustrated above one can compare the means of factor A or the means of specific cells. The extension of the method to designs with more than two factors is straightforward. For example, for a design of pqr with n subjects in each cell, the S for comparisons between category means of the factor with p levels is

$$S = \sqrt{(p-1)F_{\alpha;\,p-1,\,pqr(n-1)}}\,\sqrt{MSR\left[\sum \frac{(C_j)^2}{n_j}\right]} \tag{8.6}$$

where all terms are as defined after formula (8.3). Note that the degrees of freedom for the numerator of the tabled F ratio in (8.6) are $p-1$, or the degrees of freedom associated with the factor from which the comparisons are made. The degrees of freedom for the denominator of the tabled F ratio are $pqr(n-1)$, or the degrees of freedom for the residual term.

The Regression Equation

It was noted earlier [see Chapter 7, particularly formula (7.19) and the related discussion] that when the independent variables, or the coded vectors, are orthogonal, each β is equal to the zero-order correlation with which it is associated. Accordingly, the eight β's for the eight coded vectors of Table 8.2 are equal to the correlations given in the last line of the table. To calculate the b's, we apply formula (7.20), which is repeated here with a new number:

$$b_j = \beta_j \frac{s_y}{s_j} \tag{8.7}$$

where b_j = regression coefficient for the jth independent variable; β_j = standardized coefficient for the jth independent variable; s_y = standard deviation of the dependent variable, Y; s_j = standard deviation of the jth independent variable.

The calculations of the b's for the data of Table 8.2 are summarized in Table 8.4. The intercept, a, is obtained using formula (3.4), which is repeated here with a new number:

$$a = \bar{Y} - b_1\bar{X}_1 - \cdots - b_k\bar{X}_k \tag{8.8}$$

where a = intercept; $\bar{Y}$ = mean of the dependent variable, Y; b = regression coefficient; k = the kth independent variable. Since the mean of each of the

TABLE **8.4** SUMMARY OF CALCULATIONS OF b'S FOR DATA OF TABLE 8.2[a]

Vector	β_j	s	$b_j = (\beta_j)(s_y)/(s_j)$
1	.37574	.84017	2.00003
2	.65079	1.45521	2.00000
3	−.28180	.84017	−1.49999
4	.48810	1.45521	1.50002
5	.03834	.68599	.24995
6	−.06642	1.18818	−.25000
7	−.06642	1.18818	−.25000
8	.11504	2.05798	.24999
Y		4.47214	

[a]$\beta_j = r_{yj}$, and s = standard deviation, were obtained from Table 8.2.

independent variables in the present example (the means of vectors 1–8 of Table 8.2) equals zero, it follows that the application of formula (8.8) results in a being equal to the mean of the dependent variable, $\bar{Y}$. In the present case, $a =$ 11.00. From Table 8.4 the regression equation to two decimals is

$$Y' = 11.00 + 2.00X_1 + 2.00X_2 - 1.50X_3 + 1.50X_4 + .25X_5 - .25X_6$$
$$- .25X_7 + .25X_8$$

The application of the regression equation to the scores of the coded vectors for any subject of Table 8.2 will, of course, yield the mean of the cell, or factor combination, to which the subject belongs. Thus, for example, the predicted score for the first subject of Table 8.2 is

$$Y' = 11.00 + 2.00(1) + 2.00(1) - 1.50(1) + 1.50(1) + .25(1) - .25(1)$$
$$- .25(1) + .25(1) = 15$$

Note that 15 is the mean of A_1B_1, that is, suburban school taught by method B_1. Predicting in a similar manner the scores for each subject of Table 8.2 and subtracting each predicted score from the observed score, Y, one obtains the residual $(Y - Y')$ for each subject. Squaring and summing all the residuals yields the residual sum of squares, which in the present case is 32.00 (see Table 8.3).

Orthogonal Comparisons of Means

The data of Table 8.2 were treated as if planned comparisons were not intended by the researcher. The orthogonal coding was only used for ease of calculation. Suppose, however, that the orthogonal coding of Table 8.2 in fact represents comparisons that are of interest to the researcher and that were formulated prior to the analysis. If this is the case, it is necessary to test each comparison for significance. This can be accomplished in one of the following ways: (a) testing the significance of the b coefficients obtained from the regression analysis with orthogonal coding, and (b) testing the significance of the regression sum of squares due to each vector or comparison. Both methods are applied to the data of Table 8.2.

Significance Testing of the b's

It was shown in Chapter 7 that when orthogonal coding is used, testing the significance of a b coefficient amounts to testing the significance of the comparison reflected by the vector with which it is associated. Thus, for example, testing the significance of b_1 is the same as testing the significance of the comparison between $\bar{Y}_{A_1}$ and $\bar{Y}_{A_2}$ (the comparison reflected by vector 1 of Table 8.2).

In the case of orthogonal coding, or orthogonal independent variables, the standard error of a b weight can be calculated by formula (7.21), which is repeated here with a new number:

$$s_{b_j} = \sqrt{\frac{s^2_{y.12...k}}{\sum x_j^2}} \qquad (8.9)$$

where s_{b_j} = standard error of b for the jth variable, or coded vector; $s^2_{y.12...k}$ = variance of estimate, or the mean square residuals; $\sum x_j^2$ = sum of squares for the jth variable. And the t ratio for a b weight is

$$t = \frac{b_j}{s_{b_j}} \qquad (8.10)$$

where t = t ratio; b_j = b coefficient for the jth variable, or coded vector; s_{b_j} = standard error of b for the jth variable. The degrees of freedom for the t ratio equal the degrees of freedom for the variance of estimate, or the mean square residuals ($s^2_{y.12...k}$).

Now calculate the t ratio for b_1. From Table 8.4, $b_1 = 2.00003$. From Table 8.2, $\sum x_1^2 = 12$. And from Table 8.3, $s^2_{y.12...8} = 3.55564$.

$$s_{b_1} = \sqrt{\frac{s^2_{y.12...8}}{\sum x_1^2}} = \sqrt{\frac{3.55564}{12}} = .54434$$

$$t = \frac{b_1}{s_{b_1}} = \frac{2.00003}{.54434} = 3.67, \quad \text{with 9 degrees of freedom,} \quad p < .01$$

It is concluded that $\bar{Y}_{A_1}$ is significantly different from $\bar{Y}_{A_2}$ (at the .01 level).

In this way one calculates t ratios for each of the b's. These calculations are summarized in Table 8.5. Note that b_1, b_2, and b_4 are significant beyond the .01 level, while b_3 is significant beyond the .05 level. None of the remaining b's ($b_5 - b_8$) is significant. Consequently, the comparisons reflected by vectors 1, 2, and 4 of Table 8.2 are significant at the .01 level, while the comparison reflected by vector 3 is significant at the .05 level. Specifically, for the comparison reflected by vector 1: the mean of suburban schools ($\bar{Y}_{A_1} = 15$) is significantly different from the mean of urban schools ($\bar{Y}_{A_2} = 11$). For vector 2: the mean of suburban and urban schools [$(\bar{Y}_{A_1} + \bar{Y}_{A_2})/2 = 13$] is significantly different from the mean of rural schools ($\bar{Y}_{A_3} = 7$). For vector 3: the mean of experimental

TABLE **8.5**   TESTS OF SIGNIFICANCE OF THE b'S FOR THE DATA OF TABLE 8.2

Vector	$\sum x^2$	b	s_b	t	p
1	12	2.00003	.54434	3.67	.01
2	36	2.00000	.31427	6.36	.01
3	12	−1.49999	.54434	−2.76	.05
4	36	1.50002	.31427	4.77	.01
5	8	.24995	.66667	.37	n.s.
6	24	−.25000	.38490	−.65	n.s.
7	24	−.25000	.38490	−.65	n.s.
8	72	.24999	.22222	1.12	n.s.

teaching method B_1 ($\bar{Y}_{B_1} = 11$) is significantly different from the mean of experimental teaching method B_2 ($\bar{Y}_{B_2} = 14$). Note that when a post hoc comparison between these two methods was made they were declared to be not significantly different. This is because orthogonal comparisons are more sensitive than post hoc comparisons.[3] For vector 4: the mean of the two experimental teaching methods [($\bar{Y}_{B_1} + \bar{Y}_{B_2})/2 = 12.5$] is significantly different from the mean of the traditional method ($\bar{Y}_{B_3} = 8$).

Significance Testing of Regression Sums of Squares

When orthogonal coding reflects orthogonal comparisons, the squared zero-order correlation of each vector with the dependent variable, Y, indicates the proportion of variance accounted for by the comparison reflected by the coded vector. While it is possible to test the significance of each proportion thus obtained, we choose instead to first express each proportion as a component of the total sum of squares due to each vector. It is thus possible to see clearly the partitioning of the total sum of squares into orthogonal components of regression sums of squares and a component due to residuals. Each component regression sum of squares is then tested for significance.

To obtain the regression sum of squares due to a coded vector, or a comparison, one multiplies the squared zero-order correlation of the vector with the dependent variable, Y, by the total sum of squares, Σy^2. Thus, for example, to obtain the regression sum of squares due to vector 1 of Table 8.2, it is noted that $r_{y1} = .37574$, and that $\Sigma y^2 = 340$. The regression sum of squares due to vector 1 is

$$ss_{\text{reg}(1)} = (r_{y1}^2)(\Sigma y^2)$$
$$= (.14118)(340.00000) = 48.00120$$

In a similar manner one obtains the regression sum of squares for each of the coded vectors, or orthogonal comparisons of Table 8.2. These components of the regression sum of squares are reported in Table 8.6, with accompanying mean squares. Dividing each regression mean square by the residual mean square yields an F ratio for each comparison. The results of this analysis are also summarized in Table 8.6.

Note that the same results were obtained in Table 8.5 where the tests of significance for the b's were reported. In fact, since each F ratio in Table 8.6 has one degree of freedom for its numerator and 9 degrees of freedom its denominator, it follows that each F of Table 8.6 is equal to the t^2 for the same comparison given in Table 8.5. Thus, for example, the t for b_1 of Table 8.5 (the b for vector 1) is 3.67. $3.67^2 = 13.47$, which is, within rounding error, equal to the F ratio for the same comparison in Table 8.6, row 1. All other comparisons are similar. The interpretation of the analysis summarized in Table 8.6 is the same as that given for the tests of significance of the b's (Table 8.5).

---

[3]For a discussion of this point, see Chapter 7.

Source	Prop. of Variance	ss	df	ms	F	p
1	.14118	48.00120	1	48.00120	13.50	.01
2	.42353	144.00020	1	144.00020	40.50	.01
3	.07941	26.99940	1	26.99940	7.59	.05
4	.23824	81.00160	1	81.00160	22.78	.01
5	.00147	.49980	1	.49980	.14	n.s.
6	.00441	1.49940	1	1.49940	.42	n.s.
7	.00441	1.49940	1	1.49940	.42	n.s.
8	.01323	4.49820	1	4.49820	1.27	n.s.
Residual	.09412	32.00080	9	3.55564		
Total	1.00000	340.00000	17			

[a]Numbers 1 through 8 under "Source" are the eight orthogonal comparisons reflected by the orthogonal coding of Table 8.2; proportion of variance for each vector equals the squared zero-order correlation of the vector with the dependent variable, Y. $ss = (r^2_{yj})(\Sigma y^2)$ where j = a coded vector, and Σy^2 = total sum of squares of the dependent variable.

Computer Analysis

To this point the calculations in this chapter were done with a desk calculator. One of the reasons for using orthogonal coding was to simplify the calculations and demonstrate that a regression analysis can be done with relative ease without the aid of a computer. Nevertheless, we agree with Green (1966) who, commenting on the computer revolution, says: "Today there is no need to find ways of simplifying calculations so that they may be done by desk calculators. Nor is there any excuse for avoiding procedures solely on the basis of computational difficulty (p. 437)."

There are several obvious advantages to the use of computers in data analysis. In the first place, the use of a computer saves time and energy that can be devoted to *thinking* about the analyses and the results. Researchers become so involved in calculations that by the time they are finished they are mentally exhausted and may not give careful consideration to the meaning of the results. Furthermore, in an attempt to avoid complex calculations researchers frequently resort to analyses that are not appropriate for their data. Consequently, they cannot obtain adequate answers to research questions. The computer enables one to use appropriate analyses regardless of their complexity.

Lastly, computers are, in general, highly accurate — within their finite capacities — and greatly reduce the probability of computational error, the sort of error that desk calculators and slide rules are prone to in complex multivariate computation. This is not to say that one should accept computer output without question. Errors occur even with computer analyses, although most such errors can be traced to human errors in reading data, providing the wrong instructions in the computer program, making the wrong choices, and the like. There is no substitute for thinking and for careful scrutiny of results. When looking at the results of an analysis, whether obtained from a computer or by other means, one should always pose the question: *Do the results make sense*? When a researcher is at a loss in attempting to answer this question, the chances are he does not understand the analytic method he has used. If this is indeed the case, he should not have used the method in the first place.

Hereafter, in this chapter and succeeding ones, we present results obtained by computer. This will enable us to concentrate more on the meaning of the methods presented rather than on the mechanics of their calculations.[4] Only the results necessary for the analysis and interpretation of a problem under consideration are reported. At each stage of the reporting, a discussion of the pertinent computer printout is provided. Before analyzing the data of the 3×3 design of Table 8.1 with other coding methods, we report part of the results of an analysis of these data with orthogonal coding obtained by computer.

Computer Output for the 3 × 3 Problem, Orthogonal Coding

The data for the 3×3 design presented in Table 8.2 were analyzed by a computer program for multiple regression analysis. Following are some of the pertinent results of this analysis. The first piece of relevant information (labeled "Coefficient of Determination") is

$$R^2_{y.12345678} = .90589$$

This, of course, is the same as the value obtained in the analysis presented earlier. The program then prints an analysis of variance table, which is reproduced here as Table 8.7. Note that the regression sum of squares and the residual sum of squares reported in Table 8.7 are, within rounding errors, the same as those obtained in the earlier analysis. The F ratio of 10.83 with 8 and 9 degrees of freedom was also obtained earlier, and of course refers to the significance of the R^2, or the proportion of variance accounted for by the independent variables.

The printout then provides a summary table, parts of which are reproduced

---

[4]Although there are many computer programs for regression analysis, we have used two programs, BMD03R (Dixon, 1970) and MULR. BMD03R is part of a set of programs available at many computer installations. MULR, a program written by one of us to do certain analyses not ordinarily done by other programs, is given in Appendix C. The BMD03R results are used in this chapter and in most of the remaining chapters of Part II.

TABLE 8.7   ANALYSIS OF VARIANCE FOR THE MULTIPLE
LINEAR REGRESSION

Source of Variation	df	ss	ms	F
Due to regression	8	308.00000	38.50000	10.82811
Deviation about regression	9	32.00000	3.55556	
Total	17	340.00000		

as Table 8.8. Each row in Table 8.8 refers to one "independent variable." These variables are printed in the order in which they were read into the computer. In the present case, each row refers to one of the coded vectors of Table 8.2. Thus, for example, the row for variable 1 of Table 8.8 refers to the first coded vector of Table 8.2, the vector in which category A_1 is contrasted with category A_2. Since eight coded vectors as shown in Table 8.2 were read in to represent the independent variables, there are eight rows in Table 8.8.

The b for vector 1 is reported as 2.00000 with a standard error of .54433. Dividing b by s_b yields a t ratio, which for b_1 is 3.67423. Compare the b's, the s_b's, and the t's of Table 8.8 with those of Table 8.5 and note that they are the same, within rounding errors. The interpretation of these terms is, of course, also the same as given earlier. Note that when the orthogonal coding does in fact reflect planned comparisons, one obtains directly from the computer output the t ratio for each comparison.

The regression sum of squares and the proportion of variance due to a coded vector are reported in Table 8.8 in the row corresponding to the given vector. Thus, the sum of squares due to vector 1 is 48.00, and the proportion of variance for this vector is .14118 (see Table 8.8, row 1). Since all the vectors in the present analysis are orthogonal, the regression sum of squares due to each vector, as well as the proportion of variance accounted for by each vector, are independent components respectively of the regression sum of squares and the proportion of variance accounted for by the independent variables. Consequently, the proportion of variance accounted for by a coded vector equals the squared zero-order correlation of the coded vector with the dependent variable. Thus, $r_{y1}^2 = .14118$.

Dividing the sum of squares due to a vector by the mean square error

TABLE **8.8**   REGRESSION COEFFICIENTS AND THEIR STANDARD
ERRORS, t RATIOS, SUMS OF SQUARES, AND PROPORTIONS OF VARIANCE.
ORIGINAL DATA OF TABLE 8.2.[a]

Variable	b	s_b	t	ss	Prop. of Variance
1	2.00000	.54433	3.67423	48.00000	.14118
2	2.00000	.31427	6.36396	144.00000	.42353
3	−1.50000	.54433	−2.75568	27.00000	.07941
4	1.50000	.31427	4.77297	81.00000	.23824
5	.25000	.66667	.37500	.50000	.00147
6	−.25000	.38490	−.64952	1.50000	.00441
7	−.25000	.38490	−.64952	1.50000	.00441
8	.25000	.22222	1.12500	4.50000	.01324
Σ:				308.00000	.90589

[a]Variables 1 through 8 = eight vectors of orthogonal coding reported in Table 8.2; b = regression coefficient; s_b = standard error of b; t = t ratio; ss = sum of squares added; prop. of variance = proportion of variance added.

reported in Table 8.7, one obtains an F ratio, which in the present case is equal to the square of the t ratio listed in the same row (note that the sum of squares for each row has 1 degree of freedom and is therefore equal to the mean square regression for the row). For example, the regression sum of squares for row 1 is 48.00000 and the mean square error is 3.55556 (see Table 8.7). Therefore, $F =$ 48.00000/3.55556 = 13.499998, with 1 and 9 degrees of freedom, which is equal to the square of the t ratio associated with b_1 $(3.67423)^2$, with 9 degrees of freedom. Compare the sums of squares and the proportions of variance reported in Table 8.8 with those reported in Table 8.6, and note again that they are the same, within rounding error.

Assuming the researcher did not plan comparisons and is instead interested in the sums of squares for the main effects and the interaction, these terms are also easily obtainable from Table 8.8. For each factor one adds the sums of squares that are associated with the rows that represent the factor. Thus, vectors 1 and 2 of Table 8.2, or rows 1 and 2 of Table 8.8, are associated with the two degrees of freedom for factor A. Adding the sum of squares for these rows (48.00 and 144.00) yields 192.00, which is the sum of squares for A with 2 degrees of freedom. Rows 3 and 4 of Table 8.8 are associated with factor B. The sums of squares for these rows are 27.00 and 81.00 respectively. Their sum, 108.00, is the sum of squares for factor B, with 2 degrees of freedom. Rows 5 through 8 of Table 8.8 are associated with the interaction. Adding the sums of the squares for rows 5 through 8 (.50 + 1.50 + 1.50 + 4.50) one obtains 8.00, which is the sum of squares due to interaction.

Compare the figures obtained above for the main effects and the interaction with those reported in Table 8.3. As in Table 8.3, each sum of squares is divided by its degrees of freedom to obtain a mean square. Dividing each mean square by the residual mean square yields an F ratio for the factor under consideration. Since these calculations were done in Table 8.3, and were followed by a discussion of their interpretation, they are not repeated here.

In conclusion, it will be noted that the sum of the elements of the proportion of variance associated with the rows reflecting a given factor indicates the proportion of variance accounted by the factor. Thus, for factor A, the sum of the two proportions in rows 1 and 2 of Table 8.8 is .56471 (.14118 + .42353), indicating that about 56 percent of the variance is accounted for by factor A. For factor B the sum of the proportions of rows 3 and 4 of Table 8.8 is .31765 (.07941 + .23824). Factor B accounts for about 32 percent of the variance. The sum of the proportions of rows 5 through 8 of Table 8.8 is .02353 (.00147 + .00441 + .00441 + .01324), indicating that the interaction accounts for about 2 percent of the variance. Obviously, the sum of all the elements in the column of the proportion of variance equals R^2, which in the present case is .90589.

Effect Coding

The data for the 3×3 design introduced in Table 8.1 and analyzed subsequently with orthogonal coding are now analyzed with effect coding. Effect coding was introduced in Chapter 7, where it was noted that the coding is $\{1,0,-1\}$. In each coded vector, members of the category being identified are assigned 1's, all others are assigned 0's, except that the members of the last category are assigned -1's. The procedure is the same for factorial designs, where each factor is coded separately as if it were the only independent variable in the design. For each factor, the number of coded vectors equals the number of categories in the factor minus one, the number of degrees of freedom associated with the factor. The vectors for the interaction are obtained by cross multiplying, in succession, each coded vector of one factor by each of the coded vectors of the other factor, just as in orthogonal coding. One thus obtains a number of vectors equal to the product of the number of vectors associated with the factors whose interaction is being represented. Since the number of vectors for each factor equals the degrees of freedom associated with it, the number of vectors for the interaction obtained in the manner described above equals the degrees of freedom associated with the interaction.

The data for the 3×3 design, with the coded vectors for effect coding, are given in Table 8.9. Note that in vector 1 subjects belonging to category A_1 are assigned 1's, subjects in category A_2 are assigned 0's, while subjects in category A_3 are assigned -1's. In vector 2, subjects in category A_3 are still assigned -1's, but now subjects in category A_1 are assigned 0's, while subjects in A_2 are assigned 1's. Consequently, vectors 1 and 2 represent factor A (residential regions). Vectors 3 and 4 represent factor B (teaching methods). In these two vectors category B_3 is assigned -1's, while B_1 is assigned 1's in vector 3, and category B_2 is assigned 1's in vector 4. Vectors 5 through 8 represent the interaction of A and B.

As noted earlier, the overall results obtained from a regression analysis are the same regardless of the method used for coding the categorical independent variables. Therefore, the results of the analysis of the 3×3 design with effect coding, as shown in Table 8.9, are the same as those obtained in the analysis of these data with orthogonal coding. Instead of repeating in detail the results and their interpretation, we focus on those aspects of the analysis and results that are specific to effect coding.

Pertinent results, as obtained from a computer analysis of the data of Table 8.9, are given in Table 8.10. The sum of the column labeled ss (308.00) is equal to regression sum of squares due to the eight coded vectors of Table 8.9. Furthermore, the sum of the column labeled proportion of variance (.90589) is equal to the proportion of variance in the dependent variable, Y, accounted for by the eight coded vectors of Table 8.9, or $R^2_{y.12...8}$. Compare the above totals of the regression sums of squares and the proportions of variance with those obtained in Table 8.8. They are identical. The values of the regression sum of squares and the proportion of variance associated with each row, however, are not the same in the two tables. The differences are due to the different coding methods. We turn now to a detailed treatment of the results reported in Table 8.10.

TABLE **8.9** EFFECT CODING FOR A 3×3 DESIGN, DATA OF TABLE 8.1[a]

Cell	Y	1	2	3	4	5 (1×3)	6 (1×4)	7 (2×3)	8 (2×4)
A_1B_1	16	1	0	1	0	1	0	0	0
	14	1	0	1	0	1	0	0	0
A_2B_1	12	0	1	1	0	0	0	1	0
	10	0	1	1	0	0	0	1	0
A_3B_1	7	-1	-1	1	0	-1	0	-1	0
	7	-1	-1	1	0	-1	0	-1	0
A_1B_2	20	1	0	0	1	0	1	0	0
	16	1	0	0	1	0	1	0	0
A_2B_2	17	0	1	0	1	0	0	0	1
	13	0	1	0	1	0	0	0	1
A_3B_2	10	-1	-1	0	1	0	-1	0	-1
	8	-1	-1	0	1	0	-1	0	-1
A_1B_3	10	1	0	-1	-1	-1	-1	0	0
	14	1	0	-1	-1	-1	-1	0	0
A_2B_3	7	0	1	-1	-1	0	0	-1	-1
	7	0	1	-1	-1	0	0	-1	-1
A_3B_3	4	-1	-1	-1	-1	1	1	1	1
	6	-1	-1	-1	-1	1	1	1	1
ss:	340	12	12	12	12	8	8	8	8
M:	11	0	0	0	0	0	0	0	0
s:	4.47214	.84017	.84017	.84017	.84017	.68599	.68599	.68599	.68599

[a]Y = dependent variable; vectors 1 and 2 represent factor A; vectors 3 and 4 represent factor B; vectors 5 through 8 represent the interaction of A and B.

174

TABLE **8.10**   REGRESSION COEFFICIENTS, SUMS OF SQUARES, AND
PROPORTIONS OF VARIANCE, DATA OF TABLE 8.9[a]

Variable	b	ss	Prop. of Variance
1	4.00000	192.00000	.56471
2	.00000	.00000	.00000
3	.00000	27.00000	.07941
4	3.00000	81.00000	.23824
5	.00000	.50000	.00147
6	.00000	1.50000	.00441
7	.00000	1.50000	.00441
8	1.00000	4.50000	.01324
Σ:		308.00000	.90589

[a]Variables 1 through 8: eight vectors for effect coding reported in
Table 8.9, where 1 and 2 represent factor A, 3 and 4 represent factor
B, and 5 through 8 represent $A \times B$; b = regression coefficient;
ss = regression sum of squares.

Proportions of Variance

To understand the properties of the results obtained from a regression analysis
with effect coding, we first take a close look at the column labeled "Prop. of
variance" in Table 8.10. For this purpose, it is necessary to refer to the discussion in Chapter 5, where R^2 was expressed as the sum of a set of squared
semipartial correlations. Formula (5.10) is restated with a new number:

$$R^2_{y.12...k} = r^2_{y1} + r^2_{y(2.1)} + \cdots + r^2_{y(k.12...k-1)} \qquad (8.11)$$

where $R^2_{y.12...k}$ = squared multiple correlation of Y with k independent variables;
r^2_{y1} = squared zero-order correlation of Y with variable 1; $r^2_{y(2.1)}$ = squared semipartial correlation of Y with variable 2, partialing variable 1 from variable 2;
$r^2_{y(k.12...k-1)}$ = squared semipartial correlation of Y with variable k, partialing the
remaining independent variables $(k-1)$ from variable k. In words, formula
(8.11) states that R^2 is equal to the squared zero-order correlation of the first
independent variable with the dependent variable, Y, plus the squares of all the
subsequent semipartial correlations, at each step partialing from the variable
being entered into the equation all the variables that preceded it.

Each squared semipartial correlation indicates the proportion of variance
accounted for by an independent variable, after taking into account the proportion of variance accounted for by the variables that preceded it in the equation. In other words, a squared semipartial correlation indicates an increment
in the proportion of variance accounted for by a variable under consideration,
after having noted the proportion of variance accounted for by the variables
preceding it in the equation. The column "Prop. of variance" in Table 8.10 is, in
effect, an expression of formula (8.11) for the data of Table 8.9. Thus, the correlation between Y and vector 1 of Table 8.9 is .75147, and its square is .56471,
which is the proportion of variance accounted for by vector 1 (see row 1). To

show now that the proportion of variance associated with vector 2 (row 2) is equal to a squared semipartial correlation, it is necessary to calculate $r^2_{y(2.1)}$.

We restate the formula for a first-order semipartial correlation [formula (5.7)] with a new number:

$$r_{y(2.1)} = \frac{r_{y2} - r_{y1}r_{12}}{\sqrt{1 - r^2_{12}}} \qquad (8.12)$$

where $r_{y(2.1)}$ = semipartial correlation of Y with 2, partialing variable 1 from variable 2. For the data of Table 8.9: $r_{y1} = .75147$; $r_{y2} = .37573$; $r_{12} = .50000$. Therefore,

$$r_{y(2.1)} = \frac{(.37573) - (.75147)(.50000)}{\sqrt{1 - (.50000)^2}} = \frac{.00000}{.75000} = .00000$$

Obviously $r^2_{y(2.1)} = .00000$, which is the proportion of variance reported in row 2 of Table 8.10.

Recall, however, that vectors 1 and 2 of Table 8.9, or rows 1 and 2 of Table 8.10, represent factor A. The sum of the proportions accounted for by these two vectors is $.56471 + .00000 = .56471$. The same value was obtained for factor A when the data were analyzed with orthogonal coding (see Table 8.8, rows 1 and 2). The two analyses differ in the manner in which the proportion of variance accounted for by a given factor is sliced into separate components. In the case of orthogonal coding, the proportion of variance accounted for by a vector is equal to the square of its correlation with the dependent variable, Y. In effect coding, however, it is necessary to take into account the correlations among the coded vectors. Thus, while the correlation between vector 2 of Table 8.9 and the dependent variable is .37573, the proportion of variance attributed to this vector is reported in Table 8.10 as .00000. This is because vector 1 entered first into the analysis, and $r_{12} = .50$. The important thing to remember, however, is that regardless of the manner in which the proportion of variance accounted for is sliced into components associated with each coded vector, the sum of the components always equals the proportion of variance accounted for by the factor that the set of coded vectors represents.

In effect coding, the vectors representing the main effects and the interactions are mutually orthogonal. This means that while the coded vectors representing a given factor or an interaction are correlated, there is no correlation between coded vectors across factors or interactions. Stated differently, the coded vectors of one factor are not correlated with the coded vectors of the other factors, nor are they correlated with the coded vectors representing interactions. In the present analysis, for example, vectors 1 and 2 of Table 8.9 represent factor A, while vectors 3 and 4 represent factor B. Consequently, $r_{13} = r_{14} = r_{23} = r_{24} = .00$.

Because the vectors for main effects and interactions are mutually orthogonal, in the present example $r_{y(3.12)} = r_{y3}$ (vectors 1 and 2 represent factor A, while 3 is one of the vectors representing factor B). $r_{y3} = .28180$, and its square is .07941, the value given in row 3 of Table 8.10. $r_{y4} = .56360$, and $r_{34} = .50000$.

The reader may verify that $r^2_{y(4.3)} = .23824$, the proportion of variance attributed to vector 4 after partialing variable 3 from variable 4 (see Table 8.10, row 4). The sum of these two proportions of variance $.07941 + .23824 = .31765$, is the proportion of variance accounted for by factor B. The same value was obtained for factor B when the data were analyzed with orthogonal coding (see Table 8.8, rows 3 and 4).

Applying the procedure outlined above to the vectors describing the interaction between A and B (columns 5 through 8 of Table 8.9) will yield the proportions of variance listed in rows 5 through 8 of Table 8.10. To verify the results, one would, of course, have to obtain the intercorrelations among the four coded vectors, as well as the correlation of each coded vector with the dependent variable, Y. Again, the sum of the proportions listed in rows 5 through 8 of Table 8.10 equals the proportion of variance accounted for by the interaction.

Partitioning the Sum of Squares

Since each value in the last column of Table 8.10 is the proportion of variance accounted for by a given vector, one can readily obtain the regression sum of squares due to each vector. Simply multiply the proportion by the total sum of squares of the dependent variable (Σy^2), which in the present example is 340.00 (see Table 8.9). For example, for row 1 of Table 8.10 (the row which represents vector 1 of Table 8.9), we obtain $(.56471)(340.00000) = 192.00$ as the regression sum of squares due to vector 1. The remaining sums of squares listed in Table 8.10 are similarly obtained.

The sum of the regression sums of squares for rows 1 and 2 of Table 8.10 is 192.00, which is the regression sum of squares due to factor A. The sums of squares for rows 3 and 4 (27.00 and 81.00 respectively) add to 108.00, the regression sum of squares due to factor B. For rows 5 through 8 the sums of squares are .50, 1.50, 1.50, and 4.50, Their sum is 8.00, the regression sum of squares due to the interaction ($A \times B$).

Since, as noted above, the main effects and the interactions are mutually orthogonal, it is possible to obtain the same results by first adding the proportions of variance associated with the vectors representing a given factor and then multiplying by the sum of squares of the dependent variable (Σy^2). For example, for factor B, the proportions of variance are .07941 and .23824 (rows 3 and 4 of Table 8.10). Therefore, the regression sum of squares due to factor B is

$$ss_{\text{reg}(B)} = (.07941 + .23824)(340.00000) = 108.00$$

The discussion and calculations presented above were meant to clarify the meaning of some of the elements in Table 8.10 and the relations between them. With computer output of the kind reported in Table 8.10, the simplest approach for the purpose of analysis is to add the regression sums of squares associated with the coded vectors of a given factor. Each sum of squares thus obtained is divided by its degrees of freedom (the number of coded vectors from which it is derived) to obtain a regression mean square. Each mean square due to regression is then divided by the mean square residuals, resulting in an F ratio for each factor. This procedure is summarized in Table 8.11.

Note that the results reported in Table 8.11 are identical with those reported in Table 8.3 for the analysis of the same data with orthogonal coding. Consequently, the interpretation of the results is the same as that given earlier.

F Ratios via Proportions of Variance

It should be obvious that instead of working with regression sums of squares, it is possible to obtain the same F ratios by working with proportions of variance accounted for by each factor. To demonstrate this we repeat first formula (4.16) with a new number:

$$F = \frac{(R^2_{y.12...k_1} - R^2_{y.12...k_2})/(k_1 - k_2)}{(1 - R^2_{y.12...k_1})/(N - k_1 - 1)} \tag{8.12}$$

where $R^2_{y.12...k_1}$ = squared multiple correlation coefficient for the regression of Y on k_1 variables; and $R^2_{y.12...k_2}$ = squared multiple correlation coefficient for the regression of Y on k_2 variables, where k_2 is any set of variables selected from the set of variables k_1. The degrees of freedom for the F ratio are $k_1 - k_2$ and $N - k_1 - 1$ for the numerator and the denominator, respectively.

Recalling that the coded vectors for the main effects and the interaction are mutually orthogonal, the proportion of variance accounted for by a given factor equals R^2 of the coded vectors representing the factor with the depen-

TABLE **8.11**   SUMMARY OF MULTIPLE REGRESSION ANALYSIS, EFFECT CODING, DATA OF TABLE 8.9[a]

Source	ss	df	ms	F
Vector 1	192.00000	1		
Vector 2	.00000	1		
Factor A	192.00000	2	96.00000	27.00
Vector 3	27.00000	1		
Vector 4	81.00000	1		
Factor B	108.00000	2	54.00000	15.19
Vector 5	.50000	1		
Vector 6	1.50000	1		
Vector 7	1.50000	1		
Vector 8	4.50000	1		
Interaction ($A \times B$)	8.00000	4	2.00000	< 1
Residual	32.00000	9	3.55556	
Total	340.00000	17		

[a]Vectors 1 through 8 are the coded vectors describing the main effects and the interaction, as given in Table 8.9. The values for the ss are obtained from Table 8.10.

dent variable, Y. Accordingly, in the present example (from Table 8.10):

For factor A: $R^2_{y.12}$ = .56471 + .00000 = .56471
For factor B: $R^2_{y.34}$ = .07941 + .23824 = .31765
For $A \times B$: $R^2_{y.5678}$ = .00147 + .00441 + .00441 + .01324 = .02353

To test, for example, the proportion of variance accounted for by factor A, we note that $R^2_{y.12...8} = .90589$, with 8 degrees of freedom. In the context of formula (8.12), $R^2_{y.12...8} = R^2_{y.12...k_1}$, k_1 being 8. We note further that $R^2_{y.34} + R^2_{y.5678} = .34118$, with 6 degrees of freedom. This value is, in the context of formula (8.12), $R^2_{y.12...k_2}$, k_2 being 6. The difference between $R^2_{y.12345678}$ and $R^2_{y.345678}$ is obviously the proportion of variance accounted for by vectors 1 and 2, or $R^2_{y.12}$, which is the proportion of variance accounted for by factor A.

Applying formula (8.12),

$$F = \frac{(.90589 - .34118)/(8-6)}{(1-.90589)/(18-8-1)} = 27.00$$

with 2 and 9 degrees of freedom. The same F ratio for factor A was obtained when the calculations were done with the regression sum of squares (see Table 8.11). One may similarly obtain the F ratios for B and $A \times B$.

The purpose of this demonstration was to enhance the understanding of the analysis, as well as to prepare for future applications, when the approach outlined above becomes crucial. One can obtain the F ratios from computer output that reports only R^2's using the method described above. It is also possible to calculate regression sums of squares, as shown in an earlier section ("Partitioning the Regression Sum of Squares").

The Regression Equation

In Chapter 7 it was shown that the regression equation for effect coding with one categorical independent variable reflects the linear model. The same is true for the regression equation for effect coding in factorial designs. For two categorical independent variables, the linear model is

$$Y_{ijk} = \mu + \alpha_i + \beta_j + \alpha\beta_{ij} + \epsilon_{ijk} \tag{8.13}$$

where Y_{ijk} = the score of subject k in row i and column j, or the treatment combination α_i and β_j; μ = the population mean; α_i = the effect of treatment i; β_j = the effect of treatment j; $\alpha\beta_{ij}$ = the effect of the interaction between treatments α_i and β_j; ϵ_{ijk} = the error associated with the score of individual k under treatment combination α_i and β_j.

Formula (8.13) is expressed in parameters. In statistics, the linear model for two categorical independent variables is

$$Y_{ijk} = \bar{Y} + a_i + b_j + ab_{ij} + e_{ijk} \tag{8.14}$$

where the terms on the right are estimates of the respective parameters of Equation (8.13). Thus, for example, $\bar{Y}$ = the grand mean of the dependent variable, and is an estimate of μ in formula (8.13) — and similarly for the remaining terms. The score of a subject is conceived as composed of five components: the grand mean, the effect of treatment a_i, the effect of treatment b_j, the interaction between treatment a_i and b_j, and error.

In the light of the above, we turn our attention to the regression equation for the 3×3 design analyzed with effect coding (the original data are given in Table 8.9). From the computer output we obtain a (intercept) $= 11$. The b's for this analysis are given in Table 8.10. Accordingly, the regression equation is

$$Y' = 11 + 4X_1 + 0X_2 + 0X_3 + 3X_4 + 0X_5 + 0X_6 + 0X_7 + 1X_8$$

Note that a is equal to the grand mean of the dependent variable, $\bar{Y}$. Of the eight b's, the first four are associated with the vectors representing the main effects. Specifically, b_1 and b_2 are associated with vectors 1 and 2 of Table 8.9, the vectors representing factor A. Similarly, b_3 and b_4 are associated with the main effects of factor B, and b_5 through b_8 are associated with the interaction $(A \times B)$. We deal separately with the regression coefficients for the main effects and those for the interaction.

Regression Coefficients for the Main Effects

In order to facilitate the understanding of the regression coefficients for the main effects, the means of the treatment combinations (cells), as well as the treatment means and the treatment effects, are given in Table 8.12. From the table it can be noted that each b is equal to the treatment effect with which it is associated.[5] Thus, in vector 1 of Table 8.9 subjects belonging to category A_1 were assigned 1's. Accordingly, the coefficient for vector 1, b_1, is equal to the effect of category, or treatment, A_1. That is, $b_1 = \bar{Y}_{A_1} - \bar{Y} = 15 - 11 = 4$. Vector

---

[5]The example was contrived so that the means of the cells, the main effects, and the grand mean are integers. Although results of this kind are rarely obtained in actual data analysis, it was felt that avoiding fractions would facilitate the presentation and the discussion. It will be noted, further, that the proportion of variance accounted for in this example is very large compared to that generally obtained in behavioral research. Again this was contrived so that the results would be significant despite the small number of subjects involved.

TABLE **8.12**   CELL AND TREATMENT MEANS AND TREATMENT EFFECTS
FOR DATA OF TABLE 8.9[a]

Regions	Teaching Methods			$\bar{Y}_A$	$\bar{Y}_A - \bar{Y}$
	B_1	B_2	B_3		
A_1	15	18	12	15	4
A_2	11	15	7	11	0
A_3	7	9	5	7	-4
$\bar{Y}_B$	11	14	8	$\bar{Y} = 11$	
$\bar{Y}_B - \bar{Y}$	0	3	-3		

[a]$\bar{Y}_A$ = means for the three categories of factor A; $\bar{Y}_B$ = means for the three categories of factor B; $\bar{Y}$ = grand mean; $\bar{Y}_A - \bar{Y}$ and $\bar{Y}_B - \bar{Y}$ = treatment effects for a category of factor A and a category of factor B, respectively.

2 identifies category A_2, and the coefficient associated with this vector, b_2, indicates the effect of category A_2: $b_2 = \bar{Y}_{A_2} - \bar{Y} = 11 - 11 = 0$. Similarly, the coefficients of vectors 3 and 4, b_3 and b_4, indicate the treatment effects of B_1 and B_2 respectively (see Table 8.12 and note that $\bar{Y}_{B_1} - \bar{Y} = 11 - 11 = 0 = b_3$, and that $\bar{Y}_{B_2} - \bar{Y} = 14 - 11 = 3 = b_4$).

The remaining treatment effects – that is, those associated with the categories that are assigned -1's (in the present example these are A_3 and B_3) – can be easily obtained in view of the constraint that $\Sigma a_i = \Sigma b_j = 0$. That is, the sum of the effects of any factor equals zero. Therefore, the effect for $A_3 = -\Sigma a_i = -(4 + 0) = -4$. The effect for $B_3 = -\Sigma b_j = -(0 + 3) = -3$. Compare these values with those of Table 8.12.

Before dealing with the regression coefficients for the interaction, we digress for a brief discussion of the meaning of the interaction.

The Meaning of Interaction

The concept of interaction is probably best understood when viewed from the frame of reference of prediction. In order to minimize errors of prediction, is it necessary to resort to terms other than the main effects? When the treatment effects are independent of each other, they provide all the information necessary for optimal prediction. If, on the other hand, the effects of the treatments of one factor depend on their specific combinations with treatments of another factor, it is necessary to note in what manner the factors interact in order to achieve optimal prediction.

The above may be clarified by providing a formal definition of the interaction, which for the case of two factors takes the following form:

$$AB_{ij} = (\bar{Y}_{ij} - \bar{Y}) - (\bar{Y}_{A_i} - \bar{Y}) - (\bar{Y}_{B_j} - \bar{Y}) \tag{8.15}$$

where AB_{ij} = interaction of treatments A_i and B_j; $\bar{Y}_{ij}$ = mean of treatment combination A_i and B_j, or the mean of cell ij; $\bar{Y}_{A_i}$ = mean of category, or treatment, i of factor A; $\bar{Y}_{B_j}$ = mean of category, or treatment, j of factor B; $\bar{Y}$ = grand mean. Note that $\bar{Y}_{A_i} - \bar{Y}$ in formula (8.15) is the effect of treatment A_i, and that $\bar{Y}_{B_j} - \bar{Y}$ is the effect of treatment B_j. From formula (8.15) it follows that when the deviation of a cell mean from the grand mean is equal to the sum of the treatment effects related to the given cell, then the interaction term for the cell is zero. Stated differently, in order to predict the mean of such a cell it is sufficient to know the grand mean and the treatment effects.

Using formula (8.15) we calculate the interaction term for each of the treatment combinations. These are reported in Table 8.13. For example, the term for the cell $A_1 B_1$ is obtained as follows:

$$
\begin{aligned}
A_1 \times B_1 &= (\bar{Y}_{A_1 B_1} - \bar{Y}) - (\bar{Y}_{A_1} - \bar{Y}) - (\bar{Y}_{B_1} - \bar{Y}) \\
&= (15 - 11) - (15 - 11) - (11 - 11) \\
&= 4 - 4 - 0 = 0
\end{aligned}
$$

The other terms of Table 8.13 are similarly obtained. In five of the cells of Table 8.13 the interaction terms are zero, which means that for these cells it is pos-

TABLE **8.13**   INTERACTION EFFECTS FOR THE DATA OF TABLE 8.9

Regions	Treatments			
	B_1	B_2	B_3	Σ
A_1	0	0	0	0
A_2	0	1	−1	0
A_3	0	−1	1	0
Σ:	0	0	0	

sible to express each individual's score as a composite of the grand mean, main effects, and error.

The four remaining cells of Table 8.13 have nonzero interaction terms, indicating that part of each individual's score in these cells is due to an interaction between factors A and B. Whether nonzero interactions are sufficiently large to be attributed to other than random fluctuation is determined by statistical tests of significance. If the increment in the proportion of variance accounted for by the interaction is not significant, the interaction may be ignored; it is sufficient to speak of main effects only. Recall that the interaction in the present example is not significant (see Table 8.11).

When, however, the interaction is significant, it is necessary to study the way the variables interact. Instead of testing differences between the treatments of one variable across the treatments, or categories, of the other variable (that is, main effects), differences between treatments of one variable at each category of the other variable are tested. Such tests are referred to as tests of simple effects. Furthermore, a graphic presentation of a significant interaction may be a useful aid in interpreting the results. For graphic representations of interactions, and tests of significance subsequent to a significant interaction, the reader is referred to Kirk (1968), Winer (1971), and Marascuilo and Levin (1970).

It was shown earlier how one may do tests of significance between individual cell means. This type of analysis, too, may be used subsequent to finding a significant interaction.

Regression Coefficients for the Interaction

We repeat the regression equation for the 3×3 design of the data given in Table 8.9.

$$Y' = 11 + 4X_1 + 0X_2 + 0X_3 + 3X_4 + 0X_5 + 0X_6 + 0X_7 + 1X_8$$

The first four b weights in this equation were discussed earlier. It was shown that b_1 and b_2 refer to factor A, and that b_3 and b_4 refer to factor B. The remaining four b weights refer to the interaction. Specifically, each b refers to the cell with which it is associated. Look back at Table 8.9 and note that vector 5 is obtained as a product of vectors 1 and 3, that is, the vectors identifying A_1 and

B_1. Therefore, the regression coefficient associated with vector 5, b_5, is associated with cell A_1B_1. Note that $b_5 = 0$, as is the cell for A_1B_1 in Table 8.13.

Similarly, b_6, b_7, and b_8 refer to cells A_1B_2, A_2B_1, and A_2B_2, respectively. As with the main effects, the remaining terms for the interaction can be obtained in view of the constraint that $\Sigma ab_{ij} = 0$. That is, the sum of the interaction terms for each row or column equals zero (see Table 8.13). Thus, for example, the b for $A_2B_3 = -\Sigma(0+1) = -1$.

Applying the Regression Equation

The discussion of the properties of the regression equation for effect coding, as well as the overall analysis of the data of Table 8.9, can best be summarized by using the regression equation to predict the scores of the subjects. Applying the regression equation obtained above to the coding of the first row of Table 8.9, that is, the coding for the first subject, we obtain

$$Y' = 11 + 4(1) + 0(0) + 0(1) + 3(0) + 0(1) + 0(0) + 0(0) + 1(0)$$
$$= 11 + 4 + 0 + 0 = 15$$

Note that 15 is the mean of the cell to which the first subject belongs, A_1B_1. The regression equation will always yield the mean of the cell to which a subject belongs. Note, too, that in arriving at a statement about the predicted score of the subject, we collected terms that belong to a given factor or to the interaction. For example, the second and third terms refer to factor A, and they were therefore collected, that is, $4(1) + 0(0) = 4$. Similarly, the fourth and the fifth terms were collected to express factor B, and the last four terms were collected to express the interaction.

The residual, or error, for the first subject is $Y - Y' = 16 - 15 = 1$. It is now possible to express the score of the first subject as a composite of the five components of the linear model. To demonstrate this, we repeat formula (8.14) with a new number:

$$Y_{ijk} = \bar{Y} + a_i + b_j + ab_{ij} + e_{ijk} \tag{8.16}$$

where $Y_{ijk} =$ the score of subject k in row i and column j, or in treatment combination a_i and b_j; $\bar{Y} =$ grand mean; $a_i =$ the effect of treatment i of factor A; $b_j =$ the effect of treatment j of factor B; $ab_{ij} =$ the effect of the interaction between a_i and b_j; $e_{ijk} =$ the error associated with the score of individual k under treatments a_i and b_j. For the first subject in cell A_1B_1 the expression of formula (8.16) takes the following form:

$$16 = 11 + 4 + 0 + 0 + 1$$

where $11 =$ the grand mean; $4 =$ the effect of treatment A_1; $0 =$ the effect of treatment B_1; $0 =$ the effect of the interaction for cell A_1B_1; $1 =$ the residual, $Y - Y'$.

As another example, the regression equation is applied to the last subject of Table 8.9:

$$Y' = 11 + 4(-1) + 0(-1) + 0(-1) + 3(-1) + 0(1) + 0(1) + 0(1) + 1(1)$$

$$= 11 - 4 - 3 + 1 = 5$$

Again, the predicted score, Y', is equal to the mean of the cell to which the subject belongs, A_3B_3. The residual for this subject is $Y - Y' = 6 - 5 = 1$. Expressing the scores of the last subject of Table 8.9 in the components of the linear model,

$$6 = 11 - 4 - 3 + 1 + 1$$

In this way the scores for all the subjects of Table 8.9 are reported in Table 8.14 as components of the linear model. A close study of Table 8.14 will enhance understanding of the analysis of these data. Note that squaring and summing the elements in the column for the main effects of factor A (a_i) yield a sum of squares of 192. This is the same sum of squares obtained earlier for factor A (see, for example, Table 8.11). Similarly, the sum of the squared elements for the effects of factor B (b_j), the interaction between $A \times B$ (ab_{ij}), and

TABLE **8.14**   DATA FOR A 3×3 DESIGN EXPRESSED AS COMPONENTS OF THE LINEAR MODEL[a]

Cell	Y	$\bar{Y}$	a_i	b_j	ab_{ij}	Y'	$Y - Y'$
A_1B_1	16	11	4	0	0	15	1
	14	11	4	0	0	15	-1
A_2B_1	12	11	0	0	0	11	1
	10	11	0	0	0	11	-1
A_3B_1	7	11	-4	0	0	7	0
	7	11	-4	0	0	7	0
A_1B_2	20	11	4	3	0	18	2
	16	11	4	3	0	18	-2
A_2B_2	17	11	0	3	1	15	2
	13	11	0	3	1	15	-2
A_3B_2	10	11	-4	3	-1	9	1
	8	11	-4	3	-1	9	-1
A_1B_3	10	11	4	-3	0	12	-2
	14	11	4	-3	0	12	2
A_2B_3	7	11	0	-3	-1	7	0
	7	11	0	-3	-1	7	0
A_3B_3	4	11	-4	-3	1	5	-1
	6	11	-4	-3	1	5	1
Σ^2:			192	108	8		32

[a] Y = observed score; $\bar{Y}$ = grand mean; a_i = effect of treatment i of factor A; b_j = effect of treatment j of factor B; ab_{ij} = interaction between a_i and b_j; Y' = predicted score, where in each case it is equal to the sum of the elements in the four columns preceding it; $Y - Y'$ = residual, or error.

the residuals, $Y - Y'$, are 108, 8, and 32. The same values were obtained earlier. Adding the four sum of squares obtained in Table 8.14, one gets the total sum of squares for Y:

$$\Sigma\, y^2 = 192 + 108 + 8 + 32 = 340$$

Dummy Coding

The presentation thus far has been devoted to factorial design with orthogonal and effect coding. It is of course possible to do the analysis of the 3×3 design with dummy coding. Some general comments about the method will suffice.

First, the method of coding the main effects with dummy coding is the same as with effect coding, except that instead of assigning -1's to the last category of each factor, 0's are assigned. As in the previous analyses, the vectors for the interaction are obtained by cross multiplying the vectors for the main effects.

Second, the overall results obtained with dummy coding are the same as those obtained with orthogonal and effect coding. The regression equation, however, is different. The a (intercept) equals the mean of the cell that as a result of the dummy coding has 0's in all the vectors. Using as an example the 3×3 design analyzed with the other methods of coding, the cell that will have 0's in all the vectors is A_3B_3. Without going into a lengthy explanation about the b's, it is pointed out that their determination, too, is related to the cell that is assigned 0's in all the vectors.

Third, while the vectors of the main effects for one factor are not correlated with the vectors of the main effects for the other factors, there is a correlation between the vectors for the interaction and those for the main effects. With orthogonal and effect coding there is no correlation between the vectors for the interaction and the vectors for the main effects. Using the 3×3 design as an example, it should be noted that, unlike orthogonal and effect coding, with dummy coding,

$$R^2_{y.12345678} \neq R^2_{y.12} + R^2_{y.34} + R^2_{y.5678}$$

where $R^2_{y.12\ldots8}$ = squared multiple correlation of Y with eight dummy vectors for a 3×3 design; $R^2_{y.12}$ = squared multiple correlation of Y with the dummy vectors for factor A; $R^2_{y.34}$ = squared multiple correlation of Y with the dummy vectors for factor B; $R^2_{y.5678}$ = squared multiple correlation of Y with the vectors for the interaction.

When doing the calculations, it is important to make the adjustment for the intercorrelations between the coded vectors. In the 3×3 example, the calculation of all the necessary terms can be done as follows:

For factor A, calculate: $R^2_{y.12}$
For factor B, calculate: $R^2_{y.34}$
For $A, B, A \times B$, calculate: $R^2_{y.12345678}$
For $A \times B$, calculate: $R^2_{y.12345678} - (R^2_{y.12} + R^2_{y.34})$
For residuals, calculate: $1 - R^2_{y.12345678}$

In order to convert the above terms to sums of squares, all that is necessary is to multiply each term by the total sum of squares, Σy^2.

In general it is preferable to use orthogonal or effect coding for factorial designs. As shown in detail in the preceding sections, the properties of these coding systems have much to recommend them.

Analyses with More than Two Categorical Variables

Analyses of results from an experiment with two categorical variables, each with three categories, were presented with two coding methods. It should be stressed that the same approach applies to data from nonexperimental research. As long as one is dealing with categorical variables, it is possible to code them according to one's needs or preferences. Moreover, the procedure presented in this chapter can be extended to any number of variables with any number of categories. All that is needed is to apply one of the coding methods and to generate for each of the variables a number of coded vectors equal to the number of categories of the variable minus one. The interactions are then obtained by multiplying the vectors of the variables involved.

In designs with more than two variables, higher-order interactions are of course calculated. The vectors for such interactions are also obtained by cross multiplying the vectors of the pertinent variables. Suppose, for example, that one has a design with three variables as follows: A with two categories, B with three categories, and C with four categories; then variable A will have one coded vector (say vector number 1), variable B will have two vectors (2 and 3), and variable C will have three vectors (4, 5, and 6). The first-order interactions, $A \times B$, $A \times C$, and $B \times C$, are obtained in the manner described earlier: by cross-multiplying vectors 1 and 2, 1 and 3, and so on. The second-order interaction, that is, $A \times B \times C$, is obtained by cross multiplying the vectors associated with these variables as follows: $1 \times 2 \times 4$; $1 \times 3 \times 4$; $1 \times 2 \times 5$; $1 \times 3 \times 5$; $1 \times 2 \times 6$; $1 \times 3 \times 6$. Altogether, six vectors are generated to represent the 6 degrees of freedom associated with this interaction (the degrees of freedom for A, B, and C, respectively, are 1, 2, and 3. The degrees of freedom for the interaction $A \times B \times C$ are therefore $1 \times 2 \times 3 = 6$).[6] Having generated the necessary vectors one does a multiple regression analysis using the coded vectors as the independent variables and the scores on the dependent measure as the dependent variable. As shown earlier, when a computer program that gives the vectors' sums of squares is used for the analysis, the sums of squares associated with given variables are obtained by adding the sums of squares associated with the vectors of the variables.

[6]One of the virtues of the BMD programs is that by using a transgeneration feature one can do various operations, like addition, subtraction, multiplication, raising to powers, of vectors that are read in from cards. Consequently, one can, for example, generate the interaction vectors without punching them on cards. MULR, the program given in Appendix C, requires the user to punch the coded vectors for the main effects only. The interaction vectors are generated automatically by the program.

Note the flexibility of the coding approach. Researchers frequently encounter difficulties in obtaining computer programs that meet their specific needs. A researcher may, for example, have a four-variable design and discover to his chagrin that the computer center to which he has access has only a three-variable program. With coding, any multiple regression program can be used for the analysis with fair ease.

Categorical Variables with Unequal Frequencies

When the frequencies in the treatment combinations of categorical variables are equal, the partition of the total sum of squares is unambiguous. It is for this reason that one should always try to have equal frequencies in the different treatment combinations. An experimenter may start with equal frequencies but for some reason may lose subjects while an experiment is in progress. In an experiment conducted in a school, for example, subject attrition may be caused by illness, moving out of the neighborhood, simple forgetfulness to appear at a session, unwillingness to continue, and many other reasons. In experiments with animals, attrition may be due to illness, death, or other causes.

Whenever there are unequal frequencies in treatment combinations, the partition of the total sum of squares becomes ambiguous. The reason is that the treatment effects and interactions are no longer orthogonal. In other words, the treatment effects and their interactions are correlated. This makes it difficult to determine what portion of the sum of squares is to be attributed to each of the treatments and to their interactions.

To understand the approach taken in the least squares solution with unequal frequencies, it is necessary to review briefly some aspects of regression analysis presented earlier in Chapter 5 and in the present chapter. Recall that when the independent variables are correlated, one can orthogonalize them by using semipartial correlations. For convenience we restate formula (8.11) as it would apply, for example, to four independent variables:

$$R^2_{y.1234} = r^2_{y1} + r^2_{y(2.1)} + r^2_{y(3.12)} + r^2_{y(4.123)} \tag{8.17}$$

where $R^2_{y.1234}$ = squared multiple correlation of Y with four independent variables; r^2_{y1} = squared zero-order correlation of Y with variable 1; $r^2_{y(2.1)}$ = squared semipartial correlation of Y with variable 2, partialing 1 from 2; the remaining terms are similarly defined.

Each squared semipartial in formula (8.17) indicates the proportion of variance accounted for by a given variable after partialing out from it what it shares with all the variables preceding it. It is obvious that when the variables are correlated, the order in which they enter into the calculations is crucial. Note, in formula (8.17), that because variable 1 is entered first, it is shown to account for a proportion of variance equal to the square of the zero-order correlation between it and the dependent variable. Had variable 1 entered later in the calculations it would have accounted for more or less of the variance depending on the signs and magnitudes of the correlations among the variables involved.

Since the order in which the variables are introduced determines how the variance due to regression will be apportioned among them, how does one decide on the order? As shown below, one of the methods for the least squares solution is based on an a priori ordering of variables derived from a theoretical formulation. At this point, suffice it to note that the basic notion behind the least squares solutions as applied to the case of unequal cell frequencies is the use of semipartial correlations. It is sometimes more convenient, however, to express squared semipartials as combinations of squared multiple correlations. The information obtained from either expression is, of course, the same. To illustrate this point, formula (8.17) is expressed in squared multiple correlation terms:

$$R^2_{y.1234} = R^2_{y.1} + (R^2_{y.12} - R^2_{y.1}) + (R^2_{y.123} - R^2_{y.12}) + (R^2_{y.1234} - R^2_{y.123}) \qquad (8.18)$$

where the first term on the right is simply the square of the correlation of y with 1. The second term on the right, $R^2_{y.12} - R^2_{y.1}$, is equal to $r^2_{y(2.1)}$, that is the proportion of variance accounted for by variable 2, after having partialed out variable 1 from it. In other words, $R^2_{y.12} - R^2_{y.1}$ indicates the proportion of the variance in Y that is explained by that portion of variable 2 that is not related to variable 1, or the increment that is due to variable 2. The other terms in (8.18) are similarly interpreted.

Several approaches to the application of a least squares solution to data from unequal cell frequencies are possible. Overall and Spiegel (1969), for example, describe three different approaches, and discuss the conditions under which each of them may be appropriately used. The presentation here is limited to two approaches.

The first approach is appropriate when one wishes to make statements about main effects and interaction in the conventional manner. This type of analysis is referred to here as the *Experimental Design Approach*. The second type of analysis, here called the *A Priori Ordering Approach*, is appropriate for analysis of data from nonexperimental designs, when the researcher can specify, on the basis of theory, the order in which the variables should enter into the regression analysis.

The Experimental Design Approach

Assume an experiment on attitude change toward the use of marijuana. The experiment consists of two factors, each with two treatments, as follows. Factor A refers to source of information, where A_1 = a former addict, and A_2 = nonaddict. Factor B refers to fear arousal, where B_1 = mild fear arousal, and B_2 = intense fear arousal. Without going into the details of the design, assume further, for the sake of illustration, that five subjects are randomly assigned to each treatment combination.

In short, the experiment consists of four treatment combinations, namely A_1B_1, A_1B_2, A_2B_1, and A_2B_2. This is, of course, a 2×2 factorial design.[7] Assume that the experiment has been in progress for several sessions and that subject attrition has occurred. During the final session, measures of attitude change were available for only 14 of the 20 original subjects. The scores for these subjects, the cell means, and the unweighted treatment means are given in Table 8.15.

TABLE **8.15**   FICTITIOUS DATA FROM AN EXPERIMENT
ON ATTITUDE CHANGE

Source of Information	Fear Arousal		Unweighted Means
	B_1	B_2	
	4	8	
A_1	3	10	
	2		
	$\bar{Y} = 3.00$	$\bar{Y} = 9.00$	6.00
	3	5	
	2	4	
A_2	5	5	
	6	6	
	4		
	$\bar{Y} = 4.00$	$\bar{Y} = 5.00$	4.50
Unweighted Means	3.50	7.00	5.25

Outline of the Analysis

The methods of coding categorical independent variables in a design with unequal cell frequencies are the same as in designs with equal cell frequencies. For the present example, we use effect coding. In Table 8.16 the data originally given in Table 8.15 are repeated, together with the effect coding. Note that vector 1 of Table 8.16 identifies factor A (since there are two categories in factor A, one coded vector is necessary). Similarly, vector 2 of Table 8.16 identifies factor B. Vector 3, obtained by the multiplication of vectors 1 and 2, represents the interaction.

Thus far we have followed the same procedures that were used earlier in the chapter. It is not appropriate, however, to do a regression analysis in the usual way, because the unequal cell frequencies introduce correlations among the coded vectors. Consequently, as noted earlier, the order in which the vectors are introduced into the analysis affects the proportion of variance attributed to each of them.

[7]A 2×2 design is used for simplicity of presentation. The same approach can be extended to as many factors and at as many levels as necessary.

TABLE **8.16**   EFFECT CODING FOR DATA FROM AN EXPERIMENT
ON ATTITUDE CHANGE[a]

	Y	1	2	3 (1×2)
A_1B_1	4 3 2	1 1 1	1 1 1	1 1 1
A_2B_1	3 2 5 6 4	-1 -1 -1 -1 -1	1 1 1 1 1	-1 -1 -1 -1 -1
A_1B_2	8 10	1 1	-1 -1	-1 -1
A_2B_2	5 4 5 6	-1 -1 -1 -1	-1 -1 -1 -1	1 1 1 1
Σy^2:	64.35714			

[a]Y = data originally given in Table 8.15; 1 = coded vector for factor A; 2 = coded vector for factor B; 3 = coded vector for the interaction between A and B.

The solution to the problem in the case of the experimental design approach is to adjust the proportion of variance attributed to a factor for the correlation of the factor with all the other factors in the design. This boils down to noting the increment in the proportion of variance due to each factor when it is entered last in the analysis of the main effects only. The interaction is then adjusted for its correlation with the main effects. In other words, one notes the increment in the proportion of variance due to the interaction when it is entered last in the analysis.

For the present example, the method takes the form indicated by the following formulas:

$$\text{Incr.}\,A = R^2_{y.12} - R^2_{y.2} \tag{8.19}$$

$$\text{Incr.}\,B = R^2_{y.12} - R^2_{y.1} \tag{8.20}$$

$$\text{Incr.}\,A \times B = R^2_{y.123} - R^2_{y.12} \tag{8.21}$$

$$\text{Prop. of variance due to residuals} = 1 - R^2_{y.123} \tag{8.22}$$

where Incr. = increment; Y = dependent variable; 1 = factor A; 2 = factor B; $3 = A \times B$. The reasoning behind formulas (8.19) through (8.21) is the same. Note, for example, that formula (8.19) is, in effect, another form for expressing the squared semipartial correlation of Y with 1, partialing 2 from 1, that is, $r^2_{y(1.2)}$. Note, too, that in the case of equal cell frequencies there is, of course, no correlation between the coded vectors, that is $r_{12} = 0$, and therefore $r^2_{y(1.2)} = r^2_{y.1}$. Obviously, the application of this method to a design with equal cell frequencies results in exactly the same analysis as that used earlier in the chapter.

In formula (8.20) an adjustment is made for factor A, while in (8.21) the adjustment is for factors A and B.[8] The proportion of variance due to residuals is obtained in the same manner as in the case of equal cell frequencies. This is because despite the unequal cell frequencies there is no ambiguity as to what proportion of variance is not accounted for by the information available.

While the discussion thus far has dealt with proportions of variance, it should be obvious that it applies also to sums of squares. Multiplying each proportion obtained in formulas (8.19) through (8.22) by the total sum of squares of the dependent variable, Σy^2, yields the sum of squares for the respective component.

Having obtained proportions of variance, or sums of squares, the analysis proceeds now in the same manner as in the case of equal cell frequencies. That is, mean squares are obtained by dividing the sum of squares for each factor and for the interaction by the appropriate degrees of freedom. F ratios are then calculated by dividing each regression mean square by the residual mean square. The method outlined above is now applied to the numerical example of Table 8.16.

Analysis of the Numerical Example

The various terms necessary for the analysis of the data of Table 8.16 are given in Table 8.17. Using the appropriate values from Table 8.17, and noting

[8]In the present example there are only two factors. The procedure is the same for more than two factors. If, for example, there are three factors, A, B, and C, the proportion of variance for each of them is adjusted for the correlations among them. Thus, to obtain, for example, the proportion of variance due to factor A, it is necessary to subtract R^2 of Y with B and C from R^2 of Y with A, B, and C.

TABLE **8.17**   ZERO-ORDER AND SQUARED MULTIPLE CORRELATIONS
FOR DATA FROM AN EXPERIMENT ON ATTITUDE CHANGE[a]

(I)	1	2	3	Y
1	—	.04303	.14907	.21355
2	.00185	—	−.28868	−.62512
3	.02222	.08334	—	−.29983
Y	.04560	.39077	.08990	—
(II)	$R^2_{y.123} = .75138$		$R^2_{y.12} = .44869$	

[a]Original data given in Table 8.16. Y = dependent variable; 1 = coded vector for factor A; 2 = coded vector for factor B; 3 = coded vector for $A \times B$. In part (I) of the table, the values above the principal diagonal are zero-order correlations, while those below the diagonal are the squared zero-order correlations.

that $\Sigma y^2 = 64.35714$ (see Table 8.16), calculate the regression sums of squares and the residual sum of squares. Adapting formula (8.19) for the calculation of the regression sum of squares due to factor A, we obtain

$$ss_{reg(A)} = \Sigma y^2 (R^2_{y.12} - R^2_{y.2})$$

$$= 64.35714(.44869 - .39077)$$

$$= 64.35714(.05792) = 3.72757$$

The sums of squares for the remaining terms are similarly obtained by adapting formulas (8.20)–(8.22). For each sum of squares thus obtained a mean square is calculated. Each mean square is then divided by the mean square residuals to obtain F ratios for the main effects and the interaction.

These steps are summarized in Table 8.18, where for convenience the terms are expressed symbolically as well as numerically. Note, first, that the sum of the separate components of the sums of squares of Table 8.18 is not equal to the total sum of squares of the dependent variable, Σy^2. The former is 65.15002, while the latter is 64.35714. When the experimental design approach is applied to data with unequal cell frequencies, the sum of the sums of squares for the separate components does not necessarily equal the total sum of squares. This is a consequence of the adjustments in the proportions of variance attributed to each of the main effects.

In the present analysis the F ratio for the difference between the two sources of information is not significant (factor A: $F = 2.33$, with 1 and 10 degrees of freedom). There is a significant difference between the two conditions of fear arousal (factor B: $F = 16.21$, with 1 and 10 degrees of freedom). Since, how-

TABLE **8.18**   ANALYSIS OF VARIANCE SUMMARY TABLE FOR DATA OF TABLE 8.16[a]

Source	ss	df	ms	F	p
A	$\Sigma y^2 (R^2_{y.12} - R^2_{y.2})$ $64.35714(.44869 - .39077) = 3.72757$	1	3.72757	2.33	n.s.
B	$\Sigma y^2 (R^2_{y.12} - R^2_{y.1})$ $64.35714(.44869 - .04560) = 25.94172$	1	25.94172	16.21	.01
$A \times B$	$\Sigma y^2 (R^2_{y.123} - R^2_{y.12})$ $64.35714(.75138 - .44869) = 19.48026$	1	19.48026	12.17	.01
Residual	$\Sigma y^2 (1 - R^2_{y.123})$ $64.35714(1 - .75138) = 16.00047$	10	1.60005		
$\Sigma y^2 = 64.35714$	$\Sigma: 65.15002$	13			

[a]A = Source of Information; B = Fear Arousal; Σy^2 = total sum of squares of the dependent variable, Y; 1 = coded vector for factor A; 2 = coded vector for factor B; 3 = coded vector for $A \times B$.

ever, the interaction is also significant ($F = 12.17$, with 1 and 10 degrees of freedom), one would test the simple effects. In the present case one would want to test the differences between the two fear arousal conditions (B_1 and B_2) at each category of source of information (A_1 and A_2). In other words, one would test the difference between the mean of cell A_1B_1 and that of cell A_1B_2, and the difference between the mean of cell A_2B_1 and that of cell A_2B_2. This is not done here, since the only concern is with the problem of unequal cell frequencies.

The A Priori Ordering Approach

Unequal cell frequencies in experimental research result in correlations among the coded vectors that represent the independent variables. The reverse is true in nonexperimental research, where unequal cell frequencies are generally obtained *because* of correlations among independent variables. Consequently, orthogonalizing correlated independent variables in nonexperimental research by subject selection or by statistical adjustments makes no theoretical sense, since it may lead to a distortion of reality, or (borrowing a phrase from Hoffman, 1960) *dismemberment of reality*. Take, for example, a study of the effects of ethnicity and social class on risk taking. Ethnicity and social class tend to be correlated, and orthogonalizing them when analyzing risk-taking amounts to pretending that they are not correlated.

The absence of manipulation of independent variables and the absence of randomization of subjects in nonexperimental research invalidate attempts to determine independent main effects and interactions. In nonexperimental research an independent variable may be dependent, in part or entirely, on one or more independent variables introduced explicitly into the design by the researcher or on variables implicitly introduced by the process of the selection of subjects.

What, then, is the appropriate analysis and interpretation of data from nonexperimental research? There is not one appropriate approach, but several approaches depending on theory, knowledge, and purpose. The present approach is appropriate when the researcher can make a decision as to the order in which the variables should enter into the regression analysis. A discussion of the various aspects of arriving at such a decision is beyond the scope of this chapter.[9] Suffice it to say that the researcher may give precedence to one variable over another because it comes earlier in time, or because he believes that it "causes" the other.

Suppose we take the data presented earlier in the experiment on attitude change. This time, however, we conceive of the study as nonexperimental. The first independent variable (A) is race, (A_1 = black; A_2 = white), The second independent variable (B) is education (B_1 = high school; B_2 = college). Assume that the dependent variable is attitudes toward birth control. The first thing to

---

[9]See Chapter 11 for such a discussion, and for approaches to the analysis when the researcher has no theoretical basis to decide on an order.

note is that since there is a correlation between race and education, one is bound to obtain unequal category combinations when drawing representative samples of blacks and whites from defined populations. It is quite possible that the samples will contain more college educated whites than blacks. The contrived frequencies in Table 8.15 illustrate such a situation, although they should by no means be taken as representative of the "true" state of the relation between race and education. If a researcher were interested in studying the relations of race and education to attitudes toward birth control, he has to decide on the order in which the variables should enter the regression analysis. In the present case, it seems reasonable that race be given preference, since the race of a person may determine to some extent the level of education he may achieve. The reverse is obviously not the case. Consequently, the researcher will enter race first, followed by education and the interaction between race and education. It should be stressed that when the a priori ordering approach is used, one does not speak of main effects and interactions in the same way as one does with the experimental design approach. A priori ordering is used in an attempt to study the proportion of variance accounted for by each of the variables and their interactions, when the variables are taken in an order specified by the researcher.[10]

The a priori ordering approach is an ordinary multiple regression analysis in which the independent variables are coded vectors that represent categorical variables and their interactions. The need for a priori ordering arises when in nonexperimental research the categorical independent variables are correlated, resulting in unequal cell frequencies.

Analysis of the Numerical Example

For the purpose of comparison, the data previously analyzed by the experimental design approach are now analyzed with the a priori approach. Using the data of Table 8.16, vector 1 (coded vector for factor A) is entered first, followed by vector 2 (coded vector for factor B), and vector 3 (coded vector for $A \times B$). The results of the analysis are summarized in Table 8.19.

Unlike the analysis of these data with the experimental design approach, the sum of the sum of squares for the various components of Table 8.19 equals the total sum of squares for the dependent variable, Σy^2. This is to be expected, since in the a priori ordering there is a process of successive orthogonalizations, such that each term is orthogonalized for the ones preceding it. The difference between the two approaches results, in the present case, in different sums of squares for factor A. In the experimental design approach the sum of squares for A is 3.72757 (see Table 8.18), while in the present analysis the sum of squares for A is 2.93492 (see Table 8.19). The reason for the discrepancy is that in the experimental design approach the sum of squares for A was adjusted for the correlation between A and B, and vice versa. In the a priori approach,

---

[10]The total amount of variance accounted for is, of course, the same regardless of the order in which the variables are entered into the analysis. It is only the proportions of variance attributed to each factor that are affected by the variable order.

TABLE **8.19**   SUMMARY OF THE MULTIPLE REGRESSION ANALYSIS FOR
DATA OF TABLE 8.16

Variable	b	Prop. of Variance	ss	df	ms	F	p
A	.75000	.04560	2.93492	1	2.93492	1.83	n.s.
B	−1.75000	.40309	25.94170	1	25.94170	16.21	.01
$A \times B$	−1.25000	.30269	19.48052	1	19.48052	12.18	.01
Residual		.24862	16.00000	10	1.60000		
Total		1.00000	64.35714	13			

on the other hand, the sum of squares for factor A is not adjusted, while the sum of squares for B is adjusted for the correlation between A and B.[11]

Since in both analyses the interaction is adjusted for both factors, the values for the interaction are the same. As noted several times earlier, the magnitude of the squared multiple correlation is unchanged regardless of the order in which the variables are entered into the analysis. $R^2_{y.123} = .75138$ is of course the same in both analyses, as are the residuals, since they always equal $1 - R^2$ (.24862).

The Regression Equation

In conclusion, we comment on the properties of the regression equations that result from either analysis when all the coded vectors are entered. The intercept, a, is 5.25. Obtaining the b's from Table 8.19, the regression equation is

$$Y' = 5.25 + .75X_1 - 1.75X_2 - 1.25X_3$$

Since effect coding was used (see Table 8.16), the regression equation has the same properties as the regression equation obtained from effect coding with equal cell frequencies. In the case of unequal cell frequencies, however, the terms in the equation refer to unweighted means. Thus, the intercept, 5.25, is equal to the unweighted grand mean, or the mean of the cell means (see Table 8.15). b_1 (.75) is equal to the difference between the unweighted mean of A_1 (see Table 8.16, where subjects belonging to category A_1 are assigned 1's) and the unweighted grand mean. In other words, b_1 is equal to the effect of treatment A_1. b_2 (−1.75) is equal to the effect of treatment B_1, that is, the difference between the unweighted mean of B_1 and the unweighted grand mean. b_3 (−1.25) is equal to the interaction term for cell A_1B_1.[12] As noted earlier, the terminology of main effects and interaction in the sense used in factorial designs is appropriate only for the experimental design approach.

[11]The size of the discrepancy between the two solutions will, of course, depend on the signs and magnitudes of the correlations among the variables.
[12]For a more extensive treatment of the interpretation of regression coefficients with effect coding, see earlier sections of the chapter.

When the regression equation obtained above is applied to the codes of any of the subjects in Table 8.16, the predicted score is equal to the mean of the cell, or treatment combination, to which the subject belongs. Thus, for example, the predicted score for the first subject of Table 8.16 is

$$Y' = 5.25 + .75(1) - 1.75(1) - 1.25(1) = 3.00$$

The mean of cell A_1B_1, the cell to which the first subject of Table 8.16 belongs, is 3.00 (see Table 8.15).

Summary

In this chapter the notion of coding categorical independent variables was extended to factorial designs. It was shown that regardless of the coding method used, the basic approach and the overall results are the same. As in the case of one categorical independent variable, the scores on the dependent variable measure are regressed on a set of coded vectors. In factorial designs, however, there are subsets of coded vectors, each subset representing the main effects of a factor, or the interaction between factors.

For the main effects, each independent variable is coded separately as if it were the only variable in the design. For each variable one generates a number of coded vectors equal to the number of categories of the variable minus one. The vectors for the interaction between any two variables, or factors, are obtained by cross multiplying each of the vectors of one factor by each of the vectors of the other factor. Vectors for higher-order interactions are similarly obtained. That is, each of the vectors of one factor is multiplied by each of the vectors of the factors whose higher-order interaction is being considered.

Tests for main effects and interaction can be made either by using the proportions of variance accounted for by each subset of coded vectors, or by using the regression sums of squares attributable to each subset of coded vectors. When the main effects are significant, and the interaction is not significant, it is possible to proceed with post hoc comparisons, for example, Scheffé tests, to determine specifically which categories, or treatments of a factor, differ significantly from each other. When the interaction is significant, it is more meaningful to do tests for simple effects, or to compare individual cell means.

Although the three coding methods — orthogonal, effect, and dummy — yield the same overall results, each of them yields unique intermediate results. The selection of a coding method depends, therefore, on the kind of intermediate results one wishes to obtain. It was shown, for example, that orthogonal coding is particularly useful when the researcher has planned orthogonal comparisons. Effect coding, on the other hand, results in a regression equation that reflects the linear model.

When the cell frequencies are unequal, the partitioning of the regression sum of squares depends on the frame of reference taken by the researcher. A distinction was made between two approaches: the experimental design approach and the a priori ordering approach.

It was shown that the experimental design approach is appropriate for the analysis of data from experiments, where one wishes to make statements about main effects and interactions in the manner in which such statements are made with factorial analysis of variance. The a priori ordering approach, on the other hand, is appropriate for analysis of data of nonexperimental research. As the name implies, the a priori ordering approach requires that the researcher specify the order in which the independent variables enter into the analysis. This order is, of course, determined on the basis of theoretical considerations.

Study Suggestions

1. What is the meaning of the term "factorial experiment"?
2. Discuss the advantages of factorial experiments as compared to single-variable experiments.
3. In a factorial experiment, factors A, B, and C have 3, 3, and 5 categories, respectively. How many coded vectors are necessary to represent the following: (a) factor A; (b) factor C; (c) $A \times B$; (d) $B \times C$; (e) $A \times B \times C$?
    (*Answers*: (a) 2; (b) 4; (c) 4; (d) 8; (e) 16.)
4. In an experiment with two factors, A with 3 categories and B with 6 categories, there are 10 subjects per cell or treatment combination. What are the degrees of freedom associated with the following F ratios: (a) for factor A; (b) for factor B; (c) for the interaction $A \times B$?
    (*Answers*: (a) 2 and 162; (b) 5 and 162; (c) 10 and 162.)
5. When b coefficients obtained with orthogonal coding are tested for significance, what is in effect being tested?
6. In a factorial analysis with orthogonal coding, there are three coded vectors, 1, 2, and 3, for factor A. The zero-order correlations between each of these vectors and the dependent variable, Y, are: $r_{y1} = .36$; $r_{y2} = .41$; $r_{y3} = .09$. The total sum of squares, Σy^2, is 436.00.
    (a) What are the β's associated with each of the three coded vectors for factor A?
    (b) What is the proportion of variance accounted for by factor A?
    (c) What is the regression sum of squares for factor A?
    (*Answers*: (a) $\beta_1 = .36$, $\beta_2 = .41$, $\beta_3 = .09$; (b) .3058; (c) 133.3288.)
7. What are the properties of the regression equation obtained in a factorial analysis with effect coding?
8. In a factorial analysis effect coding was used. Factor B consists of four categories. In the regression equation obtained from the analysis, a (intercept) $= 8.5$. The three b coefficients associated with the coded vectors for factor B are: $b_1 = .5$, $b_2 = -1.5$, $b_3 = 2.5$. What are the means of the main effects of factor B?
    (*Answers*: $\bar{Y}_{B_1} = 9$, $\bar{Y}_{B_2} = 7$, $\bar{Y}_{B_3} = 11$, $\bar{Y}_{B_4} = 7$.)
9. In a factorial analysis with three factors, the following coded vectors were used: 1 and 2 for factor A; 3, 4, and 5 for factor B; 6, 7. and 8 for factor C. There were five subjects per cell. The proportion of variance accounted for by all the factors and their interactions is .37541. The proportion of variance accounted for by factor B is .23452. What is the F ratio for factor B?
    (*Answer*: $F = 24.05$, with 3 and 192 degrees of freedom.)
10. Distinguish between two analytic approaches for unequal cell frequencies in factorial designs. Under what conditions may each of them be used?

11. Using effect coding, analyze the following data obtained from a 2×3 factorial experiment:

	B_1	B_2	B_3
A_1	3	4	7
	5	3	8
	4	2	6
A_2	1	6	4
	3	6	2
	2	3	3

(a) What are the proportions of variance accounted for by the factors and by their interaction?
(b) What is the regression equation?
(c) What are the treatment effects for each of the categories of factor A and factor B?
(d) What are the F ratios for each of the factors and for their interaction?
(*Answers*: (a) factor $A = .12500$, factor $B = .18750$, $A \times B = .43750$; (b) $Y' = 4.00 + .67 X_1 - 1.00 X_2 + .00 X_3 + .33 X_4 - 1.67 X_5$; (c) $A_1 = .67$, $A_2 = -.67$, $B_1 = -1.00$, $B_2 = .00$, $B_3 = 1.00$; (d) for factor A, $F = 6.00$, with 1 and 12 *df*; for factor B, $F = 4.50$, with 2 and 12 *df*; for $A \times B$, $F = 10.50$, with 2 and 12 *df*.)

Trend Analysis: Linear and Curvilinear Regression

It has been demonstrated that multiple regression and the analysis of variance yield identical results. We have dealt, however, with categorical independent variables. For different teaching methods, for example, the choice between regression analysis and analysis of variance is quite arbitrary and will probably depend on one's familiarity with the two methods as well as the availability of computer programs and other computing facilities.[1]

As pointed out in Chapter 6, one cannot order objects on a categorical variable. The only operation possible is the assignment of each object to one of a set of mutually exclusive categories. Even though one may assign numbers to the different categories, as is done when creating dummy variables, the numbers are used solely for identification of group or category membership. Needless to say, however, experiments in the behavioral sciences are not limited to categorical variables. Many independent variables are continuous. In learning experiments, for example, one encounters continuous independent variables such as hours of practice, schedules of reinforcement, hours of deprivation, intensity of electrical shock, and the like. In studies over time or practice, one may observe so-called growth trends, which are often referred to as growth curves. The development of moral judgment in children is an example.

When the independent variable is continuous the choice between analysis

---

[1]We believe that the multiple regression approach is preferable since it is more flexible and allows more direct interpretation. What is more important, however, is that under certain circumstances the multiple regression approach is called for even when the independent variables are categorical. This was demonstrated in Chapter 8 for the case of unequal cell frequencies. When one analyzes both categorical and continuous variables, as in analysis of covariance, the multiple regression approach must be used.

of variance and regression analysis is not arbitrary. As shown in this chapter, when regression analysis is appropriate it yields a more sensitive statistical test than does the analysis of variance applied to the same data. The presentation is limited to the analysis of data with one continuous independent variable. Linear regression is presented first, followed by curvilinear regression.

Analysis of Variance with a Continuous Independent Variable

Let us assume that 15 subjects have been randomly assigned to five treatments in a learning experiment with paired-associates. The treatments vary so that one group is given one exposure to the list, a second group is given two expo- sures, and so on to five exposures for the fifth group. The dependent variable measure is the number of correct responses on a subsequent test. In this ex- ample the independent variable is continuous: different numbers of exposures to the list.

The data for the five groups, as well as the calculations of the analysis of variance are presented in Table 9.1. The F ratio of 2.10 with 4 and 10 degrees of freedom is not significant at the .05 level. Consequently, it is concluded that the hypothesis $\bar{X}_1 = \bar{X}_2 = \bar{X}_3 = \bar{X}_4 = \bar{X}_5$ cannot be rejected.[2] In this analysis

[2] We are not concerned here with the important distinction between statistical significance and meaningfulness, but with the two statistical analyses applied to the same data. It is possible that on the basis of meaningfulness one would conclude that the experiment should be replicated with larger n's in order to increase the power of the test.

TABLE **9.1**   FICTITIOUS DATA FOR A LEARNING EXPERIMENT AND ANALYSIS OF VARIANCE CALCULATIONS

	Number of Exposures					
	1	2	3	4	5	
	2	3	3	4	4	
	3	4	4	5	5	
	4	5	5	6	6	
ΣX:	9	12	12	15	15	$\Sigma X_t = 63$
$\bar{X}$:	3	4	4	5	5	$(\Sigma X_t)^2 = 3969$
						$\Sigma X_t^2 = 283$

$$C = \frac{3969}{15} = 264.60$$

$$\text{Total} = 283 - 264.60 = 18.40$$

$$\text{Between} = \frac{9^2 + 12^2 + 12^2 + 15^2 + 15^2}{3} - 264.60 = 8.40$$

Source	df	ss	ms	F
Between	4	8.40	2.10	2.10 (n.s.)
Within	10	10.00	1.00	
Total	14	18.40		

the continuous independent variable was treated as if it were categorical; that is, as if there were five distinct treatments. We now analyze the data treating the independent variable as continuous.

Linear Regression Analysis of the Learning Experiment

When doing a linear regression analysis, one should first establish that the data follow a linear trend. For linear regression to be applicable, the means of the arrays (or the five treatments in the present case) should fall on the regression line. It is possible, however, that even though the means of the population fall on the regression line, the means of the samples do not fall exactly on it, but are sufficiently close to describe a linear trend. The question is therefore whether there is a linear trend in the data, or, in other words, whether the deviation from linearity is statistically significant. If it is not statistically significant, linear regression analysis is appropriate. If, on the other hand, the deviation from linearity is statistically significant, one can still do an analysis in which the continuous variable is treated as a categorical variable—that is, an analysis of variance.[3] In what follows the methods of regression and analysis of variance are applied to the data in order to determine which of the two is more appropriate.

The data presented in Table 9.1 can be displayed for the purpose of regression analysis as in Table 9.2. Following the procedures outlined in Chapter 2

[3]The alternative of applying curvilinear regression analysis is dealt with later in the chapter.

TABLE **9.2**   DATA FROM THE LEARNING EXPERIMENT, LAID OUT FOR REGRESSION ANALYSIS

X	Y	XY
1	2	2
1	3	3
1	4	4
2	3	6
2	4	8
2	5	10
3	3	9
3	4	12
3	5	15
4	4	16
4	5	20
4	6	24
5	4	20
5	5	25
5	6	30
Σ: 45	63	204
Σ^2: 165	283	

we obtain

$$\Sigma y^2 = \Sigma Y^2 - \frac{(\Sigma Y)^2}{N} = 283 - \frac{(63)^2}{15} = 18.40$$

$$\Sigma x^2 = \Sigma X^2 - \frac{(\Sigma X)^2}{N} = 165 - \frac{(45)^2}{15} = 30.00$$

$$\Sigma xy = \Sigma XY - \frac{(\Sigma X)(\Sigma Y)}{N} = 204 - \frac{(63)(45)}{15} = 15.00$$

$$ss_{\text{reg}} = \frac{(\Sigma xy)^2}{\Sigma x^2} = \frac{(15.00)^2}{30.00} = 7.50$$

$$b = \frac{\Sigma xy}{\Sigma x^2} = \frac{15.00}{30.00} = .50$$

$$a = \bar{Y} - b\bar{X} = 4.20 - (.50)(3.00) = 2.70$$

Look back at Table 9.1 in which the between treatments sum of squares was found to be 8.40. The sum of squares due to deviation from linearity is calculated by subtracting the regression sum of squares from the between treatments sum of squares.

$$ss_{\text{dev}} = ss_{\text{treat}} - ss_{\text{reg}}$$

$$ss_{\text{dev}} = 8.40 - 7.50 = .90$$

The Meaning of the Deviation Sum of Squares

Before interpreting the results, the meaning of the sum of squares due to deviation from linearity should be explained. This is done with the aid of a figure as well as by direct calculation. In Figure 9.1 the 15 scores of Table 9.2 are plotted. The regression line is drawn following the procedures discussed in Chapter 2 and using the regression equation calculated above. The mean of each of the five arrays is symbolized by a circle. Note that while the circles are close to the regression line, none of them is actually on the line. The vertical distance between the mean of an array and the regression line is the deviation of that mean from linear regression. Since the regression line is expressed by the formula $Y' = 2.7 + .5X$, this equation can be used to calculate the predicted Y's for each of the X's:

$$Y'_1 = 2.7 + .5X_1 = (2.7) + (.5)(1) = 3.2$$

$$Y'_2 = 2.7 + .5X_2 = (2.7) + (.5)(2) = 3.7$$

$$Y'_3 = 2.7 + .5X_3 = (2.7) + (.5)(3) = 4.2$$

$$Y'_4 = 2.7 + .5X_4 = (2.7) + (.5)(4) = 4.7$$

$$Y'_5 = 2.7 + .5X_5 = (2.7) + (.5)(5) = 5.2$$

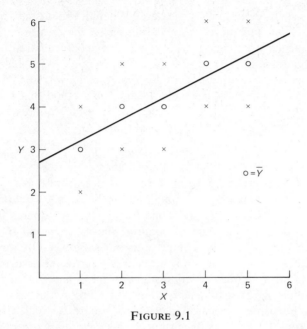

FIGURE 9.1

The five predicted Y's fall on the regression line and it is the deviation of the mean of each from its Y' that describes the deviation from linearity. They are

$$3.00 - 3.2 = -.2$$

$$4.00 - 3.7 = +.3$$

$$4.00 - 4.2 = -.2$$

$$5.00 - 4.7 = +.3$$

$$5.00 - 5.2 = -.2$$

In each case the predicted Y of a given array is subtracted from the mean of the Y's of that array. For example, the predicted Y for the array of X's with the value of 1 is 3.2, while the mean of the Y's of that array is 3.00 $[(2+3+4)/3]$. Squaring each deviation listed above, weighting the result by the number of scores in its array, and summing all the values yields the sum of squares due to deviation from regression:

$$(3)(-.2^2) + (3)(+.3^2) + (3)(-.2^2) + (3)(+.3^2) + (3)(-.2^2)$$

$$= .12 + .27 + .12 + .27 + .12 = .90$$

The same value (.90) was obtained by subtracting the regression sum of squares from the treatment sum of squares. When calculating the sum of squares due to deviation from linearity one is really asking the question: What is the difference between putting a restriction on the data so that they conform to a linear trend and putting no such restriction on the data? When the between treatments sum of squares is calculated there is no restriction on the trend of the treatment means. If the means are in fact on a straight line, the between treatments sum of squares will be equal to the regression sum of squares. With departures from linearity the between treatments sum of squares will always be larger than the regression sum of squares. What is necessary is a method that enables one to decide when the difference between the two sums of squares is sufficiently small to warrant the use of linear regression analysis.

Test of the Deviation Sum of Squares

The method of testing the significance of deviation from linearity is straight-forward. Instead of the one F ratio obtained for the between treatments sum of squares, two F ratios are obtained—one for the sum of squares due to linear regression and one for the sum of squares due to deviation from linearity. These two sums of squares are components of the between treatments sum of squares, as shown above. The sum of squares due to linear regression has 1 degree of freedom, while the deviation sum of squares has $k-2$ degrees of freedom (k = number of treatments). Dividing each sum of squares by its degrees of freedom yields a mean square. Each mean square is divided by the mean square error from the analysis of variance, thus yielding two F ratios. If the F ratio associated with the sum of squares due to deviation from linearity is not significant, one may conclude that the data describe a linear trend, and that the application of linear regression analysis is appropriate. If, on the other hand, the F ratio associated with the sum of squares due to deviation from linearity is significant, one may apply the analysis of variance.[4] This procedure, as applied to the data from the learning experiments is summarized in Table 9.3.

Note how the treatments sum of squares is broken down into two components, as are the degrees of freedom associated with the treatments sum of squares. The mean square for deviation from linearity (.30) is divided by the mean square error (1.00) to yield an F ratio of .30, which is not significant. One therefore concludes that the deviation from linearity is not significant, and that linear regression analysis is appropriate. The F ratio associated with the linear trend is 7.50. Since it is significant at the .01 level the researcher may conclude

[4]See footnote 3.

TABLE **9.3**   ANALYSIS OF VARIANCE TABLE: TEST FOR LINEARITY OF LEARNING EXPERIMENT DATA

Source	df	ss		ms	F	
Between Treatments	4	8.40				
Linearity	1		7.50	7.50	7.50	($< .01$)
Deviation from Linearity	3		.90	.30	< 1	
Within Treatments	10	10.00		1.00		
Total	14	18.40				

that there is a significant difference between the treatments. In effect, this conclusion means that the b weight is significant.

With an increase of a unit in X (the independent variable) there is a significant increment in Y (the dependent variable). It is important to note in the present example that the two analyses led to different conclusions. The analysis of variance led one not to reject the null hypothesis (Table 9.1), while the regression analysis led one to reject it (Table 9.3). This is because the two F ratios have the same denominators but different numerators. In Table 9.1 the numerator of the F ratio was obtained by dividing a sum of squares of 8.40 by 4 degrees of freedom, while in Table 9.3 a sum of squares of 7.50 was divided by 1 degree of freedom. Although it is true that the application of analysis of variance to data with continuous independent variables is valid, regression analysis applied to the same data will result in a more sensitive test and may thus yield significant results when the analysis of variance does not.

Multiple Regression Analysis of the Learning Experiment

The analysis of variance and the regression analysis of the same data were presented in detail in order to show clearly the process and meaning of testing deviation from linearity. We now demonstrate how the same results may be obtained in the context of multiple regression analysis. The necessary calculations have already been done. The basic approach is displayed in Table 9.4. Look first at the vectors Y and X. These are the same as the Y and X vectors in

TABLE **9.4**   DATA FROM LEARNING EXPERIMENT, LAID OUT FOR
MULTIPLE REGRESSION ANALYSIS

Treatment	Y	X	1	2	3	4	5
	2	1	1	0	0	0	0
1	3	1	1	0	0	0	0
	4	1	1	0	0	0	0
	3	2	0	2	0	0	0
2	4	2	0	2	0	0	0
	5	2	0	2	0	0	0
	3	3	0	0	3	0	0
3	4	3	0	0	3	0	0
	5	3	0	0	3	0	0
	4	4	0	0	0	4	0
4	5	4	0	0	0	4	0
	6	4	0	0	0	4	0
	4	5	0	0	0	0	5
5	5	5	0	0	0	0	5
	6	5	0	0	0	0	5

Table 9.3. Consequently we know (see calculations following Table 9.3) that $\Sigma xy = 15.00$; $\Sigma x^2 = 30.00$; $\Sigma y^2 = 18.40$. It is therefore possible to calculate r_{xy}.

$$r_{xy} = \frac{15.00}{\sqrt{30.00}\sqrt{18.40}} = \frac{15.00}{\sqrt{552}} = \frac{15.00}{23.4947} = .63844$$

$$r_{xy}^2 = (.63844)^2 = .40761$$

r_{xy}^2 indicates the proportion of variance in the Y scores attributed to the X scores.

Suppose, now, it is decided not to restrict the data to a linear regression. In other words, suppose a multiple regression analysis is calculated in which Y is the dependent variable, and group membership in the various treatments is the independent variable. It will be recalled from Chapters 6 and 7 that any method of coding group membership will yield the same results. Look now at Table 9.4 in which the vectors 1 through 5 describe group membership. We have used such coded vectors before. The coded vectors in Table 9.4, however, represent the different treatments of the experiment. Thus, group 1 received only one exposure to the list and is assigned a 1 in vector 1, while all other subjects are assigned a zero. Group 2 received two exposures and is assigned 2 in vector 2, and so forth for the remaining groups. It should be stressed again that it would make no difference in the overall results if we used 1's and 0's or any other coding method. There are advantages in using the present system, however, when working with a continuous independent variable.

Of the five vectors in Table 9.4, only four are independent. Suppose, therefore, that $R_{y.1234}^2$ is calculated. In effect a one-way analysis of variance can thus be done. $R_{y.1234}^2$ is equal to η_{yx}^2, or the ratio of the between treatments sum of squares to the total sum of squares. The numerical value of $R_{y.1234}^2$ is .45653. When a restriction of linearity was placed on the data it was found that the proportion of variance accounted for was .40761 (r_{yx}^2). When, on the other hand, no trend restrictions were placed on the data, the proportion of variance accounted for by X was .45653 ($R_{y.1234}^2$). It is now possible to test whether the increment in the proportion of variance accounted for is significant when no restriction is placed on the data. In other words, is the deviation from linearity significant? For this purpose we adapt formula (8.12) and restate it with a new number:

$$F = \frac{(R_{y.1234}^2 - R_{y.x}^2)/(k_1 - k_2)}{(1 - R_{y.1234}^2)/(N - k_1 - 1)} \tag{9.1}$$

where $R_{y.1234}^2$ = squared multiple correlation of the dependent variable, Y, and vectors 1 through 4 of Table 9.4; $R_{y.x}^2 = r_{yx}^2$ = squared correlation of Y with the vector X of Table 9.4, in which the independent variable is treated as continuous; k_1 = number of vectors associated with the first R^2; k_2 = number of vectors associated with the second R^2; N = number of subjects. The degrees of freedom for the F ratio are $k_1 - k_2$ and $N - k_1 - 1$ for the numerator and the denominator respectively.

Note that $R_{y.1234}^2 \geq R_{y.x}^2$. That is, $R_{y.1234}^2$ must be larger than or equal to

$R^2_{y.x}$. When the relation between Y and X is exactly linear, that is, when the Y means for all X values are on a straight line, $R^2_{y.1234} = R^2_{y.x}$. When, on the other hand, there is a deviation from linearity $R^2_{y.1234} > R^2_{y.x}$. It is this deviation from linearity that is tested by formula (9.1). For the data of Table 9.4,

$$F = \frac{(.45653 - .40761)/(4-1)}{(1-.45653)/(15-4-1)} = \frac{(.04892)/3}{(.54347)/10} = \frac{.01631}{.05435} = .30$$

The F ratio of .30 with 3 and 10 degrees of freedom is the same as that obtained before (see Table 9.3). We conclude, of course, that the deviation from linearity is not significant.

Using formula (9.1) we test for the significance of the linear trend: $F = .40761/.05435 = 7.50$, with 1 and 10 degrees of freedom. The numerator is r^2_{yx} and the denominator is the same as the error term used in calculating the F ratio for the deviation from linearity. The F ratio of 7.50 is the same as the F ratio obtained in Table 9.3 with the same degrees of freedom.

The procedure and calculation of the various sums of squares are illustrated below:

$$\text{Overall regression} = (R^2_{y.1234})(\Sigma y^2) = (.45653)(18.40) = 8.40$$

$R^2_{y.1234}$ indicates the proportion of variance accounted for by the overall regression when no restriction for trend is placed on the data. $\Sigma y^2 =$ total sum of squares of the dependent variable, Y. The sum of squares due to overall regression, 8.40, is equal to the between treatments sum of squares obtained in the analysis of variance (see Table 9.2).

$$\text{Linear regression} = (r^2_{yx})(\Sigma y^2) = (.40761)(18.40) = 7.50$$

The sum of squares due to deviation from linearity can, of course, be obtained by subtracting the regression sum of squares due to linearity from the overall regression sum of squares. Symbolically the sum of squares due to deviation from linearity is

$$(R^2_{y.1234})(\Sigma y^2) - (r^2_{yx})(\Sigma y^2) = (R^2_{y.1234} - r^2_{yx})(\Sigma y^2)$$

For the present data,

$$\text{Deviation from linearity} = (.45653 - .40761)(18.40) = .90$$

The sum of squares due to error is, as always, $(1-R^2)(\Sigma y^2)$; that is, the proportion of variance not accounted for multiplied by the total sum of squares. For the present data,

$$\text{Error} = (1 - R^2_{y.1234})(\Sigma y^2) = (1 - .45653)(18.40) = 10.00$$

All the sums of squares obtained above are identical to those obtained in Table 9.3.

The choice of the coding system used in the present example should be explained. Look at Table 9.4 and note that the X vector can be obtained by summing vectors 1 through 5. When using a computer program that enables one to generate new vectors, one need only punch the Y vector and vectors 1 through 5 of Table 9.4. With the aid of the program one can, in a single run, generate vector X by adding vectors 1 through 5, and obtain r^2_{yx}, and $R^2_{y.1234}$, as well as all the sums of squares calculated above.

Recapitulation

It has been shown that when the independent variable is continuous it is advisable to test possible departure from linearity. If the departure from linearity is significant, one can analyze the data by treating the continuous independent variable as a categorical variable. In other words, one can do multiple regression analysis with coded vectors representing the categories of the continuous independent variable. If, on the other hand, the deviation from linearity is not significant, linear regression analysis with the continuous variable is appropriate and will yield a more sensitive test.

It was also shown that one can perform the analysis either by using a combination of analysis of variance and regression analysis (as demonstrated in the beginning of the chapter) or by regression analysis only (as demonstrated in the latter part of the last section). The results are, of course, the same. Nevertheless, we recommend the use of regression analysis, since it affords clear and direct interpretation in the general context of multiple regression.

Curvilinear Regression Analysis

The presentation has been, until now, restricted to linear regression analysis. If the data depart significantly from linearity, one can do a multiple regression analysis in which the continuous variable is treated as a categorical variable. All that such an analysis can tell, however, is whether there is some trend in the data. If the researcher wants to study the nature of the trend he must use curvilinear regression analysis.

The Polynomial Equation

The method of curvilinear regression analysis is similar to linear regression analysis. The difference between the two approaches is in the regression equation used. Curvilinear regression analysis uses a polynomial regression equation. This means that the independent variable is raised to a certain power. The highest power to which the independent variable is raised indicates the degree of the polynomial. The equation

$$Y' = a + b_1 X + b_2 X^2$$

is a second-degree polynomial, since X is raised to the second power.

$$Y' = a + b_1 X + b_2 X^2 + b_3 X^3$$

is a third-degree polynomial equation.

The order of the equation indicates the number of bends in the regression curve. A first-degree polynomial, like $Y = a + bX$, describes a straight line. A second-degree polynomial describes a single bend in the regression curve, and is referred to as a quadratic equation. A third-degree polynomial has two bends and is referred to as a cubic equation. The highest order that any given equation may take is equal to $k-1$, where k is the number of distinct values in the independent variable. If, for example, a continuous independent variable consists of seven distinct values, these values may be raised to the sixth power. When this is done the regression equation will yield predicted Y's that are equal to the means of the different Y arrays, thus resulting in the smallest possible value for the residual sum of squares. In fact, when the highest-degree polynomial is used with any set of data the resulting R^2 is equal to η^2, since both analyses permit as many bends in the curve as there are degrees of freedom for the between treatments sum of squares.

One of the goals of scientific research, however, is parsimony. Our interest is not in the predictive power of the highest-degree polynomial equation possible, but rather in the highest-degree polynomial equation necessary to describe a set of data.[5]

We pointed out earlier that methods of curvilinear regression analysis are similar to those of linear regression analysis. Actually, the analysis can best be conceived as a series of steps, testing at each one whether a higher-degree polynomial adds significantly to the variance of the dependent variable accounted for. Suppose, for example, that one has a continuous variable with five distinct values. The increments in the proportion of variance accounted for at each stage are obtained as follows:

Linear:	$R^2_{y.x}$
Quadratic:	$R^2_{y.x,x^2} - R^2_{y.x}$
Cubic:	$R^2_{y.x,x^2,x^3} - R^2_{y.x,x^2}$
Quartic:	$R^2_{y.x,x^2,x^3,x^4} - R^2_{y.x,x^2,x^3}$

At each state, the significance of an increment in the proportion of variance accounted is tested with an F ratio. For the quartic element in above example the F ratio is

$$F = \frac{(R^2_{y.x,x^2,x^3,x^4} - R^2_{y.x,x^2,x^3})/(k_1 - k_2)}{(1 - R^2_{y.x,x^2,x^3,x^4})/(N - k_1 - 1)} \tag{9.2}$$

where N = number of subjects, k_1 = degrees of freedom for the larger R^2 (in the present case 4), k_2 = degrees of freedom for the smaller R^2 (in the present case 3). This type of F ratio has been used extensively in this book. Note, however, that the R^2 in the present example is based on one independent variable raised to a certain power, while in the earlier uses of the formula R^2 was based on several independent variables.

---

[5]It was demonstrated in the first part of this chapter that a linear equation was sufficient to describe a set of data. Using higher-degree polynomials on such data will not appreciably and significantly enhance the description of the data and the predictions based on regression equations derived from the data.

Curvilinear Regression: A Numerical Example

Suppose we are interested in the effect of time spent in practice on the performance of a visual discrimination task. Suppose, further, that subjects have been
randomly assigned to different levels of practice. After practice, a test of visual
discrimination is administered, and the number of correct responses recorded
for each subject. We focus attention on the relation between time practiced and
visual discrimination performance. In the fictitious data of Table 9.5 there are

TABLE **9.5**   FICTITIOUS DATA FROM A STUDY OF VISUAL
DISCRIMINATION[a]

Practice Time (in Minutes)					
2	4	6	8	10	12
4	7	13	16	18	19
6	10	14	17	19	20
5	10	15	21	20	21
Σ: 15	27	42	54	57	60
$\bar{Y}$: 5	9	14	18	19	20

[a]The dependent variable measure is the number of correct
responses.

three subjects for each of six levels of practice. Since there are six levels, the
highest-degree polynomial possible for these data is the fifth. The aim, however,
is to determine the lowest-degree polynomial that best fits the data.

In Table 9.6 the data are displayed in the form in which they were used for

TABLE **9.6**   FICTITIOUS DATA FROM VISUAL DISCRIMINATION STUDY
LAID OUT FOR CURVILINEAR REGRESSION[a]

Y	X	X^2	X^3	X^4	X^5
4	2	4	8	16	32
6	2	.	.	.	.
5	2	.	.	.	.
7	4	.	.	.	.
10	4	.	.	.	.
10	4	.	.	.	.
13	6	.	.	.	.
14	6	.	.	.	.
15	6	.	.	.	.
16	8	.	.	.	.
17	8	.	.	.	.
21	8	.	.	.	.
18	10	.	.	.	.
19	10	.	.	.	.
20	10	.	.	.	.
19	12	.	.	.	.
20	12	.	.	.	.
21	12	144	1728	20736	248832

[a]Y = Visual Discrimination. X = Practice time in minutes.

computer analysis.[6] The first and last values of X raised successively to higher powers are listed for illustrative purposes only. Only the vectors for Y and X are necessary. The remaining ones are generated by the computer. The pertinent results are now reported and discussed.

$R^2_{y.x,x^2,x^3,x^4,x^5} = .95144$. This means, of course, that 95 per cent of the variance in the dependent variable is accounted for by the fifth-degree polynomial of the independent variable. The analysis of variance is reported in Table 9.7. The F ratio of 47.0143, with 5 and 12 degrees of freedom, is signifi-

TABLE **9.7**   SUMMARY OF REGRESSION ANALYSIS FOR DATA OF
TABLE 9.6

Source of Variation	df	ss	ms	F
Due to Regression	5	548.50000	109.70000	47.0143
Deviation about Regression	12	28.00000	2.33333	
Total	17	576.50000		

cant beyond the .01 level. It should be noted that the same F ratio will be obtained if the data are subjected to a one-way analysis of variance. In fact, η^2 is equal to R^2 of the highest-degree polynomial possible, that is, .95144. At this stage, all we know is that there is a significant trend in the data. In order to see what degree polynomial fits these data, we report, in Table 9.8, another part of the results.

Tables of output similar to Table 9.8 were presented and discussed extensively in Chapter 8. Recall that the column labeled "Prop. of variance" indicates the variance accounted for by each variable after taking into account the variables preceding it. In other words, the proportions of variance are squared semipartial correlations, or the differences between squared multiple correlations. For example, the value of the proportion of variance in the third row, the one associated with the cubic term (X^3), is .00350. This means that the cubic term accounts for .35 percent of the variance in the dependent variable. The same statement can be made by saying that $R^2_{y.x,x^2,x^3} - R^2_{y.x,x^2} = .00350$. Each

[6]We chose to use one of the BMD (Dixon, 1970) programs, BMD03R, to show that any multiple regression program can be used, even though there is a special BMD program for polynomial regression (BMD05R).

TABLE **9.8**   REGRESSION COEFFICIENTS, SUM OF SQUARES ADDED, AND
PROPORTION OF VARIANCE ORIGINAL DATA OF TABLE 9.6[a]

Variable	b	ss	Prop. of Variance
X	5.12500	509.18571	.88324
X^2	−1.71875	34.32143	.05953
X^3	.40104	2.01667	.00350
X^4	−.03906	2.67857	.00465
X^5	.00130	.29762	.00052

[a]X = Practice time in minutes.

term can now be tested for significance by using the F test for the significance of an increment. The formula is the same as (9.1), above.

The error term is calculated first, since it is common to all stages of the testing. The error term is equal to $(1 - R^2_{y.k})/(N - k - 1)$. In the present example $R^2_{y.x,x^2,x^3,x^4,x^5} = .95144$. Therefore, $\text{Error} = (1 - .95144)/(18 - 5 - 1) = .04856/12 = .00405$. For the linear component, $F = .88324/.00405 = 218.08395$, which, with 1 and 12 degrees of freedom, is highly significant. For the quadratic component, $F = .05953/.00405 = 14.69876$, which, with 1 and 12 degrees of freedom, is significant beyond the .01 level. It is obvious that all other terms are not significant. We conclude, therefore, that the linear and the quadratic terms are significant, while the other terms are not. It should be noted that the F ratios associated with each term can also be obtained by using the column of the sum of squares of Table 9.8. This is done by dividing each sum of squares by the mean square for error. In the present case, this mean square is 2.33333 (see Table 9.7). The sum of squares for the linear component is 509.18571 (Table 9.8, row 1). For the linear component, $F = 509.18571/2.33333 = 218.22276$ (1/12 df). For the quadratic component: $F = 34.32143/2.33333 = 14.70920$ (1/12 df).

In the behavioral sciences it is rare to find significant trends beyond the quadratic. Moreover, the higher the degree of the polynomial the more it is affected by the unreliability of the measure involved, and the more difficult it is to interpret. The procedure that is therefore sometimes recommended is to test for the linear trend, the quadratic trend, and then combine all the remaining regression sum of squares and test for deviation from a quadratic trend. If the deviation is not significant, the analysis is terminated. If the deviation from the quadratic trend is significant one can still test succeeding components. The entire procedure for testing linear, quadratic, and deviation from quadratic is illustrated for the present example in Table 9.9. Even though the deviation from the quadratic trend was not significant, all other polynomial terms were tested to illustrate the complete set of tests.

TABLE **9.9**   ANALYSIS OF VARIANCE FOR TREND, VISUAL DISCRIMINATION DATA[a]

Source	df		ss	ms	F
Linear	1		509.18571	509.18571	218.22
Quadratic	1		34.32143	34.32143	14.71
Deviation from Quadratic	3		4.99286	1.66429	< 1
Cubic		1	2.01667	2.01667	< 1
Quartic		1	2.67857	2.67857	1.48
Quintic		1	.29762	.29762	< 1
Residual	12		28.00000	2.33333	
Total	Total	17	576.50000		

[a]The sums of squares for the different trend components were obtained from Table 9.8. The residual sum of squares was obtained from Table 9.7.

The Regression Equation

One may obtain a regression equation and use it for prediction. The equation will include only those terms of the polynomial that are found to be significant. In the present case only the linear and the quadratic trends were significant; therefore, only the regression weights associated with them will appear in the equation. It was shown earlier that when certain variables are deleted from the analysis, the regression coefficients change. Consequently, in order to obtain the regression equation one must do the analysis again with the linear and quadratic terms only.[7] In the recalculation of the regression equation the higher-order polynomials that were found to be not significant are relegated to the error term. In fact, whenever higher-order polynomials are not significant one may pool their sums of squares with the residual sum of squares. This generally results in a smaller error term for testing the terms retained in the equation, because the relatively small increase in the residual sum of squares is offset by an increase in the degrees of freedom. For the data of Table 9.9, the pooled residual sum of squares is 32.99286 ($28.00000 + 2.01667 + 2.67857 + .29762$), with 15 degrees of freedom. The residual mean square is therefore 2.19952, compared to 2.33333 in the original analysis. Consequently, the F ratios associated with the linear and the quadratic components are larger than those obtained in the earlier analysis: 231.50 for the linear, and 15.60 for the quadratic (compare with the corresponding values of Table 9.9).

The quadratic equation for the present data is

$$Y' = -1.90000 + 3.49464X - .13839X^2$$

For subjects practicing for 2 minutes,

$$Y' = -1.90000 + 3.49464(2) + (-.13839)(4) = 4.54$$

In other words, for subjects practicing for 2 minutes, the prediction is 4.54 correct responses on the visual discrimination test. For subjects practicing for 8 minutes,

$$Y' = -1.90000 + 3.49464(8) + (-.13839)(64) = 17.20$$

With the regression equation it is possible to interpolate and make predictions for values of the independent variable not used in the study, as long as these values are within the range of the values originally used. If one wanted, for example, to make a prediction for 5 minutes of practice (a condition not used in the experiment),

$$Y' = -1.90000 + 3.49464(5) + (-.13839)(25) = 12.11$$

For 5 minutes, the prediction is a score of 12.11 on the visual discrimination test. The same procedure may be applied to other intermediate values. One

---

[7]With some computer programs, it is possible to perform several analyses in one run by deleting variables. It is therefore possible to run simultaneously the linear and the quadratic analysis, as well as the highest possible degree of the polynomial.

should not, however, extrapolate beyond the range of values of the independent variable. That is, one should not make predictions for values of the independent variable that are outside the range used in the study. To indicate the danger of extrapolation, the scores of the present example are plotted in Figure 9.2. The open circles indicate the means of the arrays. Note that for the values 2, 4, 6, and 8 of the independent variable the trend is virtually linear. Had only these values been used in the study, one might have been led to believe that the trend is generally linear. As can be seen from Figure 9.2, however, the curve is quadratic. In sum, then, if one is interested in the effect of values outside the range of those under consideration, they should be included in the study or in a subsequent study.

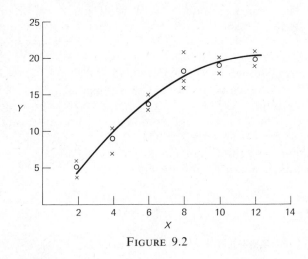

FIGURE 9.2

Curvilinear Regression Analysis and Orthogonal Polynomials

Curvilinear regression analysis may be done by using a set of orthogonal vectors coded to reflect the various degrees of the polynomials. The coefficients in such vectors are called *orthogonal polynomials*. The calculations involved in curvilinear regression analysis are considerably reduced when orthogonal polynomials are used instead of the original values of the continuous independent variable. Moreover, even though the results are the same in both approaches, their interpretation is simpler and clearer with orthogonal polynomials. The underlying principle of orthogonal polynomials is basically the same as that discussed with the orthogonal coefficients coding method (see Chapter 7). Orthogonal coefficients were constructed to contrast groups, whereas now the orthogonal coefficients are constructed to describe the different degrees of the polynomials.

When the levels of the continuous independent variable are equally spaced, and there is an equal number of subjects at each level, the construction of

orthogonal polynomials is simple. Rather than construct them himself, however, the researcher can obtain them from tables of orthogonal polynomials
reproduced in many statistics books. An extensive set of such tables can be
found in Fisher and Yates (1963). The magnitude of the difference between the
levels is immaterial as long as it is equal between all levels. In other words, it
makes no difference whether one is dealing with levels such as 2, 4, 6, and 8, or
5, 10, 15, and 20, or 7, 14, 21, and 28, or any other set of levels. The orthogonal
polynomial coefficients obtained from the tables apply equally to any set,
provided they are equally spaced and there is an equal number of subjects at
each level. Since the experimenter is interested in studying a trend, he is in a
position to make the levels of the continuous independent variable equally
spaced and to assign an equal number of subjects randomly to each level.[8]

Analysis of Visual Discrimination Data

To illustrate the method, the fictitious data of the visual discrimination experiment are now reanalyzed using orthogonal polynomials. These data and a set
of orthogonal vectors, whose coefficients were obtained from Fisher and
Yates (1963) are given in Table 9.10. Note that the sum of the coefficients in

[8]It is possible, although somewhat complicated, to construct orthogonal polynomial coefficients
for continuous variables that are not equally spaced, or when the number of subjects at each level
are not equal. For a treatment of this subject see Myers (1966) and Kirk (1968).

TABLE **9.10**   FICTITIOUS DATA FROM VISUAL DISCRIMINATION STUDY.
LAID OUT FOR ANALYSIS WITH ORTHOGONAL POLYNOMIALS

Y	1	2	3	4	5
4	−5	5	−5	1	−1
6	−5	5	−5	1	−1
5	−5	5	−5	1	−1
7	−3	−1	7	−3	5
10	−3	−1	7	−3	5
10	−3	−1	7	−3	5
13	−1	−4	4	2	−10
14	−1	−4	4	2	−10
15	−1	−4	4	2	−10
16	1	−4	−4	2	10
17	1	−4	−4	2	10
21	1	−4	−4	2	10
18	3	−1	−7	−3	−5
19	3	−1	−7	−3	−5
20	3	−1	−7	−3	−5
19	5	5	5	1	1
20	5	5	5	1	1
21	5	5	5	1	1
M: 14.16667	0	0	0	0	0
s: 5.82338	3.51468	3.85013	5.63602	2.22288	6.66863

each vector is zero, as is the sum of the cross products of any two vectors. These conditions, recall, are necessary to satisfy orthogonality. Look now at the pattern of the signs of the coefficients in each column. In column 1, Table 9.10, the signs change once from −5 to +5. In column 2, the signs change twice from +5 to +5. In column 3 they change three times from −5 to +5. These changes in signs correspond to the degree of the polynomial. Vector 1 has one sign change; it describes the linear trend. Vector 2 has two sign changes; it describes the quadratic trend. The other vectors are handled similarly.

It is now possible to calculate the multiple correlation of Y with the coded vectors as the independent variables. $R^2_{y.12345} = .95144$, the same value obtained in previous calculations. Relevant results are reported in Table 9.11. The sum of squares and the proportion of variance accounted for by each degree of the polynomial are equal to those obtained earlier (see Table 9.8). In the present case, however, since the correlations between all the coded vectors are zero, the proportion of variance accounted for by each vector is equal to the square of the zero-order correlation of a given vector with the dependent variable. R^2 is, of course, equal to the sum of the squared zero-order correlations of the coded vectors with the dependent variable. It will be recalled (see Chapter 7) that when the vectors are orthogonal to each other each t ratio associated with a given b weight is independently interpretable. For the linear component, for example, the t ratio is 14.77235 (Table 9.11, row 1). The degrees of freedom associated with each of the t ratios are equal to the degrees of freedom associated with the error term; in the present case $12 (N - k - 1 = 18 - 5 - 1)$. Since $t^2 = F$, then squaring the t ratio for the linear component should yield an F ratio equal to the one obtained in the previous analysis. $14.77235^2 = 218.22232$, with 1 and 12 degrees of freedom is indeed equal to the F ratio obtained earlier for the linear trend. Squaring each of the t ratios in Table 9.11 will yield the F ratios reported in Table 9.9. Note, however, that in the present analysis the t ratios are obtained directly, while in the previous analysis additional calculations were necessary to obtain the F ratios.

TABLE **9.11**   REGRESSION COEFFICIENTS, STANDARD ERRORS OF REGRESSION COEFFICIENTS, t RATIOS, SUM OF SQUARES ADDED, PROPORTION OF VARIANCE, AND ZERO-ORDER CORRELATIONS. VISUAL DISCRIMINATION DATA

Variable	b	s_b	t	ss	Prop. of Variance	r[a]
Linear (1)[b]	1.55714	.10541	14.77235	509.18571	.88324	.9398
Quadratic (2)	−.36905	.09623	−3.83526	34.32143	.05953	−.2440
Cubic (3)	−.06111	.06573	−.92967	2.01667	.00350	−.0591
Quartic (4)	.17857	.16667	1.07143	2.67857	.00465	.0682
Quintic (5)	.01984	.05556	.35714	.29762	.00052	.0227

[a]r = the correlation of each coded vector with the dependent variable.
[b]The numbers in the parentheses correspond to the column numbers in Table 9.10.

The Regression Equation

In the previous analysis it was noted that in order to obtain a regression equation that includes only the significant components of the trend, one has to reanalyze the data with the number of terms to be included in the regression equation. This was necessary because b weights change when the variables that are deleted are correlated with those remaining in the equation. In the case of orthogonal vectors, however, the b weights do not change regardless of the number of variables deleted. Furthermore, the intercept, a, is also not affected by deletion of variables — when the variables are orthogonal. Now,

$$a = \bar{Y} - b_i \bar{X}_i \qquad (9.3)$$

Since each orthogonal vector has, by definition, a mean of zero, a will always be equal to the mean of the dependent variable. In the present case the mean of Y is 14.66667 (see Table 9.10), and this is the value of the intercept, a. It is clear, therefore, that each dependent variable score is expressed as a composite of the mean of the dependent variable and the contribution of those components of the trend that are included in the regression equation.

To obtain the regression equation for any degree of the polynomial it is sufficient to read from Table 9.11 the appropriate b weights. The quadratic equation for the present data is therefore

$$Y' = 14.16667 + 1.55714 X_1 - .36905 X_2$$

Note that in the equation X_1 and X_2 are used to represent vectors 1 and 2 of Table 9.10. When using the regression equation for the purpose of prediction, the values inserted in it are the coded values that correspond to a given level and a given degree of the polynomial. For example, subjects who practiced for 2 minutes were assigned a -5 in the first vector (linear), and a $+5$ in the second vector (quadratic). For such subjects one would therefore predict

$$Y' = 14.16667 + 1.55714(-5) + (-.36905)(5) = 4.54$$

For subjects practicing for 8 minutes,

$$Y' = 14.1667 + 1.55714(1) + (-.36905)(-4) = 17.20$$

The same predicted values were obtained in the earlier calculations, when the original values of the independent values were inserted in the regression equation.

In sum, then, when coding is not used, the regression equation needs to be recalculated with the degree of polynomial desired. The values inserted in such an equation are those of the original variable. When orthogonal polynomials are used, on the other hand, one can obtain the regression equation at any degree desired without further calculation. The values inserted in the regression equation thus obtained are the coefficients that correspond to a given level and a given degree of the polynomial.

Calculation without a Computer

It should be evident that when using orthogonal polynomials it is fairly easy to do a curvilinear regression analysis even when a computer is not available. Simply calculate the zero-order correlation of the coded vectors with the dependent variable, the mean of the dependent variable and the standard deviations of the dependent variable and the coded vectors. Each squared zero-order correlation indicates the proportion of variance accounted for by a given component. R^2 is, as indicated earlier, the sum of the squared zero-order correlations. The b weights can be easily obtained by using the formula

$$b = r_{xy} \frac{s_y}{s_x} \tag{9.4}$$

For example, the b weight for the linear trend

$$b_{\text{lin}} = .9398 \frac{5.82338}{3.51468} = 1.55713$$

is the same value that was obtained above and reported in Table 9.11. The intercept is, as indicated in the previous section, equal to the mean of the dependent variable.

Trend Analysis with Repeated Measures

In the study of the effects of practice time on visual discrimination, the subjects were randomly assigned to different levels of practice. It is possible, however, to conduct a study in which all subjects practice at all levels. Assuming one has equivalent forms of the test, the subjects can be tested at the end of each practice period. Such a procedure is often preferable because it requires a smaller number of subjects and, more important, it provides better control. Each subject serves as his own control. The analysis of this type of design looks quite complicated when presented in the context of the analysis of variance. When considered in the context of curvilinear regression, however, it is fairly simple. All that is needed is to extend the type of analysis presented in this chapter by including coded vectors to identify the subjects.[9]

A Numerical Example

To illustrate the use of orthogonal polynomials with a repeated measures design, let us look again at the visual discrimination study. But this time let us assume that each subject practiced at all levels, and that tests were adminis-

[9]The statistical tests for repeated measured designs are based on some restrictive assumptions that are frequently not met. Consequently, some authors (for example, Greenhouse & Geisser, 1959) have proposed methods of adjusting α when the assumptions are violated. Other authors recommend that multivariate rather than univariate analysis be applied to repeated measures designs. Our purpose here is not to deal with the assumptions, but rather to show how one can apply coding to repeated measures designs for the purpose of the analysis. For a review of methods used in repeated measures designs, see Namboodiri (1972). For comparisons between univariate and multivariate analyses of repeated measures designs, see Davidson (1972).

tered at the conclusion of each practice period. The data reported in Table 9.5 were considered as collected from 18 subjects randomly assigned to six levels of practice. They are now considered, however, as collected from three subjects who experienced all the levels of the practice.[10]

Each row in Table 9.5 represents one subject. In Table 9.12 the original data from Table 9.5 are displayed for an analysis with repeated measures. Note that the Y vector and vectors 1 through 5 of Table 9.12 are identical to the vectors in Table 9.10. What makes Table 9.12 different is that it includes two additional vectors (6 and 7) that identify the subjects. Vector 6 identifies the first subject by assigning 1's to his scores and 0's to the scores of the other subjects. Similarly, vector 7 identifies the second subject. Since there are only three subjects, two vectors are necessary (the number of vectors necessary is always equal to the number of subjects minus one). The data of Table 9.12 were analyzed using Y as the dependent variable and the seven coded vectors as the independent variables. The coefficient of determination (R^2) is .98439. Look back now at the analysis with orthogonal polynomials and note that R^2 was .95144. Including information about the subjects increased the proportion

---

[10]We are using the same example because it was thoroughly analyzed and it therefore shows what happens when one controls for variance due to subjects. This kind of design is frequently used in learning experiments. The same group of subjects may, for example, be exposed to a number of trials in learning a task. Measures of performance are taken after each trial, or after blocks of trials. A trend analysis with repeated measures is then done.

TABLE **9.12**   FICTITIOUS DATA FROM VISUAL DISCRIMINATION STUDY LAID OUT FOR ANALYSIS WITH REPEATED MEASURES

Y	1	2	3	4	5	6	7
4	-5	5	-5	1	-1	1	0
6	-5	5	-5	1	-1	0	1
5	-5	5	-5	1	-1	0	0
7	-3	-1	7	-3	5	1	0
10	-3	-1	7	-3	5	0	1
10	-3	-1	7	-3	5	0	0
13	-1	-4	4	2	-10	1	0
14	-1	-4	4	2	-10	0	1
15	-1	-4	4	2	-10	0	0
16	1	-4	-4	2	10	1	0
17	1	-4	-4	2	10	0	1
21	1	-4	-4	2	10	0	0
18	3	-1	-7	-3	-5	1	0
19	3	-1	-7	-3	-5	0	1
20	3	-1	-7	-3	-5	0	0
19	5	5	5	1	1	1	0
20	5	5	5	1	1	0	1
21	5	5	5	1	1	0	0

[a]Y = dependent variable measures; 1 through 5 = orthogonal polynomials; 6 through 7 = coded vectors for subjects.

of variance accounted for by .03295 (.98439 − .95144). This may not seem to be a dramatic increase, but the important thing is what happens to the error term and consequently to the rest of the analysis.

In Table 9.13 the overall analysis of variance is reported. Compare Tables 9.7 and 9.13, and note particularly the difference in the error terms. While the

TABLE **9.13** ANALYSIS OF VARIANCE FOR THE MULTIPLE REGRESSION. VISUAL DISCRIMINATION DATA TREATED AS REPEATED MEASURES

Source of Variation	df	ss	ms	F
Due to Regression	7	567.50000	81.07143	90.08
Deviation about Regression	10	9.00000	.90000	
Total	17	576.50000		

earlier error term was 2.33333, with 12 degrees of freedom (Table 9.7), the error term now is .90000, with 10 degrees of freedom. When only one measure was available for each subject, the variability among subjects was part of the error term. With repeated measures, however, it is possible to identify the variance due to subjects and separate it from the error variance. In the present case there are three subjects, and therefore the variance due to subjects has 2 degrees of freedom. Hence, the change from an error term with 12 degrees of freedom to one with 10 degrees of freedom.

The difference between the two analyses is even more evident when one studies Table 9.14. Look at the first five rows of the table: the regression coefficients, the sums of squares, and the proportions of variance associated with the various components of the trend are identical to those obtained in the previous analysis (compare with Table 9.11). The sum of squares due to subjects is 19.00000 (Table 9.14, rows 6 and 7), with 2 degrees of freedom. The error sum of squares in the previous analysis was 28.00000, with 12 degrees of freedom (see Table 9.7). Since the sum of squares due to subjects is identified as being 19.00000, the error sum of squares in the present analysis becomes

TABLE **9.14** REGRESSION COEFFICIENTS, STANDARD ERRORS OF REGRESSION COEFFICIENTS, t RATIOS, SUM OF SQUARES ADDED, AND PROPORTION OF VARIANCE. VISUAL DISCRIMINATION DATA TREATED AS REPEATED MEASURES

Variable	b	s_b	t	ss	Prop. of Variance
Linear (1)[a]	1.55714	.06547	23.78575	509.18571	.88324
Quadratic (2)	−.36905	.05976	−6.17535	34.32143	.05953
Cubic (3)	−.06111	.04082	−1.49691	2.01667	.00350
Quadratic (4)	.17857	.10351	1.72516	2.67857	.00465
Quintic (5)	.01984	.03450	.57505	.29762	.00052
Subjects (6)	−2.50000	.54772	−4.56435	16.00000	.02775
Subjects (7)	−1.00000	.54772	−1.82574	3.00000	.00520

[a]The numbers in the parentheses correspond to the columns in Table 9.12.

9.00000, with 10 degrees of freedom. It is obvious that the difference between the two analyses is that in the repeated measures analysis the error sum of squares is reduced by extracting from it a systematic source of variance, namely the variance due to subjects. Even though this results in a loss of 2 degrees of freedom, the loss is more than offset by the considerable reduction in the error term. This is clearly seen in the t ratios associated with the first five b weights of Table 9.14. Although the b weights are identical with those obtained earlier, the t ratios are not. When the data are treated as repeated measures, the t ratios are in every instance larger. Take, for example, the t ratios for the linear trend. In the first analysis this t ratio was 14.77235, with 12 degrees of freedom (see Table 9.11), while in the present analysis the t ratio is 23.78575, with 10 degrees of freedom. The same results, of course, are obtained when working with the sums of squares or the R^2's. This is left as an exercise for the reader. In each case the obtained F ratio will be equal to the square of the corresponding t ratio reported in Table 9.14.[11]

We turn now to the sum of squares due to subjects, which is 19.00000, with 2 degrees of freedom. The mean square for subjects is therefore 9.5 (19/2). The mean square error is .9 (see Table 9.13). The F ratio due to subjects is therefore, $F = 9.5/.9 = 10.55555$, with 2 and 10 degrees of freedom. But the same F can be obtained by using the R^2's. Recall that R^2 was .95144 when the variance due to subjects was not extracted as a separate component. In the repeated measures analysis, R^2 is .98439. The F ratio for the between subjects is therefore

$$F = \frac{(.98439 - .95144)/(7 - 5)}{(1 - .98439)/(18 - 7 - 1)} = \frac{.03295/2}{.01561/10} = \frac{.016475}{.001561} = 10.55413$$

with 2 and 10 degrees of freedom. Within rounding errors, the same F ratio was obtained above. The F ratio between subjects is significant beyond the .01 level. This means, of course, that there are significant differences among subjects, or that having each subject serve as his own control significantly increases the proportion of variance accounted for.

It was pointed out above that the most dramatic effect of treating the data as repeated measures was a considerable reduction in the error term, thus making the analysis more powerful or more sensitive. This reduction in error is possible because people generally exhibit a certain amount of consistency in their responses to different treatment conditions, thus yielding variance that can be identified and separated from the sum of squares attributed to error.[12]

---

[11]This applies only to the first five t ratios, since in each case the corresponding F ratio has 1 degree of freedom for the numerator.

[12]If one is dealing with a categorical independent variable and repeated measures (instead of a continuous independent variable as in the present example), the analysis is similar. Coded vectors are generated for the categorical variable and for the subjects, and a regression analysis done. Needless to say, a trend analysis is not appropriate for categorical data. Instead, the sum of squares associated with the vectors reflecting the categorical variable are combined, as are the sums of squares for subjects. The same type of analysis applies to the randomized block design, where columns are coded for the categorical variable and rows for matched subjects.

Trend Analysis in Nonexperimental Research

The discussion and numerical examples presented thus far dealt with data obtained in experimental research. The method of studying trends, however, is equally applicable to data obtained in nonexperimental, or ex post facto, research. When, for example, a researcher studies the regression of one attribute variable on another, he must determine the trend of the regression. Using linear regression analysis only can lead to erroneous conclusions. A researcher may, for example, conclude on the basis of a linear regression analysis that the regression of variable Y on X is weak, when in fact it is strong but curvilinear.

A Numerical Example

A researcher was interested in studying the regression of satisfaction with a given job on mental ability. He selected a random sample of 40 employees to whom he administered an intelligence test. In addition, each employee was asked to rate his satisfaction with the job, using a 10-point scale, 1 indicating very little satisfaction, 10 indicating a great deal of satisfaction. The data (fictitious) are presented in Table 9.15, where Y is job satisfaction and X is intelligence.

TABLE **9.15**   FICTITIOUS DATA FOR INTELLIGENCE AND JOB SATISFACTION, $N = 40$[a]

Y	X	X^2	X^3	Y	X	X^2	X^3
2	90	8100	729000	9	104	10816	1124864
2	90	.	.	10	105	.	.
3	91	.	.	10	105	.	.
4	92	.	.	9	107	.	.
4	93	.	.	9	107	.	.
5	94	.	.	10	110	.	.
5	94	.	.	9	110	.	.
6	95	.	.	8	112	.	.
5	96	.	.	9	112	.	.
6	96	.	.	10	115	.	.
5	97	.	.	8	117	.	.
5	98	.	.	8	118	.	.
6	98	.	.	7	120	.	.
7	100	.	.	7	120	.	.
6	100	.	.	7	121	.	.
7	102	.	.	6	124	.	.
8	102	.	.	6	124	.	.
9	103	.	.	6	125	.	.
9	103	.	.	5	127	.	.
10	104	10816	1124864	5	127	16129	2048383

[a]Y = job satisfaction; X = intelligence.

In the examples presented earlier in the chapter, there were several distinct values of the independent variable. Moreover, the researcher was able to select values that were equally spaced with an equal number of subjects at each level. Consequently, the study of trends was simplified by the use of orthogonal polynomials. The researcher could fit the highest-degree polynomial possible or go to any desired level. In nonexperimental research, however, attribute variables may have many distinct values, unequally spaced,[13] with unequal numbers of subjects at each level. The procedure, therefore, is to raise the independent variable successively to higher powers, to calculate at each stage the squared multiple correlation of Y with the independent variable raised to a given power, and to test at each stage whether a higher-degree polynomial adds significantly to the proportion of variance accounted for in Y. This procedure is now applied to the fictitious data of Table 9.15. Note that the first and last values of X^2 and X^3 in each column are given for illustrative purposes. Only vectors Y and X are necessary. The remaining vectors are generated by the computer program.

We test first whether the regression is linear. This is accomplished by testing $r^2_{yx} = R^2_{y.x}$ in the usual manner. $R^2_{y.x} = .13350$. Therefore,

$$F = \frac{.13350/1}{(1-.13350)/(40-1-1)} = \frac{.13350}{.02280} = 5.86$$

with 1 and 38 degrees of freedom, significant at the .05 level. If we were to terminate the analysis at this point, we would have concluded that the regression of job satisfaction on intelligence is linear, and that about 13 percent of the variance in satisfaction is accounted for by intelligence. Furthermore, since the sign of r_{yx} is positive, we would have concluded that the higher the intelligence of a subject the more he tends to be satisfied with the job.

Calculating the squared multiple correlation for a second-degree polynomial, however, we obtain $R^2_{y.x,x^2} = .89141$, a dramatic increase in the proportion of variance accounted for. The increment due to the quadratic component is

$$R^2_{y.x,x^2} - R^2_{y.x} = .89141 - .13350 = .75791$$

This increment can of course be tested for significance in the usual manner:

$$F = \frac{(.89141-.13350)/(2-1)}{(1-.89141)/(40-2-1)} = \frac{.75791}{.00293} = 258.67$$

which, with 1 and 37 degrees of freedom, is a highly significant F ratio.

The squared multiple correlation for a third-degree polynomial is .89194.

---

[13]"Unequally spaced" does not mean that the measure used is not an interval scale, but rather that not all values within a given range are observed in a given sample. Consequently, the observed values are not equally spaced. Trend analysis relies heavily on the assumption that the independent variable measures form an interval scale. When one knows, or even suspects, serious departures from this assumption, it is advisable not to do a trend analysis. At the least, extra caution must be used with the interpretation of the data.

The increment due to the cubic component is

$$R^2_{y.x,x^2,x^3} - R^2_{y.x,x^2} = .89194 - .89141 = .00053$$

Even though this increment is minute, we test it for significance for illustrative purposes:

$$F = \frac{(.89194 - .89141)/(3-2)}{(1-.89194)/(40-3-1)} = \frac{.00053}{.00300} < 1$$

The F ratio is less than 1. Even if the F ratio were significant, a possibility that may occur when N is very large, one would not consider such a small increment meaningful and would not include the cubic trend despite its level of statistical significance.

For reasons indicated below, it is advisable to test one more polynomial beyond the first nonsignificant one. In the present example we do not report the increment due to the quartic component, which is neither significant nor meaningful. As pointed out earlier, trends beyond the quadratic are rarely observed in the behavioral sciences. When the reliabilities of the measures are not high, moreover, it is advisable not to go beyond the quadratic trend.

Revised F Ratios

The F ratios obtained in the preceding analysis were based on different error terms, because with the successive appropriations of variance by higher-order polynomials the error term is decreased successively. Proceeding with the analysis in the manner outlined above may result in declaring a component to be nonsignificant because the error term used at the given stage includes variance that may be due to higher-order polynomials yet to be identified. The error term for the linear component, for example, includes variance that may be appropriated by higher-order polynomials. It is for this reason that it was suggested above that an additional degree of polynomial be tested beyond the first nonsignificant one.

When the analysis outlined above is terminated and a decision is made about the degree of polynomial that best fits the data, it is necessary to recalculate the F ratios for the components retained in the equation, using a common error term for each component. This procedure is illustrated for the example.

Recall that the trends beyond the quadratic were not significant. $R^2_{y.x,x^2}$ is .89141. The common error term for the linear and quadratic components is therefore $(1-.89141)/(40-2-1) = .00293$, with 37 degrees of freedom. The revised F ratio for the proportion of variance accounted for by the linear component is

$$F = \frac{.13350}{.00293} = 45.56$$

with 1 and 37 degrees of freedom. Compare this F ratio with the F ratio obtained in the earlier analysis (5.86, with 1 and 38 degrees of freedom). The difference, of course, results from the great reduction in the error term due to the identification of a quadratic trend. As noted above, a given component that may be found nonsignificant in the initial analysis may be shown to be significant when the revised F ratio is based on the error term associated with the highest-degree polynomial that best fits the data. It is not necessary to recalculate the F ratio for the quadratic term, since it was based on the common error term obtained above. This always applies to the last component retained in the analysis. It is the F ratios for all components but the last retained that need to be recalculated.

The Regression Equations

The regression equations for curvilinear regression analysis are obtained in the manner shown in earlier chapters for linear regression (see, in particular, Chapter 4). When doing the analysis by computer the user may do several analyses in one run. Some computer programs (for example, BMD03R) enable the user to raise a variable to any power desired and do several analyses in one run so that each successive analysis includes a higher-order polynomial.

Parts of the computer output for the present analysis are summarized in Table 9.16. The minor discrepancies between the F ratios reported in Table 9.16 and those reported earlier are due to rounding errors.

The intercept, a, for the linear regression equation is $-.84325$ and the a for the quadratic equation is -199.03392. Obtaining the b's from Table 9.16 it is possible now to write the two regression equations. For the linear regression

TABLE **9.16**   LINEAR AND QUADRATIC REGRESSION ANALYSIS OF JOB SATISFACTION DATA[a]

I: Linear Regression

	b	Prop. of Variance	df	ss	ms	F
Linear (X)	.07197	.13350	1	25.95249	25.95249	5.85
Residual		.86650	38	168.44751	4.43283	
Total		1.00000	39	194.40000		

II: Quadratic Regression

	b	Prop. of Variance	df	ss	ms	F
Linear (X)	3.77115	.13350	1	25.95249	25.95249	45.49
Quadratic (X^2)	$-.01707$	.75791	1	147.33780	147.33780	258.25
Residual		.10859	37	21.10971	.57053	
Total		1.00000	39	194.40000		

[a]Original data in Table 9.15.

analysis only, the regression equation is

$$Y' = -.84325 + .07197X$$

The quadratic equation is

$$Y' = -199.03392 + 3.77115X - .01707X^2$$

The data for the present example are plotted in Fig. 9.3, along with the two regression curves. Note how the curve for the quadratic trend fits the data much better than does the one for the linear regression analysis. On the basis of the present analysis it can be concluded that subjects of relatively low or relatively high intelligence tend to be less satisfied with the job, as compared with subjects of average intelligence. One may speculate that the type of job under study is moderately demanding intellectually and therefore people of average intelligence seem to be most satisfied with it.

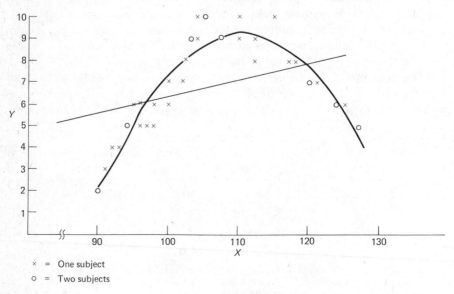

× = One subject
○ = Two subjects

FIGURE 9.3

Three Research Examples

We now summarize three studies in which trend analysis was used. You are urged to search the literature in your field of interest for additional examples. When reviewing the literature you will probably encounter instances in which data lent themselves to trend analysis, but were subjected instead to cruder analyses. We showed in the beginning of this chapter that data subjected to trend analysis may be found statistically significant, whereas a cruder analysis of the same data may yield results that are not statistically significant.

The Effect of Induced Muscular Tension on Heart Rate and Performance on a Learning Task

Wood and Hokanson (1965) tested an aspect of the theory of physiological activation, which states that subjects under moderate levels of tension will perform better than subjects under no tension. Under high levels of tension, however, the theory predicts a decrement in performance. Wood and Hokanson hypothesized that there is a linear relation between muscular tension and heart rate: increased muscular tension leads to increased heart rate. The researchers hypothesized further that there is a quadratic relation between muscular tension and performance on a simple learning task (a digit symbol task). In other words, increased muscular tension will lead to higher performance on a learning task up to an optimal point, after which further increase in such tension will lead to a decline in performance on the task.

Subjects were assigned to five different levels of induced muscular tension. Changes in heart rate and the learning of digit symbols were subjected to trend analyses. Both hypotheses were supported. Specifically, for the heart rate only the linear trend was significant, while for the digit symbols only the quadratic trend was significant.

Compare these results with those that would have been obtained had Wood and Hokanson done only an analysis of variance treating muscular tension as a categorical variable with five categories. For digit symbols, for example, the authors report an F ratio of 4.5417, with 4 and 76 degrees of freedom, for the differences between the five levels of induced tension. When the analysis for trend is done, however, the F ratio for the linear trend is less than one. The F ratio for the quadratic trend is 16.0907, with 1 and 76 degrees of freedom. The F ratio for the cubic trend is 0.00, and the F ratio for the quartic trend is slightly larger than one. It is thus seen clearly that the trend for the digit symbol data is, as predicted by the authors, quadratic.

Sex, Age, and the Perception of Violence

Moore (1966) used a stereoscope to present a viewer with pairs of pictures simultaneously. One eye was presented with a "violent" picture, and the other eye was presented with a "nonviolent" picture. One of the pairs, for example, was a mailman and a man who had been knifed. Under such conditions of binocular rivalry, binocular fusion takes place: the subject sees only one picture. Various researchers have demonstrated that binocular fusion is affected by cultural and personality factors. Moore hypothesized that when presented with pairs of violent–nonviolent pictures in a binocular rivalry situation, males will see more violent pictures than will females. Moore further hypothesized that there is a positive relation between age and the perception of violent pictures, regardless of sex.

Subjects in the study were males and females from grades 3, 5, 7, 9, 11, and college freshmen. (Note that grade is a continuous independent variable with six levels.) As predicted, Moore found that males perceived significantly more violent pictures than did females, regardless of the grade level. Furthermore, within each sex there was a significant linear trend between grade (age) and the perception of violent pictures. Moore interpreted his findings within the context of differential socialization of sex roles across age.

Involvement, Discrepancy of Information, and Attitude Change

There is a good deal of evidence that relates attitude change to discrepancy of new information about the object of the attitude. That is, in some studies it has been found that the more discrepant new information about an attitude object is from the attitude held by an individual, the more change there will be in his attitude toward the object. In addition, other studies have considered the initial involvement of the individual with the object of the attitude. Freedman (1964) hypothesized that under low involvement the relation between the discrepancy of information and attitude change is monotonic. This means, essentially, that as the discrepancy between the information and the attitude held increases, there is a tendency toward an increase in attitude change. In any event, an increase in the discrepancy will not lead to a decrease in attitude change. With high involvement, however, Freedman hypothesized that the relation is non-monotonic: with increased discrepancy between information and attitude there is an increase in attitude change up to an optimal point. Further increase in discrepancy will lead to a decrease in attitude change, or what has been called a "boomerang" effect.

Freedman induced the conditions experimentally and demonstrated that in the low involvement group only the linear trend was significant, whereas in the high involvement group only the quadratic trend was significant. In the high involvement group moderate discrepancy, as predicted, resulted in the greatest attitude change. Earlier we warned against extrapolating from a curve. Freedman maintains that the relation between discrepancy and attitude change is nonmonotonic also when the level of involvement is low. In other words, he claims that in the low involvement group the relation is also quadratic. That he obtained a linear relation in the low involvement group Freedman attributes to the range of discrepancy employed in his study. He claims that with greater discrepancy a quadratic trend will emerge in the low involvement group also. Be this as it may, we repeat that one should not extrapolate from the linear trend. To test Freedman's notions one would have to set up the appropriate experimental conditions.

Summary

It is a sign of relatively sophisticated theory when predictions derived from it are not limited to statements about differences between conditions or treatments, but also address themselves to the pattern of the differences. At the initial stages of formulating a theory one can probably state only that the phenomena under study differ under different conditions. A relatively sophisticated theory generally provides more specific predictions about relations among variables.

The methods presented in this chapter provide the means for testing predicted trends in the data. When cruder analyses are applied to data for which a trend analysis is appropriate, the consequences may be failure to support the hypothesis being tested or, at the very least, a loss of information.

The studies summarized in the previous section make it clear that the relation between theory and analytic method is close. It was the use of trend analysis that enabled the researchers to detect relations predicted by theory. Trend analysis is, of course, also useful when a researcher has no hypothesis about the specific pattern of relations among the variables under study, but wishes to learn what the pattern is. The discovery of trends may lead the researcher to reformulate theory and conduct subsequent studies to test such reformulations.

It was pointed out that for trend analysis the independent variable has to be continuous. In addition, the measurement of the variable should have relatively high reliability, or trends may seem to appear when in fact they do not exist, or trends that do exist may be overlooked.

In sum, trend analysis is a powerful technique that, when appropriately applied, can enhance the predictive and explanatory power of scientific inquiry.

Study Suggestions

1. Why is it not advisable to transform a continuous variable into a categorical variable?
2. What conditions must be satisfied for linear regression analysis to be applicable?
3. When a continuous independent variable has seven distinct values, how many degrees of freedom are associated with the sum of squares due to deviation from linearity?
    (*Answer*: 5.)
4. Under what conditions is $\eta_{yx}^2 = r_{yx}^2$?
5. In a study with a continuous independent variable consisting of eight distinct values, the following results were obtained: proportion of variance due to overall regression = .36426; proportion of variance due to linear regression = .33267. The total number of subjects was 100.
    Calculate F ratios for the following: (a) overall regression; (b) linear regression; (c) deviation from linearity.
    (*Answers*: (a) $F = 7.53$, with 7 and 92 *df*; (b) $F = 48.14$, with 1 and 92 *df*; (c) $F = .76$, with 6 and 92 *df*.)
6. When a continuous independent variable has six distinct values, what is the highest-degree polynomial that can be fitted to the data?
    (*Answer*: 5.)
7. In a study with a continuous independent variable, a third-degree polynomial was fitted. Some of the results are: $R_{y.x}^2 = .15726$; $R_{y.x,x^2}^2 = .28723$; $R_{y.x,x^2,x^3}^2 = .31626$. The total number of subjects was 150. Calculate the F ratios for the following components: (a) linear; (b) quadratic; (c) cubic.

(*Answers*: (a) $F = 33.60$, with 1 and 146 df; (b) $F = 27.77$, with 1 and 146 df; (c) $F = 6.20$, with 1 and 146 df.)

8. Why should one not extrapolate from a curve?
9. Discuss the advantages of using orthogonal polynomials.
10. A continuous independent variable consists of seven distinct values equally spaced. There is an equal number of subjects for each value of the independent variable. Using an appropriate table, indicate the orthogonal polynomials for the following components: (a) linear; (b) quadratic; (c) cubic.

    (*Answers*: (a) linear: $-3\ -2\ -1\ \ 0\ \ 1\ \ 2\ \ 3$; (b) quadratic: $5\ \ 0\ -3\ -4$ $-3\ \ 0\ \ 5$; (c) cubic: $-1\ \ 1\ \ 1\ \ 0 -1 -1\ \ 1$.)

11. Discuss the advantages of using repeated measures on the same subjects. In what types of studies are repeated measures particularly useful?
12. A researcher studied the regression of risk-taking on ego-strength. To a sample of 25 subjects he administered a measure of risk-taking and one of ego-strength. The data (fictitious) are as follows:

Risk-Taking	Ego-Strength	Risk-Taking	Ego-Strength
2	1	10	6
3	1	11	6
4	2	11	7
4	2	12	7
5	2	12	7
5	3	12	8
5	3	12	8
6	3	11	8
8	4	12	9
8	4	12	9
9	5	12	10
10	5	12	10
10	5		

What are the proportions of variance accounted for by the following components: (a) linear; (b) quadratic; (c) cubic?

What are the initial F ratios associated with the following components: (d) linear; (e) quadratic; (f) cubic? (g) What is the degree of polynomial that best fits the data?

What are the revised F ratios for the following components: (h) linear; (i) quadratic?

Interpret the results.

(Answers: (a) linear $= .88542$; (b) quadratic $= .08520$; (c) cubic $= .00342$; (d) linear: $F = 177.80$ with 1 and 23 df; (e) quadratic: $F = 63.58$ with 1 and 22 df; (f) cubic: 2.76 with 1 and 21 df; (g) quadratic; (h) linear: $F = 660.76$ with 1 and 22 df; (i) quadratic: $F = 63.58$ with 1 and 22 df.)

Continuous and Categorical Independent Variables, Interaction, and Analysis of Covariance

In the preceding chapters analysis of data derived from categorical and continuous independent variables was discussed and illustrated, but the two kinds of variables were treated separately. In the present chapter they are treated together: we now discuss the multiple regression analysis of data with both continuous and categorical independent variables. The present chapter thus serves as an integration of methods that have been treated by some researchers as distinct and by others as even incompatible.

The chapter begins with an example of an analysis in which one independent variable is continuous and another is categorical. Next, the use of a continuous variable to achieve better control or to study aptitude-treatment interaction is discussed. This is followed by discussion and illustration of tests for trends when the data are derived from continuous and categorical independent variables. The chapter concludes with a discussion and an illustration of the analysis of covariance, which is shown to be a special case of the general methods presented.

Analysis of Data from an Experiment in Retention

A researcher is studying the effect of an incentive on the retention of subject matter. He is also interested in the effect on retention of time devoted to study. Another question he wishes to pursue is whether there is an interaction between the two variables in their effect on retention. Subjects are randomly assigned to two groups, one receiving and the other not receiving an incentive. Within these groups, subjects are randomly assigned to 5, 10, 15, or 20 minutes of study of a passage specifically prepared for the experiment. At the end of the study period, a test of retention is administered. A fictitious set of data from such an experiment is reported in Table 10.1. Note that one of the variables, Incentive–No Incentive, is categorical, while the other variable, Study Time, is continuous.

TABLE **10.1**   FICTITIOUS DATA FROM A RETENTION EXPERIMENT WITH ONE CONTINUOUS AND ONE CATEGORICAL VARIABLE

Treatments	Study Time (in Minutes)				
	5	10	15	20	
No Incentive	3	4	5	7	
	4	5	6	8	
	5	6	8	9	$\bar{Y}_{\text{No Inc.}} = 5.83$
Incentive	7	9	8	10	
	8	10	11	11	
	9	11	12	13	$\bar{Y}_{\text{Inc.}} = 9.92$
$\bar{Y}$:	6.00	7.50	8.33	9.67	$\bar{Y}_t = 7.875$

In order to lay the groundwork for discussion of the analysis, the data have been plotted in Figure 10.1. Open circles identify subjects who received an incentive (I), while crosses identify subjects who received no incentive (NI). The regression lines of retention on study time for the two groups are also drawn in the figure. Two questions may be asked about these regression lines. The first is whether the slopes of the two lines (indicated by the b coefficients)

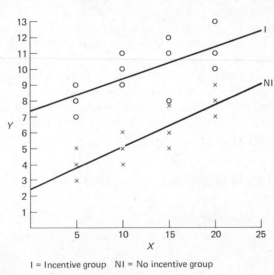

I = Incentive group   NI = No incentive group
X = Study time   Y = Retention
FIGURE 10.1

are equal. Stated differently, are the two lines parallel? Equality of slopes means that the effect of the continuous variable (Study Time) is the same in both groups. Assuming the b's to be equal, one can ask the second question: Are the intercepts (a's) of the two regression lines equal? The second question is addressed to the elevation of the regression lines. Equality of intercepts means that a single regression line fits the data for both groups, so that there is really no difference between them. If, on the other hand, the b's are equal while the a's are not, this indicates that one group is superior to the other group along the continuum of the continuous variable.

From Figure 10.1 it is evident that the regression lines are not parallel. It is possible, however, that the departure from parallelism is due to chance. This hypothesis can be tested by testing the significance of the difference between the b's. If the b's are not significantly different, one can then test the significance of the difference between the a's.

The calculations of the regression equations for the Incentive and No Incentive group are summarized in Table 10.2. The regression equation for the Incentive group is

$$Y' = 7.33330 + .20667X$$

and for the No Incentive group

$$Y' = 2.49996 + .26667X$$

While the b's are quite alike, there is a marked difference between the a's. The procedures for testing differences between regression coefficients and differences between intercepts are outlined below and applied to the data.

Tests of Differences between Regression Coefficients

As discussed in earlier chapters (see, in particular, Chapter 6), a test of significance can be conceived as an attempt to answer the question: Does additional information add significantly to the "explanation" of the variance of the dependent variable? Applied to the topic under discussion, the question is: Does using separate regression coefficients for each group add significantly to the regression sum of squares, as compared to the regression sum of squares obtained when a common regression coefficient is used?

A common regression coefficient for several groups may be calculated by the following formula:

$$b_c = \frac{\Sigma xy_1 + \Sigma xy_2 + \cdots + \Sigma xy_k}{\Sigma x_1^2 + \Sigma x_2^2 + \cdots + \Sigma x_k^2} \tag{10.1}$$

where b_c = common regression coefficient; Σxy_1 = sum of the products in group 1, and similarly for all other terms in the numerator; Σx_1^2 = sum of the squares in group 1, and similarly for all other terms in the denominator. Note that the numerator in (10.1) is the pooled sum of products within groups, while the denominator is the pooled sum of squares within groups. For the present

TABLE **10.2**   CALCULATION OF REGRESSION EQUATIONS FROM THE
RETENTION EXPERIMENT

No Incentive			Incentive		
Y	X	XY	Y	X	XY
3	5	15	7	5	35
4	5	20	8	5	40
5	5	25	9	5	45
4	10	40	9	10	90
5	10	50	10	10	100
6	10	60	11	10	110
5	15	75	8	15	120
6	15	90	11	15	165
8	15	120	12	15	180
7	20	140	10	20	200
8	20	160	11	20	220
9	20	180	13	20	260
Σ: 70	150	975	119	150	1565
M: 5.83333	12.500		9.91667	12.500	
Σ^2: 446	2250		1215	2250	

$$\sum xy = 975 - \frac{(70)\,(150)}{12} = 100 \qquad \sum xy = 1565 - \frac{(119)\,(150)}{12} = 77.5$$

$$\sum x^2 = 2250 - \frac{(150)^2}{12} = 375 \qquad \sum x^2 = 2250 - \frac{(150)^2}{12} = 375$$

$$b = \frac{\Sigma xy}{\Sigma x^2} = \frac{100}{375} = .26667 \qquad b = \frac{\Sigma xy}{\Sigma x^2} = \frac{77.5}{375} = .20667$$

$$a = \overline{Y} - b\overline{X} = 5.83333 \qquad a = \overline{Y} - b\overline{X} = 9.91667$$
$$\quad - (.26667)(12.5) = 2.49996 \qquad \quad - (.20667)(12.5) = 7.33330$$

$$Y' = 2.49996 + .26667X \qquad Y' = 7.33330 + .20667X$$

$$ss_{reg} = \frac{(\Sigma xy)^2}{\Sigma x^2} = \frac{(100)^2}{375} = 26.66667 \qquad ss_{reg} = \frac{(\Sigma xy)^2}{\Sigma x^2} = \frac{(77.5)^2}{375} = 16.01667$$

example (see Table 10.2)

No Incentive group: $\Sigma xy = 100.00$; $\Sigma x^2 = 375.00$

Incentive group: $\Sigma xy = 77.50$; $\Sigma x^2 = 375.00$

$$b_c = \frac{77.50 + 100.00}{375.00 + 375.00} = .23667$$

Recall that the calculation of a regression coefficient is based on the principle of least squares: b is calculated so that the sum of the squared residuals is minimized. This, of course, results in maximizing the regression sum of squares. Now, when regression lines are parallel the b's are obviously identical. Consequently, the sum of the regression sums of squares obtained from using each b

for its own group is the same as the regression sum of squares obtained from using a common b for all groups.

When, however, regression lines are not parallel, the common b is not equal to the separate b's. Since the b for each group provides the best fit for the group data, the sum of the regression sums of squares obtained from using the separate b's is larger than the regression sum of the squares obtained from using a common b. The discrepancy between the sum of the regression sums of squares obtained from separate b's and the regression sum of squares obtained from a common b is due to the departure from parallelism of the regression lines of the separate groups. When the increment in the regression sum of squares due to the use of separate b's is not significant, it is concluded that there are no significant differences between the b's. In other words, the common b is tenable for all the groups.

In the calculations of Table 10.2 separate b's were used for each group. The regression sum of squares for the No Incentive group is 26.66667 and for the Incentive group: 16.01667. The sum of these regression sums of squares is 42.68334. The regression sum of squares due to a common b may be obtained as follows:

$$ss_{\text{reg}} \text{ for common } b = \frac{(\text{pooled } \Sigma xy)^2}{\text{pooled } \Sigma x^2} \qquad (10.2)$$

For the present data we obtain

$$\frac{(77.50 + 100.00)^2}{375.00 + 375.00} = \frac{177.50^2}{750.00} = 42.00833$$

The discrepancy between the sum of the regression sums of squares for the separate b's and the regression sum of squares for the common b is $42.68334 - 42.00833 = .67501$. It is this value that is tested for significance.

The foregoing presentation was meant as an explanation of the approach to testing the difference between the b's. Although the procedure presented can, of course, be used to do the calculations, we demonstrate now how the analysis is done in the context of the procedures of preceding chapters.

In Table 10.3 the data from the retention experiment are displayed for such an analysis. Vector 1 in the table identifies group membership: No Incentive (NI) and Incentive (I). In vector 2, the values of the continuous variable Study Time (ST) are recorded for the NI subjects, while the I subjects are assigned 0's. The reverse is true of vector 3: the NI subjects are assigned 0's, and the I subjects are assigned the values of ST. Vector 4 is obtained by adding vectors 2 and 3. In other words, vector 4 contains the values of the continuous variable of all subjects.

We now calculate $R^2_{y.123}$ and obtain the proportion of variance accounted for by group membership (NI and I) and separate vectors for Study Time for each of the groups. $R^2_{y.123} = .82679$. In addition, we calculate $R^2_{y.14}$ and obtain the proportion of variance accounted for by group membership (NI and I) and a single vector in which Study Time of all subjects is contained. $R^2_{y.14} = .82288$.

The difference between these two R^2's is the increment in the proportion of variance accounted for by using separate vectors for the continuous variable as compared to the use of a single vector. This increment is $.82679 - .82288 = .00391$. Recall that a proportion of variance accounted for is expressed as a regression sum of squares when it is multiplied by the sum of squares for the dependent variable, Σy^2. From Table 10.3, $\Sigma y^2 = 172.625$. The increment in regression sum of squares due to the use of the separate vectors for Study Time is

$$(.00391)(172.625) = .67496$$

Note that the same value, within rounding error, was obtained in the earlier calculations.

That the present analysis accomplishes the same purpose as the previous analysis can also be seen from a comparison of the regression coefficients of

TABLE **10.3**   DATA FROM THE RETENTION EXPERIMENT, LAID OUT FOR REGRESSION ANALYSIS[a]

Treatment	Y	1	2	3	4
	3	1	5	0	5
	4	1	5	0	5
	5	1	5	0	5
	4	1	10	0	10
	5	1	10	0	10
No Incentive	6	1	10	0	10
	5	1	15	0	15
	6	1	15	0	15
	8	1	15	0	15
	7	1	20	0	20
	8	1	20	0	20
	9	1	20	0	20
	7	−1	0	5	5
	8	−1	0	5	5
	9	−1	0	5	5
	9	−1	0	10	10
	10	−1	0	10	10
Incentive	11	−1	0	10	10
	8	−1	0	15	15
	11	−1	0	15	15
	12	−1	0	15	15
	10	−1	0	20	20
	11	−1	0	20	20
	13	−1	0	20	20

Σy^2: 172.625

[a]Y = measures of retention originally presented in Table 10.1; 1 = coded vector for Incentive–No Incentive; 2 = Study Time of subjects under No Incentive; 3 = Study Time of subjects under Incentive; 4 = Study Time of all subjects.

the two analyses. When $R^2_{y.123}$ is calculated $b_2 = .26667$ and $b_3 = .20667$. The same regression coefficients for No Incentive and Incentive respectively were obtained in the calculations of Table 10.2. When $R^2_{y.14}$ is calculated $b_4 = .23667$, which is equal to b_c (common regression coefficient) obtained above.

The increment in the proportion of variance accounted for by using separate b's as compared to using a common b can, of course, be tested by the F test used frequently in earlier chapters. Adapting formula (9.1) for the present problem,

$$F = \frac{(R^2_{y.123} - R^2_{y.14})/(3-2)}{(1 - R^2_{y.123})/(24-3-1)}$$

$$= \frac{(.82679 - .82288)/(3-2)}{(1 - .82679)/(24-3-1)} = \frac{.00391/1}{.17321/20} = .45$$

As noted above, the increment in the proportion of variance accounted for by using separate b's, as compared to a common b, is .00391. The F ratio associated with this increment is smaller than one and therefore not significant. We conclude that the b's are not significantly different. In other words, the effect of Study Time on retention is not significantly different in the No Incentive and the Incentive groups. The increment due to the separate b's (.00391) is relegated to the error term.

Test of the Common Regression Coefficient

Having demonstrated that the b weights do not differ significantly and that one can therefore use a common b weight, the question is whether this b is significant. In the present context, the question is whether the continuous variable, Study Time, adds significantly to the variance in the dependent variable over and above the variance due to group membership (NI and I). This is done by comparing the regression of Y on variables 1 and 4, both group membership and the continuous variable, to the regression of Y on variable 1, group membership only. For the present data,

$$R^2_{y.14} = .82288 \qquad R^2_{y.1} = .57953$$

$$F = \frac{(.82288 - .57953)/(2-1)}{(1 - .82288)/(24-2-1)} = \frac{.24335/1}{.17712/21} = \frac{.24335}{.00843} = 28.87$$

With 1 and 21 degrees of freedom, this is a highly significant F ratio. It is concluded that the continuous variable adds significantly to the proportion of variance accounted for. Or, the addition of variable 4, Study Time, adds significantly to the regression of Y, retention, on variable 1, group membership (NI and I).

Test of the Difference between Intercepts

A test of the difference between intercepts is performed only after it has been established that the b weights do not differ significantly.[1] Only then does it

---

[1]When the b's are significantly different an interaction between the independent variables is indicated. This topic is treated later in the chapter.

make sense to ask whether one of the treatments is more effective than the other along the continuum of the continuous variable. Testing the difference between intercepts amounts to testing the difference between the treatment effects of the categorical variable. This test, too, is accomplished by testing the difference between two R^2's, or between two proportions of variance. It is done, in effect, by noting whether there is a significant difference in the proportion of variance accounted for by fitting a single regression line to the data as compared to fitting separate regression lines.

In the present context, we wish to determine whether knowledge of the categorical variable (NI and I) adds significantly to the proportion of variance accounted for over and above the proportion accounted for by the continuous variable (ST). We therefore test the difference between $R^2_{y.14}$ and $R^2_{y.4}$. (1 is the vector for the categorical variable; 4 is the vector for the continuous variable. See Table 10.3.) For the present data we obtain

$$R^2_{y.14} = .82288 \qquad R^2_{y.4} = .24335$$

$$F = \frac{(.82288 - .24335)/(2-1)}{(1 - .82288)/(24-2-1)} = \frac{.57953/1}{.17712/21} = \frac{.57953}{.00843} = 68.75$$

with 1 and 21 degrees of freedom, a highly significant F ratio. It is concluded that the two groups (NI and I) do not have a common intercept. In other words, there is a significant difference between the two intercepts. The F ratio of 68.75 indicates that the difference between the means of the No Incentive and the Incentive groups, 5.83 and 9.92, respectively, is significant.

In the foregoing analysis the categorical variable had only two categories. The same procedure applies when the categorical variable consists of more than two categories. When this is the case, it is obviously necessary to generate a number of coded vectors equal to the number of categories minus one, or the number of degrees of freedom associated with the categorical variable. The analysis is then done in the manner shown above. First, one tests whether there is a significant departure from parallelism among the regression lines. If the lines can be considered parallel, one tests whether there are significant differences among their intercepts.

When differences among intercepts are significant, it is necessary to do multiple comparisons to determine which of the categories, or the treatments, differ significantly from each other. The multiple comparisons are done in the manner described in chapter 7 [formulas (7.15) and (7.16)], except that the MSR (mean square residual) in the kind of analysis used here is based on $N - k - 2$ degrees of freedom, instead of $N - k - 1$ of formula (7.16), where k = degrees of freedom associated with the categorical variable. The loss of an additional degree of freedom in the MSR is due to the use of a continuous variable in the present analysis. It is also possible to do orthogonal comparisons among the treatments of the categorical variable in the manner shown in Chapter 7. Recall, however, that such comparisons are appropriate only when they are formulated prior to the analysis.

Relative Contributions to the Variance

Since the present study consisted of two manipulated variables with equal cell frequencies, the independent variables are orthogonal. It is therefore possible to state unambiguously the proportion of variance accounted for by each of the independent variables. Both variables account for about 82 percent of the variance ($R^2_{y.14} = .82288$). Of this 82 percent, the variable Incentive–No Incentive accounts for about 58 percent, while the variable Study Time accounts for about 24 percent. (See the above calculations. Also $r^2_{y1} = .57953$ and $r^2_{y4} = .24335$.)

It will be recalled that the researcher also wanted to know whether there was a significant interaction between the two variables. This question was answered when it was found that the b's did not differ significantly. In fact, it was found that using two separate b weights adds less than 1 percent to the variance, as compared to the variance accounted for by the common b weight. (See the earlier discussion and the test of significance for the difference between the b's.)

Categorizing Continuous Variables

Behavioral scientists frequently partition continuous independent variables into dichotomies, trichotomies, and so on, and then analyze the data as if they came from discrete categories of an independent variable. Several broad classes of research in which there is a tendency to partition continuous variables can be identified. The first class may be found in experiments in which a manipulated variable is continuous and is treated in the analysis as categorical. The retention experiment discussed in this chapter is an illustration. One variable was categorical (Incentive–No Incentive), while the other was continuous (Study Time) with four levels. Instead of doing an analysis of the kind discussed earlier, some researchers will treat the continuous variable as composed of four distinct categories and analyze the data with a 2×4 factorial analysis of variance.[2]

The second class of research studies in which one frequently encounters categorization of a continuous variable is what is referred to as the treatments-by-levels design. In such designs the continuous variable is primarily a control variable. For example, a researcher may be interested in the difference between two methods of instruction. Since the subjects differ in intelligence, he may wish to control this variable. One way of doing this is to block or create groups with different levels of intelligence and to randomly assign an equal number of subjects from each level to each of the treatments. The levels are then treated as distinct categories and a factorial analysis of variance is done. The purpose of introducing the levels into the design is to decrease the error term. This is done by identifying the sum of squares due to intelligence, thereby reducing the error sum of squares. A more sensitive F test for the main effect, methods of instruc-

---

[2]Whether one uses the conventional calculating methods or performs the calculations by the methods presented in this book (see Chapter 8), the results will of course be the same.

tion, is thereby obtained. The reduction in the error term depends on the correlation between the continuous variable and the dependent variable. The larger the correlation the greater the reduction will be in the error term.

The third class of studies in which researchers tend to categorize continuous variables is similar to the second class just discussed. While the categorization in the second class was primarily motivated by the need for control, the categorization in this class of studies is motivated by an interest in the possible interaction between the independent variables. This approach is referred to as Aptitude-Treatment Interaction (ATI). It is important in the behavioral sciences, since it may help identify relations that will otherwise go unnoticed. Behavioral scientists and educators have frequently voiced concern that while lip service is paid to the fact that people differ and that what may be appropriate for some may not be appropriate for others, little has been done to search for and identify the optimal conditions of performance for different groups of people. (For a discussion and analysis of studies from the frame of reference of ATI, See Bracht, 1970; Berliner & Cahen, in press; Cronbach & Snow, 1969.)

If, for example, a researcher who is studying the effectiveness of different teaching methods includes in his study the variable of intelligence not only for the purposes of control but also for the purpose of seeking possible interactions between teaching methods and intelligence, he is working in the ATI framework. Note that the point of departure of the researcher is a search for optimal methods of teaching subjects with different levels of intelligence. The analysis is the same as in the treatments-by-levels design, that is, a factorial analysis of variance. In both cases the researcher studies the main effects and the interaction. The difference between the two approaches is in the conceptualization of the research. Does the researcher include the continuous variable primarily for the purpose of control or for the purpose of studying interaction?

One may point to various areas in which categorization of variables in the framework discussed above is done. With achievement motivation, for example, researchers measure need achievement, dichotomize it into high and low need achievement, and treat it as a categorical variable. Researchers also tend to partition variables like authoritarianism, dogmatism, cognitive style, self-esteem, and ego-defense similarly.

There are two major questions on the categorization of continuous variables: On what basis does one categorize? and What effect does the categorization have on the analysis? The first question cannot be easily answered, since categorization is a somewhat arbitrary process. One researcher may choose to split his group at the median and label those above the median as high and those below the median as low. All subjects within a given subgroup are treated as if they had identical scores on what is essentially a continuous variable. This is particularly questionable when the variability of the continuous measure is relatively large. Moreover, in a median split, a difference of one unit on the continuous variable may result in labeling a subject as high or low. In order to avoid this possibility, some researchers create a middle group and use it in the analysis or ignore it altogether. There are other variations on the theme. One can take the continuum of intelligence, for example, and create as many categories as one fancies or believes to be appropriate. There are no solid principles to guide the categorization of continuous variables. It can be said, however, that one should not categorize a continuous variable because there is nothing to be gained; indeed there is danger of loss.

It is possible that some of the conflicting evidence in the research literature of a given area may be attributed to the practice of categorization of continuous variables. For example, let us assume that two researchers are independently studying the relation between dogmatism and susceptibility to a prestigious source. Suppose, further, that the researchers follow identical procedures in their research designs. That is, they have the same type of prestigious source, the same type of suggestion, the same number of subjects, and so on. They administer the Dogmatism Scale to their subjects, split them at the median and create a high Dogmatism and a low Dogmatism group. Subjects from each of these groups are then randomly assigned to a prestigious or a nonprestigious source. The result is a 2×2 factorial analysis of variance—two sources and two levels of Dogmatism. Note, however, that the determination of "high" and "low" is entirely dependent on the type of subjects involved. It is true that in relation to one's group it is appropriate to say that a subject is "high" or "low." But it is possible that the "highs" of the first researcher may be more like the "lows" of the second researcher due to differences in samples. When the two studies are reported, it is likely that little specific information about the subjects is offered. Instead, reporting is generally restricted to statements about high and low dogmatists, as if this were determined in an absolute fashion rather than relative to the distribution of dogmatism of a given group. It should therefore not come as a surprise that under such circumstances the "highs" in one research study behave more like the "lows" in the other study, thus leading to conflicting results in what are presumably similar studies.

The answer to the second question—What effects does the categorization have on the analysis?—is clear-cut. Categorization leads to a loss of information, and consequently to a less sensitive analysis. For example, in the research illustration on dogmatism and susceptibility to a prestigious source, the researcher is interested in the relation between these two variables. Having dichotomized dogmatism into high and low, he can make statements about differences between the two subgroups. His attempts to estimate the relation between the variables, however, will be limited due to the reduction in the variability of dogmatism resulting from the categorization.

As mentioned above, all subjects within a category are treated alike even though they may have originally been quite different on the continuous variable. For example, if one's cutting score for the high group on intelligence is 115, then all subjects above this score are considered alike on intelligence. In the subsequent analysis no distinction is made between a subject whose score was 115, and another whose score was, say, 130. But if, in the first place, the choice of the continuous variable (intelligence) was made because of its relation to the dependent variable, then one would expect a difference in performance between the two subjects, even though they were given the same treatment. It is this loss of information about the differences between subjects, or the reduction in the variability of the continuous variable, that leads to a reduction in the sensitivity of the analysis.

A Numerical Example

In order to illustrate the decrease in the sensitivity of the analysis caused by the categorization process, we reanalyze the data of the retention experiment reported in Table 10.1. Now, however, the continuous variable, Study Time, is treated as a categorical variable with four partitions. Think of the data as a treatment-by-levels design, or an ATI design. For example, instead of the variable Incentive–No Incentive (the categorical variable in the retention experiment reported in Table 10.1) think of two methods of teaching, or two sources of information, or two methods of attitude change. Instead of Study Time (the continuous variable in the retention experiment) think of four levels of intelligence, or four levels of authoritarianism.[3]

The data from Table 10.1 are displayed in Table 10.4 for an analysis in which both independent variables are treated as categorical variables. The procedures for such an analysis were discussed and illustrated in Chapter 8 and are therefore not repeated here. Note that vector 1 in Table 10.4 identifies the Incentive–No Incentive variable. Vectors 2, 3, and 4 express the variable Study Time (treated now as a categorical variable). Vectors 5, 6, and 7 express the interaction between the independent variables. In Chapter 8 it was demonstrated that when the frequencies in each cell are equal, the vectors identifying each variable and the interaction are mutually orthogonal, when orthogonal or effect coding is used. Note that effect coding is used in Table 10.4. For the present data,[4]

$$R^2_{y.1234567} = R^2_{y.1} + R^2_{y.234} + R^2_{y.567}$$
$$.83780 = .57953 + .24596 + .01231$$

The sum of squares associated with each term can be easily obtained by multiplying the proportion of variance by the total sum of squares ($R^2\Sigma y^2$). The analysis of the data is summarized in Table 10.5.

Note that while the proportion of variance accounted for by Incentive–No Incentive is identical to the one obtained in the earlier analysis (.57953), the F ratios associated with this proportion are different. In the earlier analysis the F ratio was 68.75, with 1 and 21 degrees of freedom. In the present analysis the F ratio associated with the same variable is 57.17, with 1 and 16 degrees of

[3]That the values associated with the continuous variable in this example are 5, 10, 15, and 20 should pose no problem. If you wish, add, for example, a constant of 100 to each of the values, and you may now think of the scores as IQ's. The results will not be affected by the addition of a constant.
[4]There is a slight discrepancy between the total R^2 obtained here (.83780) and the one obtained earlier between the dependent variable and the categorical and continuous variables (.82288). The source of this discrepancy is discussed later in the chapter.

TABLE **10.4** DATA FROM THE RETENTION EXPERIMENT, LAID OUT FOR ANALYSIS IN WHICH THE INDEPENDENT VARIABLES ARE TREATED AS CATEGORICAL VARIABLES[a]

Y	1	2	3	4	5 (1×2)	6 (1×3)	7 (1×4)
3	1	1	0	0	1	0	0
4	1	1	0	0	1	0	0
5	1	1	0	0	1	0	0
4	1	0	1	0	0	1	0
5	1	0	1	0	0	1	0
6	1	0	1	0	0	1	0
5	1	0	0	1	0	0	1
6	1	0	0	1	0	0	1
8	1	0	0	1	0	0	1
7	1	-1	-1	-1	-1	-1	-1
8	1	-1	-1	-1	-1	-1	-1
9	1	-1	-1	-1	-1	-1	-1
7	-1	1	0	0	-1	0	0
8	-1	1	0	0	-1	0	0
9	-1	1	0	0	-1	0	0
9	-1	0	1	0	0	-1	0
10	-1	0	1	0	0	-1	0
11	-1	0	1	0	0	-1	0
8	-1	0	0	1	0	0	-1
11	-1	0	0	1	0	0	-1
12	-1	0	0	1	0	0	-1
10	-1	-1	-1	-1	1	1	1
11	-1	-1	-1	-1	1	1	1
13	-1	-1	-1	-1	1	1	1

Σy^2: 172.625

[a]Y = measures of retention originally given in Table 10.1; 1 = coded vector for Incentive–No Incentive; 2, 3, and 4 = coded vectors for Study Time; 5, 6, and 7 = vectors expressing the interaction between Incentive–No Incentive and Study Time.

freedom. Not only is there a decrease in the size of the F ratio, but there is also a decrease in the degrees of freedom associated with the denominator. The difference between the two analyses is even more dramatic when one considers the variable Study Time. When treated as a continuous variable, Study Time accounts for .24335 of the variance. When treated as a categorical variable, it

TABLE **10.5** ANALYSIS OF VARIANCE SUMMARY TABLE.
RETENTION EXPERIMENT DATA. BOTH VARIABLES TREATED AS
CATEGORICAL[a]

Source	Prop. of Variance	ss	df	ms	F
Incentive—No Incentive $R^2_{y.1}$	.57953	100.04137	1	100.04137	57.17
Study Time $R^2_{y.234}$	.24596	42.45884	3	14.15295	8.09
Interaction $R^2_{y.567}$	.01231	2.12501	3	.70834	< 1
Error $(1 - R^2_{y.1234567})$	.16220	27.99978	16	1.74999	
Total	1.00000	172.62500	23		

[a]Original data given in Table 10.4.

accounts for much the same proportion of the variance, .24596. The F ratios for these two proportions, however, are quite different. In the earlier analysis, the F ratio was 28.87, with 1 and 21 degrees of freedom, while in the present analysis the F ratio is 8.09, with 3 and 16 degrees of freedom. This difference is mainly due to the difference in the degrees of freedom associated with each of the numerators of the F ratios. In the earlier analysis, the numerator of the F ratio was obtained by dividing R^2 by 1 degree of freedom. In the present analysis, on the other hand, almost the same R^2 is divided by 3 degrees of freedom, thus yielding a smaller numerator for the F ratio.

To demonstrate this more clearly, we calculate the F ratio for Study Time using the proportion of variance accounted for by this variable. For the present analysis, $R^2_{y.1234567} = .83780$. $R^2_{y.1567}$ (Incentive–No Incentive and Interaction) = .59184. For Study Time,

$$F = \frac{(.83780 - .59184)/(7-4)}{(1 - .83780)/(24 - 7 - 1)} = \frac{.24596/3}{.16220/16} = \frac{.08199}{.01014} = 8.09$$

Not surprisingly, the same F ratio was obtained when the sums of squares were used.[5]

While in both analyses the F ratios associated with the two variables are significant beyond the .01 level, it is evident that the analysis in which the continuous variable was categorized is less sensitive. It is quite possible, therefore, that treating a continuous variable in its original form in an analysis may result

[5]For the calculation of the F ratio when Study Time was treated as a continuous variable, see earlier parts of this chapter. Compare also the two error terms for the two F ratios and note that while the proportion due to error in the present analysis is slightly smaller than in the former analysis (.16620 and .17712, respectively), the denominator of the F ratio in the present analysis is larger than the one in the former analysis (.01014 and .00843, respectively). This is due to the loss of a larger number of degrees of freedom for the denominator in the present analysis (16, as compared to 21 in the former analysis).

in a significant F ratio at a prespecified level of significance, while the F ratio for the same variable may fail to reach the prespecified level of significance when the continuous variable is categorized.

Treating a continuous variable as categorical leads to loss of information. The analysis just completed, in which a continuous variable, Study Time, was transformed to a four-way partition, was less sensitive than the earlier analysis in which all the values of the continuous variable were used. The obvious conclusion is, of course: Do not partition continuous variables.

The Study of Interaction

To ask whether independent variables interact is, in effect, to ask about the model that best fits the data. When an interaction is not significant, an additive model is sufficient to describe the data. This means that a subject's score on the dependent variable is conceived as a composite of several additive components. In the most general case, these are: an intercept, treatment effects, and an error term. For the retention experiment, the additive model is

$$Y = a + b_1X_1 + b_4X_4 + e \tag{10.3}$$

where X_1 is group membership (NI and I), X_4 is the continuous variable (ST. See Table 10.3), and b_1 and b_4 are the regression coefficients associated with these vectors.

If, on the other hand, the interaction is significant, this means that an additive model is not adequate to describe the data. One needs to add terms that reflect the interaction. For the present example this may take the form

$$Y = a + b_1X_1 + b_4X_4 + b_5X_1X_4 \tag{10.4}$$

The difference between formulas (10.3) and (10.4) is that in the latter a term that is the product of the values of the two independent variables has been added (that is, X_1X_4). It is, of course, possible to determine whether this new term adds significantly to the proportion of variance accounted for, thereby testing whether the additive model is adequate or not. Before demonstrating how this is done, we discuss two types of interactions.

Ordinal and Disordinal Interaction

In the retention experiment analyzed above the interaction was found to be not significant. When, however, an interaction is significant, it is necessary to study it carefully in order to decide on a proper course of analytic action. A significant interaction per se does not tell the whole story. Lindquist (1953) and Lubin (1961) distinguish between ordinal and disordinal interactions. An *ordinal interaction* is one in which the "rank order of the treatments is constant," whereas a *disordinal interaction* is one in which the "rank order of the treatments changes" (Lubin, 1961, p. 808).

This distinction can best be illustrated graphically, as in Figure 10.2. In Figure 10.2(a), a situation with no interaction is depicted. The two regression

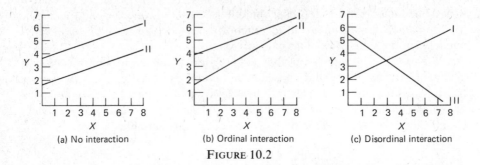

(a) No interaction   (b) Ordinal interaction   (c) Disordinal interaction

FIGURE 10.2

lines are parallel. There is a constant difference between Treatments I and II along the continuum of the continuous variable. In other words, the b weights for the two regression lines are identical, and the difference between the treatments is entirely accountable by the difference between the intercepts of the regression lines. In (b), while Treatment I is still superior to Treatment II along the continuum, it is relatively more effective at the lower end of the X variable than at the upper end. Note, however, that in all cases the rank order of the means of Treatment I is higher than that of the means of Treatment II. Thus this is an ordinal interaction.

In Figure 10.2(c), the regression lines cross. This is an example of a disordinal interaction. Treatment II is superior at the lower levels of X (up to 3), while Treatment I is superior at the upper regions of X (from 3 and up). At the value of $X = 3$, the two treatments seem to be equally effective. When the interaction is disordinal [as in (c)], it is not meaningful to speak of main effects (or differences between intercepts). One needs to qualify one's statements and specify at what levels of X Treatment I is superior to Treatment II, and at what levels of X the reverse is true.

It is obvious that if the regression lines of Figure 10.2(b) are extended they will cross each other as in a disordinal interaction. Therefore, the question is: When is an interaction considered ordinal and under what conditions is it considered disordinal? The answer lies in the range of interest of the researcher.

The Research Range of Interest

The research range of interest is defined by the values of the continuous variable (X) of relevance to the purposes of the research. For example, if the continuous variable is intelligence, the researcher may be interested in the IQ range of 90 to 110. In other words, it is for subjects within this range of intelligence that he wishes to make statements about the effectiveness of teaching methods or some other treatment.

The decision as to whether an interaction is ordinal or disordinal is based on the point at which the regression lines cross each other. If this point is outside the range of interest, the interaction is considered ordinal. If, on the other hand, the point at which the lines intersect is within the range of interest, then the interaction is considered disordinal. To illustrate, let us assume that for

Figure 10.2 the researcher's range of interest is from 1 to 8 on the X variable. It is evident that the regression lines in (b) do not intersect within the range of interest, while those in (c) cross each other well within the range of interest (at the point where $X = 3$).

Determining the Point of Intersection

It is possible to calculate the point at which the regression lines intersect. Note that at the point of intersection the predicted Y for Treatment I is equal to the predicted Y for Treatment II. When the regression lines are parallel a prediction of equal Y's for two treatments at a given value of X will not occur. The regression equations for two parallel lines consist of different intercepts and identical b weights. For example, assume that in a given research study consisting of two treatments (A and B) the regression lines are parallel. Assume further that the intercept for Treatment A is 7 while the intercept for Treatment B is 2, and that the b weights for each of the regression lines is .8. The two regression equations are

$$Y'_A = 7 + .8X$$
$$Y'_B = 2 + .8X$$

For any value of X the value of Y'_A will be 5 points higher than the value of Y'_B (this is the difference between the intercepts, and it is constant along the continuum of X).

Suppose, however, that the two equations are

$$Y'_A = 7 + .3X$$
$$Y'_B = 2 + .8X$$

An inspection of these two equations indicates that when the values of X are relatively small Y'_A will be larger than Y'_B. The reason is that the intercept plays a more important role relative to the b weight in the prediction of Y. But as X increases the b weight plays an increasingly important role thus offsetting the difference between the intercepts, until a point is reached where a balance is struck and $Y'_A = Y'_B$. Beyond that point, Y'_B will be larger than Y'_A. The point of intersection can be calculated with the following formula:

$$\text{Point of intersection } (X) = \frac{a_1 - a_2}{b_2 - b_1} \qquad (10.5)$$

where the a's are the intercepts of the regression lines, and the b's are the regression coefficients. For the above example, $a_1 = 7$, $a_2 = 2$, $b_1 = .3$, $b_2 = .8$.

$$X = \frac{7 - 2}{.8 - .3} = \frac{5}{.5} = 10$$

The point at which the lines intersect is at the value of $X = 10$. This is illustrated in Figure 10.3. If the range of interest in the research depicted in Figure 10.3 is from 3 to 15, then the interaction is disordinal since the lines intersect within this range.

The regression equations for the two lines of Figure 10.3 are now applied to several values of X in order to illustrate the major points made in the discussion above. For $X = 10$,

$$Y'_A = 7 + (.3)(10) = 10$$
$$Y'_B = 2 + (.8)(10) = 10$$

The same value is predicted for subjects under treatments A or B. This is because the lines intersect at $X = 10$. For $X = 5$,

$$Y'_A = 7 + (.3)(5) = 8.5$$
$$Y'_B = 2 + (.8)(5) = 6$$

The value of $X = 5$ is below the point of intersection, and the regression equation for Treatment A leads to higher predicted value of Y than does the regression equation for Treatment B. The reverse is true for values of X that are above the point of intersection. For example, for $X = 12$,

$$Y'_A = 7 + (.3)(12) = 10.6$$
$$Y'_B = 2 + (.8)(12) = 11.6$$

It is important that the point of intersection of the regression lines be quite removed from the range of interest in order for the researcher to be confident in his choice of treating the interaction as ordinal rather than disordinal. For example, suppose that in the retention experiment analyzed in the beginning of this chapter the interaction was significant. We repeat first the two regression equations obtained earlier.

$$Y'_I = 7.33330 + .20667X$$
$$Y'_{NI} = 2.49996 + .26667X$$

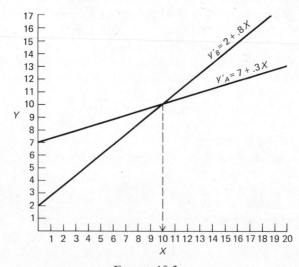

FIGURE 10.3

Recall that the range of interest in the retention experiment was from 5 to 20 minutes of study time. The point of intersection of the two regression lines is

$$\frac{7.33330 - 2.49996}{.26667 - .20667} = \frac{4.83334}{.06} = 80.56$$

At about 80 minutes of study time the lines will intersect. This is far removed from the range of interest. Had the interaction been significant, one could have concluded with confidence that it is ordinal. Predicted Y's at the point of intersection are

$$Y'_I = 7.33330 + (.20667)(80.56) = 23.98$$
$$Y'_{NI} = 2.49996 + (.26667)(80.56) = 23.98$$

In the context of the retention experiment one might have speculated that as study time increases considerably the differential effects of the treatments (Incentive–No Incentive) tend to disappear. To reiterate, however, the interaction in the retention experiment was not significant. The discussion and the calculations were presented for illustrative purposes only.

An Example with Interaction

It has been maintained that students' satisfaction with the teaching styles of their teachers depends, among other variables, on the students' tolerance of ambiguity. Specifically, students whose tolerance of ambiguity is relatively low prefer teachers whose teaching style is largely directive, while students whose tolerance of ambiguity is relatively high prefer teachers whose style is largely nondirective. To test this hypothesis students were randomly assigned to "directive" and "nondirective" teachers. In the beginning of the semester the students were administered a measure of tolerance of ambiguity on which the higher the score the greater the tolerance. At the end of the semester students rated their teacher on a 7-point scale, 1 indicating very little satisfaction, 7 indicating a great deal of satisfaction. Fictitious data for two classes, each consisting of 20 students, are reported in Table 10.6.

In Chapter 8 it was shown that cross-product vectors representing categorical variables indicate the interaction of such variables. The procedure is the same for the cases of categorical and continuous variables. In Table 10.6 we use dummy coding (vector 1) and effect coding (vector 2) for the categorical variable, teaching styles.[6] It is, of course, not necessary to use both coding methods. We use them here to show some of their special properties. Vector 3 in Table 10.6 represents the continuous variable, tolerance of ambiguity. Vector 4 is a product of vectors 1 and 3, while vector 5 is a product of vectors 2 and 3. (Note that these vectors need not be punched on cards, since they can be generated by the computer.)

---

[6]When the categorical variable consists of two categories, effect coding and orthogonal coding are indistinguishable. For more than two categories, the methods presented here also apply with orthogonal coding.

TABLE **10.6**   TOLERANCE OF AMBIGUITY AND RATING OF TEACHING
STYLES, FICTITIOUS DATA[a]

Teaching Styles	Y	1	2	3	4 (1×3)	5 (2×3)
	2	1	1	5	5	5
	1	1	1	7	7	7
	2	1	1	10	10	10
	2	1	1	15	15	15
	3	1	1	17	17	17
	3	1	1	20	20	20
	3	1	1	25	25	25
Nondirective	4	1	1	23	23	23
	4	1	1	27	27	27
	5	1	1	30	30	30
	6	1	1	35	35	35
	5	1	1	37	37	37
	5	1	1	40	40	40
	5	1	1	42	42	42
	6	1	1	45	45	45
	6	1	1	47	47	47
	6	1	1	50	50	50
	7	1	1	55	55	55
	7	1	1	60	60	60
	6	1	1	62	62	62
	7	0	−1	5	0	−5
	7	0	−1	7	0	−7
	7	0	−1	9	0	−9
	6	0	−1	13	0	−13
	5	0	−1	12	0	−12
	6	0	−1	16	0	−16
	5	0	−1	18	0	−18
	5	0	−1	21	0	−21
	5	0	−1	22	0	−22
	4	0	−1	27	0	−27
Directive	4	0	−1	26	0	−26
	3	0	−1	32	0	−32
	3	0	−1	35	0	−35
	2	0	−1	40	0	−40
	3	0	−1	45	0	−45
	2	0	−1	47	0	−47
	2	0	−1	49	0	−49
	1	0	−1	53	0	−53
	2	0	−1	58	0	−58
	1	0	−1	63	0	−63

[a]Y = Ratings of teachers; vector 1 = teaching styles, where 1 is for nondirective and 0 is for
directive; vector 2 = teaching styles, where 1 is for nondirective and −1 is for directive; 3 =
tolerance of ambiguity.

Testing the Interaction

In order to determine whether there is a significant interaction between the categorical variable and the continuous variable, one tests whether the increment in the proportion of variance of Y accounted for by the interaction vectors is significant. Recall that a significant interaction indicates that the b weights are significantly different. In other words, the regression lines for the separate categories of the categorical variable, or the separate groups, are not parallel.

With the present example we can do two analyses, one using dummy coding and one using effect coding. Referring to the coded vectors of Table 10.6, the increment due to the interaction can be expressed in two ways:

$$R^2_{y.134} - R^2_{y.13} = R^2_{y.235} - R^2_{y.23}$$

The R^2's obtained with either of the coding methods, of course, will be the same. Some of the intermediate statistics, however, will differ for different coding methods. For the data of Table 10.6 we obtain

$$R^2_{y.134} = R^2_{y.235} = .90693$$

$$R^2_{y.13} = R^2_{y.23} = .01581$$

The increment due to the interaction is therefore

$$R^2_{y.134} - R^2_{y.13} = R^2_{y.235} - R^2_{y.23} = .90693 - .01581 = .89112$$

The interaction accounts for about 89 percent of the variance of Y. This increment is tested for significance in the usual manner:

$$F = \frac{(.90693 - .01581)/(3-2)}{(1-.90693)/(40-3-1)} = \frac{.89112/1}{.09307/36} = \frac{.89112}{.00259} = 344.06$$

with 1 and 36 degrees of freedom, a highly significant F ratio. It is concluded that the interaction of teaching styles and tolerance of ambiguity is significant. Stated differently, this means that the b weights for the regression of teacher ratings on tolerance of ambiguity are significantly different in the two groups that were exposed to different teaching styles.

As noted earlier, when the interaction is significant it is necessary to determine whether it is ordinal or disordinal. For this purpose one has to calculate the regression equation for each group and then determine the point of intersection of the regression lines. In the retention experiment presented earlier in the chapter we calculated the regression equation for each group separately (see Table 10.2). It is possible, however, to obtain the separate regression equations from the overall regression analysis that includes the coded vectors for the categorical variable, the vectors for the continuous variables, and the interaction vectors.

Obtaining Separate Regression Equations from
Overall Analysis

The regression equation for each group can be obtained from the overall analysis regardless of the method used for coding the categorical variable. We demonstrate how this is accomplished with the dummy and effect coding used in Table 10.6.

Dummy Coding. The regression equation for the overall analysis with dummy coding is

$$Y' = 7.19321 - 5.99972X_1 - .10680X_3 + .20516X_4$$

where X_1, X_3, and X_4 refer to vectors 1, 3, and 4 of Table 10.6. This equation can be interpreted in an analogous manner to the interpretation of the equation obtained for dummy coding with a categorical variable only (see Chapter 7).

The intercept, a, is equal to the intercept of the regression equation for the group that is assigned 0's in all the coded vectors. In the present example the group taught by a "directive" teacher was assigned 0's (see Table 10.6). Consequently, the intercept of the regression equation for this group is 7.19321, the intercept obtained from the overall regression equation.

Each b coefficient for a coded vector in the overall regression equation is equal to the difference between the intercept of the regression equation for the group identified by the vector (that is the group assigned 1's in the vector) and the intercept of the regression equation for the group assigned 0's in all the coded vectors. Therefore, in order to obtain the intercept for a given group it is necessary to add the b weight for the coded vector in which it is assigned 1's and the a for the overall regression equation. In the present example there is only one dummy vector (vector 1 of Table 10.6) for which the b coefficient was reported above as -5.99972. The intercept, a, of the regression equation for the group taught by the "nondirective" teacher (the group assigned 1's in vector 1) is therefore

$$7.19321 + (-5.99972) = 1.19349$$

The method of obtaining the regression coefficient of the regression equation for each group is similar to the method outlined above for obtaining the separate intercepts. Specifically, the regression coefficient, b, for the continuous vector in the overall regression equation is equal to the b for the continuous variable of the regression equation for the group assigned 0's throughout. In the present example the b for vector 3 representing tolerance of ambiguity (see Table 10.6) was reported above as $-.10680$. This, then, is the b coefficient for tolerance of ambiguity for the regression equation for the group taught by the "directive" teacher.

The b coefficient associated with the interaction vector in the overall regression equation is equal to the difference between the b for the group assigned 1's in the coded vector that was used to generate the cross-product vector and the b for the group assigned 0's throughout. In the present example vector 1 was multiplied by vector 3 to obtain vector 4. In vector 1 the group taught by the "nondirective" teacher was assigned 1's. Accordingly, the b associated with vector 4 in the overall regression equation (.20516) is equal to the difference between the b for the group taught by the "nondirective" teacher and the b for the group taught by the "directive" teacher (the group assigned

0's). It was shown above that the b for the group assigned 0's is $-.10680$. The b for the group assigned 1's is therefore

$$-.10680 + .20516 = .09836$$

On the basis of the above calculations we now write the regression equations for the two groups. They are

$$Y'_{(D)} = 7.19321 - .10680X$$

$$Y'_{(ND)} = 1.19349 + .09836X$$

where D = directive; ND = nondirective; X = tolerance of ambiguity. The same regression equations would be obtained if one calculated them separately for each group. The reader is advised to calculate the regression equations for the two groups of Table 10.6 using the method presented in earlier chapters (for example, Chapter 2. This method was also used in Table 10.2) and compare them to the two regression equations obtained above.

Effect Coding.   We now show how one can obtain the separate regression equations from the overall regression analysis when effect coding is used for the categorical variable. The overall regression equation for the data of Table 10.6 is

$$Y' = 4.19336 - 2.99986X_2 - .00422X_3 + .10258X_5$$

where X_2, X_3, and X_5 refer to vectors 2, 3, and 5 of Table 10.6.

With effect coding, the intercept, a, of the overall regression equation is equal to the average of the intercepts of the separate regression equations. It was found above that the two a's of the regression equations for the present example are 7.19321 and 1.19349. The average of these two values is 4.193350, which is, within rounding errors, equal to the a of the overall regression equation obtained with effect coding. The b associated with a coded vector of the overall regression equation is equal to the difference, or deviation, of the intercept for the group assigned 1's in the given vector from the average of the intercepts. Consequently, the a for the group assigned 1's in a given vector is equal to the b associated with this vector plus the average of the intercepts, or the a of the overall regression equation. In the present example, the group taught by the "nondirective" teacher was assigned 1's in vector 2. The b for this vector (the b for X_2) is -2.99986. Accordingly, the intercept for the group taught by the "nondirective" teacher is

$$4.19336 + (-2.99986) = 1.19350$$

where 4.19336 is the a of the overall regression equation reported above.

The intercept for the group assigned -1's in all the coded vectors is equal to the intercept obtained from the overall regression equation minus the sum of the b's of the overall regression equation associated with all the coded vectors. In the present example there is only one coded vector (X_2) whose coefficient is -2.99986. The intercept for the group taught by the "directive" teacher (that is, the group assigned −1's in vector 2 of Table 10.6) is therefore

$$4.19336 - (-2.99986) = 7.19322$$

When effect coding is used for the categorical variable, the b coefficient associated with the continuous variable (X_3 in the present example) is equal to the average of the regression coefficients, for the separate regression equations.[7] The b for X_3 was reported above as −.00422.

The b associated with each interaction vector is equal to the difference, or deviation, of the b for the group assigned 1's in the vector that was used to generate the product or interaction vector from the average of the b's. Consequently, the b for the group assigned 1's in a given vector is equal to the b for the interaction vector with which it is associated plus the average of the b's. In the present example, the product vector, 5, was generated by multiplying vector 2, in which the group taught by the "nondirective" teacher was assigned 1's, by the vector of the continuous variable (vector 3 of Table 10.6). The b for vector 5 was reported above as .10258. The average of the b's was shown to be −.00422. Therefore the b for the group taught by the "nondirective" teacher is

$$-.00422 + .10258 = .09836$$

The b for the group assigned −1's is equal to the average of the regression coefficients minus the sum of the b's for all the interaction vectors in the overall regression equation. In the present example there is only one interaction vector, X_5, whose coefficient is .10258. The b for the group assigned −1's (that is, the group taught by the "directive" teacher) is therefore

$$-.00422 - (.10258) = -.10680$$

The values obtained from the analysis with effect coding are the same as those obtained with dummy coding.

While the method of obtaining the separate regression equations from an overall analysis was illustrated for the case of a categorical variable with two categories, it can be extended to a categorical variable with any number of categories.[8] Moreover, while in the present example only one continuous variable was used, the same method of analysis as well as the same method of obtaining the separate regression equations applies with any number of con-

---

[7]Note that the average of the b's is not the common regression coefficient, b_c, discussed earlier in the chapter. To obtain b_c use formula 10.1, or the following formula:

$$b_c = \frac{\Sigma x_1^2 b_1 + \Sigma x_2^2 b_2 + \cdots + \Sigma x_k^2 b_k}{\Sigma x_1^2 + \Sigma x_2^2 + \cdots + \Sigma x_k^2}$$

From this formula it can be seen that when the Σx^2 for all the groups are equal the average of the b's equals b_c. As noted earlier, b_c can be obtained from the regression analysis in which the interaction vectors are not included. In such an analysis, the regression coefficient associated with the continuous vector is the b_c. In the present example b_c is the coefficient associated with vector 3 in the regression equation for $R_{y.13}^2$ or $R_{y.23}^2$. The value of this coefficient is −.00721.

[8]For an example of an analysis in which the categorical variable consists of three categories, see "Analysis of Covariance" later in the chapter.

tinuous variables and one categorical variable. Although in the present example a linear regression analysis was done, the method outlined above also applies to curvilinear regression analysis.

The Point of Intersection

Having obtained the separate regression equations one can calculate the point of intersection of the two regression lines. For convenience we repeat the two regression equations:

$$Y'_{(D)} = 7.19321 - .10680X$$

$$Y'_{(ND)} = 1.19349 + .09836X$$

where D = directive; ND = nondirective; X = tolerance of ambiguity. Applying formula (10.5) to the values of these equations, obtain the point of intersection:

$$X = \frac{7.19321 - 1.19349}{(.09836) - (-.10680)} = \frac{5.99972}{.20516} = 29.24$$

The value of X at which the regression lines intersect is well within the range of scores of the continous variable (the scores range from 5 to 63). The interaction is therefore disordinal.

    The data of Table 10.6 are plotted in Figure 10.4, along with the two regression lines. For the group taught by a "nondirective" teacher the regression of teacher ratings on tolerance of ambiguity is positive. The situation is

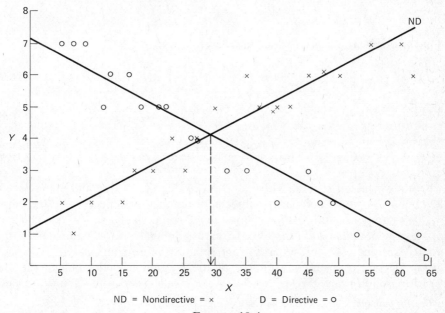

ND = Nondirective = ×   D = Directive = o

FIGURE 10.4

reversed for the group taught by a "directive" teacher. This, of course, is also evident from the regression coefficients of the separate regression equations. It appears, then, that students who are more tolerant of ambiguity prefer a "nondirective" teacher, while students who are less tolerant of ambiguity prefer a "directive" teacher.

Regions of Significance and the Johnson-Neyman Technique

Inspection of Figure 10.4 indicates that students whose scores on tolerance of ambiguity are closer to the point of intersection of the regression lines (29.24) differ less in their satisfaction with the teachers, as compared with students whose scores on tolerance are farther from the point of intersection. In other words, the differential effects of teaching styles are more marked for students whose scores on tolerance of ambiguity are relatively high or low compared to students whose scores on ambiguity are in the middle of the range.

For any level of the independent variable, X, one can determine whether subjects from different groups differ significantly on the dependent variable, Y.[9] A technique developed by Johnson and Neyman (1936), however, is generally preferable because it enables the researcher to establish regions of significance, thereby making it possible to state within what ranges of the X scores subjects from different groups differ significantly on Y, and within what range of X subjects from different groups do not differ significantly on Y.

The application of the Johnson-Neyman technique is now demonstrated for the above example. In order to establish the regions of significance it is necessary to solve for the two values of X in the following formula[10]:

$$X = \frac{-B \pm \sqrt{B^2 - AC}}{A} \tag{10.6}$$

The terms of formula (10.6) are defined as follows:

$$A = \frac{-F_\alpha}{N-4} (ss_{res}) \left(\frac{1}{\sum x_1^2} + \frac{1}{\sum x_2^2} \right) + (b_1 - b_2)^2 \tag{10.7}$$

$$B = \frac{F_\alpha}{N-4} (ss_{res}) \left(\frac{\bar{X}_1}{\sum x_1^2} + \frac{\bar{X}_2}{\sum x_2^2} \right) + (a_1 - a_2)(b_1 - b_2) \tag{10.8}$$

$$C = \frac{-F_\alpha}{N-4} (ss_{res}) \left(\frac{N}{n_1 n_2} + \frac{\bar{X}_1^2}{\sum x_1^2} + \frac{\bar{X}_2^2}{\sum x_2^2} \right) + (a_1 - a_2)^2 \tag{10.9}$$

where F_α = tabled F ratio with 1 and $N-4$ degrees of freedom at a selected level of α; N = total number of subjects, or number of subjects in both groups; n_1, n_2 = number of subjects in groups 1 and 2 respectively; ss_{res} = residual sum of squares obtained from the overall regression analysis, or, equivalently, the pooled residual sum of squares from separate regression analyses for each group; $\sum x_1^2$, $\sum x_2^2$ = sum of squares of the continuous independent variable (X) for groups 1 and 2, respectively; $\bar{X}_1$, $\bar{X}_2$ = means of groups 1 and 2 respec-

[9]See, for example, Johnson and Jackson (1959, pp. 435–437).
[10]The formulas used here were adapted from formulas given by Walker and Lev (1953, p. 401).

tively on the continuous independent variable, X; b_1, b_2 = regression coefficients of the regression equations for groups 1 and 2, respectively; a_1, a_2 = intercepts of the regression equations for groups 1 and 2, respectively.

The values necessary for the application of the above formulas to the data of Table 10.6 are[11]

$$ss_{res} = 13.06758 \qquad \Sigma x_1^2 = 5776.80 \qquad \Sigma x_2^2 = 6123.80$$
$$\bar{X}_1 = 32.60 \qquad \bar{X}_2 = 29.90$$
$$a_1 = 1.19349 \qquad a_2 = 7.19321$$
$$b_1 = .09836 \qquad b_2 = -.10680$$

The tabled F value with 1 and 36 degrees of freedom at the .05 level is 4.11.

$$A = \frac{-4.11}{36}(13.06758)\left(\frac{1}{5776.80}+\frac{1}{6123.80}\right)+(.20516)^2 = .04160$$

$$B = \frac{4.11}{36}(13.06758)\left(\frac{32.60}{5776.80}+\frac{29.90}{6123.80}\right)+(-5.99972)(.20516) = -1.21521$$

$$C = \frac{-4.11}{36}(13.06758)\left(\frac{40}{400}+\frac{32.60^2}{5776.80}+\frac{29.90^2}{6123.80}\right)+(-5.99972)^2 = 35.35519$$

$$X = \frac{1.21521 \pm \sqrt{1.21521^2 - (.04160)(35.35519)}}{.04160}$$

$$X_1 = 31.07 \qquad X_2 = 27.36$$

The two X values are now used to establish the region of nonsignificance. Values of Y for subjects whose scores lie within the range of 27.36 and 31.07 on X are not significantly different across groups. There are two regions of significance, one for X scores above 31.07 and one for X scores below 27.36. In other words, the ratings indicated by students whose scores on tolerance of ambiguity are above 31.07 or below 27.36 are significantly different in the two groups. Note that in the present example the region of nonsignificance is narrow: about 4 points on X. Practically all the X values are in the regions of significance. On the basis of the analysis we conclude that students whose scores on tolerance of ambiguity are above 31.07 are more satisfied with a "nondirective" teacher, while students whose scores on tolerance of ambiguity are below 27.36 are more satisfied with a "directive" teacher.

Applications and Extensions of the Johnson-Neyman Technique

In the example analyzed above the interaction was disordinal. The Johnson-Neyman technique is equally applicable to ordinal interactions. The procedure

[11]The residual sum of squares, which is part of the computer output, was shown several times earlier to be equal to $\Sigma y^2(1-R^2)$. Σy^2 for the present example is 140.40. $R^2_{y.134} = .90693$, $ss_{res} = (140.40)(1-.90693) = 13.06703$, which, within rounding errors, is equal to the residual sum of squares obtained from the computer output.

Note that the first two terms in formulas (10.7), (10.8), and (10.9) are the same, except that in formulas (10.7) and (10.9) their sign is negative, while in formula (10.8) the sign is positive.

is the same except that with ordinal interactions one of the regions of significance is outside the research range of interest.[12]

The technique is not limited to the case of a categorical variable with two categories, or two groups, nor is it limited to one continuous independent variable. For extensions to more than two categories, or groups, and more than one continuous variable, see Johnson and Fay (1950), Abelson (1953), Walker and Lev (1953, pp. 404–411), Johnson and Jackson (1959, pp. 438–441), Potthoff (1964).

The examples analyzed above were taken from experimental research. The procedures shown for testing the significance of the difference between regression coefficients and between intercepts, as well as the Johnson-Neyman Technique, are equally applicable to the analysis of data from nonexperimental research. One may, for example, wish to compare regression equations for blacks and whites,[13] or for males and females, or for Catholics, Jews, Moslems, and Protestants. Furthermore, in the examples analyzed the number of subjects in the groups was equal. The same analysis applies when the number of subjects in groups is unequal.

Recapitulation

The procedure for analyzing data with continuous and categorical independent variables is now summarized to clarify the analytic sequence. The steps to be followed are presented in the form of a set of questions. Depending on the nature of the answer to a given question, one may either have to go to another step or terminate the analysis and summarize the results.

Create a vector Y that will include the measures of the dependent variable for all subjects. Create coded vectors to indicate membership in the categories of the categorical variable. Create a vector, or vectors, that will include the values of the continuous variable, or variables, for all subjects. Generate new vectors by multiplying the vectors for the categorical variable by the vector for the continuous variable, or variables. These product vectors represent the interaction terms.

To make the following presentation more concise we stay with the example of a categorical variable with two categories (for example, male, female) and one continuous variable (for example, motivation). The vector representing the categorical variable is symbolized by A, the vector for the continuous variable is symbolized by B, and the product of vectors A and B is symbolized by C.

1. *Is the proportion of variance accounted for meaningful?* Calculate

---

[12]For a discussion of the research range of interest, see earlier sections of the chapter. For an example of regions of significance when the interaction is ordinal, see Study Suggestion 9 at the end of the chapter.

[13]See, for example, Cleary (1968), Duncan (1969), and Chapter 16. For an example of the comparison of regression equations for males and females, see Study Suggestion 9.

$R^2_{y.abc}$. This indicates the proportion of variance accounted for by the main effects and the interaction. If $R^2_{y.abc}$ is too small to be meaningful within the context of your theoretical formulation and your knowledge of the findings in the field of study, terminate the analysis. Whether R^2 is significant or not, if, in your judgment, its magnitude has little substantive meaning, there is no point in going further. If $R^2_{y.abc}$ is meaningful, go to step 2.

2. *Is there a significant interaction?*   Calculate $R^2_{y.ab}$. Test

$$F = \frac{(R^2_{y.abc} - R^2_{y.ab})/(3-2)}{(1 - R^2_{y.abc})/(N-3-1)}$$

A nonsignificant F ratio indicates that the interaction is not significant. If the interaction is not significant go to step 3. If it is significant, go to step 5.

3. *Is the common regression weight significant?*   In the present context, this is the same as asking whether the continuous variable accounts for a significant increment in the proportion of the variance in the dependent variable. Calculate $R^2_{y.a}$. Test

$$F = \frac{(R^2_{y.ab} - R^2_{y.a})/(2-1)}{(1 - R^2_{y.ab})/(N-2-1)}$$

Note that in the analysis in which the interaction vectors are deleted, the b coefficient for the continuous variable is the common b, or b_c. Consequently, the t ratio for b_c obtained in such an analysis is equal to $\sqrt{F}$, which is obtained from the application of the above formula. Go to step 4.

4. *Are the intercepts significantly different?*   That is, is one treatment, or group, superior in an equal amount over the other treatment, or group, along the continuum of the continuous variable? Calculate $R^2_{y.b}$. Test

$$F = \frac{(R^2_{y.ab} - R^2_{y.b})/(2-1)}{(1 - R^2_{y.ab})/(N-2-1)}$$

At this stage the analysis is terminated. From the regression analysis in which the interaction vector was deleted, calculate the regression equation for each group. Note that the b for the continuous variable will be the same for both regression equations, that is, a common b, while the a's will differ. Using the regression equations, plot the parallel regression lines.

If the F ratio obtained above is not significant, this indicates that there is no significant difference between the methods or groups. In this case, report the regression equation that is common to both groups. Interpret the results.

5. *Establish regions of significance.*   If the F ratio calculated in step 2, above, is significant, calculate the separate regression equations and the point of intersection of the regression lines. Plot the regression lines. Apply the Johnson-Neyman technique to establish regions of significance. Interpret the results.

Testing for Trends

In the examples presented thus far the data were linear (see Figures 10.1 and 10.4). In many instances, however, the trend may not be as obvious. Moreover, one should not rely on visual inspection alone, although study of plotted data is always valuable. It is the application of tests for trends that enables one to make a decision about the analysis that is most appropriate for a given set of data.

Tests with Orthogonal Polynomials

The method of fitting polynomials was discussed and demonstrated in Chapter 9. One tests whether successively higher degrees of polynomials add significantly to the variance accounted for. The procedure is now applied to the data from the retention experiment presented in the beginning of the chapter. It will be recalled that the continuous variable, Study Time, had four levels. The highest-degree polynomial for these data is therefore cubic (number of levels minus one). This may be obtained by raising the values of the continuous variable to the second and the third powers. Since the experiment also involved a categorical variable (Incentive–No Incentive), we must study the interactions on the linear, quadratic, and cubic levels. To perform the entire analysis it is necessary to have the following vectors for the independent variables: (1) Incentive–No Incentive, (2) linear trend in Study Time, (3) quadratic trend (Study time squared), (4) cubic trend (Study Time raised to the third power), (5) linear interaction (product of vectors 1 and 2), (6) quadratic interaction (product of vectors 1 and 3), and (7) cubic interaction (product of vectors 1 and 4).

Because the retention experiment consisted of equal cell frequencies and equal intervals for the continuous variable, it is possible to simplify the analysis by the use of orthogonal polynomials.[14] The data from the experiment are repeated in Table 10.7, along with the necessary vectors to test for trends and interactions. Vector Y, as in earlier tables, identifies the dependent variable. Vector 1 identifies Incentive–No Incentive. Vectors 2, 3, and 4 represent the linear, quadratic, and cubic components for Study Time. The coefficients for these vectors were obtained from a table of orthogonal polynomials (for example, Fisher & Yates, 1963). Vectors 5, 6, and 7 are obtained by multiplying, in turn, vector 1 by vectors 2, 3, and 4. Thus vector 5 represents the linear interaction, and is a product of vectors 1 and 2. Vectors 6 and 7 represent the quadratic and cubic interactions, respectively. It will be noted that all the vectors are orthogonal to each other. The analysis is therefore straightforward, and the results directly indicate the proportion of variance accounted for by each vector. Table 10.8 reports a summary of the regression analysis.

Look first at column 7 of Table 10.8. The first vector (Incentive–No Incentive) accounts for .57953 of the variance, a finding identical to the ones

[14]See Chapter 9.

TABLE **10.7** DATA FROM THE RETENTION EXPERIMENT, LAID OUT
FOR TREND ANALYSIS[a]

Y	1	2	3	4	5 (1 × 2)	6 (1 × 3)	7 (1 × 4)
3	1	−3	1	−1	−3	1	−1
4	1	−3	1	−1	−3	1	−1
5	1	−3	1	−1	−3	1	−1
4	1	−1	−1	3	−1	−1	3
5	1	−1	−1	3	−1	−1	3
6	1	−1	−1	3	−1	−1	3
5	1	1	−1	−3	1	−1	−3
6	1	1	−1	−3	1	−1	−3
8	1	1	−1	−3	1	−1	−3
7	1	3	1	1	3	1	1
8	1	3	1	1	3	1	1
9	1	3	1	1	3	1	1
7	−1	−3	1	−1	3	−1	1
8	−1	−3	1	−1	3	−1	1
9	−1	−3	1	−1	3	−1	1
9	−1	−1	−1	3	1	1	−3
10	−1	−1	−1	3	1	1	−3
11	−1	−1	−1	3	1	1	−3
8	−1	1	−1	−3	−1	1	3
11	−1	1	−1	−3	−1	1	3
12	−1	1	−1	−3	−1	1	3
10	−1	3	1	1	−3	−1	−1
11	−1	3	1	1	−3	−1	−1
13	−1	3	1	1	−3	−1	−1

Σy^2:   172.625

[a]Y = measures of retention originally given in Table 10.1; 1 = coded vector for Incentive–No Incentive; 2 = vector for linear component of Study Time; 3 = quadratic; 4 = cubic; 5, 6, and 7 = interaction between Incentive–No Incentive and Study Time.

obtained in earlier calculations. The same applies to vector 2, the linear component of Study Time (.24335). The quadratic component is very small, as is the cubic component: .00024 and .00237 respectively. (See rows 3 and 4, Table 10.8.) The linear-by-treatment interaction is .00391. This value is identical to the one obtained in the beginning of this chapter, where the significance of the

TABLE **10.8**   SUMMARY OF REGRESSION ANALYSIS WITH ORTHOGONAL POLYNOMIALS. RETENTION EXPERIMENT DATA[a]

(1) Vector	(2) M	(3) s	(4) b	(5) s_b	(6) t	(7) Prop. of Variance
1	0	1.02151	−2.04167	.27003	−7.56086	.57953
2	0	2.28416	.59167	.12076	4.89947	.24335
3	0	1.02151	−.04167	.27003	−.15430	.00024
4	0	2.28416	.05833	.12076	.48305	.00237
5	0	2.28416	.07500	.12076	.62106	.00391
6	0	1.02151	.20833	.27003	.77152	.00603
7	0	2.28416	−.05833	.12076	−.48305	.00237
Y	7.875	2.73960				

Σ: .83780

[a]Original data given in Table 10.7. Vector 1 = Incentive–No Incentive; 2 = linear component of Study Time; 3 = quadratic; 4 = cubic; 5, 6, and 7 = interaction between Incentive–No Incentive and Study Time; Y = measure of retention.

difference between the b weights for the Incentive and the No Incentive groups was tested. The interactions of the treatment with the quadratic and the cubic components are both very small: .00603 and .00237 respectively (rows 6 and 7 of Table 10.8). The sum total of column 7 is .83780. This is in effect $R^2_{y.1234567}$, and is identical to the R^2 obtained earlier, when the continuous variable was categorized. In both cases, the vectors exhaust all the available information.

What, then, is the difference between the two analyses, namely the present one and the one in which the continuous variable was partitioned? (See Table 10.5.) The difference is in the specific breakdown of the variance such as the one reported in Table 10.8, which permits the researcher to note that other than the Incentive–No Incentive variable and the linear trend of Study Time, little else is meaningful. Adding these two components we obtain .57953 + .24335 = .82288. This compares well with the proportion of variance accounted for when all the information is exhausted (.83780); that is, when the categorical variable and all the trends and the interactions are used. The small discrepancy between the two is clearly unimportant. It is this discrepancy, which results from using partial information as compared to all the information, that was alluded to earlier.

What about significance? Column 6 of Table 10.8 consists of the t ratios associated with the b's of column 4. It was shown in Chapter 8 that when the vectors are orthogonal to each other, as they are in the present case, each t ratio is separately interpretable and indicates whether the sum of squares (and the proportion of variance associated with it) is significant. For example, the t ratio associated with b_1 (vector 1 in Table 10.8) is −7.56086, with 16 degrees of freedom (degrees of freedom associated with the overall error term). Squaring −7.56086 yields an F ratio of 57.17, with 1 and 16 degrees of freedom. The same F ratio was obtained earlier (see Table 10.5). It is obvious that the only other significant t ratio in Table 10.8 is the one associated with the linear trend of Study Time (4.89947, row 2, column 6). All other t ratios are smaller than one.

To demonstrate clearly the distinctions and the similarities between the analysis in which the continuous variable was categorized and the present analysis, we combine some of the information of Table 10.8 and report it in Table 10.9. Compare Table 10.9 with Table 10.5 and note that the data reported in columns 1, 3, and 5 of Table 10.9 are identical to the data reported in Table 10.5. The difference between the two tables is that, in Table 10.9, the sum of squares due to Study Time is divided into two components, linear trend and deviation from linearity. This results in a larger F ratio for Study Time: 24.00 compared to 8.09. The interaction sum of squares is similarly divided.

Looking now at the F ratios associated with the various components of Study Time and the interactions, we see that the only significant F ratio is the one for the linear trend of Study Time. In addition, the F ratio for the categorical variable, Incentive–No Incentive, is significant. We therefore take these two significant terms only, and relegate all the remaining terms to the error term. This results in $R^2_{y.12} = .82288(.57953 + .24335)$, and an error term of $.17712$ $(1 - .82288)$. The degrees of freedom for the F ratio associated with $R^2_{y.12}$ are 2 for the numerator and 21 $(24 - 2 - 1)$ for the denominator. These are the figures obtained in the beginning of this chapter, where the b weights for the two groups were found to be not significantly different. We therefore did an analysis that included one vector for Incentive–No Incentive and another

TABLE **10.9**   SUMMARY OF REGRESSION ANALYSIS WITH ORTHOGONAL POLYNOMIALS, TESTS FOR TREND, RETENTION EXPERIMENT DATA[a]

	1	2	3	4	5	6	7
Source	Prop. of Variance	Prop. of Variance	ss	ss	df	ms	F
I–NI	.57953		100.04137		1	100.04137	57.17
ST	.24596		42.45884		3	14.15295	8.09
Linear		.24335		42.00829	1	42.00829	24.00
Deviation from Linearity		.00261		.45055	2	.22528	<1
Interaction	.01231		2.12501		3	.70834	<1
I – NI × ST Linear		.00391		.67496	1	.67496	<1
Deviation from Linearity		.00840		1.45005	2	.72502	<1
Error	.16220		27.99978		16	1.74999	
Total	1.00000		172.62500		23		

[a]I–NI = Incentive–No Incentive. ST = Study Time.

vector for Study Time. The difference between that analysis and the present one is that here we actually tested for deviations from linearity whereas in the earlier analysis a linear trend was assumed.

Tests for Trend when the Continuous Variable Is an Attribute Variable

The above illustration dealt with a continuous variable which was manipulated by the researcher. In such cases, it is the researcher who chooses the values of the continuous variable, and he can therefore determine that they be equally spaced along the continuum. This, it will be recalled, makes the application of orthogonal polynomials simple and straightforward.

In many designs, however, the researcher employs an attribute variable.[15] In a treatments-by-levels design, or in an Aptitude-Treatment-Interaction design, the attribute variable may be, for example, IQ, motivation, anxiety, cognitive style, and the like. In such situations the intervals between the values of the continuous variable may not be equal. Furthermore, the continuous variable may consist of many values with unequal numbers of subjects for the different values. One still must test deviation from linearity. In certain studies a trend other than linear may even be part of the hypothesis. In other words, on the basis of theoretical formulations a researcher may hypothesize a quadratic or a cubic trend. Obviously, such hypotheses need to be tested.

The procedures for testing the deviation from linearity, or given trends, with attribute variables follow a sequence of steps. At each step a decision needs to be made on the next appropriate one to be taken. These procedures are now illustrated.

Sequence of Testing for Trends

For the purpose of illustration, let us assume that we have two methods of teaching (A_1 and A_2) and that the attribute variable (B) is intelligence. Suppose we wish to test whether the trend is linear or quadratic. The sequence of steps, formulated as questions to be answered, is outlined below:

1.   *What is the proportion of variance accounted for by the teaching methods, the linear and quadratic trends, and the interaction of the two variables?*   Calculate $R^2_{y.a,b,b^2,ab,ab^2}$.

Note that B (intelligence) is squared and the product vectors of A and B and A and B^2 are generated.

2.   *Is the quadratic trend significant?*   Calculate $R^2_{y.a,b,ab}$. Test

$$F = \frac{(R^2_{y.a,b,b^2,ab,ab^2} - R^2_{y.a,b,ab})/(5-3)}{(1 - R^2_{y.a,b,b^2,ab,ab^2})/(N-5-1)}$$

If F is significant, go to step 3. If F is not significant, proceed with the sequence of steps shown earlier in the section "Recapitulation." Briefly, test first whether there is a significant linear interaction. If the interaction is significant, use the

---

[15]See Kerlinger (1973) for a discussion of active (manipulated) variables and attribute variables.

Johnson-Neyman technique to establish regions of significance, and interpret the results. If the linear interaction is not significant, test the difference between the intercepts. If the intercepts differ significantly, two parallel lines fit the data. That is, one method is superior to the other along the continuum of the continuous variable. If the intercepts do not differ significantly, a single regression line describes the data adequately. In other words, the methods do not differ significantly.

3. *Is there a significant quadratic interaction?*   If the F ratio under step 2 above is significant, test

$$F = \frac{(R^2_{y.a,b,b^2,ab,ab^2} - R^2_{y.a,b,b^2})/(5-3)}{(1 - R^2_{y.a,b,b^2,ab,ab^2})/(N-5-1)}$$

If the F ratio is significant, calculate the regression equation for each group (this can be done by using the overall regression equation in the manner shown earlier in the chapter). Plot the two regression curves and interpret the results. Note that when the interaction is significant several alternatives are possible. It may be, for example, that the two curves intersect, or that the difference between the curves is not constant, or that the regression for one group is linear while that of the other group is curvilinear. If the F ratio for the interaction is not significant, go to step 4.

4. *Is there a constant difference between the curves?*   In other words, is one method superior to the other along the continuum, which is described by two quadratic curves? Test

$$F = \frac{(R^2_{y.a,b,b^2} - R^2_{y.b,b^2})/(3-2)}{(1 - R^2_{y.a,b,b^2})/(N-3-1)}$$

If the F ratio is significant, one method is superior to the other. From the regression equation in which the interaction vectors were deleted, that is, $R^2_{y.a,b,b^2}$, calculate the regression equations for the two groups. Plot the two regression curves. If the F ratio is not significant, one quadratic curve is sufficient to describe the data from both groups. One concludes that the two methods do not differ significantly.

It should be obvious that the same process outlined above is followed when one wishes to study the cubic trend. In the above example, one will need two more vectors, that is, b^3 and ab^3. The testing sequence will be the same as that outlined above, starting with tests for the cubic trend.

Analysis of Covariance

In traditional statistics books the analysis of covariance is presented as a separate topic. Students who are not familiar with regression analysis are frequently baffled by the analysis of covariance, and tend to do calculations blindly. Needless to say, little understanding is gained. The picture becomes even more complicated when more than one covariate is used, or when one is dealing with a factorial analysis of covariance.

The Uses of the Analysis of Covariance

The behavioral researcher usually thinks of the analysis of covariance when he has to deal with intact groups, which may differ on a concomitant variable related to the dependent variable. Suppose that one is studying the effects of different teaching methods on achievement. The school in which the study is conducted does not permit the breaking up of classes and insists that intact classes be used. The researcher suspects, or knows, that the classes differ in intelligence. Under such conditions, it is possible that the classes higher in intelligence will perform better regardless of the teaching method to which they are assigned. This may therefore lead to an erroneous conclusion that a given method is superior, when its apparent superiority is due to the mental ability of the subjects assigned to it. To avoid such a blunder, the researcher attempts to "equalize" the groups on intelligence by the use of the analysis of covariance. In other words, he attempts to take into account, or adjust for, initial differences in intelligence. The same logic applies when the researcher believes that the groups differ in motivation as well as in intelligence. The adjustment is then made for both variables.

It should be noted, however, that one of the assumptions underlying the analysis of covariance is randomization. In other words, for the conclusions to be valid it is necessary that subjects be randomly assigned to groups and that groups be randomly assigned to treatments. When it is not possible to randomly assign subjects to groups, one should, at the very least, randomly assign treatments to groups. When randomization is not possible at all, analysis of covariance can be misleading and should therefore be used with caution.

Analysis of covariance is useful because it enables the researcher to identify and take into account sources of variance due to concomitant variables, thereby providing greater control. If the variability of a given concomitant variable is relatively large, and this variable is correlated with the dependent measure, it is possible to use the subjects' scores on the concomitant variable as a covariate. An adjustment for the covariate will lead to a reduction in the error term, and consequently to a more sensitive analysis. In this application the analysis of covariance is not unlike the treatments-by-levels, or randomized blocks design.[16]

The Logic of Analysis of Covariance

It will be recalled that when a variable is residualized, the correlation between the predictor variable and the residuals is zero (see Chapter 5). In other words, the residualized variable is one from which whatever it shared with the predictor variable has been purged. Suppose now that one were studying the effects of different teaching methods on achievement and wished to adjust the achievement score for differences in intelligence. The independent variable is teaching methods, the dependent variable is achievement, and the covariate is

---

[16]For a discussion of this point, see Feldt (1958). For the analysis of covariance and its uses and assumptions, see Cochran (1957) and Elashoff (1969).

intelligence. One can first use the subjects' intelligence scores to predict their achievement on basis of the regression of achievement on intelligence. If Y_{ij} is the actual achievement of individual i in group j, then Y'_{ij} is his predicted score. $Y_{ij} - Y'_{ij}$ is, of course, the residual. Calculating the residuals for all subjects, one arrives at a set of scores (residuals) which have zero correlations with intelligence. A test of significance between the residuals of the various groups will indicate whether the groups differ significantly *after* their scores have been adjusted for possible differences in intelligence. This is the logic behind the analysis of covariance. It can be summarized by the following formula:

$$Y_{ij} = \bar{Y} + T_j + b(X_{ij} - \bar{X}) + e_{ij} \qquad (10.10)$$

where Y_{ij} = the score of subject i under treatment j; $\bar{Y}$ = the grand mean on the dependent variable; T_j = the effect of treatment j; b = a common regression coefficient for Y on X; X_{ij} = the score on the covariate for subject i under treatment j; $\bar{X}$ = the grand mean of the covariate; e_{ij} = the error associated with the score of subject i under treatment j. Formula (10.10) can be rewritten as

$$Y_{ij} - b(X_{ij} - \bar{X}) = \bar{Y} + T_j + e_{ij} \qquad (10.11)$$

which clearly shows that after adjustment $[Y_{ij} - b(X_{ij} - \bar{X})]$, a score is conceived as composed of the grand mean, a treatment effect, and an error term. The right-hand side of formula (10.11) is an expression of the linear model presented in Chapter 7. In fact, if b were zero, that is, if the covariate were not related to the dependent variable, formula (10.11) would be identical to formula (7.11).

Homogeneity of Regression Coefficients

The process of adjustment for the covariate (X) in formula (10.10) involves the application of a common regression coefficient (b) to the deviation of X from the grand mean of X $(\bar{X})$. The use of a common b weight is based on the assumption that the b weights for the regression of Y on X in each group are not significantly different. This assumption is also referred to as the homogenity of regression coefficients. The testing of the assumption proceeds in exactly the same manner as the testing of the differences between regression coefficients, which was discussed and illustrated in the beginning of the chapter. Essentially, one tests whether the use of separate regression coefficients adds significantly to the proportion of variance accounted for, as compared to the proportion of variance accounted for by the use of a common regression coefficient.

Having established that the use of a common regression coefficient is appropriate, one can determine whether there is a significant difference between the means of the treatment groups after adjusting the scores on the dependent variable for possible differences on the covariate. As in previous work, this test attempts to answer the question whether additional information adds significantly to the proportion of variance accounted for. In the present context the question may be phrased: Does knowledge about treatments add significantly to the proportion of variance accounted for by the covariate? This is the test of significance between intercepts presented earlier.

When there are significant differences between the b's, that is, when there is an interaction between the covariate and the treatment, analysis of covariance should not be used. One can instead study the pattern of the interaction in the manner described earlier in the chapter, that is, by establishing regions of significance.

The logic of the analysis of covariance was presented as an analysis of residuals for the purpose of clarifying what in effect is being accomplished by such an analysis. One need not, however, actually calculate the residuals. From the foregoing discussion it should be clear that the calculations of the analysis of covariance follow the same pattern described in the present chapter.

A Numerical Example

Let us assume that a researcher is studying the effects of three different teaching methods on achievement. Assume, further, that he must work with intact classes and that he is only able to randomly assign the classes to the different treatments. The researcher therefore decides to do an analysis of covariance using a pretest and a posttest of achievement, where the pretest is the covariate. The analysis will adjust for possible initial differences in achievement among the groups. A set of fictitious data for such an experiment is reported in Table 10.10. X symbolizes the pretest, or the covariate, while Y symbolizes the posttest at the conclusion of the study. These data are repeated in Table 10.11, this time displayed for a regression analysis. Vector Y in Table 10.11 represents the dependent measure, or the final achievement scores. Vector X represents the pretest, or the covariate, and vectors 1 and 2 are effect coding for the treatments. Vectors 3 and 4 are obtained by multiplying, in turn, vector X by vectors 1 and 2.

TABLE **10.10**   FICTITIOUS DATA FROM AN EXPERIMENT WITH THREE TEACHING METHODS[a]

			Treatments			
	I		II		III	
X	Y	X	Y	X	Y
1	5	4	9	6	10
2	5	5	8	7	10
3	6	6	8	8	13
4	6	7	10	9	11
5	9	8	11	10	12
6	8	9	11	11	13
Σ: 21	39	39	57	51	69
M: 3.5	6.5	6.5	9.5	8.5	11.5

[a]X = pretest (covariate); Y = posttest.

TABLE **10.11**   ANALYSIS OF COVARIANCE OF AN EXPERIMENT WITH TEACHING METHODS, LAID OUT FOR REGRESSION ANALYSIS[a]

Treatments	Y	X	1	2	3 $(X \times 1)$	4 $(X \times 2)$
	5	1	1	0	1	0
	5	2	1	0	2	0
	6	3	1	0	3	0
I	6	4	1	0	4	0
	9	5	1	0	5	0
	8	6	1	0	6	0
	9	4	0	1	0	4
	8	5	0	1	0	5
	8	6	0	1	0	6
II	10	7	0	1	0	7
	11	8	0	1	0	8
	11	9	0	1	0	9
	10	6	−1	−1	−6	−6
	10	7	−1	−1	−7	−7
	13	8	−1	−1	−8	−8
III	11	9	−1	−1	−9	−9
	12	10	−1	−1	−10	−10
	13	11	−1	−1	−11	−11
Σ:	165	111				
Σy^2:	108.50	Σx^2: 128.50				

[a]Y = posttest; X = pretest; 1 and 2 = coded vectors for treatments.

Following the method presented earlier in the chapter we now obtain the regression equations for the three groups. For this purpose we use the regression equation with effect coding for the overall regression analysis. That is, the regression equation in which the covariate (X of Table 10.11), the teaching methods (vectors 1 and 2 of Table 10.11), and the product vectors for the two variables (vectors 3 and 4 of Table 10.11) are used. The overall regression equation is

$$Y' = 5.42857 + .63810X - 1.62857(V_1) + .17143(V_2) + .13333(V_3) - .03810(V_4)$$

where X = the covariate; V_1–V_4 = vectors 1 through 4 of Table 10.11.

As noted earlier, the intercept, a, of the overall regression equation with effect coding is equal to the average of the intercepts of the separate regression equations. The intercepts for group I, II, and III are therefore[17]

$$a_I = 5.42857 + (-1.62857) = 3.80000$$
$$a_{II} = 5.42857 + (.17143) = 5.60000$$
$$a_{III} = 5.42857 - [(-1.62857 + .17143)] = 6.88571$$

[17]See the section "Obtaining Separate Regression Equations from Overall Analysis" for an explanation of the method.

The average of the b's is .63810, the b for the continuous vector (X) in the overall regression equation. The b's for groups I, II, and III are

$$b_{\mathrm{I}} = .63810 + .13333 = .77143$$
$$b_{\mathrm{II}} = .63810 + (-.03810) = .60000$$
$$b_{\mathrm{III}} = .63810 - [(.13333) + (-.03810)] = .54287$$

Accordingly, the regression equations for the three groups are

$$Y'_{\mathrm{I}} = 3.80000 + .77143X$$
$$Y'_{\mathrm{II}} = 5.60000 + .60000X$$
$$Y'_{\mathrm{III}} = 6.88571 + .54287X$$

Test first the homogeneity of the b's. For the data of Table 10.11 this amounts to testing the difference between $R^2_{y.x1234}$ and $R^2_{y.x12}$.

$$R^2_{y.x1234} = .90203, \qquad R^2_{y.x12} = .89747$$

$$F = \frac{(.90203 - .89747)/(5-3)}{(1-.90203)/(18-5-1)} = \frac{.00456/2}{.09797/12} = \frac{.00228}{.00816} = < 1$$

The F ratio is less than one. We conclude that the b weights are homogeneous, or that they do not differ significantly. The use of a common regression coefficient is therefore justified. Note that the increment in the proportion of variance accounted for by using separate regression coefficients is indeed minute (.00456).

We now test whether the teaching methods add significantly to the proportion of variance accounted for after allowance is made for the covariate. This is done by testing the significance of the difference between $R^2_{y.x12}$ and $R^2_{y.x}$, where x stands for the covariate; 1 and 2 are coded vectors representing treatments (see Table 10.11). For the present data

$$R^2_{y.x12} = .89747, \qquad R^2_{y.x} = .85999$$

$$F = \frac{(.89747 - .85999)/(3-1)}{(1-.89747)/(18-3-1)} = \frac{.03748/2}{.10253/14} = \frac{.01874}{.00732} = 2.56$$

with 2 and 14 degrees of freedom. This F ratio is not significant, and it is concluded that there are no significant differences between the treatments, after having adjusted for the covariate. Note that while the covariate accounts for about 86 percent of the variance, the treatments add about 4 percent. Evidently the groups started out with quite different scores on the initial test. This is clearly seen by an inspection of the means of the three treatment groups on the covariate, which are 3.5, 6.5, and 8.5 (see Table 10.10). It is obvious that the variance accounted for is due mostly to the differences on the initial measure. To illustrate this point further, the data are plotted in Fig. 10.5. Crosses identify subjects in Treatment I, open circles identify subjects in Treatment II, and closed circles identify subjects in Treatment III. The three separate regression lines are also drawn in Figure 10.5. Note that these lines are close to each other.

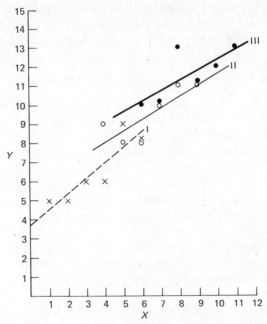

X = pretest; Y = posttest;   I, II, III, = Treatments I, II, and III

FIGURE 10.5

It is not difficult to visualize a common regression line adequately describing the data. In other words, the regression of Y (dependent variable) on X (covariate) is sufficient to describe the data.

One might have reached very different conclusions had no attention been paid to possible differences in initial scores — in other words, had no adjustment been made for the covariate. This can be easily seen by calculating $R^2_{y.12}$, which is the squared multiple correlation between the dependent measure and the coded vectors representing the treatments (see Table 10.11). For the present data, $R^2_{y.12} = .70046$.

$$F = \frac{.70046/2}{(1 - .70046)/(18 - 2 - 1)} = \frac{.35023}{.29954/15} = \frac{.35023}{.01997} = 17.54$$

with 2 and 15 degrees of freedom. This F ratio is significant beyond the .01 level. Furthermore, it is associated with a considerable proportion of variance — about 70 percent. As pointed out above, however, these differences almost disappear when initial differences between the groups are taken into account. The nature of the adjustment becomes even clearer when it is applied to the means of the groups on the final scores.

Adjustment of Means and Tests of Significance

The means for the three treatments groups on the posttest were reported in Table 10.10. They are 6.5, 9.5, and 11.5 for Treatments I, II, and III, respectively. These means reflect not only possible differences in treatment effects but also differences between the groups that are due to covariate differences. It is possible to adjust each of the means and observe the differences after the effect of the covariate is removed. The general formula is

$$\bar{Y}_{j(\text{adj})} = \bar{Y}_j - b(\bar{X}_j - \bar{X}) \tag{10.12}$$

where $\bar{Y}_{j(\text{adj})}$ = the adjusted mean of treatment j; $\bar{Y}_j$ = the mean of treatment j before adjustment; b = the common regression coefficient; $\bar{X}_j$ = the mean of the covariate for treatment group j; $\bar{X}$ = the grand mean of the covariate.

To appreciate what is accomplished by formula (10.12), let us assume that in a study in which the analysis of covariance is used subjects are randomly assigned to treatments. Assume, further, that because of randomization all treatment groups end up having identical means on the covariate. This means that $\bar{X}_j = \bar{X}$ in formula (10.12). The application of formula (10.12) under such circumstances will result in no adjustments of the means of $\bar{Y}_j$. This makes sense, since the groups are equal on the covariate. Note that the case described above brings us back to the designs presented earlier in the chapter where one variable was continuous and another was categorical. The function of the covariate is to identify a systematic source of variance and thereby reduce the error term. Although randomization of subjects will generally not result in equal group means on the covariate, the differences between such means will probably be small. Consequently, the adjustments of means by formula (10.12) will be relatively small.

In the example being analyzed here, groups were randomly assigned to treatments. The means of the three groups on the covariate are quite different: 3.5, 6.5, and 8.5 for the three treatment groups (see Table 10.10). We now calculate the adjusted means. The common regression coefficient is .63810.[18] The grand mean on the covariate $(\bar{X})$ is 6.16667. The adjusted means for the three treatment groups are

$$\bar{Y}_{\text{I(adj)}} = 6.5 - (.63810)(3.5 - 6.16667) = 8.20$$
$$\bar{Y}_{\text{II(adj)}} = 9.5 - (.63810)(6.5 - 6.16667) = 9.29$$
$$\bar{Y}_{\text{III(adj)}} = 11.5 - (.63810)(8.5 - 6.16667) = 10.01$$

The adjusted means are closer to each other than are the unadjusted ones. This is because the initially lower mean was adjusted upward, while the initially higher means were adjusted downward. The group that started with a handicap is compensated, so to speak.

We illustrate now that testing the differences among adjusted means is the same as testing differences among intercepts of equations in which the common regression coefficient for the covariate, b_c is used. As shown earlier, it is possible to calculate the regression equations with a common b by using the re-

---

[18] In the present example the common coefficient, b_c, is equal to the average of the coefficients for the separate regression equations. This is because in the present example $\Sigma x_1^2 = \Sigma x_2^2 = \Sigma x_3^2$. For an explanation, see footnote 7.

gression equation in which the product vectors are deleted. For the present example this regression equation is

$$Y' = 5.23175 + .63810X - .96508(V_1) + .12063(V_2)$$

where $X =$ the covariate of Table 10.11; V_1, $V_2 =$ vectors 1 and 2 of Table 10.11.

The intercepts for groups I, II, and III are

$$a_I = 5.23175 + (-.96508) = 4.26667$$
$$a_{II} = 5.23175 + (.12063) = 5.35238$$
$$a_{III} = 5.23175 - [(-.96508) + (.12063)] = 6.07620$$

Note that the differences among the intercepts are the same as the differences among the adjusted means reported above. For example, the adjusted means for groups I and II are 8.20 and 9.29, respectively. The intercepts for groups I and II are 4.26667 and 5.35238, respectively. Accordingly

$$9.29 - 8.20 = 1.09$$
$$5.35238 - 4.26667 = 1.09$$

and similarly for all other differences.

In the present example, it was found that the treatments did not differ significantly. When, however, there are significant differences among treatments, it is necessary to determine specifically which of them differ significantly from each other. This is done by multiple comparisons between adjusted means. As usual, the kind of test used depends on whether the comparisons are a priori or post hoc.

As shown in Chapter 7, when the researcher uses orthogonal comparisons, the treatments can be coded to reflect such comparisons. The analysis of covariance with orthogonal coding is done in the same manner as illustrated above with effect coding. Subsequent to the overall analysis, however, the increment in the proportion of variance accounted for by each vector, after allowance is made for the covariate, is tested for significance in the manner shown in Chapter 7. The error term for the F ratio for each comparison is the residual mean square of the analysis of covariance. In the example analyzed above there were three treatments. Consequently two orthogonal vectors could have been used to reflect two possible orthogonal comparisons. In the foregoing analysis it was found that, after allowing for the covariate, the treatments accounted for about 4 percent of the variance. Had orthogonal coding been used this increment would have been the same. The use of orthogonal coding, however, would have resulted in partitioning the 4 percent increment due to treatments into two orthogonal components. It is these components that are tested for significance in the manner shown in Chapter 7.

Instead of orthogonal comparisons we can test a priori nonorthogonal comparisons between means. For comparisons of this kind in an analysis of covariance, t ratios are calculated for differences between the adjusted means of the treatments to which the a priori comparisons refer. The formula for the t ratio for a test between two adjusted means is

$$t = \frac{\bar{Y}_{1(adj)} - \bar{Y}_{2(adj)}}{\sqrt{MSR \left(\frac{1}{n_1} + \frac{1}{n_2}\right) \left[1 + \frac{ss_{reg(c)}}{k ss_{res(c)}}\right]}} \tag{10.13}$$

where $\bar{Y}_{1(adj)}$ and $\bar{Y}_{2(adj)}$ = adjusted means for treatments 1 and 2 respectively; MSR = residual mean square of the analysis of covariance; n_1, n_2 = number of subjects in groups 1 and 2, respectively; $ss_{reg(c)}$ = regression sum of squares of the covariate when it is regressed on the treatments; $ss_{res(c)}$ = residual sum of squares of the covariate when it is regressed on the treatments; k = number of coded vectors for treatments, or the degrees of freedom for treatments. The degrees of freedom for the t ratio of formula (10.13) equal the degrees of freedom for the residual mean square of the analysis of covariance.

It was said above that when all treatment groups have equal means on the covariate no adjustment of means takes place. Therefore the numerator of formula (10.13) will consist of unadjusted means when all treatment groups have equal means on the covariate. Furthermore, the regression sum of squares of the covariate ($ss_{res(c)}$) will equal zero. Consequently, formula (10.13) will reduce to the conventional t ratio formula, except that the MSR is the one of the analysis of covariance. It is when treatment groups differ on the covariate that the adjustments indicated in formula (10.13) are necessary. For designs with equal number of subjects in the groups, the denominator of formula (10.13) is constant for comparisons between any two adjusted means.

In order to illustrate the application of formula (10.13) let us assume that a priori comparisons were formulated for the example analyzed above, and that one of the comparisons is between Treatments I and II. It is first necessary to regress the covariate on the treatments. In the present example this means the regression of X of Table 10.11 on vectors 1 and 2 (the vectors representing the treatments). In other words, it is necessary to calculate $R^2_{x.12}$ and then obtain the regression sum of squares and the residual sum of squares of X, the covariate. For the present example we obtain

$$ss_{reg(c)} = 76.00$$
$$ss_{res(c)} = 52.50$$

From earlier calculation we have $\bar{Y}_{I(adj)} = 8.20$ and $\bar{Y}_{II(adj)} = 9.29$. The MSR (mean square residuals) is .79456 (see Table 10.12); $k = 2$. Applying formula (10.13) to the comparison between the adjusted means for treatments I and II,

$$t = \frac{9.29 - 8.20}{\sqrt{.79456 \left(\frac{1}{6} + \frac{1}{6}\right)\left[1 + \frac{76.00}{(2)(52.50)}\right]}}$$

$$= \frac{1.09}{\sqrt{.79456 \left(\frac{2}{6}\right)(1.72381)}} = \frac{1.09}{\sqrt{.45656}} = \frac{1.09}{.67569} = 1.61$$

with 14 degrees of freedom.

The Scheffé method for post hoc comparisons between means was discussed and illustrated in Chapters 7 and 8. This method also applies to post hoc comparisons in the analysis of covariance. Formula (7.15) is adapted for the case of the analysis of covariance:

$$D = C_1\bar{Y}_{1(\text{adj})} + C_2\bar{Y}_{2(\text{adj})} + \cdots + C_j\bar{Y}_{j(\text{adj})} \qquad (10.14)$$

where D = difference or contrast; C = coefficient by which a given adjusted mean, $\bar{Y}_{(\text{adj})}$, is multiplied; j = number of means involved in the comparison. For the kind of comparisons dealt with here it is required that $\Sigma C_j = 0$. That is, the sum of the coefficients in a given contrast must be zero.

A comparison is considered statistically significant if $|D|$ (the absolute value of D) exceeds a value S, which is defined as follows:

$$S = \sqrt{kF_\alpha; k, N-k-2} \sqrt{MSR\left[\sum \frac{(C_j)^2}{n_j}\right]} \qquad (10.15)$$

where k = number of coded vectors for treatments, or number of treatments minus one; $F_\alpha; k, N-k-2$ = tabled value of F with k and $N-k-2$ degrees of freedom at a prespecified α level; MSR = residual mean square of the analysis of covariance; C_j = coefficient by which the mean of group j is multiplied; n_j = number of subjects in group j.

Even though no significant differences between treatments were found in the above analysis, we illustrate the application of the Scheffé method to the comparison of the adjusted means of Treatments I and II. The information necessary for formulas (10.14) and (10.15) is

$$\bar{Y}_{I(\text{adj})} = 8.20; \quad \bar{Y}_{II(\text{adj})} = 9.29; \quad k = 2; \quad MSR = .79456; \quad N-k-2 = 14$$

The tabled F for 2 and 14 degrees of freedom for the .05 level is 3.74.

$$D = (1)(\bar{Y}_{I(\text{adj})}) + (-1)(\bar{Y}_{II(\text{adj})})$$
$$= (1)(8.20) + (-1)(9.29) = -1.09$$

$$S = \sqrt{(2)(3.74)} \sqrt{.79456\left[\frac{(1)^2}{6} + \frac{(-1)^2}{6}\right]}$$

$$= \sqrt{7.48} \sqrt{.79456\left(\frac{2}{6}\right)} = \sqrt{7.48}\sqrt{.26485}$$

$$= \sqrt{1.98108} = 1.41$$

Since $|D| = 1.09$ is smaller than $S = 1.41$, we conclude that the difference between the adjusted means of Treatments I and II is not significant at the .05 level. We repeat that this test was done for illustrative purposes only. In the present analysis the overall test for the differences among treatments was not significant, and therefore post hoc comparisons should not have been made.

Tabular Summary of the Analysis of Covariance

The major results of the foregoing analysis are reported in Table 10.12, thus providing a succinct summary of the procedures followed in the analysis of

TABLE **10.12**   SUMMARY OF THE ANALYSIS OF COVARIANCE FOR AN
EXPERIMENT WITH THREE TEACHING METHODS[a]

I:	Source	Prop. of Variance	ss	df	ms	F
	Covariate $R^2_{y.x}$	.85999	93.30934	1		
	Treatments (after adjustment) $R^2_{y.x12} - R^2_{y.x}$	.03748	4.06685	2	2.03342	2.56
	Error $(1 - R^2_{y.x12})$	.10253	11.12381	14	.79456	
	Total	1.00000	108.50000	17		

II:		Treatments		
		I	II	III
	Original Mean:	6.50	9.50	11.50
	Adjusted Mean:	8.20	9.29	10.01

[a]Original data given in Table 10.11. Y = dependent variable; X = covariate; 1 and 2 = coded vectors for treatments.

covariance. Part (I) of Table 10.12 reports the analysis of covariance. Part (II) of Table 10.12 reports the original and adjusted means of the dependent variable.

Analysis with Multiple Covariates

The procedures presented thus far can be easily extended to include more than one covariate. One needs to calculate the squared multiple correlation of the dependent measure with the covariates and the vectors representing treatments. Another squared multiple correlation is then calculated between the dependent variable and the covariates only. The difference between these two squared multiple correlations indicates the proportion of variance accounted for by the treatments after adjusting for the covariates. This difference is tested for significance with the F test. If, for example, in the experiment with teaching methods there were another covariate, motivation (X_2), then the basic analysis for differences between treatments would have been

$$F = \frac{(R^2_{y.x_1x_212} - R^2_{y.x_1x_2})/(4-2)}{(1 - R^2_{y.x_1x_212})/(N-4-1)}$$

where $R^2_{y.x_1x_212}$ = squared multiple correlation of Y with covariates X_1 and X_2 and two coded vectors (1 and 2) for the three treatments; $R^2_{y.x_1x_2}$ = squared multiple correlation of Y with covariates X_1 and X_2; N = total number of subjects.

Analysis of Covariance with Multiple Categorical Variables

In the teaching experiment analyzed above there was only one categorical variable, teaching methods. It is conceivable, naturally, that more than one independent variable will be used. For example, another variable might have been the educational background of the teachers; for instance, having a degree from a liberal arts school or from a school of education. In this case, there would have been a covariate, premeasures, and two categorical variables, teaching methods and teachers' educational backgrounds. This is factorial analysis of covariance. In the context of the present chapter, an additional coded vector is all that is needed to represent the teachers' educational backgrounds. Product vectors between the categorical variables are then generated to represent their interaction. The analysis is: calculate the difference between the squared multiple correlation that includes the covariate(s) and all the coded vectors and the squared multiple correlation with the covariate(s) only. Such an analysis is in effect a combination of the methods presented in this chapter and in Chapter 8.

In conclusion, analysis of covariance is seen to be a special case of the methods discussed and illustrated earlier in the chapter. It can be used either for better control (reduction in the error term) or for adjustment for initial differences on a variable related to the dependent variable, or both. The basic analysis consists of testing differences between a squared multiple correlation that includes all vectors, and one that includes the covariate(s) only. Various computer programs enable one, by successive deletions of variables, to calculate in a single run all the R^2's one needs.

Summary

The collection and analysis of data in scientific research are guided by hypotheses derived from theoretical formulations. Needless to say, the closer the fit between the analytic method and the hypotheses being tested, the more one is in a position to draw appropriate and valid conclusions. The overriding theme of this chapter was that certain analytic methods considered by some researchers as distinct or even incompatible are actually part of the multiple regression approach. To this end, methods introduced separately in preceding chapters were brought together.

Whether the independent variables are continuous, categorical, or a combination of both, the basic approach is to bring all the information to bear on the explanation of the variance of the dependent variable. Used appropriately, this approach can enhance the researcher's efforts to explain phenomena. It was shown, for example, that the practice of categorizing continuous variables leads to loss of information. More important, however, is the loss of explanatory power in that the researcher is not able to study the trend of the relation between the continuous variable that was categorized and the dependent variable.

The analytic methods presented in this chapter were shown to be particularly useful for testing hypotheses about trends and interactions between continuous and categorical variables. The application of these methods not only enables the researcher to test specific hypotheses but also to increase the general sensitivity of the analysis. Finally, it was shown that the use of continuous variables to achieve better control, as in the analysis of covariance, is an aspect of the overall regression approach.

Study Suggestions

1. Distinguish between categorical and continuous variables. Give examples of each.
2. In a study of the relation between X and Y in three separate groups, some of the results were: $\Sigma x_1 y_1 = 72.56$; $\Sigma x_2 y_2 = 80.63$; $\Sigma x_3 y_3 = 90.06$; $\Sigma x_1^2 = 56.71$; $\Sigma x_2^2 = 68.09$; $\Sigma x_3^2 = 75.42$.
    Calculate: (a) the three separate b coefficients; (b) the common b coefficient; (c) regression sum of squares when the separate b's are used; (d) regression sum of squares when the common b is used.
    (*Answers*: (a) $b_1 = 1.28$; $b_2 = 1.18$; $b_3 = 1.19$; (b) $b_c = 1.21$; (c) 295.86; (d) 295.53.)
3. Distinguish between ordinal and disordinal interaction.
4. What is meant by "the research range of interest"?
5. In a study with two groups, A and B, the following regression equations were obtained:
$$Y'_A = 22.56 + .23X$$
$$Y'_B = 15.32 + .76X$$

At what value of X do the two regression lines intersect?
(*Answer*: 13.66.)
6. What is meant by "aptitude-treatment interaction"? Give examples of research problems in which the study of ATI may be important.
7. Discuss the uses of analysis of covariance. Concentrate on its functions in experimental research. Give examples.
8. Why is it important to determine whether the b's are homogeneous when doing an analysis of covariance?
9. A researcher wished to determine whether the regression of achievement on achievement motivation is the same for males and females. For a sample of males ($N = 12$) and females ($N = 10$) he obtained measures of achievement and achievement motivation. Following are the data (fictitious):

	Males		Females	
		Achievement		Achievement
	Achievement	Motivation	Achievement	Motivation
	25	1	22	1
	29	1	24	1
	34	2	24	2
	35	2	22	3
	35	3	29	3
	39	3	30	3
	42	4	30	4
	43	4	33	4
	46	4	32	5
	46	5	35	5
	48	5		
	50	5		

Analyze the data.

(a) What is the proportion of variance accounted for by sex, achievement motivation, and their interaction?

(b) What is the proportion of variance accounted for by the interaction of sex and achievement motivation?

(c) What is the F ratio for the interaction?

(d) What is the overall regression equation for achievement motivation, sex, and the interaction (when effect coding $\{1, -1\}$ is used for sex)?

(e) What are the regression equations for the two groups?

(f) What is the point of intersection of the regression lines?

(g) What type of interaction is there in the present example?

(h) What is the region of significance at the .05 level?

Plot the regression lines and interpret the results.

(*Answers*: (a) .93785; (b) .03629; (c) $F = 10.52$, with 1 and 18 df; (d) $Y' = 21.06903 + 3.95617(AM) + 1.64575(S) + 1.15723(INT.)$; (e) $Y'_{(M)} = 22.71478 + 5.11340X$, $Y'_{(F)} = 19.42328 + 2.79894X$; (f) $= -1.42214$; (g) ordinal; (h) males and females whose scores on achievement motivation are larger than .52 differ significantly in achievement.)

10. An educational researcher studied the effects of three different methods of teaching on achievement in algebra. He randomly assigned 25 students to each method. At the end of the semester he obtained achievement scores on a standardized algebra test. In order to increase the sensitivity of his analysis, the researcher decided to use the students' IQ as a covariate. The data (fictitious) for the three groups are as follows:

Method A		Method B		Method C	
IQ	Algebra	IQ	Algebra	IQ	Algebra
90	42	90	48	90	58
92	40	91	48	92	58
93	42	93	50	93	58
94	42	94	50	94	60
95	42	95	50	95	60
96	44	96	52	96	62
97	44	97	52	97	62
98	46	98	52	99	62
99	44	99	52	100	63
100	46	100	54	102	64
102	46	101	54	103	66
103	46	102	52	104	64
104	48	103	54	105	66
106	48	104	56	107	64
107	48	105	54	108	65
108	50	106	55	110	66
109	50	108	56	111	64
111	50	109	56	112	66
113	50	111	56	113	68
114	52	112	58	115	68
116	53	114	60	116	70
118	54	115	60	118	70
119	54	116	60	118	68
120	56	118	60	120	72
121	56	120	62	121	74

Analyze the data.

(a) What are the b's for the separate groups?

(b) What is the common b?

(c) What is the F ratio for the test of homogeneity of regression co-efficients?

(d) What is the proportion of variance accounted for by teaching methods and the covariate?

(e) What is the proportion of variance accounted for by the covariate?

(f) What is the F ratio for teaching methods without covarying IQ?

(g) What is the F ratio for teaching methods after covarying IQ?

(h) What are the adjusted means for the three methods?

Interpret the results.

(*Answers*: (a) $b_A = .48096$; $b_B = .44152$; $b_C = .42577$; (b) $b = .44990$; (c) $F = 1.62$, with 2 and 69 *df*; (d) .98422; (e) .28058; (f) $F = 98.16$, with 2 and 72 *df*; (g) $F = 1599.18$, with 2 and 71 *df*; (h) $\bar{Y}_A = 47.64$; $\bar{Y}_B = 54.86$; $\bar{Y}_C = 64.38$.)

Explanation and Prediction

In Part I of this book it was noted that the interpretation of results from a multiple regression analysis may become complex and perplexing. It was pointed out, for example, that the question of the relative importance of variables is so complex that it almost seems to elude a solution. Although different approaches toward a solution were presented (for example, the magnitude of the β's and squared semipartial correlations), we now need to elaborate points made earlier as well as to develop new points.

Regression analysis can play an important role in predictive and explanatory research frameworks. Prediction and explanation reflect different research concerns and emphases. In prediction studies the main emphasis is on practical application. On the basis of knowledge of one or more independent variables, the researcher wishes to develop a regression equation to be used for the prediction of a dependent variable, usually some criterion of performance or accomplishment. The choice of independent variables in the predictive framework is determined primarily by their potential effectiveness in enhancing the prediction of the criterion.

In an explanatory framework, on the other hand, the basic emphasis is on the explanation of the variability of a dependent variable by using information from one or more independent variables. The choice of independent variables is determined by theoretical formulations and considerations. Stated differently, when the concern is explanation, the emphasis is on formulating and testing explanatory models or schemes. It is within this context that questions about the relative importance of independent variables become particularly meaningful. Explanatory schemes may, under certain circumstances, be enhanced by inferences about causal relations among the variables under study.

The basic analytic techniques of regression analysis are the same when used in studies primarily concerned with prediction or with explanation. The interpretation of the results, however, may differ depending on whether the emphasis is on one or the other. Consequently, prediction and explanation are dealt with separately in this chapter. Some applications of multiple regression analysis are of course appropriate in either a predictive or an explanatory framework.

We begin with a brief discussion of shrinkage of the multiple correlation and the method of cross validation. This is followed by a treatment of three types of solutions useful in the predictive framework. Basic problems of explanatory models are then presented and discussed. In this context two approaches are presented, namely commonality analysis and path analysis. While the former is an extension of some of the ideas and methods presented in earlier chapters, the latter requires further elaboration of these methods, as well as a treatment of the complex problems related to attempts of making causal inferences in non-experimental research.

Shrinkage of the Multiple Correlation and Cross-Validation

The choice of a set of weights in a regression analysis is designed to yield the highest possible correlation between the independent variables and the dependent variable. Recall that the multiple correlation can be expressed as the correlation between the predicted scores based on the regression equation and the observed criterion scores. If one were to apply a set of weights derived in one sample to the predictor scores of another sample and then correlate these predicted scores with the observed criterion scores, the resulting R will almost always be smaller than the R obtained in the sample for which the weights were originally calculated. This phenomenon is referred to as the shrinkage of the multiple correlation. The reason for shrinkage is that in calculating the weights to obtain a maximum R, the zero-order correlations are treated as if they were error-free. This is of course never the case. Consequently, there is a certain amount of capitalization on chance, and the resulting R is biased upwards.

The degree of the overestimation of R is affected, among other things, by the ratio of the number of independent variables to the size of the sample. Other things equal, the larger this ratio, the greater the overestimation of R. Some authors recommend that the ratio of independent variables to sample size be at least 30 subjects per independent variable. This is a rule of thumb that does not satisfy certain researchers who say that samples should have at least 400 subjects. Needless to say, the larger the sample the more stable the results. It is therefore advisable to work with fairly large samples.

Even though it is not possible to determine exactly the degree of over-estimation of R, it is possible to estimate the amount of shrinkage by applying the following formula:

$$\hat{R}^2 = 1 - (1 - R^2)\left(\frac{N-1}{N-k-1}\right) \tag{11.1}$$

where $\hat{R}^2$ = estimated squared multiple correlation in the population; R^2 = obtained squared multiple correlation; N = size of the sample; k = number of independent variables. For a detailed discussion and an unbiased estimator of R^2, see Olkin and Pratt (1958).

The application of formula (11.1) is demonstrated for three different sample sizes. Assume that the squared multiple correlation between three independent variables and a dependent variable is .36. What will $\hat{R}^2$ be if the ratios of the independent variable to the sample size were $1:5, 1:30, 1:50$? In other words, what will $\hat{R}^2$ be if the sample sizes for which the R was obtained were 15, 90, 150?

For a sample of 15 (1 : 5 ratio),

$$\hat{R}^2 = 1 - (1 - .36)\left(\frac{15-1}{15-3-1}\right) = 1 - (.64)\left(\frac{14}{11}\right) = 1 - .81 = .19$$

For a sample of 90 (1 : 30 ratio)

$$\hat{R}^2 = 1 - (1 - .36)\left(\frac{90-1}{90-3-1}\right) = 1 - (.64)\left(\frac{89}{86}\right) = 1 - .66 = .34$$

For a sample of 150 (1 : 50 ratio),

$$\hat{R}^2 = 1 - (1 - .36)\left(\frac{150-1}{150-3-1}\right) = 1 - (.64)\left(\frac{149}{146}\right) = 1 - .65 = .35$$

Note that with a ratio of $1:5$, $\hat{R}^2$ is about half the size of R^2 (.19 and .36 respectively), when the ratio is $1:30$, the estimated shrinkage of R^2 is about .02 (from .36 to .34), and with a ratio of $1:50$ it is about .01 (from .36 to .35).

The above discussion and formula (11.1) apply to the case when all the independent variables are used in the analysis. When a selection procedure is applied to the independent variables as, for example, in a stepwise solution, the capitalization on chance is even greater. This is because the "best" set of variables selected from a larger pool is bound to have errors due to the correlations of these variables with the criterion, as well as errors due to the intercorrelations among the predictors. In an effort to offset some of these errors, one sould have large samples (about 500) whenever a number of variables is to be selected from a larger pool of variables.

Cross-Validation

Probably the best method for estimating the degree of shrinkage is to perform a cross-validation (Mosier, 1951; Lord & Novick, 1968, pp. 285 ff.; Herzberg, 1969). This is done by using two samples. For the first sample a regular regression analysis is performed, and R^2 and the regression equation are calculated. The regression equation is then applied to the predictor variables of the second sample, thus yielding a Y' for each subject. The first sample is referred to as the *screening sample*, and the second as the *calibration sample* (Lord & Novick, 1968, p. 285). (If a selection of variables is used in the screening sample, the regression equation is applied to the same variables in the calibration sample.) A Pearson r is then calculated between the observed criterion scores (Y) in the calibration sample and the predicted criterion scores (Y'). This $r_{yy'}$ is analogous to a multiple correlation in which the equation used is the one obtained in the screening sample.

The difference between R^2 of the screening sample and R^2 of the calibration sample is an estimate of the amount of shrinkage. If the shrinkage is small and the R^2 is considered meaningful by the researcher, he can apply the regression equation obtained in the screening sample to future predictions. As pointed out by Mosier (1951), however, a regression equation based on the combined samples (the screening and calibration samples) has greater stability due to the larger number of subjects on which it is based. It is therefore recommended that after deciding that the shrinkage is small, the two samples be combined and the regression equation for the combined samples be used in future predictions.[1]

Cross-validation, then, needs two samples. Sometimes, long delays in assessing the findings in a study may occur due to difficulties in obtaining a second sample. In such circumstances, an alternative approach is recommended. A large sample (say 500) is randomly split into two sub-samples. One subsample is used as the screening sample, and the other is used for calibration.

Double Cross-Validation

Some researchers are not satisfied with cross-validation and insist on double cross-validation (Mosier, 1951). The procedure outlined above is applied twice. For each sample (or random subsample of a given sample), R^2 and the regression equation are calculated. Each regression equation obtained in one sample is then applied to the predictor variables of the other sample, and R^2 is calculated by using $r_{yy'}$. One thus has two R^2's calculated directly in each sample, and two R^2's calculated on the basis of regression equations obtained from alternate samples. It is then possible to study the differences between the R^2's as well as the differences in the two regression equations. If the results are close, one may combine the samples and calculate the regression equation to be used in prediction. Double cross-validation is strongly recommended as the most rigorous approach to the validation of results from regression analysis in a predictive framework.

---

[1]Note that it is not possible to estimate the shrinkage of R^2 for the combined sample, unless one were to obtain another calibration sample. For a discussion of this point, see Mosier (1951).

It is always necessary to be alert to possible future changes in situations that may diminish the usefulness of regression equations, or even make them useless. If, for example, the criterion is grade-point average in college, and there have been important changes in grading policies, a regression equation derived in a situation prior to such changes may not apply any longer.

Selecting Variables for Prediction

A researcher's primary interest is often not in hypothesis testing, or in assessing the relative importance of independent variables, but rather in making as good a prediction to a criterion as possible on the basis of several predictor variables. Under such circumstances, one's efforts are directed toward obtaining as high a squared multiple correlation as possible. Because many of the variables in the behavioral sciences are intercorrelated, it is often possible to select from a pool of variables a smaller set, which will yield an R^2 almost equal in magnitude to the one obtained by using the total set.

When variables are selected from an available pool, the aim is usually the selection of the minimum number of variables necessary to account for almost as much of the variance as is accounted for by the total set. But practical considerations, such as relative costs in obtaining measures of the variables, ease of administration, and the like, often enter into the selection process. Under such circumstances, one may end up with a larger number of variables than the minimum that would be selected when the sole criterion is to account for almost as much of the variance as does the total set of variables. A researcher may, for example, select five variables in preference to three others that would yield about the same R^2 but at much greater cost.

Because practical considerations vary with given sets of circumstances, it is not possible to formulate a systematic selection method that takes such considerations into account. The researcher must select the variables on the basis of his specific means, needs, and circumstances. When, however, his sole aim is the selection of the minimum number of variables necessary to account for much of the variance accounted for by the total set, he may use one of several selection methods that have been developed for this purpose. We present three such methods: the forward solution, the backward solution, and the stepwise solution.

Forward Solution

This solution proceeds in the following manner. The correlations of all the independent variables with the dependent variable are calculated. The independent variable that has the highest zero-order correlation with the dependent variable is entered first into the analysis. The next variable to enter is the one that produces the greatest increment to R^2, after having taken into account the variable already in the equation. In other words, it is the variable that has the highest squared semipartial correlation with the dependent variable, after partialing the variable already in the equation. The squared semipartial indicates the increment in the R^2, or the incremental variance, attributed to the second variable.[2] The third variable to enter is the one that has the highest squared semipartial correlation with the dependent variable after having partialed out the first two variables already in the equation. Some authors work with partial

---

[2]The criterion here is purely statistical. As noted above, other considerations may enter into a selection process.

rather than with semipartial correlations. The results are the same.[3] The process outlined above may be continued for as many variables as one wishes to enter. At each succeeding step, the variable with the largest squared semipartial correlation is the one to enter the equation.

Criteria for Terminating the Analysis.   Since the reason for using a forward solution is to select a smaller set of variables from those available, it is necessary to know when to terminate the analysis. In other words, there is need for a criterion when to stop entering additional variables into the equation. Basically, two kinds of criteria may be used: statistical significance and meaningfulness.

At each stage of the analysis one can test whether an increment in R^2 attributed to a given variable is statistically significant. A formula introduced early in this book, and used frequently throughout it, is most suited for this purpose. This formula is repeated here for easy reference, but without elaboration:

$$F = \frac{(R^2_{k_1} - R^2_{k_2})/(k_1 - k_2)}{(1 - R^2_{k_1})/(N - k_1 - 1)} \tag{11.2}$$

A significant F ratio indicates that the increment in R^2 is statistically significant.

With large samples even a minute increment will be statistically significant. Since the use of large samples is mandatory in regression analysis, it is also advisable to use the criterion of meaningfulness. This involves a decision by the researcher as to whether an increment is substantively meaningful. Such a decision may, for example, be based on the effort involved in obtaining the additional variable in relation to the increment it contributes to the R^2. Meaningfulness is specific to particular situations. What is considered a meaningful increment in one situation may not be considered meaningful in another situation.

In sum, then, one can terminate the analysis on purely statistical grounds, or one can use, in addition, the criterion of meaningfulness. It is recommended that meaningfulness be given the primary consideration. What good is a statistically significant increment if it is not meaningful?

A Numerical Example.   The forward solution is now illustrated with a numerical example using three independent variables. Assume that one wishes to predict the grade-point average (GPA) of college students, so that a selection procedure may be established. On the basis of a review of the literature, and theoretical and practical considerations, the researcher selects the following independent variables: socioeconomic status (SES), intelligence (IQ), and need

---

[3]A squared semipartial is equal to the product of a squared partial and a residual. For example, dealing with variable 2 after having entered variable 1 in the equation,

$$r^2_{y(2.1)} = r^2_{y2.1} (1 - r^2_{y1})$$

The left side of the equation is a squared semipartial. The right side of the equation is the product of a squared partial correlation and the residual, after accounting for variable 1. Either the left or the right side of the above equation indicates the increment in R^2 due to variable 2. Working with semipartials seems to be a more direct approach.

achievement (n Ach).[4] The researcher wishes now to determine whether the three variables are needed, or whether one or two will yield almost as good prediction as all three.

A correlation matrix based on fictitious data for 100 subjects is reported in Table 11.1. Note that variable 2 (IQ) has the highest zero-order correlation with Y (GPA). This variable is therefore the first to enter into the equation. It is now necessary to calculate the semipartials for the remaining variables. We repeat formula (5.7), which for the semipartial of Y with variable 1 takes the following form:

$$r_{y(1.2)} = \frac{r_{y1} - r_{y2}r_{12}}{\sqrt{1 - r_{12}^2}} \tag{11.3}$$

$$r_{y(1.2)} = \frac{(.33) - (.57)(.30)}{\sqrt{1 - (.30)^2}} = \frac{.1590}{\sqrt{.91}} = \frac{.1590}{.9539} = .1667$$

$$r_{y(3.2)} = \frac{(.50) - (.57)(.16)}{\sqrt{1 - (.16)^2}} = \frac{.4088}{\sqrt{.9744}} = \frac{.4088}{.9871} = .4141$$

Since $r_{y(3.2)}$ is the larger of the two semipartials, variable 3 (n Ach) is the next to enter the equation.

$$R_{y.23}^2 = r_{y2}^2 + r_{y(3.2)}^2 = (.57)^2 + (.4141)^2 = .4964$$

Testing now for the significance of the increment to R^2 attributed to variable 3,

$$F = \frac{(R_{y.23}^2 - R_{y.2}^2)/(2-1)}{(1 - R_{y.23}^2)/(100 - 2 - 1)}$$

$$= \frac{(.4964 - .3249)/(2-1)}{(1 - .4964)/(100 - 2 - 1)} = \frac{.1715/1}{.5036/97} = \frac{.1715}{.00519} = 33.04$$

Need achievement accounts for about 17 percent of the variance in GPA, over and above the proportion of variance accounted for by IQ (32 percent). This increment is obviously meaningful. It is also significant beyond the .001 level ($F = 33.04$, with 1 and 97 df).

[4]Ordinarily, one would start with a larger number of variables. The choice of three in the present example is for illustrative purposes only.

TABLE 11.1   CORRELATION MATRIX FOR THREE INDEPENDENT VARIABLES AND A DEPENDENT VARIABLE. $N = 100$

	1 SES	2 IQ	3 n Ach	Y GPA
1	1.00	.30	.41	.33
2		1.00	.16	.57
3			1.00	.50
Y				1.00

Since there is only one variable left (SES), the question is whether it will add meaningfully and significantly to the R^2. For this purpose we calculate

$$r_{y(1.23)} = \frac{r_{y(1.2)} - r_{y(3.2)}r_{1(3.2)}}{\sqrt{1 - r_{1(3.2)}^2}}$$

$$= \frac{.1667 - (.4141)(.3667)}{\sqrt{1 - (.3667)^2}} = \frac{.01485}{\sqrt{.86553}} = \frac{.01485}{.93034} = .01596$$

$$r_{y(1.23)}^2 = (.01596)^2 = .0002$$

SES will add about .02 percent to the proportion of variance in GPA over and above the proportion of variance already accounted for by IQ and n Ach. It is clear that if the sole criterion is prediction, SES adds practically nothing to the prediction and will therefore not be used. Nevertheless, for the purpose of illustration, we test for significance of the increment due to SES:

$$R_{y.231}^2 = r_{y2}^2 + r_{y(3.2)}^2 + r_{y(1.23)}^2$$

$$= .3249 + .1715 + .0002 = .4966$$

$$F = \frac{(.4966 - .4964)/(3-2)}{(1 - .4966)/(100-3-1)} = \frac{.0002/1}{.5034/96} = \frac{.0002}{.0052} = .04$$

The F ratio is less than one. We thus have an increment that is neither meaningful nor statistically significant.

Using only IQ and n achievement will account for about 50 percent of the variance of GPA. The regression equation expressed in z scores is

$$z_y' = .5029z_2 + .4195z_3$$

where z_2 is the standard score on IQ, and z_3 is the standard score on n achievement. It is generally more economical to express the regression equation in raw score form, since one can thereby obtain the predicted score without having first to convert the raw scores on the predictor variables to standard scores. In the present example, however, we started with correlations and therefore cannot express the regression equation in raw score form.

It should be noted that in the forward solution no allowance is made for studying the effect the introduction of new variables may have on the usefulness of the variables already in the equation. It is possible, due to the combined contribution of variables introduced at a later stage of the analysis, and the relations of these variables with those already in the equation, that a variable introduced at an earlier stage may be disposed of with very little loss in R^2. In the forward solution, however, the variables are "locked in" in the order in which they were introduced into the equation. The only course open to the researcher is to note whether the addition of a variable is meaningful or significant.

Backward Solution

The backward solution starts out with the squared multiple correlation of all independent variables with the dependent variable. Each independent variable is deleted from the regression equation one at a time, and the loss to R^2 due to the deletion of the variable is studied. In other words, each variable is treated as if it were entered last in the equation. It is thus possible to observe which variable adds the least when entered last. The loss in R^2 that occurs as a result of the deletion of a variable may be assessed against a criterion of meaningfulness as well as significance. A variable considered not to add meaningfully or significantly to prediction is deleted.

If no variable is deleted, the analysis is terminated. Evidently all the variables contribute meaningfully to the prediction of the criterion. If, on the other hand, a variable is deleted, then the process described above is repeated for the remaining variables. That is, each of the remaining variables is deleted, in turn, and the one with the smallest contribution to R^2 is studied. Again, it may be either deleted or retained on the basis of the criteria used by the researcher. If the variable is deleted, one repeats the process described above to determine whether an additional variable may be deleted. The analysis continues as long as one deletes variables that produce no meaningful or significant loss to R^2. When the deletion of any one variable produces a meaningful or significant loss to R^2, the analysis is terminated.

A Numerical Example. The backward solution is now applied to the data used to illustrate the forward solution. It will be recalled that the dependent variable was grade-point average, while the three independent variables were socioeconomic status, intelligence, and need achievement (see Table 11.1). $R^2_{y.123}$ was .4966. It is now necessary to delete, in turn, each of the independent variables and to observe the loss caused to the R^2. In other words, it is necessary to calculate $R^2_{y.23}$, $R^2_{y.13}$, and $R^2_{y.12}$, and note which of these is the least discrepant from $R^2_{y.123}$. The actual calculations of the three multiple correlations are left as an exercise for the reader. A summary of the losses in R^2 as a result of the deletion of each of the variables is provided in Table 11.2. The deletion of variable 1 results in a loss of .0002 in the R^2, or the proportion of variance accounted for. The deletion of variable 2 results in a loss of .2278, and the deletion of variable 3 results in a loss of .1439. It is clear that variable 1 can

TABLE **11.2**   FIRST STEP IN THE BACKWARD SOLUTION. DATA FROM
TABLE 11.1. $N = 100$

Variable Deleted	Operation	Prop. of Variance Lost	df	F
1–SES	$R^2_{y.123} - R^2_{y.23}$	$.4966 - .4964 = .0002$	1	< 1
2–IQ	$R^2_{y.123} - R^2_{y.13}$	$.4966 - .2688 = .2278$	1	43.47
3–n Ach	$R^2_{y.123} - R^2_{y.12}$	$.4966 - .3527 = .1439$	1	27.46
Error	$(1 - R^2_{y.123})$	.5034	96	

be deleted without meaningful loss. The F ratios for the deletion of each of the variables are also reported in Table 11.2. Each F ratio indicates whether the deletion of the variable with which it is associated results in a significant loss in R^2. The F ratio associated with variable 1 is smaller than one, while the F ratios associated with the other two variables are significant beyond the .001 level. On both counts, meaningfulness and significance, variable 1 can be deleted.

The next step is to note whether another variable can be deleted. In the present case there are only two variables left, variables 2 and 3. $R^2_{y.23} = .4964$. Deleting variable 2,

$$R^2_{y.23} - r^2_{y3} = .4964 - (.50)^2 = .2464$$

Deleting variable 3,

$$R^2_{y.23} - r^2_{y2} = .4964 - (.57)^2 = .1715$$

In view of the fact that the deletion of either variable 2 or variable 3 results in a meaningful loss in the variance accounted for (about 25 percent for variable 2, and about 17 percent for variable 3), it is decided not to delete any of the remaining variables. Variables 2 and 3 are retained, the regression equation is calculated, and the analysis is terminated.

In the present example both solutions (the forward and the backward) yielded the same results. It is possible, however, for the two solutions to select different sets of independent variables. As noted above, in the forward solution a variable already in the equation is not deleted even though its usefulness may be diminished by variables entering at subsequent stages. In the backward solution, on the other hand, each variable is viewed in the light of the contribution of all the other variables. Thus a variable may be deleted by the backward solution while it is retained in the forward solution.

The backward solution is more laborious than the forward solution. It seems desirable to have a method with some of the advantages of both solutions. Such a method is the stepwise solution.

Stepwise Solution

The stepwise solution is a variation on the forward solution. One of the shortcomings of the latter is that variables entered into the equation are retained despite the fact that they may have lost their usefulness in the light of the contributions made by variables entered at later stages. In the stepwise solution tests are performed at each step to determine the contribution of each variable already in the equation if it were to enter last. It is thus possible to discard a variable that was initially a good predictor. Criteria for removal of a variable may be meaningfulness or significance.

A Numerical Example. The application of the stepwise solution is illustrated with a set of fictitious data. The criterion measure is the grade-point average (GPA) earned in a graduate psychology department. Four predictors are used. Of these, three are measures administered to each student at the time of application. They are: (1) Graduate Record Examination-Quantitative

(GRE-Q); (2) Graduate Record Examination-Verbal (GRE-V); and (3) Miller Analogies Test (MAT). In addition, each applicant is interviewed by three professors, each of whom rate the applicant on a five-point scale, five indicating a very promising candidate. The average rating (AR) given by the three professors is the fourth predictor. A stepwise regression analysis is to be done to select the set of variables that best predicts GPA and that eliminates superfluous variables. A set of fictitious data for 30 subjects[5] on the five variables is reported in Table 11.3.

The calculations of a stepwise regression analysis are quite laborious. It is therefore recommended that they be done with a computer. Several computer programs for the stepwise regression analysis are available. For the present example we used BMD02R (Dixon, 1970). In this program one specifies, among other things, a level of F ratio for entering a variable into the equation and a level of F for removing a variable from the equation. At each step, the variable that makes the greatest increment to R^2 is entered, provided the F ratio associated with it — labeled "F to enter" — exceeds the prespecified F for entering a variable. Equivalently, it is the variable that has the highest partial correlation with the criterion, after having partialed all the variables already in the equation.

The contribution of each of the variables in the equation is then re-examined. This is done by treating, in turn, each variable as entering last in the equation. F ratios are calculated for each variable when it is entered last. These F ratios are labeled "F to remove," since they indicate the significance level associated with the removal of the variable. If a variable has an "F to remove" smaller than the prespecified F ratio for removal, it is removed from the equation. The next step is then taken. The analysis is terminated when no variable not in the equation has an "F to enter" that exceeds the prespecified F for entering, and no variable in the equation has an "F to remove" smaller than the prespecified F for removal.

At each step of the analysis, various calculations are performed and printed. Among these are: R, the regression equation, partial correlations, regression and residual sums of squares, "F to enter" for each variable not in the equation, and "F to remove" for each variable in the equation. What follows does not report all the results, but only those most pertinent for the presentation and interpretation of the stepwise regression analysis.

The intercorrelation matrix (zero-order correlations) for the five variables is reported in Table 11.4. A summary of the various steps taken in the present analysis is provided in Table 11.5. This table is divided into two major parts: one for variables in the equation and one for variables not in the equation. At each step there is an indication of the variable that was entered, the R between the variables in the equation (including the one entered in the step) and the dependent variable, and "F to remove" for each variable in the equation. For

---

[5] As discussed earlier, a larger sample is required. The present set is provided for illustrative purposes only.

TABLE **11.3** FICTITIOUS DATA FOR A
STEPWISE REGRESSION ANALYSIS. $N = 30$[a]

Ss	GPA	GRE-Q	GRE-V	MAT	AR
1	3.2	625	540	65	2.7
2	4.1	575	680	75	4.5
3	3.0	520	480	65	2.5
4	2.6	545	520	55	3.1
5	3.7	520	490	75	3.6
6	4.0	655	535	65	4.3
7	4.3	630	720	75	4.6
8	2.7	500	500	75	3.0
9	3.6	605	575	65	4.7
10	4.1	555	690	75	3.4
11	2.7	505	545	55	3.7
12	2.9	540	515	55	2.6
13	2.5	520	520	55	3.1
14	3.0	585	710	65	2.7
15	3.3	600	610	85	5.0
16	3.2	625	540	65	2.7
17	4.1	575	680	75	4.5
18	3.0	520	480	65	2.5
19	2.6	545	520	55	3.1
20	3.7	520	490	75	3.6
21	4.0	655	535	65	4.3
22	4.3	630	720	75	4.6
23	2.7	500	500	75	3.0
24	3.6	605	575	65	4.7
25	4.1	555	690	75	3.4
26	2.7	505	545	55	3.7
27	2.9	540	515	55	2.6
28	2.5	520	520	55	3.1
29	3.0	585	710	65	2.7
30	3.3	600	610	85	5.0
M:	3.31	565.33	575.33	67.00	3.57
s:	.60	48.62	83.03	9.25	.84

[a]GPA = Grade-Point Average; GRE-Q = Graduate Record
Examination-Quantitative; GRE-V = Graduate Record
Examination-Verbal; MAT = Miller Analogies Test; AR =
Average Rating.

each variable not in the equation the following are reported: the partial correlation with the dependent variable, partialing all the variables in the equation
at the given step, and "F to enter," which indicates the F ratio for the increment in R^2 if the variable were to enter last into the equation.

Let us take a closer look at Table 11.5, starting with step 1. Since AR has
the highest zero-order correlation with GPA (.621; see Table 11.4) it is

TABLE **11.4**   CORRELATION MATRIX OF VARIABLES USED IN THE
STEPWISE REGRESSION ANALYSIS[a]

	GPA	GRE-Q	GRE-V	MAT	AR
GPA	1.000	.611	.581	.604	.621
GRE-Q		1.000	.468	.267	.508
GRE-V			1.000	.426	.405
MAT				1.000	.525
AR					1.000

[a]GPA = Grade-Point Average; GRE-Q = Graduate Record Examination-Quantitative; GRE-V = Graduate Record Examination-Verbal; MAT = Miller Analogies Test; AR = Average Rating. Original data in Table 11.3.

TABLE **11.5**   SUMMARY OF STEPWISE REGRESSION ANALYSIS FOR
DATA OF TABLE 11.3. $N = 30$[a]

	Variables in Equation					Variables Not in Equation			
Step	Variable Entered	R	df	F[b]	F to Remove[c]	Vari-able	Partial Correla-tion	df	F to Enter[d]
1	AR	.6207	1/28	17.550	17.550	GRE-Q	.43814	1/27	6.414
						GRE-V	.46022	1/27	7.256
						MAT	.41727	1/27	5.692
2	GRE-V	.7180	2/27	14.363					
	AR		1/27		9.886	GRE-Q	.34032	1/26	3.406
	GRE-V		1/27		7.256	MAT	.34113	1/26	3.424
3	MAT	.7562	3/26	11.577					
	AR		1/26		4.761	GRE-Q	.40029	1/25	4.770
	GRE-V		1/26		4.831				
	MAT		1/26		3.424				
4	GRE-Q	.8003	4/25	11.134					
	AR		1/25		1.629				
	GRE-V		1/25		2.105				
	MAT		1/25		4.789				
	GRE-Q		1/25		4.770				
5	AR-REMOVED	.7855	3/26	13.964					
	GRE-V		1/26		2.311	AR	.24735	1/25	1.629
	MAT		1/26		8.949				
	GRE-Q		1/26		8.389				

[a]See footnote a, Table 11.4, for the names of the independent variables.

[b]The F ratio for the overall R at each step. For example, at step 2 the F ratio for $R = .7180$ is 14.363 with 2 and 27 degrees of freedom.

[c]The "F to Remove" at each step is a test of the loss caused to R by removing a given variable. For example, in step 2 the proportion of variance with which R is decreased by removing the variable AR has an F ratio of 9.886 with 1 and 27 degrees of freedom.

[d]The "F to Enter" is a test of the increment in the proportion of variance accounted for by a given variable entered last in the equation. For example, in step 1, if GRE-Q is entered after AR, which is already in the equation, the increment in the proportion of variance due to GRE-Q has an F ratio of 6.414, with 1 and 27 degrees of freedom.

selected to enter first in the equation. R in this case is the same as the zero-order correlation (.6207), and the F ratio is 17.550, with 1 and 28 degrees of freedom. Note that the F to remove is the same as the F for the R, since we are dealing with one variable only. We turn now to the variables not in the equation, and demonstrate how the statistics associated with each variable are obtained. For example, the partial correlation for GPA and GRE-Q, partialing AR (which is already in the equation)[6] is

$$r_{\text{GPA, GRE-Q.AR}} = \frac{(.611) - (.621)(.508)}{\sqrt{1 - (.621)^2}\sqrt{1 - (.508)^2}} = \frac{.2955}{\sqrt{1 - .3856}\sqrt{1 - .2581}}$$

$$= \frac{.2955}{\sqrt{.6144}\sqrt{.7419}} = \frac{.2955}{(.7838)(.8613)} = \frac{.2955}{.6751} = .438$$

which is equal to the partial correlation reported in Table 11.5 (.43814).

The contribution of GRE-Q to R^2, over and above AR, which is already in the equation, can be calculated in several ways. While the most straightforward method is to calculate the squared semipartial correlation of GPA and GRE-Q, partialing AR from the latter, we demonstrate the calculation by using partial correlation, since this is the statistic reported in Table 11.5. The contribution of GRE-Q is calculated as follows[7]:

$$r^2_{\text{GPA, GRE-Q. AR}}(1 - r^2_{\text{GPA, AR}}) = .43814^2[1 - (.6207)^2]$$

$$= (.1920)(.6147) = .1180$$

Introducing GRE-Q after AR will add .1180 to the R^2. Therefore,

$$R^2_{\text{GPA.AR,GRE-Q}} = (.6207)^2 + .1180 = .3853 + .1180 = .5033$$

The F ratio for the increment due to GRE-Q is

$$F = \frac{(.5033 - .3853)/(2 - 1)}{(1 - .5033)/(30 - 2 - 1)} = \frac{(.1180)/1}{(.4967)/27} = \frac{.1180}{.0184} = 6.41$$

The same F ratio, with 1 and 27 degrees of freedom, is reported in Table 11.5.

The partial correlations and their associated F ratios for the other variables are similarly calculated. The F level to enter specified in this analysis was 3.00. All three variables have F's exceeding this value. The variable selected for the next step is the one that has the highest partial correlation, or equivalently, the one that has the highest F to enter. In step 1 this variable is GRE-V, whose partial correlation is .46022; its F ratio is 7.256. Step 2 therefore includes AR and GRE-V. The level of F ratio to remove was specified as 2.00. Both variables have F's to remove exceeding this value, and therefore none is removed. From the variables not in the equation, MAT has the higher partial correlation. The F ratio associated with MAT exceeds 3.00 (the level to enter) and is therefore entered next in step 3. In step 3 all variables in the equation have F ratios larger than 2.00 (the specified level to remove) and none is therefore removed.

[6]The zero-order correlations are obtained from Table 11.4.
[7]See footnote 3.

The only variable not in the equation in step 3 is GRE-Q. Since its F ratio is larger than 3.00, it is entered in step 4. Note now that the F to remove for AR is smaller than 2.00. AR is therefore removed in step 5. Thus a variable which was the best single predictor is shown to be the least useful after other variables have been introduced into the equation. The increment of R^2 due to AR, over and above the other three variables, is not significant ($F = 1.629$, with 1 and 25 degrees of freedom). The proportion of variance accounted for by the four variables is .6405 ($R = .8003$), while the proportion of variance accounted for after removal of AR is .6170 ($R = .7855$; see Table 11.5, steps 4 and 5). It follows, therefore, that the proportion of variance due to AR, when it is entered last is $.6405 - .6170 = .0235$, about 2 percent of the variance. Compare this with the almost 39 percent AR accounted for when it entered first.

As noted earlier, meaningfulness is more important than statistical significance. The variable AR was shown to account for about 2 percent of the variance when it was entered last in the equation. The question therefore is whether this increment is meaningful. In the present example it probably is not. It will be recalled that AR is obtained as a result of three interviews with each applicant. For predictive purposes only, the time and effort involved in obtaining AR seem not to be justified. This is not to say that interviewing applicants for graduate study is worthless, or that the only information it yields is a rating on a five-point rating scale. One may wish to retain the interviews for other purposes, despite the fact that they are of little use in predicting GPA.

In conclusion, it should be noted that on the basis of significance testing only, the contribution of GRE-V is not significant ($F = 2.311$, with 1 and 26 degrees of freedom; see Table 11.5, step 5). The significance test is affected by the size of the sample, which, as pointed out above, is not sufficient. When entered last in the analysis, GRE-V contributes slightly over 3 percent of the variance. Unlike AR, GRE-V is relatively easy to obtain. Again, GRE-V may provide important information for purposes other than the prediction of GPA. Consequently the decision to retain or not retain the test depends on the researcher's judgment of its usefulness in relation to the efforts required in obtaining the data.

Assuming it is decided to remove the variable AR and retain the rest of the predictor variables, the regression equation for the data of Table 11.3 is

$$Y' = -2.14877 + .00493(\text{GRE-Q}) + .00161(\text{GRE-V}) + .02612(\text{MAT})$$

Explanation

Thus far we have been preoccupied with prediction. As noted earlier, however, another aspect of scientific inquiry is explanation. Philosophers of science have devoted a good deal of attention to the differences as well as the relations between prediction and explanation.[8] Kaplan (1964) states that from the stand-

---

[8]See, for example, Brodbeck (1963), Braithwaite (1953), Hempel (1965), and Kaplan (1964).

point of a philosopher of science the ideal explanation is probably one that allows prediction.

> The converse, however, is surely questionable; predictions can be and often are made even though we are not in a position to explain what is being predicted. This capacity is characteristic of well-established empirical generalizations that have not yet been transformed into theoretical laws... In short, explanations provide under-standing, but we can predict without being able to understand, and we can under-stand without necessarily being able to predict. It remains true that if we can predict successfully on the basis of certain explanations we have good reason, and perhaps the best sort of reason, for accepting the explanation [pp. 349–350].

In their search for explanation of phenomena behavioral scientists have attempted to determine the relative importance of explanatory variables. Various criteria for the importance of variables have been used. In the context of multiple regression analysis the two most frequently encountered criteria are the relative contribution to the proportion of variance accounted for in the dependent variable, and the relative magnitude of the squared β's.[9] These criteria are unambiguous, however, only in experimental research with balanced designs, or when the independent variables are not correlated. It is under such circumstances that the partitioning of the regression sum of squares, or the proportion of variance accounted for, is unambiguous.

When the independent variables are not correlated, the proportion of variance attributable to a given variable is equal to the squared zero-order cor-relation between it and the dependent variable. Furthermore, under such cir-cumstances, each β is equal to the zero-order correlation between the de-pendent variable and the variable with which it is associated. Consequently, squaring β's amounts to squaring zero-order correlations.

In nonexperimental, or ex post facto, research, however, the independent variables are generally correlated, sometimes substantially. This makes it difficult, if not impossible, to untangle the variance accounted for in the dependent variable and to attribute portions of it to individual independent variables. Various authors have addressed themselves to this problem, some concluding that it is insoluble. Goldberger (1964), for example, states: "...When orthogonality is absent the concept of contribution of an individual regressor remains inherently ambiguous (p. 201)." Darlington (1968) is even more explicit, saying: "It would be better to simply concede that the notion of 'independent contribution to variance' has no meaning when predictor vari-ables are intercorrelated (p. 169)."

And yet, if we are to explain phenomena we need to continue searching for methods that will enable us to untangle the effects of independent variables on the dependent variable, or at least provide some better understanding of these effects. That such methods are urgently needed may be noted from the different, and frequently contradictory, interpretations of findings from studies

---

[9]For a review and an excellent discussion of these and other attempts to determine the relative importance of variables, see Darlington (1968).

that have important implications for public policies and for the behavior of individuals. Witness, for example, the controversy that has surrounded some of the findings of the study, *Equality of Educational Opportunity* (Coleman et al., 1966), since its publication. The controversy has not been restricted to professional journals but has also received wide, frequently oversimplified and consequently misleading, coverage in the popular press.

The controversy about some of the findings of the Coleman Report (as *Equality of Educational Opportunity* is often called) is an almost classic example of a controversy about the relative importance of variables in the context of multiple regression analysis. Coleman et al. report that students' attitudes and home background accounted for a far larger proportion of the variance in school achievement compared to the proportion of variance accounted for by the schools that the pupils attended. In fact, the differences among schools have been found to account for a negligible portion of the variance in school achievement.[10] Some researchers took these findings to mean that students' attitudes and background are far more important than the school they attend, while other researchers have challenged this interpretation of relative importance of variables in view of the intercorrelations among them. The possible consequences of one interpretation or another on public educational policies and the attitudes and behavior of individuals cannot be over-estimated. Needless to say, whatever the interpretation, it must be reached by sound methodology.

Another example of a controversy that evokes great emotional outbursts on the part of professionals and laymen alike has to do with the attempts to determine the relative importance of heredity and environment on intelligence. Note, for example, the criticisms, accusations, and counteraccusations that followed the publication of Jensen's (1969) paper on this topic. (See *Harvard Educational Review*, 1969, for a compilation that includes Jensen's paper as well as reactions to it by authorities in various disciplines.)

Being cognizant of the implications of the findings from the Coleman Report, various authors have reanalyzed some of the original data and offered their own interpretation, which frequently differed greatly from the original interpretations made by the authors of the report. Notable among those who have reanalyzed and reinterpreted data from the Coleman Report are researchers who were involved in the original analysis (for example, Mood, 1969, 1971) and researchers currently associated with the United States Office of Education (Mayeske et al., 1969; United States Office of Education, 1970). The basic analytic method used by these authors is what they have called *commonality analysis*.

Commonality Analysis

Commonality analysis is a method of analyzing the variance of a dependent variable into common and unique variances to help identify the relative in-

---

[10]For a more detailed discussion of the findings of the Coleman Report, see Chapter 16.

fluences of independent variables. Mood (1969, 1971) and Mayeske et al. (1969), who developed the method, have applied it to data of the Coleman Report. It will be noted that two researchers in England, Newton and Spurrell (1967a, 1967b), independently developed the same system and applied it to industrial problems. They referred to the method as elements analysis.[11] For comments on some earlier and related attempts, see Creager (1971).

The unique contribution of an independent variable is defined as the variance attributed to it when it is entered last in the regression equation. Thus defined, the unique contribution is actually a squared semipartial correlation between the dependent variable and the variable of interest, after partialing all the other independent variables from it.[12] With two independent variables, the unique contribution of variable 1 is defined as follows:

$$U(1) = R^2_{y.12} - R^2_{y.2} \qquad (11.4)$$

where $U(1)$ = unique contribution of variable 1; $R^2_{y.12}$ = squared multiple correlation of Y with variables 1 and 2; $R^2_{y.2}$ = squared correlation of Y with variable 2. Similarly, the unique contribution of variable 2 is defined as follows:

$$U(2) = R^2_{y.12} - R^2_{y.1} \qquad (11.5)$$

where $U(2)$ = unique contribution of variable 2. The definition of the commonality of variables 1 and 2 is

$$C(12) = R^2_{y.12} - U(1) - U(2) \qquad (11.6)$$

where $C(12)$ = commonality of variables 1 and 2. Substituting the right-hand sides of formulas (11.4) and (11.5) for $U(1)$ and $U(2)$ in formula (11.6), we obtain

$$
\begin{aligned}
C(12) &= R^2_{y.12} - (R^2_{y.12} - R^2_{y.2}) - (R^2_{y.12} - R^2_{y.1}) \\
&= R^2_{y.12} - R^2_{y.12} + R^2_{y.2} - R^2_{y.12} + R^2_{y.1} \\
&= R^2_{y.2} + R^2_{y.1} - R^2_{y.12} \qquad (11.7)
\end{aligned}
$$

As a result of determining unique and common contribution of variables, it is possible to express the correlation between any independent variable and the dependent variable as a composite of the unique contribution of the variable of interest plus its commonalities with other independent variables. Thus $R^2_{y.1}$ in the above example can be expressed as follows:

$$R^2_{y.1} = U(1) + C(12) \qquad (11.8)$$

That this is so can be demonstrated by restating formula (11.8) using the

---

[11]We believe that a more appropriate name may be components analysis, since the method partitions the variance of the dependent variable into a set of components, some of which are unique, while the others are commonalities.

[12]See Wisler (1969) for a mathematical development in which commonality analysis is expressed as squared semipartial correlations.

right-hand sides of formulas (11.4) and (11.7):

$$R_{y.1}^2 = (R_{y.12}^2 - R_{y.2}^2) + (R_{y.2}^2 + R_{y.1}^2 - R_{y.12}^2)$$

Similarly,

$$R_{y.2}^2 = \mathrm{U}(2) + \mathrm{C}(12) \qquad (11.9)$$

The commonality of variables 1 and 2 is referred to as a second-order commonality. With more than two independent variables second-order commonalities are determined for all the possible pairs of variables. In addition, third-order commonalities are determined for all possible sets of three variables, fourth-order commonalities for all sets of four variables, and so forth up to one commonality whose order is equal to the total number of independent variables. Thus, for example, with three variables, A, B, and C, there are three unique components, namely $\mathrm{U}(A)$, $\mathrm{U}(B)$, and $\mathrm{U}(C)$; three second-order commonalities, namely $\mathrm{C}(AB)$, $\mathrm{C}(AC)$, and $\mathrm{C}(BC)$, and one third-order commonality, namely $\mathrm{C}(ABC)$. Altogether, there are seven components in a three-variable problem. In general, the number of components is equal to $2^k - 1$, where k is the number of independent variables. Thus, with four independent variables there are $2^4 - 1 = 15$ components, four of which are unique, six are second-order, four are third-order, and one is a fourth-order commonality. With five independent variables there are $2^5 - 1 = 31$ components. Note that with each addition of an independent variable there is a considerable increase in the number of components, a point discussed below.

Rules for Writing Commonality Formulas

Mood (1969) and Wisler (1969) offer a rule for writing formulas for the unique and commonality components in a commonality analysis. This rule can be explained by an example. Suppose we have three independent variables, X_1, X_2, and X_3, and a dependent variable, Y. To write the formula for the unique contribution of variable X_2, for example, we first construct the following product:

$$-(1 - X_2)X_1X_3$$

The variable of interest, X_2, is substracted from one and this term is multiplied by the remaining independent variables, which in the present example are X_1 and X_3. The above product is now expanded:

$$-(1 - X_2)X_1X_3 = -(X_1X_3 - X_1X_2X_3) = -X_1X_3 + X_1X_2X_3$$

After expanding the product, each term is replaced by R^2 of the dependent variable with the variables indicated in the given term. Thus, using the above expansion, the unique contribution of variable X_2 is

$$\mathrm{U}(X_2) = -R_{y.x_1x_3}^2 + R_{y.x_1x_2x_3}^2$$

or, written more succinctly,

$$\mathrm{U}(2) = -R_{y.13}^2 + R_{y.123}^2$$

We now illustrate how the rule applies to the writing of the formula for the commonality of two variables, namely X_2 and X_3. First, construct the product. This time, however, there are two terms in which each of the variables of interest is subtracted from one. The product of these terms is multiplied by the remaining independent variable(s), which in the present example is X_1. The product to be expanded for the commonality of X_2 and X_3 is therefore

$$-(1-X_2)(1-X_3)X_1$$

After expansion,

$$-(1-X_2)(1-X_3)\,X_1 = -X_1 + X_1\,X_2 + X_1\,X_3 - X_1\,X_2\,X_3$$

Replacing each term in the right-hand side of this equation by R^2, we obtain

$$C\,(23) = -R_{y.1}^2 + R_{y.12}^2 + R_{y.13}^2 - R_{y.123}^2$$

To write the formula for the commonality of X_1 and X_3 one expands the product $-(1-X_1)(1-X_3)X_2$ and then replaces each term by the appropriate R^2.

For the commonality of all the independent variables in the above example, it is necessary to expand the following product:

$$-(1-X_1)(1-X_2)(1-X_3)$$

After expansion, the above product is equal to

$$-1 + X_1 + X_2 - X_1\,X_2 + X_3 - X_1\,X_3 - X_2\,X_3 + X_1\,X_2\,X_3$$

When the rule is applied to the writing of the formula for the commonality of all the independent variables, the expansion of the product has one term equal to -1. This term is deleted and the remaining terms are replaced by R^2's in the manner illustrated above. Accordingly, using the expansion for the product terms of X_1, X_2, and X_3, the formula for the commonality of these variables is

$$C\,(123) = R_{y.1}^2 + R_{y.2}^2 - R_{y.12}^2 + R_{y.3}^2 - R_{y.13}^2 - R_{y.23}^2 + R_{y.123}^2$$

The rule illustrated above applies to any number of independent variables. We illustrate this for some components in a problem with k independent variables. To obtain, for example, the unique contribution of variable X_1, the following product is constructed:

$$-(1-X_1)\,X_2\,X_3 \ldots X_k$$

After expanding this product, each term is replaced by R^2 of the dependent variable with the independent variables indicated in the given term. To write the formula for the commonality of variables X_1, X_2, X_3, and X_4, for example, the following product is expanded:

$$-(1-X_1)(1-X_2)(1-X_3)(1-X_4)\,X_5\,X_6 \ldots X_k$$

Again, each term after the expansion is replaced by the appropriate R^2. The formula for the commonality of all k independent variables is obtained by expanding the following product:

$$-(1-X_1)(1-X_2) \ldots (1-X_k)$$

As noted above, after expansion with all the independent variables there is one term equal to -1. This term is deleted, and all other terms are replaced by R^2's in the manner shown above.

A Numerical Example

Before discussing some aspects and problems in interpreting results from commonality analysis, we apply the method to a fictitious numerical example. The same example was used earlier in this chapter in connection with the forward and backward regression solutions. Three independent variables were used: socioeconomic status (SES), intelligence (IQ), and need achievement (n Ach). The dependent variable was the grade-point average (GPA) of college students. The intercorrelations among these variables are repeated in part I of Table 11.6. In part II of the table we report the various R^2's necessary for a commonality analysis.

The rule for writing the formulas for the various components is now applied to the present example. In order to avoid cumbersome symbolism, however, we use the following symbols: $X_1 = $ SES; $X_2 = $ IQ; $X_3 = n$ Ach; $Y = $ GPA. For the unique contribution of X_1 expand the following product:

$$- (1 - X_1)X_2X_3 = -X_2X_3 + X_1X_2X_3$$

Replacing each of the terms in the expansion by the appropriate R^2's from Table 11.6, we obtain

$$U(1) = -R^2_{y.23} + R^2_{y.123} = -.4964 + .4966 = .0002$$

The unique contributions of X_2 and X_3 are similarly obtained. They are

$$U(2) = -R^2_{y.13} + R^2_{y.123} = -.2688 + .4966 = .2278$$

$$U(3) = -R^2_{y.12} + R^2_{y.123} = -.3527 + .4966 = .1439$$

TABLE **11.6**   FICTITIOUS DATA FOR A COMMONALITY ANALYSIS[a]

I:			Correlation Matrix		
		1 SES	2 IQ	3 n Ach	Y GPA
	1	1.0000	.3000	.4100	.3300
	2	.0900	1.0000	.1600	.5700
	3	.1681	.0256	1.0000	.5000
	Y	.1089	.3249	.2500	1.0000

II:	Squared Multiple Correlations	
	$R^2_{y.123} = .4966$	$R^2_{y.12} = .3527$
	$R^2_{y.13} = .2688$	$R^2_{y.23} = .4964$

[a]The entries above the principal diagonal of the correlation matrix are zero-order correlations, while those below the diagonal are squared zero-order correlations. For example $r_{12} = .3000$, $r^2_{12} = .0900$.

For the commonality of X_1 and X_2 we expand the following product:

$$- (1-X_1)(1-X_2)X_3 = -X_3 + X_1X_3 + X_2X_3 - X_1X_2X_3$$

Replacing each term in the expansion by the appropriate R^2's from Table 11.6,

$$C(12) = -R^2_{y.3} + R^2_{y.13} + R^2_{y.23} - R^2_{y.123}$$

$$= -.2500 + .2688 + .4964 - .4966 = .0186$$

The commonality of X_1 and X_3, and that of X_2 and X_3 are similarly obtained. They are

$$C(13) = -R^2_{y.2} + R^2_{y.12} + R^2_{y.23} - R^2_{y.123}$$

$$= -.3249 + .3527 + .4964 - .4966 = .0276$$

$$C(23) = -R^2_{y.1} + R^2_{y.12} + R^2_{y.13} - R^2_{y.123}$$

$$= -.1089 + .3527 + .2688 - .4966 = .0160$$

The commonality of variables X_1, X_2, and X_3 is obtained by the following expansion:

$$- (1-X_1)(1-X_2)(1-X_3) = -1 + X_1 + X_2 - X_1X_2 + X_3 - X_1X_3$$
$$- X_2X_3 + X_1X_2X_3$$

Deleting the -1 and replacing the remaining terms by the R^2's from Table 11.6,

$$C(123) = R^2_{y.1} + R^2_{y.2} - R^2_{y.12} + R^2_{y.3} - R^2_{y.13} - R^2_{y.23} + R^2_{y.123}$$

$$= .1089 + .3249 - .3527 + .2500 - .2688 - .4964 + .4966 = .0625$$

The analysis is summarized in Table 11.7 in the manner suggested by Mayeske et al. (1969). Several observations may be made about this table. Note that each term in the last line, the line labeled Σ, is equal to the squared zero-order correlation of the variable with which it is associated and the dependent

TABLE **11.7**   SUMMARY OF COMMONALITY ANALYSIS OF DATA
OF TABLE 11.6

	Variables		
	1 SES	2 IQ	3 n Ach
Unique to 1, SES	.0002		
Unique to 2, IQ		.2278	
Unique to 3, n Ach			.1439
Common to 1 and 2	.0186	.0186	
Common to 1 and 3	.0276		.0276
Common to 2 and 3		.0160	.0160
Common to 1, 2, and 3	.0625	.0625	.0625
Σ:	.1089	.3249	.2500

variable. Thus, for example, in the last line under SES we have .1089, which is equal to the squared zero-order correlation between SES and GPA ($.3300^2$. See Table 11.6).

Reading down each column in Table 11.7 it is possible to note how the proportion of variance accounted for by a given variable is partitioned into various components. The proportion of variance accounted for by SES, for example, is partitioned as follows: .0002 unique to SES, .0186 common to SES and IQ, .0276 common to SES and n Ach, and .0625 common to SES, IQ, and n Ach. From this analysis it is evident that SES makes practically no unique contribution. Most of the variance accounted for by SES (.1089) is due to its commonalities with the other independent variables. In contrast, intelligence and need achievement show relatively large unique contributions, about 23 and 14 percent respectively.

The squared multiple correlation can be written as a composite of all the unique and common components. Thus, for the present problem,

$$R^2_{y.123} = U(1) + U(2) + U(3) + C(12) + C(13) + C(23) + C(123) \qquad (11.10)$$

$$.4966 = .0002 + .2278 + .1439 + .0186 + .0276 + .0160 + .0625$$

From this form of partitioning of the variance it appears that the unique contributions of IQ and n Ach comprise about 37 percent of the variance accounted for, while all the commonalities account for the remaining 13 percent.

Does this analysis enable us to answer the question about the relative importance of variables? Can we conclude, for example, that SES is not important, since it makes almost no unique contribution? Or, does the larger proportion of unique variance associated with IQ, as compared to n Ach, indicate that it is the more important variable of the two? There are no simple answers to these questions, considering that unique and common contributions are affected by the intercorrelations among the independent variables.

Problems in the Interpretation of Commonality Analysis

The unique contribution of a variable was defined above as the increment in the proportion of variance accounted for when it is entered last in the regression equation. It is therefore important to note that the uniqueness of variables depends on the relations among the specific set of variables under study. Addition or deletion of variables may change drastically the uniqueness attributed to some or all the remaining variables. Moreover, the higher the correlations among the variables, the larger the commonalities and the smaller the unique components. Directing attention to the difficulty that arises when commonalities are large and unique components are small, Mood (1971) attributes it not to the method of commonality analysis, but to "our state of ignorance about educational variables (p. 197)." He further maintains that commonality analysis helps us "identify indicators which are failing badly with respect to specificity (p. 197)."

It seems to us that Mood's argument has greater validity when considered in a predictive rather than an explanatory framework. Commonality analysis can be used as an alternative to other methods for the selection of variables in a predictive framework (for example, stepwise solution). In fact, Newton and Spurrell (1967a, 1967b) maintain that commonality analysis is superior to other selection methods currently used, and give empirical evidence to support this claim. Among several rules they formulate for the selection of variables from a larger pool, one is that variables with small commonalities and large unique components are preferred. This rule makes good sense in a predictive framework. Its application in an explanatory framework, however, may be misleading. (Incidentally, Newton and Spurrell recommend that their rules be used in what we have called a predictive framework.)

While it is true that large commonalities are a consequence of high correlations among variables, it does not necessarily follow that a high correlation reflects lack of specificity of variables. Creager (1971) comments on this point, saying: "Correlations between sets may be of substantive importance and not solely artifacts of the inadequacy of the proxy variables (p. 675)." It is possible, for example, that two variables are highly correlated because one of them is a cause of the other. Commonality analysis, however, does not differentiate between situations in which variables lack specificity and those in which there are causal relations. Applying commonality analysis in the latter situation may sometimes lead to the erroneous conclusion that a presumed cause lacks specificity or is unimportant because it has little or no uniqueness, on the one hand, and large commonalities with other variables, on the other hand.

Another problem with commonality analysis, alluded to earlier, is the proliferation of higher-order commonalities that results from the addition of independent variables. Even with five independent variables only, there are 31 components, 26 of which are commonalities. While it may be possible, although by no means always easy, to explain a second- or a third-order commonality, it is extremely difficult, and even impossible, to explain commonalities of higher orders. Mood and Mayeske et al. recognize this difficulty and suggest as a remedy that independent variables be grouped and that commonality analysis be done on the grouped variables. Wisler (1969), for example, maintains that, "It is by grouping variables and performing commonality analyses that one can begin to discern the structure in nonexperimental, multivariate data (p. 359)." While admittedly simpler, commonality analysis with grouped variables may still lead to results that are difficult to interpret. One can find examples of such difficulties in the reanalysis of data from the Coleman Report by Mayeske et al. (1969). For example, Mood (1971) reproduces one such analysis from Mayeske et al., in which two grouped independent variables, peer quality and school quality, were used. Each of these variables was obtained as a result of grouping about 30 indicators. In an analysis of achievement in grades 3, 6, 9, and 12 it was found that the unique contributions of each of the grouped variables ranged from .04 to .11, while their commonalities ranged from .45 to .75. Mood (1971) concludes: "The overlap between peer quality and school quality is so large that there seems hardly any point in referring to them as different factors; or perhaps the problem is that we are so ignorant about specificity of indicators that ours have almost no specificity at all (p. 198)."

Still another problem encountered in commonality analysis is that some of the commonalities can have negative signs. It should be pointed out that the unique components are always positive. Negative commonalities can be obtained in situations where one of the variables is a suppressor, or when correlations among independent variables are negative. There seems no need to elaborate the conceptual difficulties that arise when a negative proportion of variance is attributed to the commonality of a set of variables.

In conclusion, it seems to us that commonality analysis can make a greater contribution in a predictive than in an explanatory framework. Furthermore, commonality analysis can be useful in early and exploratory stages of research when, perhaps, not much is known about the relations among the independent variables. We believe, however, that at its present stage of development commonality analysis cannot be of much help in testing hypotheses derived from relatively sophisticated theories.

We turn now to a treatment of path analysis, an analytic method that in recent years has been gaining wide currency, particularly among sociologists.

Path Analysis

Path analysis was developed by Sewall Wright as a method for studying the direct and indirect effects of variables taken as causes of variables taken as effects. It is important to note that path analysis is not a method for discovering causes, but a method applied to a causal model formulated by the researcher on the basis of knowledge and theoretical considerations. In Wright's words:

> ... the method of path coefficients is not intended to accomplish the impossible task of deducing causal relations from the values of the correlation coefficients. It is intended to combine the quantitative information given by the correlations with such qualitative information as may be at hand on causal relations to give a quantitative interpretation (Wright, 1934, p. 193).

> In cases in which the causal relations are uncertain, the method can be used to find the logical consequences of any particular hypothesis in regard to them (Wright, 1921, p. 557).

In other words, path analysis is useful in testing theory rather than in generating it. In fact, one of the virtues of the method is that in order to apply it the researcher is required to make explicit the theoretical framework within which he operates.

From these introductory remarks it is evident that causal thinking plays an important role in the application of path analysis. Consequently, to better understand what path analysis can and what it cannot do, it is necessary to discuss, although briefly, the concept of causation.

Causation

The concept of causation has stirred a great deal of controversy among philosophers and scientists alike. We do not intend to take sides in this controversy nor to review it.[13] In the work of scientists, even in the work of those who are strongly opposed to the use of the term causation, one encounters the frequent use of terms that indicate or imply causal thinking. When behavioral scientists, for example, speak about the effects of child rearing practices on the development of certain personality patterns, or the effect of reinforcement on subsequent behavior, or the reasons for delinquent behavior, or the influence of attitudes on perception, there is implication of causation. This tendency to imply causation, even when refraining from using the term, is reflected also in some of the methods employed by behavioral scientists. For example, proportions of variance are attributed to certain independent variables; or a presumed cause is partialed from two variables in order to observe whether the relation between them is spurious. "Thus, the difference between true and spurious correlations resolves into a difference between causal and noncausal connections (Brodbeck, 1963, p. 73)."[14]

Nagel (1965) summed up the status of the concept of causation, saying: "Though the *term* may be absent the *idea* for which it stands continues to have wide currency (p. 11)." Drawing attention to the fact that causal statements are frequent in the behavioral as well as the physical sciences, Nagel concludes: "In short, the idea of cause is not as outmoded in modern science as is sometimes alleged (p.11)." That this is so is not surprising since the scientist's question of why a certain event has occurred carries with it an implication of causality. Moreover, when a behavioral scientist wishes to bring about desired changes in human behavior, he must be able to identify the factors affecting the behavior or, more plainly, the causes of the behavior.

In sum, scientists, qua scientists, seem to have a need to resort to causal frameworks, even though on philosophical grounds they may have reservations about the concept of causation.

Causation in Experimental and Nonexperimental Research

In experimental research the experimenter manipulates variables of interest and observes the manner in which the manipulation affects the variation of the dependent variable. In order to be reasonably sure that the observed variation in the dependent variable is indeed due to the manipulated variables, the experimenter must control other relevant variables. One of the most powerful methods of control is randomization.[15] Being in a position to manipulate and randomize, the experimenter may feel reasonably confident in making state-

---

[13]For discussions see, for example, Blalock (1964, 1971), Braithwaite (1953), Feigl and Brodbeck (1953), and Lerner (1965). For a discussion of causality in relation to multiple regression, see Wold and Jureen (1953) and Wold (1970).

[14]For further discussions of this point, see Blalock (1968), Brodbeck (1963), and Simon (1954).

[15]See Kerlinger (1973) for a discussion of the role of randomization in experiments.

ments about the kinds of actions that need be taken in order to produce desired changes in the dependent variable.

The situation is considerably more complex and more ambiguous in non-experimental research because the researcher can neither manipulate nor randomize. While it is possible to resort to statistical controls in lieu of randomization, the researcher must be constantly alert to the pitfalls inherent in the interpretation of analyses of data from nonexperimental research. This need for caution is probably best expressed in the oft-repeated admonition: "Correlation is no proof of causation." Nor does any other index prove causation, regardless of whether the index was derived from data collected in experimental or nonexperimental research. Covariations or correlations among variables may be suggestive of causal linkages. Nevertheless, an explanatory scheme is not arrived at on the basis of the data, but rather on the basis of knowledge, theoretical formulations and assumptions, and logical analysis. It is the explanatory scheme of the researcher that determines the type of analysis to be applied to data, and not the other way around.

Having completed an analysis, the researcher is in a position to determine whether the data are consistent with his explanatory scheme. If the data are inconsistent with the explanatory model, doubt is cast on the theory that generated it. Consistency of the data with an explanatory model, however, is not proof of a theory; it only lends support to it.

It is possible for the same data to be consistent with competing causal models. The decision as to which of the models is more tenable rests on considerations outside the data. For example, consider the following competing models involving three variables: (1) $X \rightarrow Y \rightarrow Z$; (2) $Y \rightarrow X \rightarrow Z$. The first model indicates that X affects Y, which in turn affects Z. The second model, on the other hand, indicates that Y affects X, which in turn affects Z. As will be shown below, observed correlations among the three variables may be consistent with both models. It is possible, however, that X precedes Y in a time sequence. If this is the case, the researcher may reject model 2 in favor of model 1.[16]

What is needed, then, is a method of analysis designed to shed light on the tenability of a theoretical model formulated by the researcher. One such method is path analysis. The following presentation is not intended to be exhaustive but rather to acquaint the reader with some of the basic principles and applications of path analysis. For more detailed treatments the reader is referred to Wright (1934, 1954, 1960a), Tukey (1954), Li (1955), Turner and Stevens (1959), Land (1969), and Heise (1969b).

Path Diagrams

The path diagram, although not essential for numerical analysis, is a useful device for displaying graphically the pattern of causal relations among a set of

---

[16]This is not to say that the temporal priority of X indicates that it is a cause of Y, but rather that on logical grounds we are not willing to accept the notion that an event that occurred later in time (Y in the present example) caused an event that preceded it (X of the present example).

variables. In the causal model, a distinction is made between exogenous and endogenous variables. An *exogenous variable* is a variable whose variability is assumed to be determined by causes outside the causal model. Consequently, the determination of an exogenous variable is not under consideration in the model. Stated differently, no attempt is made to explain the variability of an exogenous variable or its relations with other exogenous variables. An *endogenous variable*, on the other hand, is one whose variation is explained by exogenous or endogenous variables in the system. The distinction between the two kinds of variables is illustrated in Figure 11.1, which depicts a path diagram consisting of four variables.

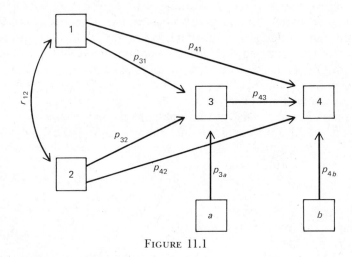

FIGURE 11.1

Variables 1 and 2 in Figure 11.1 are exogenous. The correlation between exogenous variables is depicted by a curved line with arrowheads at both ends, thus indicating that the researcher does not conceive of one variable being the cause of the other. Consequently, a relation between exogenous variables (in the present case r_{12}) remains unanalyzed in the system.

Variables 3 and 4 in Figure 11.1 are endogenous. Paths, in the form of unidirectional arrows, are drawn from the variables taken as causes (independent) to the variables taken as effects (dependent). The two paths leading from variables 1 and 2 to variable 3 indicate that variable 3 is dependent on variables 1 and 2.

The presentation in this chapter is limited to recursive models. This means that the causal flow in the model is unidirectional. Stated differently, it means that at a given point in time a variable cannot be both a cause and an effect of another variable. For example, if variable 2 in Figure 11.1 is taken as a cause of variable 3, then the possibility of variable 3 being a cause of variable 2 is ruled out.[17]

---

[17]Turner and Stevens (1959), and Wright (1960b) discuss the treatment of feedback and reciprocal causation in a system.

An endogenous variable treated as dependent in one set of variables may also be conceived as an independent variable in relation to other variables. Variable 3, for example, is taken as dependent on variables 1 and 2, and as one of the independent variables in relation to variable 4. It will be noted that in this example the causal flow is still unidirectional.

Since it is almost never possible to account for the total variance of a variable, residual variables are introduced to indicate the effect of variables not included in the model. In Figure 11.1, a and b are residual variables. As will be noted below, it is assumed that a residual variable is neither correlated with other residuals nor with other variables in the system. Thus variable a, for example, is assumed to be not correlated with b nor with 1, 2, and 4.

In order to simplify the presentation of path diagrams it is convenient not to represent the residuals in the diagram. We follow this practice in the remainder of the presentation. This, of course, does not mean that the residuals and the assumptions pertaining to them are ignored.

Assumptions

Among the assumptions that underlie the application of path analysis as presented in this chapter are:

(1) The relations among the variables in the model are linear, additive, and causal. Consequently, curvilinear, multiplicative, or interaction relations are excluded.

(2) The residuals are not correlated among themselves, nor are they correlated with the variables in the system. The implication of this assumption is that all relevant variables are included in the system. Endogenous variables are conceived as linear combinations of exogenous or other endogenous variables in the system and a residual. Exogenous variables are treated as "givens." Moreover, when exogenous variables are correlated among themselves, these correlations are treated as "givens" and remain unanalyzed.

(3) There is a one-way causal flow in the system. That is, reciprocal causation between variables is ruled out.

(4) The variables are measured on an interval scale.[18]

Path Coefficients

Wright (1934) defines a path coefficient as:

> The fraction of the standard deviation of the dependent variable (with the appropriate sign) for which the designated factor is directly responsible, in the sense of the fraction which would be found if this factor varies to the same extent as in the observed data while all others (including the residual factors...) are constant (p. 162).

[18]For a discussion of the implications of weakening these assumptions, see Land (1969), Heise (1969b), and Bohrnstedt and Carter (1971). Boyle (1970) and Lyons (1971) discuss and illustrate the application of path analysis to ordinal measures.

In other words, a path coefficient indicates the direct effect of a variable taken as a cause of a variable taken as effect.

The symbol for a path coefficient is a p with two subscripts, the first indicating the effect (or the dependent variable), and the second subscript indicating the cause (the independent variable). Accordingly, p_{32} in Figure 11.1 indicates the direct effect of variable 2 on variable 3.

The Calculation of Path Coefficients

Each endogenous (dependent) variable in a causal model may be represented by an equation consisting of the variables upon which it is assumed to be dependent, and a term representing residuals, or variables not under consideration in the given model. For each independent variable in the equation there is a path coefficient indicating the amount of expected change in the dependent variable as a result of a unit change in the independent variable. Exogenous variables, it will be recalled, are assumed to be dependent on variables not included in the model, and are therefore represented by a residual term only. The letter e or u with an appropriate subscript is used to represent residuals. As an illustration, the equations for a four-variable causal model depicted in Figure 11.2 are given. Expressing all variables in standard score form (z score), the equations are

$$z_1 = e_1 \tag{11.11a}$$

$$z_2 = p_{21}z_1 + e_2 \tag{11.11b}$$

$$z_3 = p_{31}z_1 + p_{32}z_2 + e_3 \tag{11.11c}$$

$$z_4 = p_{41}z_1 + p_{42}z_2 + p_{43}z_3 + e_4 \tag{11.11d}$$

where the residuals (e's) are also expressed in z scores. Variable 1 is exogenous and is therefore represented by a residual (e_1) only. Variable 2 is shown to be dependent on variable 1 and on e_2, which stands for variables outside the system affecting variable 2. Similar interpretations apply to the other equations. A set of equations such as (11.11) is referred to as a recursive system. It is a

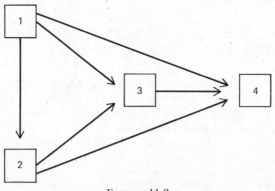

FIGURE 11.2

system of equations in which at least half of the path coefficients have been set equal to zero. Consequently, a recursive system can be organized in a triangular form, because the upper half of the matrix is assumed to consist of path coefficients which are equal to zero. For example, (11.11a) implies the following equation:

$$z_1 = e_1 + 0_{12} z_2 + 0_{13} z_3 + 0_{14} z_4$$

Similarly, for the other equations in (11.11).

If, as discussed above, it is assumed that each of the residuals in (11.11) is not correlated with the variables in the equation in which it appears, it follows that the residuals are not correlated among themselves. Under such conditions, the solution for the path coefficients takes the form of the ordinary least squares solution for β's (standardized regression coefficients) presented in earlier chapters of the book.

The process of calculating the path coefficients for the model depicted in Figure 11.2 is now demonstrated. Let us start with p_{21}, that is, the path coefficient indicating the effect of variable 1 on variable 2.

$$r_{12} = \frac{1}{N} \Sigma z_1 z_2$$

Substituting (11.11b) for z_2,

$$r_{12} = \frac{1}{N} \Sigma z_1 (p_{21} z_1 + e_2)$$

$$= p_{21} \frac{\Sigma z_1 z_1}{N} + \frac{\Sigma z_1 e_2}{N}$$

$$r_{12} = p_{21} \tag{11.12}$$

The term $\Sigma z_1 z_1 / N = \Sigma z^2_1 / N = 1$ (the variance of standard scores equals one), and the correlation between variable 1 and e_2 is assumed to be zero. Thus $r_{12} = p_{21}$. It will be recalled that when dealing with a zero-order correlation β is equal to the correlation coefficient. Accordingly, $r_{21} = \beta_{21} = p_{21}$. It was thus demonstrated that the path coefficient from variable 1 to variable 2 is equal to β_{21}, which can be obtained from the data by calculating r_{12}. A path coefficient is equal to a zero-order correlation whenever a variable is conceived to be dependent on a single cause and a residual. (Note that this is the case for variable 2 in Figure 11.2.) The same principle still applies when a variable is conceived to be dependent on more than one cause, provided the causes are independent. For example, in Figure 11.3 variables X and Z are conceived as independent causes of Y. Therefore $p_{yx} = r_{yx}$ and $p_{yz} = r_{yz}$.

Returning now to the model in Figure 11.2 we note that variable 3 is conceived to be dependent upon two variables (1 and 2) that are not independent of each other. In fact variable 2 is conceived to be dependent on 1 (in addition it is dependent on a residual not presented in the diagram). We now demonstrate the calculation of the two paths leading to variable 3, namely p_{31} and p_{32}:

$$r_{13} = \Sigma z_1 z_3$$

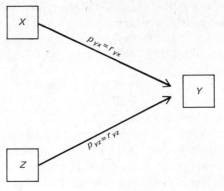

FIGURE 11.3

Substituting (11.11c) for z_3:[19]

$$r_{13} = \frac{1}{N} \sum z_1 (p_{31}z_1 + p_{32}z_2)$$

$$= p_{31} \frac{\sum z_1^2}{N} + p_{32} \frac{\sum z_1 z_2}{N}$$

$$r_{13} = p_{31} + p_{32} r_{12} \qquad (11.13a)$$

Equation (11.13a) consists of two unknowns (p_{31} and p_{32}) and therefore cannot be solved (r_{12} and r_{13} are, of course, obtainable from the data). It is possible, however, to construct another equation with the same unknowns thereby making a solution possible. To obtain the second equation,

$$r_{23} = \frac{1}{N} \sum z_2 z_3$$

Again substituting (11.11c) for z_3,

$$r_{23} = \frac{1}{N} \sum z_2 (p_{31}z_1 + p_{32}z_2)$$

$$= p_{31} \frac{\sum z_2 z_1}{N} + p_{32} \frac{\sum z_2^2}{N}$$

$$r_{23} = p_{31} r_{12} + p_{32} \qquad (11.13b)$$

We thus have two equations involving the path coefficients leading to variable 3:

$$p_{31} + p_{32} r_{12} = r_{13} \qquad (11.13a)$$

$$p_{31} r_{12} + p_{32} = r_{23} \qquad (11.13b)$$

---

[19]The presentation is simplified by dropping the residual terms, since they are assumed not to be correlated.

Equations (11.13) are similar to the normal equations used in earlier chapters for the solution of β's.[20] In fact, the above equations can be rewritten as follows:

$$\beta_{31.2} + \beta_{32.1}r_{12} = r_{13} \tag{11.14a}$$

$$\beta_{31.2}r_{12} + \beta_{32.1} = r_{23} \tag{11.14b}$$

Except for the fact that path coefficients are written without the dot notation, it is obvious that equations (11.13) and (11.14) are identical. It is therefore possible to solve for the path coefficients in the same manner that one would solve for the β's; that is, by applying a least squares solution to the regression of variable 3 on variables 1 and 2. Each path coefficient is equal to the β associated with the same variable. Thus, $p_{31} = \beta_{31.2}$ and $p_{32} = \beta_{32.1}$. Note, however, that $p_{31} \neq p_{13}$. As discussed earlier, p_{31} indicates the effect of variable 1 on variable 3, while p_{13} indicates the effect of variable 3 on variable 1. In the type of causal models under consideration in this chapter — models with one-way causation — it is not possible to have both p_{31} and p_{13}. The path coefficients that are calculated are those that reflect the causal model formulated by the researcher. If, as in the present example, the model indicates that variable 1 affects variable 3, then p_{31} is calculated.

Turning now to variable 4 in Figure 11.2, we note that it is necessary to calculate three path coefficients to indicate the effects of variables 1, 2, and 3 on variable 4. For this purpose three equations are constructed. This is accomplished in the same manner illustrated above. For example, the first equation is obtained as follows:

$$r_{14} = \frac{1}{N}\sum z_1 z_4$$

Substituting (11.11d) for z_4,

$$r_{14} = \frac{1}{N}\sum z_1 (p_{41}z_1 + p_{42}z_2 + p_{43}z_3)$$

$$= p_{41}\frac{\sum z^2_1}{N} + p_{42}\frac{\sum z_1 z_2}{N} + p_{43}\frac{\sum z_1 z_3}{N}$$

$$r_{14} = p_{41} + p_{42}r_{12} + p_{43}r_{13}. \tag{11.15a}$$

The two other equations are similarly obtained. They are

$$r_{24} = p_{41}r_{12} + p_{42} + p_{43}r_{23} \tag{11.15b}$$

$$r_{34} = p_{41}r_{13} + p_{42}r_{23} + p_{43} \tag{11.15c}$$

Again we have a set of normal equations (11.15) which are solved in the manner illustrated in earlier chapters.

In sum, then, when variables in a causal model are expressed in standardized form (z scores), and the assumptions discussed above are reasonably met, the path coefficients turn out to be standardized regression coefficients

[20]See Chapter 4, particularly the discussion in connection with Equations (4.2).

(β's) obtained in the ordinary regression analysis.[21] But there is an important difference between the two analytic approaches. In ordinary regression analysis a dependent variable is regressed in a single analysis on all the independent variables under consideration. In path analysis, on the other hand, more than one regression analysis may be called for. At each stage, a variable taken as dependent is regressed on the variables upon which it is assumed to depend. The calculated β's are the path coefficients for the paths leading from the particular set of independent variables to the dependent variable under consideration. The model in Figure 11.2 requires three regression analyses for the calculation of all the path coefficients. The path from 1 to 2 (p_{21}) is calculated by regressing 2 on 1, as indicated by equation (11.12). p_{31} and p_{32} are obtained by regressing variable 3 on variables 1 and 2, as indicated by equations (11.13). p_{41}, p_{42}, and p_{43} are obtained by regressing variable 4 on variables 1, 2, and 3, as indicated by equations (11.15).

We now consider advantages of path analysis as a tool for decomposing correlations among variables, thereby enhancing the interpretation of relations.

The Analysis of a Correlation

One of the important applications of path analysis is the analysis of a correlation into its components. Within a given causal model it is possible to determine what part of a correlation between two variables is due to the direct effect of a cause and what part is due to indirect effects. Indirect effects may occur in several ways. For example, when causes are correlated each cause has a direct effect on the dependent variable as well as an indirect effect through the correlations with the other causes. This is illustrated in Figure 11.4(a) for a three-variable causal model, in which the exogenous variables 1 and 2 are correlated. In this case r_{13} is due to a direct as well as an indirect effect of 1 on 3. The direct effect is indicated by the direct path from 1 to 3. The indirect effect is due to the correlation of variable 1 with variable 2, which is also a cause of variable 3. The same reasoning applies to r_{23}.

Another example of an indirect effect is illustrated in Figure 11.4(b), where variable 1 is exogenous, and variables 2 and 3 are endogenous. Note that variable 1 has a direct effect on variable 3. In addition, variable 1 affects variable 2, which in turn affects variable 3. This latter route indicates the indirect effect of variable 1 on variable 3 as mediated by variable 2. r_{23}, in Figure 11.4(b) can also be decomposed into a direct and an indirect effect of 2 on 3. Note, however, that here the indirect effect is actually a spurious one, since it is a consequence of variable 2 and 3 having a common cause, namely variable 1. In contrast to the path diagrams in Figures 11.4(a) and 11.4(b), the diagram in Figure 11.4(c) depicts a causal model with independent causes. In such a case the correlation between a cause and an effect is due solely to the direct effect of one on the other. Thus r_{13}, in Figure 11.4(c), is due to the direct effect of 1 on 3. Similarly for the correlation between 2 and 3.

---

[21]For a position that calls for the use of unstandardized path coefficient (b's) instead of β's see Tukey (1954) and Blalock (1968). For a response, see Wright (1960a).

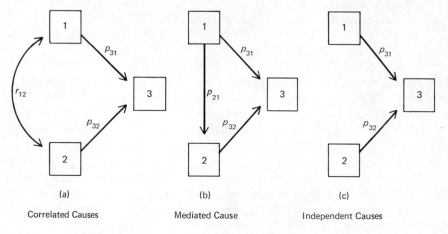

(a)

Correlated Causes

(b)

Mediated Cause

(c)

Independent Causes

FIGURE 11.4

The procedure for decomposing correlations between two variables is now illustrated by applying it to the four-variable model depicted in Figure 11.2. For convenience we repeat the necessary formulas which were developed in the previous section in connection with this model.

$$r_{12} = p_{21} \tag{11.12}$$

$$r_{13} = p_{31} + p_{32}r_{12} \tag{11.13a}$$

$$r_{23} = p_{31}r_{12} + p_{32} \tag{11.13b}$$

$$r_{14} = p_{41} + p_{42}r_{12} + p_{43}r_{13} \tag{11.15a}$$

$$r_{24} = p_{41}r_{12} + p_{42} + p_{43}r_{23} \tag{11.15b}$$

$$r_{34} = p_{41}r_{13} + p_{42}r_{23} + p_{43} \tag{11.15c}$$

Look back at Figure 11.2 and note that, except for the residuals which are not depicted in the diagram, variable 2 is affected by variable 1 only. Therefore r_{12} is due solely to the direct effect of variable 1 on 2 as indicated by p_{21} in equation (11.12).

Consider the correlation between variables 1 and 3. Since $r_{12} = p_{21}$ [equation (11.12)] it is possible to substitute p_{21} for r_{12} in equation (11.13a), obtaining

$$r_{13} = p_{31} + p_{32}p_{21} \tag{11.13a'}$$

It can now be seen that r_{13} is composed of two components. The direct effect of variable 1 on 3 is indicated by p_{31}, while the term $p_{32}p_{21}$ indicates the indirect effect of variable 1 on variable 3 via variable 2 (see Figure 11.2 for these direct and indirect paths). The same procedure can be applied to equation (11.13b) for the correlation between variables 2 and 3.

Now, variables 1 and 4. In equation (11.15a) substitute p_{21} for r_{12}. In addition, substitute for r_{13} in equation (11.15a) the right-hand term of equation

(11.13a). Making these substitutions we obtain

$$r_{14} = p_{41} + p_{42}p_{21} + p_{43}(p_{31} + p_{32}p_{21})$$
$$r_{14} = p_{41} + p_{42}p_{21} + p_{43}p_{31} + p_{43}p_{32}p_{21} \tag{11.15a'}$$

It is evident that the correlation between variable 1 and variable 4 is composed of a direct effect (p_{41}) and the following three indirect effects: $1 \to 2 \to 4$; $1 \to 3 \to 4$; $1 \to 2 \to 3 \to 4$.

Basically, then, the procedure involves first the development of the equation for the correlation between two variables, when each of them is expressed as a composite of the path coefficients leading to it (this was illustrated in detail in the previous section). The equation thus obtained is then expanded by substituting, whenever available, a compound term with more elementary terms. For example, in the equation for r_{14} (11.15a), r_{13} was substituted by two more elementary components.

We offer another example of the procedure, this time as applied to the decomposition of r_{34}.

$$r_{34} = p_{41}r_{13} + p_{42}r_{23} + p_{43} \tag{11.15c}$$

Substituting the right-hand term of (11.13a) for r_{13} and the right-hand term of (11.13b) for r_{23}, we obtain

$$r_{34} = p_{41}(p_{31} + p_{32}r_{12}) + p_{42}(p_{31}r_{12} + p_{32}) + p_{43}$$
$$= p_{41}p_{31} + p_{41}p_{31}r_{12} + p_{42}p_{31}r_{12} + p_{42}p_{32} + p_{43}$$

Substituting p_{21} for r_{12} in the above, and rearranging the terms to reflect the causal flow,

$$r_{34} = p_{43} + p_{41}p_{31} + p_{41}p_{21}p_{32} + p_{42}p_{21}p_{31} + p_{42}p_{32} \tag{11.15c'}$$

The first term on the right-hand side of (11.15c') (p_{43}) indicates the direct effect of variable 3 on variable 4. All the remaining terms indicate indirect effects due to common causes of 3 and 4. For example, part of the correlation between 3 and 4 is due to variable 1 being their common cause, as indicated by $p_{41}p_{31}$.

Wright (1934) and Turner and Stevens (1959), among others, provide algorithms that enable one to write equations such as (11.15c') by tracing the direct and indirect paths in a path diagram. While useful, such algorithms may become confusing to the novice, particularly when the path diagram is a complex one. It is therefore recommended that until one has mastered path analysis, the equations be developed in the manner demonstrated here rather than by resorting to algorithms.[22]

Total Indirect Effects

It is sometimes useful to decompose a correlation into a direct effect and Total Indirect Effects (TIE). It is then possible to study the magnitude of each

---

[22]The present procedure is further illustrated in connection with the numerical examples given below.

of these components and discern the roles they play in the system. To obtain the total indirect effects of a variable all that is necessary is to subtract its direct effect from the correlation coefficient between it and the dependent variable. The total indirect effects of variable 1 on variable 4, in the above example, are

$$\text{TIE}_{41} = r_{41} - p_{41} \tag{11.16}$$

Similarly, the TIE for variable 3 on variable 4 are

$$\text{TIE}_{43} = r_{43} - p_{43} \tag{11.17}$$

Theory Testing

Path analysis is an important analytic tool for theory testing. Through its application one can determine whether or not a pattern of correlations for a set of observations is consistent with a specific theoretical formulation. As shown in the preceding sections, a correlation between two variables can be expressed as a composite of the direct and indirect effects of one variable on the other. Using path coefficients it is therefore possible to reproduce the correlation matrix (R) for all the variables in the system. It should be noted, however, that as long as all variables are connected by paths and all the path coefficients are employed, the R matrix can be reproduced regardless of the causal model formulated by the researcher. Consequently, the reproduction of the R matrix when all the path coefficients are used is of no help in testing a specific theoretical model.

What if one were to delete certain paths from the causal model? This, in effect, will amount to setting certain path coefficients equal to zero. The implication is that the researcher conceives of the correlation between the two variables whose connecting path is deleted as being due to indirect effects only. By deleting certain paths the researcher is offering a more parsimonious causal model. If after the deletion of some paths it is possible to reproduce the original R matrix, or closely approximate it, the conclusion is that the pattern of correlations in the data is consistent with the more parsimonious model. Note that this does not mean the theory is "proven" to be "true." In fact, as is shown in the numerical examples given below, competing parsimonious models may be equally effective in reproducing the original R matrix, or closely approximating it. The crucial point is that the theoretical formulation is not derived from the analysis. All that the analysis indicates is whether or not the relations in the data are consistent with the theory.

If after the deletion of some paths there are large discrepancies between the original R matrix and the reproduced one, the conclusion is that in the light of the relations among the variables the more parsimonious theory is not tenable. Consequently, path analysis is more potent as a method for rejecting untenable causal models than for lending support to one of several competing causal models, when these models are equally effective in reproducing the correlation matrix.

What guidelines are there for the deletion of paths in a causal model? As pointed out earlier, the primary guideline is the theory of the researcher. On the basis of theory and previous research the researcher may decide that two variables in a model are not connected by a direct path. The analysis then determines whether or not the data are consistent with the theoretical formulation.

There is also a pragmatic approach to the deletion of paths. The researcher calculates first all the path coefficients in the model and then employs some criterion for the deletion of paths. Heise (1969b) refers to this approach as "theory trimming." Two kinds of criteria may be used in theory trimming. These are statistical significance and meaningfulness. It will be recalled that in the models dealt with in this chapter path coefficients are equal to β's. Therefore by testing the significance of a given β one is in effect testing the significance of the path coefficient which is equal to it. Adopting the significance criterion, one may decide to delete paths whose coefficients are not significant at a prespecified level. The problem with such a criterion, however, is that minute path coefficients may be found significant when the analysis is based on fairly large samples. Since it is recommended that one always use large samples for path analysis, the usefulness of the significance criterion is questioned.

In view of the shortcomings of the significance criterion, some researchers prefer to adopt the criterion of meaningfulness and delete all paths whose coefficients they consider not meaningful. It is not possible to offer a set of rules for determining meaningfulness. What may be meaningful for one researcher in one setting may be considered not meaningful by another researcher in another setting. In the absence of any other guidelines, some researchers (see, for example, Land, 1969) recommend that path coefficients less than .05 may be treated as not meaningful.

Subsequent to the deletion of paths whose coefficients are considered not meaningful, one needs to determine the extent to which the original R matrix can be approximated. In this case, too, there are no set rules for assessing goodness of fit. Once again the researcher has to make a judgment. Broadly speaking, if the discrepancies between the original and the reproduced correlations are small, say < .05, and the number of such discrepancies in the matrix is relatively small, the researcher may conclude that the more parsimonious model which generated the new R matrix is a tenable one.

Numerical Examples: Three-Variable Models

Suppose that a researcher has postulated a three-variable causal model as depicted in Figure 11.5. That is, variable 1 affects variables 2 and 3, while variable 2 affects variable 3. Suppose that the correlations are $r_{12} = .50$; $r_{23} = .50$; and $r_{13} = .25$. Since variable 2 is affected by variable 1 only, $p_{21} = \beta_{21} = r_{21} = .50$. To obtain p_{32} and p_{31}, calculate $\beta_{31.2}$ and $\beta_{32.1}$. Applying formula

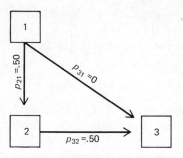

FIGURE 11.5

(6.7), which is repeated here with a new number for convenience,

$$\beta_{31.2} = \frac{r_{31} - r_{32}r_{12}}{1 - r_{12}^2}$$

(11.18)

$$\beta_{31.2} = \frac{(.25) - (.50)(.50)}{1 - .50^2} = \frac{.25 - .25}{1 - .25} = \frac{0}{.75} = 0$$

$$\beta_{32.1} = \frac{(.50) - (.25)(.50)}{1 - .50^2} = \frac{.50 - .125}{1 - .25} = \frac{.375}{.75} = .50$$

$p_{31} = \beta_{31.2} = 0$, and $p_{32} = \beta_{32.1} = .50$. Since p_{31} is zero there is no doubt that the direct path from variable 1 to variable 3 can be deleted, resulting in the more parsimonious model represented in Figure 11.6. According to this model, variable 1 has no direct effect on variable 3. Now see whether the original correlations can be reproduced. In the present case this involves the reproduction of the correlation between variables 1 and 3. It should be obvious that p_{21} is still .50 (r_{21}) and $p_{32} = .50$ (r_{32}).

According to the model in Figure 11.6,

$$z_1 = e_1$$

$$z_2 = p_{21}z_1 + e_2$$

$$z_3 = p_{32}z_2 + e_3$$

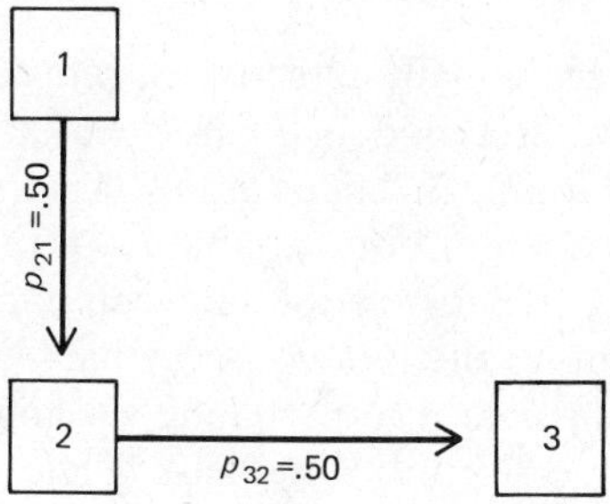

FIGURE 11.6

The residual terms are assumed not to be correlated and are therefore dropped.

$$r_{13} = \frac{1}{N}\sum z_1 z_3 = \frac{1}{N}\sum z_1 (p_{32} z_2) = p_{32} r_{12}$$

Since $r_{12} = p_{21}$,

$$r_{13} = p_{32} p_{21} = (.50)(.50) = .25$$

It was demonstrated that r_{13} can be reproduced in the more parsimonious model, thus indicating that the model is compatible with the correlations among the variables. As discussed earlier, however, there may be competing models that are just as effective in reproducing the correlation matrix. For example, suppose that one postulated that the causal model for the three variables under consideration is the one represented in Figure 11.7. In this model variable 2 is a common cause of variables 1 and 3. The path coefficients are $p_{32} = .50$ (r_{32}), and $p_{12} = .50$ (r_{12}). What about the reproduction of r_{13}?

$$z_1 = p_{12} z_2 + e_1$$

$$z_3 = p_{32} z_2 + e_3,$$

$$r_{13} = \frac{1}{N}\sum z_1 z_3 = \frac{1}{N}\sum z_1 (p_{32} z_2) = p_{32} r_{12}$$

Since $r_{12} = p_{12}$,

$$r_{13} = p_{32} p_{12} = (.50)(.50) = .25$$

Again, the correlation between variables 1 and 3 was reproduced.

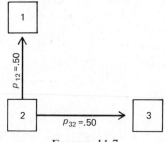

FIGURE 11.7

While both models are equally effective in reproducing r_{13}, they are substantively different. In the first model (Figure 11.6) variable 1 affects variable 2, which in turn affects variable 3. Therefore r_{13} is a consequence of variable 2 mediating the effect of variable 1 on variable 3. In the second model (Figure 11.7), on the other hand, the correlation between variables 1 and 3 is clearly spurious. It is entirely due to the two variables having a common cause (variable 2). If one were to calculate $r_{13.2}$ (partialing variable 2 from variables 1 and 3) it would be equal to zero in both models. While it makes good sense to partial variable 2 in model 11.7, it makes no sense to do so in model 11.6.[23]

[23]For discussions pertaining to spurious correlations, see Simon (1954) and Blalock (1964, 1968).

The important question, of course, is which model is more tenable. Obviously it is possible to generate more models than the two under discussion. In fact, even with as small a number as three variables it is possible to generate twelve different causal models. Again, the choice of the model is up to the researcher. The more he knows about the area in which he is working the better equipped he is to make a choice among the alternative models.

It was said in the preceding section that path analysis may be useful in rejecting a causal model when there are sizable discrepancies between the original and the reproduced R matrices. This point is illustrated by resorting again to the three variables under consideration in the present section. Suppose that the researcher has postulated the causal model to be the one represented in Figure 11.8. That is, variable 3 affects variable 1, which in turn affects variable 2. The equations therefore are

$$z_3 = e_3$$

$$z_2 = p_{21}z_1 + e_2$$

$$z_1 = p_{13}z_3 + e_1$$

$$p_{13} = r_{13} = .25 \qquad p_{21} = r_{21} = .50$$

Attempting now to reproduce r_{23},

$$r_{23} = \frac{1}{N} \sum z_3 z_2 = \frac{1}{N} \sum z_3 (p_{21}z_1) = p_{21}r_{13}$$

Since $r_{13} = p_{13}$,

$$r_{23} = p_{21}p_{13} = (.50)(.25) = .125$$

Note that there is a large discrepancy between the original r_{23} (.50) and the reproduced one (.125). It is therefore concluded that the data are not consistent with the model, and the model is rejected.

In conclusion, it is demonstrated that when all the variables are connected by paths the correlation matrix can be reproduced regardless of the causal model chosen by the researcher. Assume that the causal model is as depicted

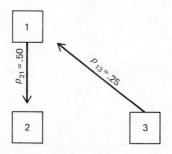

FIGURE 11.8

in Figure 11.9. The equations are

$$z_3 = e_3$$

$$z_1 = p_{13}z_3 + e_1$$

$$z_2 = p_{21}z_1 + p_{23}z_3 + e_2$$

$$p_{13} = r_{13} = .25$$

$$p_{21} = \beta_{21.3} = \frac{(.50) - (.50)(.25)}{1 - .25^2} = \frac{.50 - .125}{1 - .0625} = \frac{.375}{.9375} = .40$$

$$p_{23} = \beta_{23.1} = \frac{(.50) - (.50)(.25)}{1 - .25^2} = .40$$

$$r_{12} = \frac{1}{N} \sum z_1 z_2 = \frac{1}{N} \sum z_1 (p_{21}z_1 + p_{23}z_3) = p_{21} + p_{23}r_{13}$$

Since $r_{13} = p_{13}$,

$$r_{12} = p_{21} + p_{23}p_{13}$$

$$r_{12} = (.40) + (.40)(.25) = .50$$

The obtained r_{12} from the data is identical to the reproduced one. r_{23} can be similarly reproduced. Again, the point is that unless some paths are deleted, the correlation matrix will be reproduced regardless of the causal model employed. We turn now to a more complex model.

An Example from Educational Research

Earlier in this chapter a numerical example was presented in which the grade-point average (GPA) of college students was regressed on socioeconomic status (SES), intelligence (IQ), and need achievement (*n* Ach). A forward and a backward solution, as well as commonality analysis, were demonstrated in connection with this example. We return now to the same set of fictitious data and analyze them with path analysis. In addition to providing a demonstration of a path analysis for a more complex model, the use of the same numerical example will afford a comparison with the ordinary regression and commonality

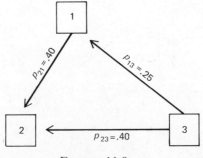

FIGURE 11.9

analyses used earlier. For convenience, we repeat in the upper half of the matrix in Table 11.8 the correlations originally presented in Table 11.1.

Assume that the causal model for the variables in this example is the one depicted in Figure 11.10. Note that SES and IQ are treated as exogenous variables. SES and IQ are assumed to affect n Ach.; SES, IQ, and n Ach are assumed to affect GPA.[24] Since it will be necessary to make frequent reference to these variables in the form of subscripts, it will be more convenient to identify them by the numbers attached to them in Figure 11.10. Accordingly, $1 = $ SES, $2 = $ IQ, $3 = n$ Ach, and $4 = $ GPA.

TABLE **11.8**   ORIGINAL AND REPRODUCED CORRELATIONS
FOR A FOUR-VARIABLE MODEL. $N = 100$[a]

	1 SES	2 IQ	3 n Ach	4 GPA
1	1.000	.300	.410	.330
2	.300	1.000	.160	.570
3	.410	.123	1.000	.500
4	.323	.555	.482	1.000

[a]The original correlations are reported in the upper half of the matrix. The reproduced correlations are reported in the lower half of the matrix. For explanation and discussion, see text below.

In order to calculate the path coefficients for the causal model depicted in Figure 11.10 it is necessary to do two regression analyses. First, variable 3 is regressed on variables 1 and 2 to obtain $\beta_{31.2} = p_{31}$ and $\beta_{32.1} = p_{32}$. Second, variable 4 is regressed on variables 1, 2, and 3 to obtain $\beta_{41.23} = p_{41}, \beta_{42.13} = p_{42}$, and $\beta_{43.12} = p_{43}$.

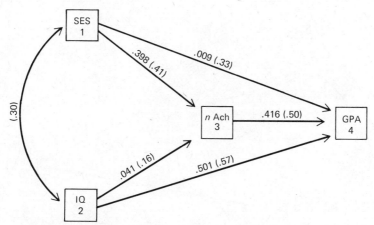

FIGURE   11.10   Numbers in parentheses indicate zero-order correlations. Other numbers are path coefficients. For example: $p_{41} = .009$, $r_{41} = .33$.

[24]The theoretical considerations that would generate this model are not discussed here, since the sole purpose of the presentation is to illustrate the analysis of such a model.

Methods for calculating β's were discussed and illustrated in earlier chapters (see particularly Chapter 4). Therefore, in the interest of space the calculations of the β's for the present problem and subsequent ones are not shown. Instead, the results are reported and applied in path analysis.[25] Regressing variable 3 on variable 1 and 2, one obtains $p_{31} = .398$ and $p_{32} = .041$. Regressing variable 4 on variables 1, 2, and 3, one obtains $p_{41} = .009$, $p_{42} = .501$, and $p_{43} = .416$.

Note that two path coefficients (p_{41} and p_{32}) are $< .05$, indicating that r_{14} and r_{23} are mainly due to indirect effects. The direct effect of variable 1 on 4 is .009, while the total of indirect effects is .321 ($r_{41} - p_{41} = .33 - .009$). In other words, SES has practically no direct effect on GPA. It does, however, affect GPA indirectly through its correlation with IQ and its effect on n Ach. The correlation between IQ and n Ach is mostly due to IQ's correlation with SES. The observations regarding p_{41} and p_{32} lead to the conclusion that the present model can be trimmed. The more parsimonious model is presented in Figure 11.11.

Are the data consistent with the new model? To answer this question, the path coefficients in the new model are calculated and used in an attempt to reproduce the original correlation matrix. In the new model, $p_{31} = r_{31} = .41$. By regressing variable 4 on variables 2 and 3, one obtains $p_{42} = .503$ and $p_{43} = .420$.

The equations that reflect the model in Figure 11.11 are

$$z_3 = p_{31}z_1 + e_3$$

$$z_4 = p_{42}z_2 + p_{43}z_3 + e_4$$

We now calculate the zero-order correlations between all the variables. Since variables 1 and 2 are exogenous, r_{12} remains unanalyzed.

$$r_{13} = p_{31} = .41$$

$$r_{23} = \frac{1}{N}\sum z_2 z_3 = \frac{1}{N}\sum z_2(p_{31}z_1) = p_{31}r_{12} = (.41)(.30) = .123$$

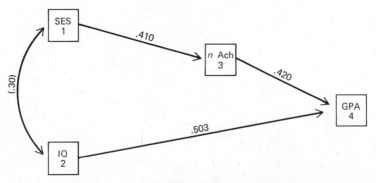

FIGURE 11.11

The original r_{23} is .160.

$$r_{14} = \frac{1}{N}\sum z_1 z_4 = \frac{1}{N}\sum z_1(p_{42}z_2 + p_{43}z_3) = p_{42}r_{12} + p_{43}r_{13}$$

Substituting p_{31} for r_{13},

$$r_{14} = p_{42}r_{12} + p_{43}p_{31}$$

$$r_{14} = (.503)(.30) + (.420)(.410) = .323$$

The original r_{14} is .330.

$$r_{24} = \frac{1}{N}\sum z_2 z_4 = \frac{1}{N}\sum z_2(p_{42}z_2 + p_{43}z_3) = p_{42} + p_{43}r_{23}$$

Substituting $p_{31}r_{12}$ for r_{23},

$$r_{24} = p_{42} + p_{43}p_{31}r_{12}$$

$$r_{24} = (.503) + (.420)(.410)(.30) = .555$$

The original r_{24} is .57.

$$r_{34} = \frac{1}{N}\sum z_3 z_4 = \frac{1}{N}\sum z_3(p_{42}z_2 + p_{43}z_3) = p_{42}r_{23} + p_{43}$$

Substituting $p_{31}r_{12}$ for r_{23},

$$r_{34} = p_{42}p_{31}r_{12} + p_{43}$$

$$r_{34} = (.503)(.410)(.30) + (.420) = .482$$

The original r_{34} is .50.

Since the discrepancies between the original and the reproduced correlations are small, it is concluded that the data are consistent with the more parsimonious model. (For ease of comparison between the original and the reproduced correlations, see Table 11.8, where the former are reported in the upper half of the matrix and the latter are reported in the lower half.)

Let us compare the present analysis with analyses done earlier on the same data, namely the forward and backward solutions and the commonality analysis. In the forward and the backward solutions it was found that SES made such a minute contribution to the proportion of variance accounted for that it was not meaningful to retain it in the equation. In the commonality analysis the unique component of SES was almost zero.[26] Consequently, one may have been led to believe that SES is not an important variable when it is considered in the context of the total pattern of relations with the other variables, namely IQ, n Ach, and GPA. In the present model, on the other hand, SES plays an important role. While it does not have a direct effect on GPA, it does affect GPA indirectly through its effect on n Ach and through its correlation with IQ. IQ and n Ach have direct as well as indirect effects on GPA. In both cases, however, the

[26]See earlier sections of this chapter.

direct effects are much larger than the indirect ones. The direct effect of IQ on GPA is somewhat larger than the direct effect of n Ach on GPA.

Assuming that the theoretical formulations with the variables under consideration are sound, the present analysis is more meaningful than analyses that are done without regard to theoretical considerations (for example, forward solution, commonality analysis). In fact, using such analyses or a single regression analysis for the present theoretical model would constitute lack of fit between theory and analysis.

Accounting for Variance in the Dependent Variable

Problems relating to the variance accounted for by each of the independent variables in a regression analysis were discussed and illustrated in an earlier section of this book. It was shown that when the independent variables are correlated, the order in which they are entered into the analysis has an effect on the proportion of variance accounted for attributed to each of them. It was therefore reasoned that, in the absence of criteria for the ordering of the independent variables, statements about the proportion of variance accounted for by each variable are of little value.

Within the framework of a causal model, however, the ordering of the independent variables is not arbitrary. On the contrary, it is determined by the theoretical considerations that generated the specific model. Two approaches may be taken to the study of increments in the proportion of variance accounted for within the context of a given model. In the first approach, one starts with the cause that is most remote from the dependent variable and successively enters variables in the direction of the causal flow, moving closer and closer to the dependent variable. It is thus possible to determine the increments in the proportion of variance accounted for by each variable when the order in which they are entered into the analysis is determined by a given causal model. Moreover, it is possible to note whether, after having entered a set of remote causes, the increment due to a cause which is closer to the dependent variable is meaningful.

In the second approach, one starts from the cause closest to the dependent variable and traces backwards to the more distant causes. In this case it is possible to note whether a more remote cause adds meaningfully to the variance accounted for, after having introduced causes closer to the dependent variable. Which of the two approaches one chooses to follow depends on one's frame of reference and specific interests.[27]

The two approaches are illustrated by applying them to the numerical example given in the preceding section (for the causal model, see Figure 11.10). Moving forward from the remotest causes, it is to be noted that since variable 1 (SES) and variable 2 (IQ) are treated as correlated exogenous variables it is not possible to determine unique proportions of variance for which each of them accounts. Instead, the two variables are treated as one unit in the analysis. The

---

[27]For a more detailed treatment of this topic, see Duncan (1970).

second unit is variable 3 (n Ach). Accordingly, $R^2_{4.12}$ indicates the proportion of variance accounted for by SES and IQ. $R^2_{4.123} - R^2_{4.12}$ indicates the increment in the proportion of variance due to n Ach.

Using the results obtained from the calculation in the preceding section,

$$R^2_{4.123} = .4966$$

$$R^2_{4.12} = .3527$$

$$R^2_{4.123} - R^2_{4.12} = .4966 - .3527 = .1439$$

When SES and IQ are entered first they jointly account for about 35 percent of the variance. The increment due to n Ach is about 14 percent.

Moving backward from the cause closest to the dependent variable,

$$R^2_{4.3} = .2500$$

$$R^2_{4.123} - R^2_{4.3} = .4966 - .2500 = .2466$$

In this approach n Ach is shown to account for 25 percent of the variance. Adding more remote causes (SES and IQ) results in an increment of about 25 percent.

Recall that in the forward, backward, and stepwise solutions the criterion is optimal prediction with a minimum number of variables. The selection of a variable is therefore determined by the increment in variance of the dependent variable it accounts for. In the present approach, on the other hand, the selection of variables is determined by a theoretical model. The choice of tracing forward from the remotest cause(s) to the dependent variable or tracing backwards from the cause(s) closest to the dependent variable depends on the question the researcher wishes to answer. Tracing backwards, for example, will answer the question whether in accounting for variance of the dependent variable it is sufficient to resort to immediate causes or whether it is necessary to include also more remote ones.

An Example from Political Science Research

In an attempt to explain the roll call behavior of Congressmen, Miller and Stokes (1963) studied the pattern of relations among the following variables: attitudes of samples of constituents in each of 116 congressional districts, attitudes of the Congressmen representing the districts, Congressmen's perceptions of the attitudes held by their constituents, and roll-call behavior of the Congressmen. In a reanalysis of the Miller and Stokes data, Cnudde and McCrone (1966) formulated the three alternative causal models presented in Figure 11.12. Cnudde and McCrone tested these alternative models by employing a technique originally developed by Simon (1954) and elaborated by Blalock (1964, 1968). The Simon-Blalock technique is similar in certain respects to path analysis, but is not as powerful.[28] The presentation here is limited to the

---

[28]See Boudon (1968) and Heise (1969b).

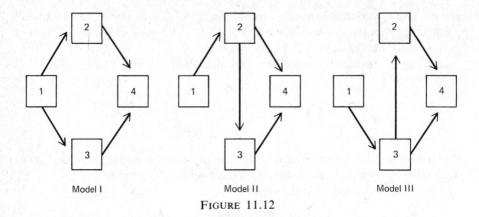

Model I   Model II   Model III

FIGURE 11.12

application of path analysis to the testing of the three alternative models proposed by Cnudde and McCrone (1966). For the theoretical considerations that generated these models the reader is referred to the papers by Miller and Stokes (1963) and Cnudde and McCrone (1966).

One of the domains dealt with in the study was attitudes and roll call behavior pertaining to civil rights. The correlations among the variables in the area of civil rights are reported in the upper half of the matrix in Table 11.9. We deal separately with each of the three models presented in Figure 11.12. First, the path coefficients for the model are reported. Second, the equations reflecting the model are stated, followed by an attempt to reproduce the correlation matrix. To facilitate the presentation, the numbers identifying the variables in Figure 11.12 are used. They are: 1 = Constituents' attitudes, 2 = Congressmen's attitudes, 3 = Congressmen's perceptions, and 4 = Roll call.

Model I.   The path coefficients for this model are

$$p_{21} = .498 \qquad p_{31} = .738 \qquad p_{42} = .721 \qquad p_{43} = .823$$

TABLE **11.9**   ORIGINAL AND REPRODUCED CORRELATIONS. ATTITUDES
AND ROLL CALL PERTAINING TO CIVIL RIGHTS[a]

	1 Constituents' Attitudes	2 Congressmen's Attitudes	3 Congressmen's Perceptions of Constituents' Attitudes	4 Roll-Call Behavior
1	1.000	.498	.738	.649
2	.475	1.000	.643	.721
3	.738	.643	1.000	.823
4	.608	.721	.823	1.000

[a]The original correlations, in the upper half of the matrix, are taken from Cnudde and McCrone (1966). In the lower half of the matrix are the correlations as reproduced by the application of Model III. For an explanation and discussion, see text.

The equations that reflect Model I are

$$z_1 = e_1$$
$$z_2 = p_{21}z_1 + e_2$$
$$z_3 = p_{31}z_1 + e_3$$
$$z_4 = p_{42}z_2 + p_{43}z_3 + e_4$$

Since all the path coefficients in this model are equal to their respective zero-order correlations, all that is necessary is to note whether r_{23} and r_{14} can be reproduced.

$$r_{23} = \frac{1}{N} \sum z_2 z_3 = \frac{1}{N} \sum z_2 (p_{31}z_1) = p_{31}r_{12}$$
$$r_{23} = p_{31}p_{21}$$
$$r_{23} = (.738)(.498) = .368. \text{ The original } r_{23} = .643$$

In view of the large discrepancy between the reproduced and the original correlation between variables 2 and 3, it is obvious that r_{14} cannot be reproduced either. Consequently, Model I is rejected.

Model II.   The path coefficients for Model II are

$$p_{21} = .498 \qquad p_{32} = .643 \qquad p_{42} = .329 \qquad p_{43} = .613$$

The equations are

$$z_1 = e_1$$
$$z_2 = p_{21}z_1 + e_2$$
$$z_3 = p_{32}z_2 + e_3$$
$$z_4 = p_{42}z_2 + p_{43}z_3 + e_4$$

Reproducing r_{13},

$$r_{13} = \frac{1}{N} \sum z_1 z_3 = \frac{1}{N} \sum z_1 (p_{32}z_2) = p_{32}r_{12}$$
$$r_{13} = p_{32}p_{21}$$
$$r_{13} = (.643)(.498) = .320. \text{ The original } r_{13} = .738$$

The discrepancy between the reproduced r_{13} and the original r_{13} is so large that it is not necessary to see whether r_{14} can be reproduced. Model II is rejected.

Model III.   The path coefficients for Model III are

$$p_{31} = .738 \qquad p_{23} = .643 \qquad p_{42} = .329 \qquad p_{43} = .613$$

The equations are

$$z_1 = e_1$$
$$z_2 = p_{23}z_3 + e_2$$
$$z_3 = p_{31}z_1 + e_3$$
$$z_4 = p_{42}z_2 + p_{43}z_3 + e_4$$

Reproducing r_{12},

$$r_{12} = \frac{1}{N} \sum z_1 z_2 = \frac{1}{N} \sum z_1 (p_{23} z_3) = p_{23} r_{13}$$

$$r_{12} = p_{23} p_{31}$$

$$r_{12} = (.643)(.738) = .475. \text{ The original } r_{12} = .498$$

There is a very small discrepancy (.023) between the original r_{12} and the reproduced r_{12}.

Reproducing r_{14},

$$r_{14} = \frac{1}{N} \sum z_1 z_4 = \frac{1}{N} \sum z_1 (p_{42} z_2 + p_{43} z_3) = p_{42} r_{12} + p_{43} r_{13} = p_{42} (p_{23} p_{31}) + p_{43} p_{31}$$

$$r_{14} = p_{42} p_{23} p_{31} + p_{43} p_{31}$$

$$r_{14} = (.329)(.643)(.738) + (.613)(.738) = .156 + .452 = .608$$

The original $r_{14} = .649$. The discrepancy between the two correlations is quite small (.041). It is concluded that the data are consistent with Model III.

Because of space limitations it is not possible to discuss in detail the implications of Model III. Suffice it to point out that according to this model there is no direct effect of the constituents' attitudes on Congressmen's attitudes. Constituents' attitudes affect Congressmen's perceptions, which in turn affect Congressmen's attitudes. Moreover, the direct effect of Congressmen's perceptions (of their constituents' attitudes) on the roll call is considerably larger than the direct effect of the Congressmen's attitudes on roll call ($p_{43} = .613$, $p_{42} = .329$). It appears that what Congressmen perceive the attitudes of their constituents to be is more important than their own attitudes in determining their roll-call behavior. The original correlations and the Model III reproduced correlations are given in Table 11.9.

Final Note

Explanation and prediction are at the core of scientific inquiry. When the primary concern is prediction, the forward, backward, and stepwise solutions and commonality analysis can be used for the selection of variables from a larger pool of variables. The choice and application of a specific method depends, of course, on the needs and interest of the researcher.[29]

The situation is more difficult and complex, however, when the primary interest is explanation. Commonality analysis and path analysis were presented as two methods intended to assist the researcher to untangle the relations among independent variables, thereby facilitating attempts to study their effects on a dependent variable. It should be noted, however, that neither commonality analysis nor path analysis are free of shortcomings. Commonality analysis is

---

[29]For a comparative study of different selection methods, see Halinski and Feldt (1970).

not too helpful when correlations among independent variables are relatively high, or when the number of independent variables is large. Path analysis, on the other hand, requires the formulation of a causal model, a requirement which frequently cannot be met at the present stage of knowledge in the behavioral sciences. In addition, path analysis imposes a set of rather restrictive assumptions, which, when seriously violated, may lead to erroneous conclusions.

Neither path analysis nor commonality analysis should be viewed as a panacea for the solution of the highly complex problems that confront the behavioral scientist. Recognizing that they are methods, it follows that they may be used judiciously or injudiciously. In the final analysis, a method is as good or as bad as the use to which it is put by a prudent or imprudent researcher. If there is any lesson to be learned from this chapter, it is that no method should be used thoughtlessly. Paraphrasing a phrase from the Jewish prayer book: The outcome of a deed depends on the thought that initiated it.

Finally, the subject matter of this chapter should indicate that we are far from solving the methodological problems facing us in our attempts to explain phenomena. A concerted effort is needed to refine and sharpen available tools, and to develop new ones that will perhaps be more in accord with the complexity of human behavior.

Study Suggestions

1. Distinguish between explanation and prediction. Give examples of studies in which the emphasis is on one or the other.
2. What is meant by "shrinkage" of the multiple correlation? What is the relation between shrinkage and the size of the sample?
3. What is cross-validation? How is it used?
4. Discuss and compare forward, backward, and stepwise solutions. What is the purpose of using such solutions?
5. Discuss two types of criteria that may be used for the termination of an analysis in which a smaller number of variables is selected from a larger pool. Which of them is more important? Why?
6. Here is a fictitious correlation matrix ($N = 300$). The dependent variable is verbal achievement. The independent variables are: race, mental ability, school quality, self-concept, and level of aspiration.

	1 Race	2 IQ	3 School Quality	4 Self- Concept	5 Level of Aspiration	6 Verbal Achievement
1	1.00	.30	.25	.30	.30	.25
2		1.00	.20	.20	.30	.60
3			1.00	.20	.30	.30
4				1.00	.40	.30
5					1.00	.40

Using the above data, do a forward solution. At each step, indicate the increment in the proportion of variance due to the variable entering the equation and the F ratio associated with the increment. Terminate the analysis when the increment in the proportion of variance is less than .01. Give the regression equation for the final solution.

(*Answers*: Step 1. $R^2_{6.2} = .3600$. $F = 167.44$, 1 and 298 *df*. Step 2. $R^2_{6.25} = .4132$. Increment $= .0532$. $F = 26.87$, 1 and 297 *df*. Step 3. $R^2_{6.253} = .4298$. Increment $= .0166$. $F = 8.61$, 1 and 296 *df*. Step 4. $R^2_{6.2534} = .4392$. Increment $= .0094$. Analysis terminated. Regression equation: $z'_6 = .511z_2 + .136z_3 + .206z_5$.)

7. What is the purpose of commonality analysis?
8. In a study with eight independent variables, how many components are there in a commonality analysis?
    (*Answer*: 255.)
9. Write the formula for the commonality of variables 3 and 5, in a study consisting of six independent variables.
    (*Answer*: $-R^2_{y.1246} + R^2_{y.12346} + R^2_{y.12456} - R^2_{y.123456}$.)
10. What is the purpose of path analysis.
11. Distinguish between exogenous and endogenous variables.
12. What is a recursive model? Give examples.
13. In studies of authoritarianism it has been found that the F scale is correlated negatively with mental ability and years of education. Assume that the correlations among these variables are as indicated on the paths in the theoretical model given in the figure below.

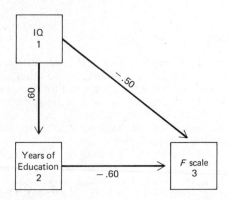

Do a path analysis. (a) What is the direct effect of mental ability on authoritarianism? (b) What is the indirect effect of mental ability on authoritarianism? (c) What is the direct effect of years of education on authoritarianism? Interpret the results.
(*Answers*: (a) $p_{31} = -.219$; (b) $-.281$; (c) $p_{32} = -.469$.)

14. Using the data of study suggestion 13 do a commonality analysis. What are the unique contributions of: (a) mental ability; (b) years of education? (c) What is the commonality of mental ability and years of education? Interpret the results and compare them with those obtained in the path analysis of the same data.
    (*Answers*: (a) .031; (b) .141; (c) .219.)
15. For the data given in the correlation matrix of study suggestion 6, con-

sider the following causal model:

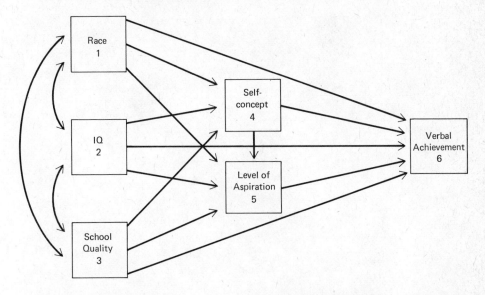

Do a path analysis. What are the path coefficients for the variables affecting: (a) self-concept; (b) level of aspiration; (c) verbal achievement?
(d) Using a criterion of .05 as a meaningful path coefficient, which paths
may be deleted in the above model? Interpret the results.
(*Answers*: (a) $p_{41} = .239$, $p_{42} = .104$, $p_{43} = .119$; (b) $p_{51} = .116$, $p_{52} = .171$,
$p_{53} = .178$, $p_{54} = .296$; (c) $p_{61} = -.019$, $p_{62} = .506$, $p_{63} = .130$, $p_{64} = .110$,
$p_{65} = .171$; (d) the path from race to verbal achievement.)

16. Use commonality analysis to determine the unique contributions to verbal achievement of the five variables in study suggestion 15. (The correlations among the variables are given in study suggestion 6.) Interpret the
results and compare them with those obtained in the path analysis of the
same data (study suggestion 15).
(*Answers*: U(1) = .0003, U(2) = .2190, U(3) = .0148, U(4) = .0097,
U(5) = .0216.)

PART 3

Multiple Regression and Multivariate Analysis

CHAPTER 2

Multiple Regression, Discriminant Analysis, and Canonical Correlation

Almost all methods of numerical data analysis are the same in one respect: they identify, partition, and control variance. Multiple regression analysis and analysis of variance, for example, break down the variance of a dependent variable according to its relations to the variances of one or more independent variables. One may even go so far as to say that all methods of analysis seek to identify and quantify variance shared by variables. Multiple regression seeks to identify and estimate the magnitude and statistical significance of the variance of the dependent variable, Y, that is shared with several independent variables. If we keep this notion firmly in mind we will have little difficulty understanding the other methods of multivariate analysis to be considered in this chapter and in Chapters 13 and 14.

Our purposes in this chapter and in Chapters 13 and 14 are, as usual, to give the reader an intuitive grasp of what these methods do, to place them in the behavioral research context, and to deepen understanding of multiple regression analysis and its scientific generality and applicability. In three chapters we cannot clearly and completely explain complex multivariate methods. Such explanation needs a whole volume. But we may be able to put the student in a position to pursue study of multivariate methods with a base of understanding that can materially aid his study.

Discriminant Analysis

A research problem that arises again and again is to classify individuals into groups on the basis of their scores on tests. The simplest case, of course, is to place people – note that instead of people we can say physical objects, geographical areas, or any other units on which we have one or more measures –

into two groups on the basis of scores on one test. We administer, say, a test of verbal aptitude or a test of creativity and then assign children to different groups on the basis of their verbal aptitude or creativity scores. In other words, by using test scores we are able to assign individuals to two (or more) groups.[1]

A more interesting procedure is to assign individuals to two or more groups on the basis of their scores on two or more tests or scales. Instead of a single test of verbal aptitude, we may have two or three aptitude tests and from them predict, for instance, success or lack of success in high school. By doing a discriminant analysis we can make such predictions. That is, on the basis of a least squares composite of k test scores, we predict success or lack of success in high school. As the reader may have guessed, this is nothing more than a multiple regression situation where the dependent variable is group membership.

Take a rather unusual but potentially fruitful example. Suppose we have three measures of administrative performance acquired through the In-Basket Test (Hemphill, Griffiths, & Frederiksen, 1962): Ability to Work with Others, X_1, Motivation for Administrative Work, X_2, and General Professional Skill, X_3. In addition, we have ratings of the same administrators on their administrative performance, as observed on the job or in simulated administrative situations. These ratings are simply "successful" and "unsuccessful." How can we assign the individuals—and other individuals not in the sample—to the "successful" and "unsuccessful" groups?

Another related and perhaps more important problem is not only to discriminate maximally between two or more groups—say males and females, Republicans and Democrats, or successful and unsuccessful—but also to be able to specify something of the nature of the discrimination. In the above administrator example, for instance, the three In-Basket measures may indeed discriminate well between the successful and unsuccessful administrators. But this may not be enough for our purposes. We may also want to "explain" the discrimination; we may want to know the relative efficacies or weights of the three tests in the discrimination.

Discriminant analysis is a way of solving such problems. The *discriminant function* is a regression equation with a dependent variable that represents group membership. With only two groups, discriminant function analysis amounts to multiple regression analysis with the dependent variable taking the values of 1 and 0. There are several measures for each individual in a sample. Using these measures as independent variables and a vector of 1's and 0's as the dependent variable, we solve the regression in the usual manner. The resulting equation, the discriminant function, maximally discriminates the members of the sample; it tells us to which group each member probably "belongs."

[1]A common reason for such classification has been to create manageable variables to use in analysis of variance and other analytic tools. By now the reader will know that this practice is questionable since it discards information. One of our points in previous chapters has been to include such variables, unaltered, in multiple regression analysis.

Again using the administrator example, we obtain a sample of administrators who had been rated as "successful" and "unsuccessful." The administrators' scores on the three tests are treated as independent variables predicting a dependent variable of 1's and 0's, 1 meaning "successful" and 0 "unsuccessful." The analysis—in this case an ordinary multiple regression analysis—will maximally discriminate between the two groups to the extent of its capabilities. It will also give information on how well the function predicts the successful and unsuccessful groups and the relative weights of the three tests in doing so.

The discriminant function can also be used for other groups of similar administrators without knowing whether they are successful or unsuccessful. One uses the function, in other words, for predictive purposes. It has to be borne in mind, however, that if the new samples differ from the original sample, the validity of the procedure is questionable. Nevertheless, with cumulative information on administrative performance and its relations to the three In-Basket tests, predictions can be relied upon. In any case, discriminant analysis can materially aid predictive use of tests.

An Educational Example

As usual, we use a fictitious example with simple numbers to show how discriminant analysis works. We use a rather ordinary example because it illustrates the discriminant function in a clearer way than a more interesting research example might. There is often the need in education to classify in-

TABLE 12.1   FICTITIOUS EXAMPLE OF DISCRIMINANT FUNCTION
ANALYSIS, WITH PREDICTED Y SCORES, GROUP MEMBERSHIP, AND
REGRESSION STATISTICS[a]

Y	X_1	X_2	Y'	Group Assignment	Satisfactory Achievement
1	8	3	1.05556	A_1	Yes
1	7	4	1.03704	A_1	Yes
1	5	5	.87963	A_1	Yes
1	3	4	.48148	A_2	Yes
1	3	2	.24074	A_2	Yes
0	4	2	.37963	A_2	No
0	3	1	.12037	A_2	No
0	3	2	.24074	A_2	No
0	2	2	.10185	A_2	No
0	2	5	.46296	A_2	No

$$Y' = -.41667 + .13889X_1 + .12037X_2$$
$$\bar{Y}'_{A_1} = .73889 \qquad \bar{Y}'_{A_2} = .26111$$

[a]Y: 1 = adequate achievement, 0 = inadequate achievement; X_1 = verbal ability; X_2 = school motivation; Y' = predicted scores; A_1 = assignment to adequate group; A_2 = assignment to inadequate group.

dividuals or to assign them to groups. Suppose, for example, that the guidance and counseling department of a high school is interested in predictions of the achievement of incoming freshmen. If the department knew which students were going to have severe achievement problems, perhaps it could do something for such students *before* the problems arise. One of the members of the department believes that a combination of a measure of verbal ability and one of motivation (which he has devised) will predict underachievement well.

To test this predictive idea, the department selects 10 sophomore students, 5 who are having considerable academic difficulty and 5 who are not having difficulty, according to the judgments of teachers. (Of course, many more than 10 students would be used.) The department has verbal ability scores on these 10 individuals, and administers the motivation measure to them. The 5 students who are achieving successfully are assigned 1's and the 5 who are having academic difficulty are assigned 0's.

The data from the 10 students are given in Table 12.1, first three columns. The rest of the data in the table will be explained as we go along. The member of the guidance department whose idea it was to use the two measures did a multiple regression analysis on the data using the Y vector of 1's and 0's as the dependent variable and the verbal ability measure, X_1, and the school motivation measure, X_2, as the independent variables.[2] The regression analysis yields the regression equation given at the bottom of Table 12.1. Using this equation, the guidance counselor calculates the predicted Y scores of each student; these scores are given in the fourth column of the table (labeled Y'). The actual achievement status of the students is given in the last column. If a student's achievement is satisfactory, Yes is recorded; if unsatisfactory, No is recorded.

How well do the two independent variables assign the individuals to the two groups? If the counselor did not know the achievement status of the students, how well could he have predicted their achievement? Keep in mind that when we predict group membership we predict 1's and 0's. Thus, we expect the predicted Y's, derived from the regression equation and the students' scores, to be close to 1 in the case of students whose achievement is satisfactory, call them A_1, and close to 0 in the case of students whose achievement is not satisfactory, call them A_2. The counselor calculated the predicted means of A_1 and A_2 (the mean of the first five Y' values, fourth column, and the mean of the second five Y' values). These means are .73889 and .26111. He then assigned the 10 students to A_1 and A_2 using the criterion of closeness to these means. The resulting classification is given in the fifth column of Table 12.1. Judging from the actual achievement status of the students, given in the last column of the table, 8 are correctly classified. All 5 whose achievement is not satisfactory have been assigned to A_2, but 2 students whose achievement is actually satisfactory have been erroneously assigned to A_2.

[2]Actual discriminant function analysis does not use the method described above. With only two groups, however, the above method works and, more important, clearly shows the multiple regression nature of discriminant analysis.

The procedure seems to indicate that verbal ability and school motivation are fairly successful in predicting achievement status. The usual F test, however, showed that $R^2_{y.12}$, which was .48, was not statistically significant.[3] If this were actually the case, the counselor cannot expect much predictive efficiency from the discriminant function.

Let us assume for the moment that the regression was statistically significant. If so, the counselor can use the discriminant function expressed by the regression equation in Table 12.1 for future students. After entering these students' X_1 and X_2 scores into the equation, he can assign them to A_1, probable satisfactory achievement, or to A_2, probable unsatisfactory achievement. He and the department, in conjunction with the administrators and teachers of the school, may want to watch the A_2 students with special care and perhaps even work out an educational program to counteract the probable unsatisfactory achievement. Such educational action must of course be handled with extreme care. Some students might be incorrectly assigned. The discriminant function only tells the *probable* future status of the students. Moreover, the discriminant function is subject to the reduced efficacy of prediction common to all regression procedures. That is, the equation maximizes the relation between the independent variables and the Y vector *of this sample*. With a new sample one does not have the full benefit of this maximization. Nevertheless, if successful, the discriminant function helps to cut down ignorance and, in doing so, points to possible remedial action.

The Discriminant Function with Three or More Groups

The above relatively simple procedure cannot be used when there are more than two classification groups. The calculations become much more complex. The essential idea behind the complexity, however, is simple: Seek the linear combination of the variables that will maximize the differences between the groups relative to the differences within the groups. This is virtually an analysis of variance way of thinking. The actual methods of calculation are beyond the scope of this book. Our main point is that discriminant analysis with more than two groups — more accurately called *multiple discriminant analysis* — can be considered a regression procedure.

Some Research Aspects of Discriminant Analysis

Although discriminant analysis seems not to have been used very much in behavioral and educational research, it has interesting potentialities. Like multiple regression in general, it can be used in two main ways: for classification and diagnosis, and to study the relations among variables in different populations and groups. The first use will probably be more common than the second. A clinical psychologist, for example, may wish to classify youths as

---

[3]This lack of statistical significance is due largely to the small number of cases used. In this example, we have sacrificed statistical significance for simplicity and realism. (The correlations between verbal ability and achievement, .62, and between school motivation and achievement, .45, are rather close to similar correlations in the literature.)

delinquent or nondelinquent. If he has measures that seem to be related to delinquency—for instance, social class, values, and personal beliefs (Jessor et al., 1968)—and also knowledge of the actual delinquency of a group of youths, the measures and the knowledge of delinquency can be used in a discriminant function. If the prediction is reasonably successful, the function can be used to assess the probable delinquency of other individuals. This amounts to the same procedure used in the earlier example. One can extend such analysis to other variables: success or not in college; school dropout or not; neurotic–not neurotic; vote for–vote against.

The second use of the discriminant function, like one use of multiple regression analysis, is more germane to basic scientific purposes. One wishes to know, for instance, something of the relations between values and career choices. Cooley and Lohnes (1962, pp. 119–123) used multiple discriminant analysis to study the prediction from knowledge of values to career plans. They used three career-plans groups: those students who enter graduate work to do basic research; an applied science group, those who continue in science and engineering, but who do not plan a research career; and a nonscience group, those who leave scientific work to enter fields that have direct involvement with people. Science and engineering majors from six eastern colleges were administered the *Study of Values* and other personality measures. The individuals were followed up over a three-year period, and a discriminant analysis performed, using the six scales of the *Study of Values* as independent variables, and group membership as the dependent variable. Cooley and Lohnes were successful in differentiating the members of the groups with the *Study of Values*, and were able to describe some of the differences. (They also outlined the calculations and discussed the results.)

This kind of analysis is important because it can lead to viable theory and research. Needless to say, the explanation of career choice is not simple. Is it possible that underlying social and personal values may to a substantial extent determine career choice? Is it likely that career choice is determined by values interacting with personality and ecological variables? The possibilities for theory development seem clear, and the use of discriminant analysis can significantly help in such development.

Canonical Correlation

Canonical correlation is the generalization of multiple regression analysis to any number of dependent variables. This is of course not a large conceptual step. It is, however, a rather large computational step. We will outline the calculations in Chapter 14. In this chapter, however, we omit computational considerations. An outline of the method is sufficient for our purposes. Moreover, canonical correlation, except for the simplest problems, is so complex as to make desk calculator calculations forbidding. Here, again, intelligent and critical reliance on the computer is necessary.

Canonical correlation analysis is multiple regression analysis with k independent variables and m dependent variables. We will call the independent variables the "variables on the left" and the dependent variables the "variables on the right," or "predictor variables" and "criterion variables." The basic idea of canonical correlation is that, through least squares analysis, two linear composites are formed, one for the independent variables, X_j, and one for the dependent variables, Y_n. The correlation between these two composites is the canonical correlation, R_c. The square of the canonical correlation, R_c^2, is an estimate of the variance shared by the two composites.[4]

The parallel with multiple regression should be apparent. Like the co-efficient of multiple correlation, the canonical coefficient is the maximum correlation possible between the two sets of variables. Like multiple regression, we have a least squares procedure that seeks the regression weights to be attached to each of the variables of both sets of variables. Multiple regression analysis can of course be considered a special case of canonical analysis. In view of the practical limitations on the use of canonical analysis, however, we prefer to consider it a generalization of multiple regression analysis. As we will see, the theoretical notion of canonical correlation is an elegant and aesthetically satisfying one, even though its actual use may leave something to be desired.

Data Matrices and Canonical Correlation

It will be useful to familiarize the reader with a conceptualization and symbolism of the raw data and correlation matrices that are used in multiple regression and canonical correlation. In multiple regression analysis, one variable, the dependent variable, is partitioned from the rest of the matrix. In canonical correlation analysis, two or more variables, the dependent variables, are partitioned from the rest of the matrix.

The basic data matrix for multiple regression analysis is simply the rectangular matrix of raw scores (or z scores), X_{ij}, where $i = 1, 2, \ldots, N$, N being the number of cases or subjects, and $j = 1, 2, \ldots, k$, k being the number of variables (tests, items, scales, and so on). Such a matrix was shown in Chapter 4 (Table 4.1) where the last or kth variable is the dependent variable. The basic data matrix for canonical correlation analysis is shown in Table 12.2. As usual, the first subscript of each X stands for rows (subjects, cases) and the second subscript for columns (variables, tests, items, and so on). Note the broken vertical line: it partitions the matrix into the k independent and the $n - k$ dependent variables.

The variables are intercorrelated and a correlation or R matrix is formed. This matrix, too, is partitioned similarly. In the multiple regression correlation matrix, the dependent variable is partitioned from the independent variables. In canonical correlation analysis the correlation matrix is partitioned as shown in Table 12.3. The partitioning is indicated by the broken lines, and the indepen-

---

[4]A *redundancy index* developed recently by Stewart and Love (1968) is also useful in the interpretation of results of a canonical correlation analysis. For an explanation of this index, as well as other approaches, see Cooley and Lohnes (1971, pp. 170 ff.).

TABLE **12.2** BASIC RAW DATA MATRIX FOR CANONICAL
CORRELATION ANALYSIS[a]

Cases	Independent Variables				Dependent Variables			
1	X_{11}	X_{12}	. . .	X_{1k}	$X_{1(k+1)}$	. . .		X_{1n}
2	X_{21}	X_{22}	. . .	X_{2k}	$X_{2(k+1)}$	. . .		X_{2n}
.		.			.			
.		.	. . . .	.	.	. . .		.
.		.	.		.			.
N	X_{N1}	X_{N2}	. . .	X_{Nk}	$X_{N(k+1)}$	. . .		X_{Nn}

[a]N = number of cases; k = number of independent variables; n = total number of variables.

dent and dependent variables are labeled. It is easier to use matrix algebra and symbols than the usual statistical symbols to indicate the solutions of canonical correlation problems. The four partitions of the correlation matrix are indicated in this way (see, for example, Cooley & Lohnes, 1971, p. 176, or Anderson, 1966, p. 166):

$$\mathbf{R} = \begin{bmatrix} \mathbf{R}_{11} & \vdots & \mathbf{R}_{12} \\ \cdots & + & \cdots \\ \mathbf{R}_{21} & \vdots & \mathbf{R}_{22} \end{bmatrix}$$

where $\mathbf{R}$ = the whole correlation matrix of the $k + (n-k)$ variables; $\mathbf{R}_{11}$ = the correlations of the k independent variables; $\mathbf{R}_{22}$ = the correlations of the $n-k$ dependent variables; $\mathbf{R}_{12}$ = the correlations between the independent and dependent variables; $\mathbf{R}_{21}$ = the transpose of $\mathbf{R}_{12}$.

In computer solutions of canonical correlation problems, the raw data matrix is read into the computer and the computer does all the basic statistics.[5] After the whole R matrix is calculated, the canonical correlation analysis begins by partitioning the matrix as indicated above. The user must of course "tell" the computer how to partition the matrix. Our purpose here is to give an intuitive feeling for the method. We use verbal description and provide examples so that the student can begin to interpret published research studies and computer output.

Canonical Correlation Process

The matrices of Table 12.3 are operated on in such a way as to produce a sort of double least squares solution. Two linear composites are formed, one of the variables on the left and one of the variables on the right. In multiple regression,

[5]Unfortunately, widely available multivariate programs do not always permit reading in a correlation matrix as well as raw data. We believe that all multiple regression, canonical correlation, and factor analysis programs should provide the options of analysis using either the raw data or the correlation matrix. It is not difficult, however, to alter existing programs to read both raw data *and* correlation matrices. A competent programmer can alter a program in an hour or two. Some programs, however, may be quite difficult to alter.

TABLE **12.3** PARTITIONED CORRELATION MATRIX FOR CANONICAL
CORRELATION ANALYSIS[a]

		Independent Variables					Dependent Variables			
		1	2	. . .	k	$k+1$		. . .	n
Independent Variables	1	r_{11}	r_{12}	. . .	r_{1k}	$r_{1(k+1)}$	. . .	r_{1n}	
	2	r_{21}	r_{22}	. . .	r_{2k}	$r_{2(k+1)}$	. . .	r_{2n}	
	.	.	.		.	.		.	
	.	.	.	.		.	.		.
	.		.	.		.		.	
	k	r_{k1}	r_{k2}	. . .	r_{kk}	$r_{k(k+1)}$	. . .	r_{kn}	
Dependent Variables	$k+1$	$r_{(k+1)1}$		. . .	$r_{(k+1)k}$	$r_{(k+1)(k+1)}$	. . .	$r_{(k+1)n}$	
	.	.		.		.		.	
	.	.	.	.		.		.	
	.	.		.		.		.	
	n	r_{n1}	r_{n2}	. . .	r_{nk}	$r_{n(k+1)}$	. . .	r_{nn}	

[a]k = number of independent variables; n = total number of variables.

one linear composite of the X's is formed taking the relations among them and the relations between them and the Y vector into account. In canonical correlation analysis, the procedure is more complex because more correlations have to be taken into account and two composites formed. At any rate, the correlation between the two composites is the canonical correlation. Its square, R_c^2, is interpreted similarly to the squared multiple correlation coefficient: it represents the variance shared by the two composites.

A canonical correlation analysis also yields weights, which, theoretically at least, are interpreted as regression weights. These weights appear to be the weak link in the canonical correlation analysis chain. Recall that regression weights in multiple regression analysis fluctuate from sample to sample and change when variables are added or subtracted from the regression equation. (They do not change, however, with different orders of entering variables in the regression equation.) The canonical correlation weights are more of a problem, particularly when more than one canonical correlation is calculated from the same set of data (see below). In other words, the weights must be interpreted with great caution and circumspection.

The value of canonical correlation analysis is enhanced by another feature of the method. More than one source of common variance can be identified and analyzed. In multiple regression analysis, although the dependent variable may contain more than one source of variance (for example, grade-point averages), there is only one regression equation. In canonical correlation, however, there can be more than one set of equations. In other words, the method systematically extracts the first and largest source of variance, and the canonical correlation coefficient is an index of the relation between the two sets of variables based on this source of variance. Then the next largest source of variance, left in the data after the first source is extracted and independent of the first source, is analyzed. The second canonical correlation coefficient, which is smaller than the first, is an index of the relation between the two sets of variables due to this second source of variance.

An example of multiple sources of variance might be the study of the relations between values and attitudes. Suppose a social psychologist wanted to know, first, how values and attitudes are related and the magnitudes of the relations, and second, the number of sources of variance in responses to value and attitude items. For example, a first source of variance may underlie religious values and religious attitudes and a second source may underlie educational values and educational attitudes. Values are the variables on the left and attitudes are the variables on the right. The canonical correlation may be .65 between religious values, on the one hand, and religious attitudes, on the other hand. This then, is a first source of variance reflected in the first canonical correlation coefficient. The canonical correlation between educational values and educational attitudes, say, is .49. This coefficient reflects the second source of variance in the data.

Although factor analysis may be a better method for investigating such problems, in some cases canonical correlation can supply useful information on relations among sets of variables and also yield tests of the statistical significance of such relations. In the foregoing example, the correlation of .49 may not be statistically significant, which would indicate that, while religious values and attitudes are significantly related, educational values and attitudes are not.

Computer programs usually do the successive analyses of canonical correlation and test the statistical significance of the successive sources of variance. If a second or third canonical R is not significant, of course, it and its weights are not interpreted. We give an example below of actual research with more than one significant canonical correlation.

Three Studies Using Canonical Correlation

One does not readily find research studies that have used canonical correlation. In earlier years the prohibitive calculations involved and general unfamiliarity with the method of course inhibited its use. Today, computer facilities and canonical correlation programs are available. Yet the method is still used only rarely. One suspects, therefore, that researchers are still unfamiliar with it. There may be a research conceptual difficulty, however. It may be that canonical correlation is not suited to most research problems. When studying the underlying relations between two sets of variables, there has to be some reasonable source of common variance in the two sets of variables. One can easily see that if one set of measures presumably reflects an underlying phenomenon or construct and the second set, similarly, reflects another related phenomenon, then canonical correlation is appropriate and valuable. In any case, the following studies illustrate quite different uses of the method.

Reading and Arithmetic versus Spelling and Language

An unusually good yet relatively simple research example has been thoroughly analyzed with different multivariate methods by Bock and Haggard (1968). In the study from which the data came (Haggard, 1957), 122 children supplied scores on achievement tests of reading, arithmetic, spelling, and language. We are only interested here in the correlations among these tests and the canonical correlation analysis. (Bock and Haggard did other analyses involving sex and grade as variables.) The correlations among the tests were substantial: from .49 to .91. Using the reading and arithmetic tests as left-hand variables and the spelling and language tests as right-hand variables, Bock and Haggard did an analysis in which they calculated two canonical correlations: .75 and .04. Only the first of these was statistically significant. (Refer to our earlier discussion of more than one source of variance and calculating canonical correlations for each source.)

The left- and right-hand weights (in standard-score form) associated with the canonical correlation of .75 were

$$.46R + .68A \qquad \text{and} \qquad -.03S + 1.02L$$

where R, A, S, and L stand for reading, arithmetic, spelling, and language. The two sets of weights are interpreted together. Taking the weights at face value, language has a substantial weight, while the weight for spelling is virtually zero. Arithmetic has a somewhat larger weight than reading. As noted above, however, these are partial coefficients whose magnitude is affected by the correlations among the variables used in the analysis.

Early Experiences and Orientation to People

In a study of sources of interests, Roe and Siegelman (1964) tested the interesting notion that early experiences produce later differences in orientation to persons and things. (They assumed that these orientations influence interests in occupations.) Their basic hypothesis was that extensive and satisfying personal relations early in life produce adults who are primarily person-oriented, while inadequate and unsatisfying relations produce adults who are primarily oriented to nonpersonal aspects of the environment (*ibid.*, pp. 4; 37–39). They used a set of independent variables that reflected early home environment and a set of dependent variables that reflected orientation toward people.[6] The analysis yielded a canonical correlation of .47. The greatest contribution to this correlation came from a measure of early social experience, an independent variable, and one of the dependent variables, a composite of questionnaire and inventory items measuring orientation toward people (*ibid.*, pp. 43–44, footnote 9). Their hypothesis was supported.

---

[6] A list of their measures and the canonical correlation analysis itself can be found in Cooley and Lohnes (1962, pp. 40–44). The original monograph does not report as much as Cooley and Lohnes do.

Social Environment and Learning

In a sophisticated study of the relations between five sets of independent variables consisting of measures of the social environment of learning, student biographical items, and miscellaneous variables (dogmatism, authoritarianism, intelligence, and so on), on the one hand, and a set of dependent variables consisting of cognitive and noncognitive measures of learning, on the other hand, Walberg (1969) used canonical correlation analysis to good effect. Separate analyses were run between each set of independent variables and the set of dependent variables. Three of the five sets predicted significantly to the learning criteria.

One interesting result was the canonical correlation between Walberg's fourteen learning or classroom environment variables—Intimacy, Friction, Formality, Democracy, and so on—and the dependent learning variables—Science Understanding, Science Interest, Physics Achievement, and so on. The canonical R was .61, indicating a fairly substantial relation between the composites of the two sets of variables. If we take R_c literally, classroom climate has some influence on achievement.

Walberg found that 15 of the independent variables (of the total 48) correlated significantly with the set of dependent variables collectively. In a separate canonical analysis of these two sets of variables, two statistically significant canonical correlations were found: $R_{c_1} = .64$ and $R_{c_2} = .60$. Since the two linear composites are orthogonal to each other, these R_c's reflect two independent sources of variance in the data. The first canonical variate or component reflected the relations between the independent variables and the cognitive variables: Physics Achievement, Science Understanding, and Science Processes. The second variate reflected the relations between the independent variables and the noncognitive variables: Science Interest, Physics Interest, and Physics Activities. In short, Walberg was able, through canonical and other analyses, to present a highly condensed generalization, as he calls it, about the relations between cognitive learning, noncognitive learning, and a variety of environmental and other variables related to learning. This is probably an important finding that helps to advance knowledge about the enormous complexity of learning and school environments and influences.[7]

We have tried to lay a foundation for understanding discriminant analysis and canonical correlation analysis in this chapter. In subsequent chapters both methods will be further explained and illustrated.

---

[7]This complexity is reflected in the difficulty of reading relatively condensed reports of such research. As multivariate analysis is used more and more in behavioral research, reading, understanding, and evaluating research studies will increase in difficulty. The problem is aggravated by the impossibility of publishing enough of the actual research data to assess studies adequately. The difficulties facing researchers who write reports, the editors who have to evaluate them, and readers of the reports are therefore considerable.

Study Suggestions

1. Describe the relation between multiple regression analysis and canonical correlation analysis. In your description, mention the similarities and differences between the two methods. Do you think that research that uses canonical correlation will gradually supplant research that uses multiple regression? Give reasons for your answer.

2. Make up a research problem that requires the use of discriminant analysis. Use entered college–did not enter college as the dependent variable. What are the basic functions of discriminant analysis? Is discriminant analysis with two groups similar to multiple regression analysis? Why?

3. Describe an area of educational research in which canonical correlation analysis can be useful. [*Hint*: Think of different kinds of school achievement.] Outline the structure of a research problem in which canonical correlation is the basic mode of analysis.

4. Suppose a psychological researcher has three measures of authoritarianism and these three measures are positively correlated. Suppose, further, that researcher has two measures of dogmatism and they are positively correlated. The researcher wants to know the overall relation between authoritarianism and dogmatism.
    (a) Can he use multiple regression analysis? If so, how would he go about it?
    (b) Can he use canonical correlation analysis? What will such an analysis tell him?
    (c) How are the two kinds of analysis alike? What is the difference between them?

5. Here are some fictitious data for a simple discriminant analysis:

X_1	X_2	Y
8	3	1
7	4	1
5	5	1
3	4	1
3	2	1
4	2	0
3	1	0
3	2	0
2	2	0
2	5	0

(a) Do a multiple regression analysis of these data. Treat the Y vector of 1's and 0's as though they were ordinary continuous scores.
(b) Does knowledge of X_1 and X_2 enable us to predict "well" to group membership (Y)?
(c) Clothe the example with variable names and interpret the results. Make Y membership in some group.
(*Answers*: $R^2 = .478$; $F = 3.202$ $(df = 2, 7)$, not significant at the .05 level; $b_1 = .139$, $b_2 = .120$, $a = -.4167$.)

6. Suppose a department of sociology in a university has worked out a prediction equation that has been useful in distinguishing successful from unsuccessful doctoral students. What is the danger in using the equation for future groups of students? Should the use of the equation be abandoned because there is danger in its use?

7. Describe how you might do a study of career choice and how discriminant analysis can be used in the analysis of the data. How might canonical correlation be used? [*Hint*: Think of different interests and abilities as independent variables and career choice as dependent variable.]

8. To learn discriminant analysis, the reader can do no better than to study the following excellent manual: M. Tatsuoka, *Discriminant Analysis: The Study of Group Differences*. Champaign, Ill.: Institute for Personality and Ability Testing, 1970. It explains the rationale, mechanics, and mathematics of the method clearly and as simply as possible. It has a good example (p. 39) that the reader should try to do. (Unfortunately, there appears to be no comparable source for canonical correlation.)

13

Multiple Regression, Multivariate
Analysis of Variance, and Factor Analysis

The two multivariate methods to be studied in this chapter, multivariate analysis of variance and factor analysis, like the two methods discussed in Chapter 12, are closely related to multiple regression analysis. As we said before, however, the relation is not as obvious as it is with discriminant analysis and canonical correlation. In this chapter, we try to explain and illustrate the relation. As a partial foundation for showing the relation between multiple regression and multivariate analysis of variance, we borrow the basic idea of generalizing multivariate analysis and tests of statistical significance. To explain the relation between multiple regression and factor analysis, on the other hand, we describe the regression nature of factor analysis and, more practically, explain and illustrate the use of factor scores, scores of individuals based on factor analysis, in regression analysis.

Multivariate Analysis of Variance and the Generalization of Research Design and Analysis

There is little doubt that the invention by Ronald Fisher in 1924 of the analysis of variance and certain important companion notions, like randomization, is one of the great achievements of our time (Hotelling, 1951; Neyman, 1967; Stanley, 1966). In addition to solving important problems of statistics and statistical inference, it laid a foundation for modern thinking on research design. Although Campbell and Stanley (1963) say that their extensive treatment of research designs is not a chapter on experimental design, it is fairly safe to say that the chapter would not have been possible without Fisher's breakthrough. The basic idea of putting multiple groups and multiple experimental treatments together in one experiment and the identification and manipulation of different sources of variance in dependent variable measures made it possible to revolutionize experimental and analytic thinking.

Until recently, most research thinking has focused on one dependent variable. The Fisherian revolution freed independent variables, so to speak, but little was done to free dependent variables. It is not strange, however, that statistical and research workers extended their thinking to more than one dependent variable. Why not multiple regression with more than one dependent variable? We have seen that this question is answered, in part at least, by canonical correlation. Why not analysis of variance with more than one dependent variable? The answer is that a large part of the statistical problem has been solved, and the analysis of variance of data from experiments with more than one dependent variable is now possible, although much more complex than analysis of variance with only one dependent variable.

Analysis of variance with any number of independent variables and any number of dependent variables is called *multivariate analysis of variance*. Like univariate analysis of variance, it was designed primarily for multivariate experimental data in which at least one of the independent variables has been manipulated. Also like univariate analysis of variance, its purpose is basically to test statistical hypotheses about experimental group means of more than one dependent variable. A simple example is a two-by-two factorial design with two dependent variables, for example, methods of teaching, A_1 and A_2, and types of incentives, B_1 and B_2, as independent variables, and arithmetic achievement and understanding of mathematical concepts, as dependent variables, as Y_1 and Y_2.[1] In univariate analysis of variance, the total sum of squares is partitioned into between groups and within groups sums of squares. In multivariate analysis of variance the total sum of products, $\Sigma y_1 y_2$, is also partitioned according to the independent variables into between groups and within groups sums of products. The test of statistical significance is used to determine whether the means of the two dependent variables, considered simultaneously, are equal. Inspection of the means and subsequent statistical tests (Hummel and Sligo, 1971; Morrison, 1967, pp. 182ff.) will show the different effects of the independent variables on the dependent variables.

If the reader will think of the dependent variable means two-dimensionally rather than one-dimensionally, he may see what is meant. A univariate F test tests the differences among means on a single continuum or dimension. A multivariate F test, however, tests the significance of mean differences k-dimensionally, in this case two-dimensionally. The multivariate test is on a plane

---

[1] One can say that the simplest possible example of multivariate analysis of variance is a one-way analysis of variance with two dependent variables. We prefer the factorial example, however, because it clearly makes both independent and dependent variables multivariate. There is no need to be overly puristic. All, or almost all, the methods of analysis can be conceived in either a multiple regression framework or in a statistical test of group means framework.

rather than on a continuum (in this case of two dependent variables only). An example will be given later if the reader is a bit confused. (It is a little clumsy to express multivariate notions verbally.)

Multivariate Analysis of Variance and Multiple Regression Analysis: Generalization of Tests of Statistical Significance

Our purpose in this section is to help the reader understand, to some extent at least, what multivariate analysis is and how it can be used. We cannot teach multivariate analysis of variance; that would take at least several chapters and a level of statistical sophistication beyond that assumed in this book. But we believe the reader can profit from seeing certain important relations between multivariate analysis of variance and multiple regression analysis. To help see these relations, we follow the lead of Rulon and Brooks (1968) and concentrate on statistical tests of group differences. In their splendid chapter, Rulon and Brooks completely generalized such tests. While we deal only with the case of one experimental independent variable and two dependent variables, it should be borne in mind that the approach and method can be extended to several independent and several dependent variables.

The most elementary parametric statistical test is the t test of two groups. If t is statistically significant, then the means are said to be significantly different. The next step up the statistical ladder is the F test applied to three or more groups — or to two groups, in which case, $t = \sqrt{F}$. The next extension is to the F test in the factorial analysis of variance. Univariate analysis of variance can of course be extended to complex factorial and other designs. We have shown in Part II how regression analysis can be applied to these designs.

When there is more than one dependent variable the ordinary t and F tests are not applicable in the usual way. They can naturally be used with each dependent variable separately, but, as Bock and Haggard (1968, p. 102) point out, because the dependent variable measures have been obtained from the same subjects and thus are correlated in some unknown way, the F tests are not independent. "No exact probability that at least one of them will exceed some critical level on the null hypothesis can be calculated (*ibid.*, p. 102)." Multivariate methods take the correlations among the dependent variables into account. Moreover, a researcher may be interested in the overall statistical significance of the differences among the dependent variables as a set. To analyze two or more dependent variables there are several tests of the statistical significance of mean differences: Hotelling's T^2, Mahalanobis' D^2, Wilks' Λ (lambda), and still others. T^2, D^2, and Λ can be tied to the F test, as Rulon and Brooks show.

For our limited purpose, we will only discuss Wilks Λ and the accompanying F test. The other tests are actually special cases of this general statistic. In the case when there are any number of independent variables and any number of dependent variables, only Λ can be used.

A Univariate Analysis of Variance Example: Λ and R^2

In Table 13.1, fictitious scores from a hypothetical experiment are given together with a conventional univariate one-way analysis of variance. Let us suppose that an experiment on changing attitudes has been done in which three kinds of appeals, A_1, A_2, and A_3, were used with prejudiced individuals.[2] A_1 was a democratic appeal or argument in which prejudice is said to be incommensurate with democracy. A_2 was a fair play appeal: the American notion of fair play demands equal treatment. And A_3 was a religious appeal: prejudice and discrimination are violations of the ethics of the major religions. The dependent variable was attitude toward blacks (higher scores indicate greater acceptance).

[2]The idea for this experiment was taken from an actual experiment by Citron, Chein, and Harding (1950).

TABLE **13.1**   FICTITIOUS EXPERIMENTAL DATA AND ANALYSIS OF VARIANCE CALCULATIONS

	A_1	A_2	A_3	
	4	5	3	
	7	6	5	
	9	3	1	
	6	8	4	
	9	3	4	
	6	2	5	
	5	5	7	
	7	6	3	
	7	7	5	
	10	5	3	
ΣX:	70	50	40	$\Sigma X_t = 160$
$\bar{X}$:	7	5	4	$(\Sigma X_t)^2 = 25{,}600$
				$\Sigma X_t^2 = 988$

$$C = \frac{25\,600}{30} = 853.333$$

$$\text{Total} = 988 - 853.333 = 134.667$$

$$\text{Between} = \frac{70^2 + 50^2 + 40^2}{10} - 853.333 = 46.667$$

Source	df	ss	ms	F
Between	2	46.667	23.334	7.160 (.01)
Within	27	88.000	3.259	
Total	29	134.667		

The analysis of variance yielded an F ratio of 7.16, statistically significant at the .01 level. (Regression analysis of these data yields an R^2 of .35.) Inspection of the means seems to indicate that A_1, the democratic appeal, was the most effective in increasing acceptance of blacks.

One formula for Wilks' Λ, a statistic we use to show something of the relation between multivariate analysis of variance and multiple regression analysis, is

$$\Lambda = \frac{ss_w}{ss_t} \tag{13.1}$$

where ss_w = within sum of squares, and ss_t = total sum of squares. Since $ss_t = ss_b + ss_w$, where ss_b = between groups sum of squares, Λ can also be written

$$\Lambda = \frac{ss_t - ss_b}{ss_t} \tag{13.2}$$

and

$$\Lambda = 1 - \frac{ss_b}{ss_t} \tag{13.3}$$

Substituting ss_w and ss_t from Table 13.1, we obtain $\Lambda = 88./134.667 = .6535$. Λ can now be used in a formula for F:

$$F = \frac{1-\Lambda}{\Lambda} \cdot \frac{N-k}{k-1} \tag{13.4}$$

where N = total number of cases, and k = number of experimental groups (in this case, 3). Substituting Λ, just calculated, and N and k,

$$F = \frac{1-.6535}{.6535} \cdot \frac{30-3}{3-1} = 7.158$$

which is the same, within errors of rounding, as the F value calculated in Table 13.1.

Now, do a multiple regression analysis of the same data using coded vectors as described in Part II. That is, string out all 30 scores in a single Y vector, code A_1 and A_2 with 1's and 0's, and do the regression analysis. Such an analysis yields $R^2 = .3465$. Use this value in a formula for F from earlier chapters:

$$F = \frac{R^2/k}{(1-R^2)/(N-k-1)} \tag{13.5}$$

where N = total number of cases, and k = number of actual independent variable vectors used in the regression analysis (in this case, 2). k and $N-k-1$ are simply the degrees of freedom, 2 and 27, given at the bottom of Table 13.1. Substituting R^2 yields

$$F = \frac{.3465/2}{(1-.3465)/(30-2-1)} = \frac{.1733}{.0242} = 7.161$$

which is the same value yielded by formula (13.4), within errors of rounding.

Look at formula (13.4) now. We can obviously write the formula as

$$F = \frac{(1 - \Lambda)/(k - 1)}{\Lambda/(N - k)} \tag{13.6}$$

If we realize that the $k - 1$ of this formula is the same as the k of formula (13.5) — both are the between groups degrees of freedom associated with $1 - \Lambda$ and R^2, respectively — then $(1 - \Lambda)/\Lambda$ must equal $R^2/(1 - R^2)$. To get to the main point, $\Lambda = 1 - R^2$, and $R^2 = 1 - \Lambda$. Recall that $1 - R^2$ is the residual variance of the regression analysis. Therefore Λ is the same residual variance. Or, it is the proportion of the total sum of squares of the dependent variable that is not associated with the regression of Y on the independent variables. And $1 - \Lambda$ is the proportion of the variance of Y due to the regression of Y on the independent variables.

A Multivariate Analysis of Variance Example

Suppose that instead of one dependent variable, attitudes toward blacks, there were two: attitudes toward blacks and attitudes toward Jews. It is clear that these attitude variables are correlated and that independent F tests are questionable, as indicated earlier. Moreover, it is possible that there may not be significant differences using either dependent variable separately in analyses of variance, but that a multivariate analysis that analyzes both dependent variables simultaneously will show a significant difference.[3] In fact, the example now to be given was expressly constructed to show this possibility.

Assume that the experiment has been done and that the data are those of Table 13.2. A_1, A_2, and A_3 are the same experimental treatments of the hypothetical experiment of the last section: democratic appeal, fair play appeal, and religious appeal. Dependent variable 1 is attitudes toward blacks and dependent variable 2 is attitudes toward Jews. For ease of calculations we have used only five subjects in each group. The cross product sums (labeled CP) are given, together with the sums and means of each column. The total sums and sums of squares of 1 and 2 are also given (on the right), as are the deviation sums of squares, total and within groups.

Although the within sums of squares, ss_{w_1} and ss_{w_2}, of the two dependent variables can be calculated directly, it is easier to calculate them with one-way analysis of variance. Separate analyses of variance were calculated for 1, attitudes toward blacks, and for 2, attitudes toward Jews. The within groups sums of squares obtained from these analyses are given at the bottom of the table. The two analyses yielded F ratios of 3.47 and 3.57, neither of which was statistically significant. Thus, the means of A_1, A_2, and A_3 for dependent variable 1, 4.6, 5.0, and 6.2, and for dependent variable 2, 8.2, 6.6, and 6.2, do not differ significantly from each other. Had we done two separate experiments and

[3]We are indebted to Li's (1964, pp. 405–410) clear demonstration of how this can happen. Later, we will repeat the essence of his demonstration. See, also, Tatsuoka's (1971b, especially pp. 22–24) clear explanation.

TABLE **13.2**   DATA FROM HYPOTHETICAL EXPERIMENT, THREE
EXPERIMENTAL TREATMENTS AND TWO DEPENDENT VARIABLES[a]

	A_1		A_2		A_3		
	1	2	1	2	1	2	
	3	7	4	5	5	5	
	4	7	4	6	6	5	
	5	8	5	7	6	6	
	5	9	6	7	7	7	
	6	10	6	8	7	8	
Σ:	23	41	25	33	31	31	$\Sigma_{t_1} = 79$
M:	4.6	8.2	5.0	6.6	6.2	6.2	$\Sigma_{t_1}^2 = 435$
CP:		194		169		196	
							$\Sigma_{t_2} = 105$
							$\Sigma_{t_2}^2 = 765$

$$ss_{t_1} = 435 - \frac{(79)^2}{15} = 18.9333$$

$$ss_{t_2} = 765 - \frac{(105)^2}{15} = 30.0000$$

$$ss_{w_1} = 12. \text{ (from ANOVA)}$$
$$ss_{w_2} = 18.80 \text{ (from ANOVA)}$$

[a]1: attitudes toward blacks; 2: attitudes toward Jews; CP: cross products sums.

obtained these results, we would conclude that the experimental treatments had had no significant effect.

This conclusion is not correct, however. Suppose we do a multivariate analysis of variance along the lines of the analysis of the preceding section, using Wilks' Λ and the F test associated with Λ. There are several similar formulas for different numbers of experimental treatments and dependent variables. Although they are considerably more complex, we give formulas that are general and fit all cases (Rulon & Brooks, 1968):

$$\Lambda = \frac{|\mathbf{W}|}{|\mathbf{T}|} \tag{13.7}$$

$$F = \frac{1 - \Lambda^{1/s}}{\Lambda^{1/s}} \cdot \frac{ms - v}{t(k-1)} \tag{13.8}$$

where $\mathbf{W}$ = the matrix of within groups sums of squares and cross products, $\mathbf{T}$ = the matrix of the total sums of squares and cross products, both defined below, m, s, and v are as defined below, t = number of dependent variables, and k = number of experimental treatments.[4] $|\mathbf{W}|$ and $|\mathbf{T}|$ indicate the determinants of the $\mathbf{W}$ and $\mathbf{T}$ matrices (see Appendix A).

[4]Rulon and Brooks (1968, pp. 72–76) explain that the formula for F, above, is an approximation by Rao which applies to any number of experimental treatments or groups and any number of dependent variables.

W and **T** are defined as

$$\mathbf{W} = \begin{pmatrix} ss_{w_1} & scp_w \\ scp_w & ss_{w_2} \end{pmatrix}$$

$$\mathbf{T} = \begin{pmatrix} ss_{t_1} & scp_t \\ scp_t & ss_{t_2} \end{pmatrix}$$

where ss_{w_1}, ss_{w_2}, ss_{t_1}, and ss_{t_2} are the within groups and total sums of squares for dependent variables 1 and 2, and scp_w and scp_t are the within groups and total sums of cross products.

m, s, and v are defined[5]

$$m = \frac{2N - t - k - 2}{2}$$

$$s = \sqrt{\frac{t^2(k-1)^2 - 4}{t^2 + (k-1)^2 - 5}}$$

$$v = \frac{t(k-1) - 2}{2}$$

where N, t, and k are as defined earlier. In our problem, $N = 15$, $t = 2$, and $k = 3$.

Two of the values of **T**, ss_{t_1} and ss_{t_2}, the total sums of squares of 1 and 2, have been calculated in Table 13.2. The other value, scp_t, the sum of the total cross products, is calculated: $(3)(7) + (4)(7) + \cdots + (7)(8) - (79)(105)/15 = 559 - 553 = 6.00$. The values of **W**, ss_{w_1}, and ss_{w_2} can be obtained by calculating the between sums of squares and subtracting them from the total sums of squares, as in one-way analysis of variance:

$$\text{Between} = \left[\frac{(23)^2}{5} + \frac{(25)^2}{5} + \frac{(31)^2}{5}\right] - \frac{(79)^2}{15} = 6.9333$$

$$ss_{w_1} = 18.9333 - 6.9333 = 12.0000 \qquad \text{(for 1)}$$

$$\text{Between} = \left[\frac{(41)^2}{5} + \frac{(33)^2}{5} + \frac{(31)^2}{5}\right] - \frac{(105)^2}{15} = 11.20$$

$$ss_{w_2} = 30.00 - 11.20 = 18.80 \qquad \text{(for 2)}$$

The within sum of cross products, scp_w, is calculated similarly, but with cross products instead of squares:

$$\text{Between} = \left[\frac{(23)(41)}{5} + \frac{(25)(33)}{5} + \frac{(31)(31)}{5}\right] - \frac{(79)(105)}{15} = -7.20$$

$$scp_w = 6.00 - (-7.20) = 13.20$$

---

[5]In certain cases the fraction 0/0 appears in s. To handle such cases, see Rulon and Brooks (1968, pp. 73–75) and Tatsuoka (1971, pp. 88–89).

The **W** and **T** matrices, then, are

$$\mathbf{W} = \begin{pmatrix} 12.00 & 13.20 \\ 13.20 & 18.80 \end{pmatrix}$$

$$\mathbf{T} = \begin{pmatrix} 18.9333 & 6.0000 \\ 6.0000 & 30.0000 \end{pmatrix}$$

Equation (13.7) calls for the determinants of the **W** and **T** matrices, $|\mathbf{W}|$ and $|\mathbf{T}|$. The calculation of two-by-two determinants is simple and is explained in Appendix A:

$$|\mathbf{W}| = (12.00)(18.80) - (13.20)(13.20) = 51.36$$

$$|\mathbf{T}| = (18.9333)(30.0000) - (6.0000)(6.0000) = 531.9990$$

Applying equation (13.7) yields

$$\Lambda = \frac{|\mathbf{W}|}{|\mathbf{T}|} = \frac{51.36000}{531.9990} = .0965$$

Calculate m, s, and t:

$$m = \frac{(2)(15) - 2 - 3 - 2}{2} = 11.5$$

$$s = \sqrt{\frac{(2)^2(3-1)^2 - 4}{(2)^2 + (3-1)^2 - 5}} = \sqrt{\frac{12}{3}} = 2$$

$$v = \frac{2(3-1) - 2}{2} = \frac{2}{2} = 1$$

Finally, calculate the F ratio with equation (13.8):

$$F = \frac{1 - .0965^{1/2}}{.0965^{1/2}} \cdot \frac{(11.5)(2) - 1}{2(3-1)} = \frac{1 - \sqrt{.0965}}{\sqrt{.0965}} \cdot 5.50$$

$$= 12.208$$

which, at $t(k-1)$ and $ms - v$, or 4 and 22, degrees of freedom, is significant at the .001 level.

This long and somewhat tedious procedure has yielded a valuable result. Evidently the experimental treatments were effective: analyzing both dependent variables simultaneously there are significant differences among the means. When considered separately, attitudes toward blacks and attitudes toward Jews were not affected differentially by the democratic, fair play, and religious appeals, but when considered together the appeals did affect the attitudes. This is not easy to understand. How can such a result happen? Li (1964, Chapter 30) has provided a strikingly clear demonstration of how it can happen. We use his explanation with the data of Table 13.2 and try to add a bit to it.

A Demonstration of Multivariate Statistical Significance

In Figure 13.1, we have plotted the paired scores of the two dependent variables, 1 and 2. The plotted pairs of A_1 are shown with open circles, those of A_2 with black circles, and those of A_3 with crosses. The means of the three experimental groups have also been plotted—indicated by circled asterisks. Notice that the plotted points overlap a good deal if viewed horizontally or vertically. If we visualize the projections of all the plotted points on the 1-axis first, we see the substantial overlap. In addition, the three means of the 1 groups have been projected on the 1-axis (circled asterisks): 4.6, 5.0, 6.2. Now visualize the projections on the 2-axis of all the points. Again, there is considerable overlap. The plotted means' projections on the 2-axis have been indicated: 8.2, 6.6, 6.2. Note when considering variable 1 alone, that there is little difference between the means of A_1, the lowest mean, and A_2, but both are different from A_3, the highest mean. When considering variable 2 alone, on the other hand, the mean of A_1, now the highest mean, is quite different from the means of A_2 and A_3—and the latter is now the lowest mean.

If, instead of regarding the plotted points one-dimensionally, we regard them two-dimensionally in the 1-2 plane, the picture changes radically. There are clear separations between the plotted points and the plotted means of A_1, A_2, and A_3. In fact, it is possible to draw straight lines to separate the clusters of plotted points. Considering the two dependent variables together, then, the groups are separated in the two-dimensional space. And the multivariate analysis faithfully reflects the separation.

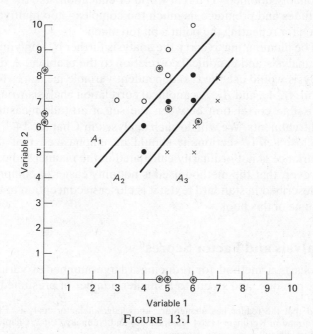

FIGURE 13.1

If the data of this little "study" are taken at face value, we have an interesting finding. A_1 is least effective in changing attitudes toward blacks and most effective in changing attitudes toward Jews, whereas the reverse is true for A_3. Although such an effect is probably rare (we know of no reported case in the literature) it *can* happen. Perhaps it is rare because its possibility has not been conceived.

The above example was carefully, in fact, almost painfully, contrived. Still, similar kinds of results will probably occur in the future as multivariate analysis and thinking are used more. With more independent variables and dependent variables, the possibilities and complexities increase enormously. It is possible that theoretical work and empirical research may be considerably enriched in the next decade by the discovery of interactions and what are now considered unusual relations. The results of the contrived attitude change and prejudice example are not far-fetched. A democratic appeal quite conceivably may have little effect on attitudes towards blacks when it does have a positive effect on attitudes toward Jews. One thinks of the situation in parts of the United States where a democratic appeal might have a negative effect if used to change attitudes toward blacks. In any case, one can see the possible enrichment of theory of attitude change and theory of prejudice and the improvement of research in both fields.

One can even see possibilities of practical applications, especially in education. Methods of teaching, for example, may work well for certain kinds of students in some subjects, while they may not work too well for other kinds of students in other subjects. Clearly, the search for a "best" universal teaching method is probably doomed. The real world of education, like the real world of changing attitudes and prejudice, is much too complex—too multivariate, if we can be forgiven for repeating the point a bit too much.

It would be illuminating to carry the analysis further by applying discriminant function analysis and canonical correlation to the problem. A discriminant function analysis would use the two dependent variable measures to predict to membership in A_1, A_2, and A_3. A canonical correlation analysis would show the maximum possible correlation between the set of attitude measures and the experimental treatments. We will do such analyses in Chapter 14.

Before closing this section, it should be emphasized that multivariate analysis of variance is not ordinarily calculated in the manner indicated above. We felt, however, that the method used is not only easier to comprehend than the method described in standard texts; it is closer in conception to the multiple regression theme of this book.

Factor Analysis and Factor Scores[6]

Factor analysis is a method for reducing a large number of variables (tests, scales, items, persons, and so on) to a smaller number of presumed underlying

---

[6]It is assumed that the reader has elementary knowledge of factor analysis. Elementary discussions can be found in Kerlinger (1964, Chapter 36) and Nunnally (1967, Chapter 9). The best all-around but more difficult reference is Harman (1967).

unities called factors. Factors are usually derived from the intercorrelations among variables. If the correlations among five variables, for instance, are zero or near-zero, no factors can emerge. If, on the other hand, the five variables are substantially correlated, one or more factors can emerge. Factors are constructs, hypothetical variables that reflect the variances shared by tests, items and scales and responses to them, and, in fact, almost any sort of stimuli and responses to stimuli.

We are interested in factor analysis not so much because it is a powerful scientific tool for discovering underlying relations but rather because it is a multivariate method related to multiple regression analysis, and because it yields so-called factor scores that can profitably be used in multiple regression analysis and in other forms of analysis.

Factor Analysis and Multiple Regression Analysis

A *factor* of a data matrix is any linear combination of the variables in the matrix. If factors have been obtained from the data matrix, there will ordinarily be fewer factors than variables, and the variables can be estimated in a multiple regression manner from the factors. There is also a second way to look at factors and variables. Rather than labor these matters verbally, however, let us use a simple example.

We give a rotated factor matrix in Table 13.3. The rows of the factor matrix are the variables $(1, 2, \ldots, 9)$ and the columns are the factors $(A, B,$ and $C)$. The entries in the table are called *factor loadings*. The entry in the A column and the first row, .80, is the factor loading of variable 1 on factor A. The magnitude of a factor loading indicates the extent to which a variable is "on that factor." The loading of .80 on A of variable 1, for instance, indicates that variable 1 is highly loaded with "A-ness," whereas its loading of .00 on B indicates that it is "not on" B. The factor loadings are in standard score form.

The h^2 column gives the communalities of the variables. Since the factors in this case are assumed to be orthogonal, the sums of squares of the factor

TABLE **13.3** FICTITIOUS ROTATED FACTOR MATRIX, THREE FACTORS, NINE VARIABLES[a]

Variables	Factors			
	A	B	C	h^2
1	*.80*	.00	.10	.65
2	*.70*	.05	−.15	.52
3	*.60*	.12	.15	.40
4	.10	*.70*	.20	.54
5	−.12	*.75*	.08	.58
6	.07	*.65*	.05	.43
7	.20	.10	*.50*	.30
8	−.15	.02	*.70*	.53
9	.00	.10	*.82*	.68

[a]Significant factor loadings are italicized.

loadings across rows equal the communalities, or $\Sigma a_j^2 = h^2$. The *communality* of a variable is the variance it shares with other variables. It is the common factor variance of a test or variable. Some part of the variance of a test may be specific to that test. This variance is the complement of the communality. It is variance not shared with other variables. Yet all tests and variables usually share something with other tests and variables. This shared "something" is the communality. In the example given above of five variables being substantially correlated, the communality of variable 2, for instance, is the variance that variable 2 shares with the other four variables.

It is said above that a factor is a linear combination of the variables in the matrix. Factor A, for instance, can be seen as the linear combination of the nine variables:

$$A = .80X_1 + .70X_2 + .60X_3 + .10X_4 - .12X_5 + .07X_6 + .20X_7 - .15X_8 + .00X_9$$

where $X_j = $ the variables. Factors B and C can be similarly written. Thus the factors can be called dependent variables and the variables independent variables. The similarity to multiple regression thinking is obvious.

But the situation can be viewed differently. Any variable can be conceived as a linear combination of the factors. Let $Y_i = $ the i variables; A, B, and $C = $ the three factors; and $a_j = $ the j factor loadings (in this case, $j = 3$). Then we write another equation to estimate any variable:

$$Y_i = a_1 A + a_2 B + a_3 C \tag{13.9}$$

Here the factors are viewed as independent variables that are used to estimate Y_i, the dependent variables. Again, the similarity to multiple regression thinking is obvious. The idea behind equation (13.9) will be used in a practical way later.

The factor loadings, a_j, in any row of the factor matrix are indices of the amount of variance each factor contributes to the estimation of the variables. This contribution can be calculated by squaring each factor loading. For example, the factor contributions to variables 1 and 4 of Table 13.3 are seen as

1:    $(.80)^2 + (.00)^2 + (.10)^2 = .64 + 0 + .01 \quad = .65$

2:    $(.10)^2 + (.70)^2 + (.20)^2 = .01 + .49 + .04 = .54$

The sums of these squares, .65 and .54, are the communalities. They indicate, as we said earlier, the common factor variance of the variables. That is, h^2 indicates the proportion of the total variance of a variable that is common factor variance. The common factor variance of a variable is that proportion of the total variance of the variable, indicated numerically by 1.00 (standard score form), that is shared with the other variables in the analysis. It is also that proportion of the total variance of the variable that the factors account for. If $h_6^2 = 1.00$, for instance, then the factors account for all the variance of variable 6. If the h^2 for variable 4 is .54, as above, then the factors account for 54 percent of the total variance of variable 4. The communalities of variables 1 and 4 are .65 and .54. Variable 1 has more common factor variance than variable 4. More important, for variable 1, $(.80)^2 = .64$, compared to $(.00)^2 = 0$ and $(.10)^2 = .01$ for factors B and C, respectively, indicating that factor A contributes most heavily to the common factor variance of variable 1. Similar reasoning applies to variable 4: factor B contributes most to the common factor variance: $(.70)^2 = .49$, compared to $(.10)^2 = .01$ and $(.20)^2 = .04$.

Much more interesting from the point of view of multiple regression is that the communalities are squared multiple regression coefficients. Recall that when the independent variables of a multiple regression problem are not correlated (that is, the correlations are zero), the formula for R^2 can be written (for three independent variables)

$$R^2_{y.123} = r^2_{y.1} + r^2_{y.2} + r^3_{y.3}$$

Similarly, if the correlations among the independent variables are all zero, then the sum of the squared beta weights equals the squared multiple correlation coefficient:

$$R^2_{y.123} = \beta_1^2 + \beta_2^2 + \beta_3^2$$

or, in general,

$$R^2_{y.12...k} = \Sigma \, \beta_j^2 \qquad (13.10)$$

Thus, if the independent variables are uncorrelated, the individual squared betas indicate the proportion of the variance accounted for by the regression of Y on the independent variables. Because the factors are orthogonal to each other (we have assumed an orthogonal solution, of course), that is, their intercorrelations are zero, the same reasoning applies:

$$h^2 = \Sigma \, a_j^2 \qquad (13.11)$$

Actually, the h^2's are squared multiple regression coefficients, and the a's are regression weights.[7]

The Purposes of Multiple Regression and Factor Analysis

While factor analysis and multiple regression analysis resemble each other in that both are regression methods, in general the two methods have quite different purposes. Multiple regression's fundamental purposes are to predict dependent variables and to test research hypotheses. It is not now necessary to discuss multiple regression and prediction; we have stressed prediction again

---

[7]See Nunnally (1967, pp. 292–296) for a complete explanation. In Chapter 4, footnote 11, we showed a mathematically elegant and conceptually simple way to calculate the β's and R^2's of a correlation matrix. One of the problems of factor analysis is what values to put into the principal diagonal of the R matrix before factoring. Experts believe (for example, Harman, 1967, pp. 86–87) that the best values to insert in the diagonal are R^2's. That is, each variable in the matrix is treated in turn as a dependent variable and the R^2 between this variable and all the other variables in the matrix as independent variables is calculated. These R^2's, called SMC's, or squared multiple correlations, are then put in the diagonal before factoring. They are lower bound estimates of the communalities, and, as such, were recommended as the best possible estimates by Guttman (1956). Note here the connection between R^2 and h^2.

and again in earlier discussions. Multiple regression's hypothesis-testing function has also been discussed.

Factor analysis' basic purpose is to discover unities or factors among many variables and thus to reduce many variables to fewer underlying variables or factors. In achieving this purpose, factors "explain" data. Note, however, that this is a different kind of explanation than the explanatory purpose of multiple regression. Multiple regression "explains" a single known, observed, and measured dependent variable through the independent variables. Factor analysis "explains" many variables, usually without independent and dependent variable distinction, by showing their basic structure, how they are similar, and how they are different. In addition, the factor analyst almost always seeks to name the components of the structure, the underlying unities or factors. This is a deep and important scientific purpose. The analyst can literally discover categories, unities, and variables. When Thurstone and Thurstone (1941), in their classic study of intelligence, factor analyzed 60 tests and found the famous six primary abilities—Verbal, Number, Spatial, Word Fluency, Memory, and Reasoning—they actually discovered some of the unities underlying intelligence. Their efforts to analyze the tests loaded on each factor, and then name the factors were productive and creative. Later evidence has shown the validity of their thinking and efforts.

The Thurstone and Thurstone research shows another important difference between factor analysis and multiple regression. In multiple regression the independent and dependent variables are observable: the X's and the Y are measures obtained from actual measurement instruments. In factor analysis, on the other hand, the factors are not observable; they are hypothetical constructs, as Harman (1967, p. 16) points out, that are estimated from the variables (or tests) and are presumed to "be there." In sum, factor analysis is more suited to testing what can be called structural hypotheses about the underlying structure of a set of variables, whereas multiple regression is more suited to testing explicit hypotheses about the relations between several independent variables and a dependent variable.[8]

Factor Scores and Multiple Regression Analysis

One of the most promising developments that has been made possible by multivariate conceptualization of research problems and multivariate analysis and the computer is the use of so-called factor scores in multiple regression equa-

[8]We hasten to add, however, that it is quite possible, perhaps desirable, to put measures of independent and dependent variables into a factor analysis. One can hypothesize the relations between them and test the hypotheses in a factor analytic manner. There is no esoteric mystery about this. Factor analysis has no built-in defects that make hypothesis-testing and conceiving variables as independent and dependent impossible. Of course, if one is only interested in the difference between means, the use of factor analysis is inappropriate. But if one's hypothesis has to do, for instance, with the way variables cluster and influence each other, then factor analysis may be appropriate and useful. See Cattell (1952, Chapter 20) and Fruchter (1966) for explorations of the hypothesis-testing possibilities of factor analysis. The Cattell chapter is a clear-sighted and pioneering probing of controlled experimentation and factor analysis. Fruchter's chapter, also clear-sighted and probing, brings the subject up-to-date.

tions. As usual with most bright and powerful ideas, the basic notion is simple: the factors found by factor analysis are used as variables in the regression equation. Instead of using what are called a priori variables — variables measured by instruments *assumed* to measure the variables — factor analysis is used to determine the variables (factors). And the variables or factors can be correlated or uncorrelated — oblique or orthogonal factors. Theoretically, uncorrelated factors are desirable because explanation and interpretation of research results are simpler, less ambiguous, and generally more straightforward, as we have seen rather plentifully throughout this book.[9]

Take professorial effectiveness as an example. Suppose that we had a large number of student ratings of professors on 15 variables. Suppose, further, that we factor analyzed the responses to the 15 variables and found three factors: I: Dynamism-Enthusiasm; II: Interest; and III: Scholarship (see Coats & Smidchens, 1966). These factors, we think, underlie student perceptions and judgments of professors and their teaching. What are the relations between these factors and an independent index of professorial effectiveness, say ratings by colleagues, or a composite measure consisting of ratings by colleagues and number of publications in the last 5 years?

The original scores on the 15 variables of the N subjects of a sample are converted to z scores — the customary thing to do in calculating factor scores. The original z scores are then converted to factor scores in standard-score form (see Harman, 1967, p. 349). In other words, the 15 standard scores of each individual are converted to three factor scores in standard score form. Call the converted matrix $\mathbf{Z}$ and the scores of individual i f_{ij}, where $i = 1, 2, \ldots, N$, and $j = 1, 2, 3$.

There is now an N-by-3 matrix of factor scores, and each individual has three factor scores, f_1, f_2, and f_3. Assume a dependent variable, Y, measuring professorial effectiveness. A multiple regression analysis can now be done and its results interpreted in the usual way. In short, the factor scores are used in the same way as original variable scores — or their z scores — in the regression equation. The regression equation is solved for the regression weights, R^2, the F ratio, and the additional analyses and statistics discussed in earlier chapters.

The point of the whole procedure is a scientific measurement one. The researcher reduces a larger number of a priori variables to a smaller number of presumably underlying variables or factors. These factors can then be used as independent variables in controlled studies of the determinants of phenomena. The factors or "factor variables" are used as X_1, X_2, and so on in regression equations and in discriminant and canonical correlation analyses. The researcher may then be better able to explain the phenomena and related phenomena. He may also extend and enrich theory in his field. For example, one of the reasons that teacher effectiveness studies have been relatively ineffectual for so

[9]Because factor analysis can yield orthogonal factors, or factors whose intercorrelations are zero, this does not mean that the factor variables will be uncorrelated. That is, the factors *are* orthogonal. But if the factor scores are incorrectly calculated, the factor scores will probably be correlated (Glass & Maguire, 1966).

many years is the almost purely ad hoc atheoretical nature of the thinking and research in the field (Getzels & Jackson, 1963; Mitzel, 1960; Ryans, 1960).

There *are* effective teachers. And something makes them more effective than other teachers. In other words, there must be an explanation of teaching effectiveness. Accepting this assumption, what is probably needed, in addition to a multivariate approach, is psychological and social psychological theory that can be developed, tested, and changed under the impact of empirical research and testing (Getzels & Jackson, 1963). With a problem so complex, with so many possible variables interacting in unknown ways, the only guide to viable research will be theory — with multivariate technical methods to back up the theory. The extraction of factors from the intercorrelations of items, the calculation of factor scores, and the application of discriminant analysis and multiple regression analysis may well help to solve the problem.

Research Use of Factor Scores

The use of factor scores in multiple regression seems to be relatively infrequent. One does not have, in other words, a plethora of studies to choose from. We briefly summarize three unusual and good studies.

Veldman and Peck: A Study of Teacher Effectiveness.   In an excellent study of teacher effectiveness, Veldman and Peck (1963) had junior and senior high school students rate student teachers on a 38-item, four-point scale. The items covered a wide variety of perceived teacher characteristics and behaviors — "Her class is never dull or boring," "She is admired by most of her students," and so on. Factor analysis yielded five factors, and Veldman and Peck calculated factor scores. These scores were found to be highly reliable. Each student teacher, then, had five factor scores, one on each of the following factors: I. "Friendly, Cheerful, Admired"; II. "Knowledgeable, Poised"; III. "Interesting, Preferred"; IV. "Strict Control"; and V. "Democratic Procedure."

The student teachers were rated for effectiveness by their university supervising professors. Although the authors did not use multiple regression, they tested the differences among the high-, medium-, and low-rated student teachers. Statistically significant differences were found on factors I, II, and IV for both male and female students. A major finding was that there was no relation between supervisor evaluations and factor III, "Interesting, Preferred." Instead, the supervisors' ratings seemed to have been a function of factors I, II, and IV. Students' and supervisors' ideas on teacher effectiveness seem quite different!

Veldman and Peck used univariate analysis of variance to test the significance of the differences of the five sets of factor scores among three levels of supervisor ratings: high, medium, and low. That is, there were three means, high effective, medium effective, and low effective, on each of the five factor dimensions. F tests then showed which sets of means were significantly different. While this procedure yielded interesting information, as reported above, a possibly better procedure would have been to use the factors as independent variables predicting to the effectiveness ratings. A good deal more information could probably have been obtained and interpreted. Nevertheless, this is an important and competent study whose findings have clear implications for theoretical development in the study of teacher effectiveness.

Khan: A Study of Affective Variables and Academic Achievement. Interested in the effects of noncognitive or nonintellective factors on academic achievement, Khan (1969) factor analyzed responses to the items of an instrument constructed to measure academic attitudes, study habits, need for achievement, and anxiety about achievement. The eight factors found (unfortunately, one cannot judge the adequacy of the analysis, for example, the legitimacy of using as many as eight factors) were used to calculate factor scores. These scores were called affective measures. Khan also obtained from his sample intellective or aptitude measures — verbal and mathematical tests — and achievement measures — reading, language, arithmetic, and the like.

To assess whether the affective variables added to the predictive power of the intellective variables, Khan used multiple regression and F tests as follows. R^2 was calculated with the intellective measures (verbal and mathematical tests), as independent variables, predicting to the separate achievement tests (reading, arithmetic, and so on), as dependent variables. R^2 was calculated with the intellective measures *and* the affective measures (study habits, attitudes toward teachers, and so on). F tests were then used to test the significance of the differences between the R^2's. All six F tests (there were six criterion variables) were statistically significant, though the magnitudes of the increments were small.

Khan also used canonical correlation to study the relations between his affective and intellective measures. He obtained a canonical correlation of .69 for males and .76 for females. He was able to identify the affective variables that contributed significantly to the canonical correlations: attitudes toward teachers and achievement anxiety for males, and achievement anxiety for females.

This study while it did not yield dramatic results, clearly points to a more sophisticated approach to complex educational and psychological problems, an approach whose sharpened conceptualization and methodology may help lead the way to theoretical and practical breakthroughs in educational and psychological research.

McGuire, Hindsman, King, and Jennings: Factor "Variables" and Aspects of Achievement. Earlier, we advised calculation of factor scores and their use as independent variables in multiple regression equations. In a large and impressive study, McGuire, Hindsman, and Jennings (1961) used a different and more elaborate procedure. They wished to set up a model of the talented behavior of adolescents in which the behavior was a function of the potentialities of the person, his expectations of the supportive behavior of others, social pressures, his sex role identification, and the institutional context and pattern of educational experiences impinging on him. They used, in essence, a large number of independent variables to measure these ideas and others and, as dependent variables, six measures of different aspects of achievement. Multiple regression was used to study the relations between each of the dependent variables and the set of independent variables. For example, the multiple regression coefficient between grade-point average and 35 independent variables in one sample was .85.

The independent variables were then factor analyzed in order to obtain a smaller set of "factor variables" that could be used in the prediction of the achievement variables. Actually, this was the most important part of the study. The authors then calculated the weighted combinations of the independent variables that most efficiently predicted the factors. Recall that we said early in this section that factor analysis can be conceived in a regression way: the underlying factors are dependent variables that are regressed on the independent variables, the variables of the analysis. This is essentially what McGuire et al. did: they calculated the regression of each of the factors found in the factor analysis on the variables. For instance, Factor I, "Cognitive Approach," was the dependent variable and 8 of the 32 independent variables were differentially weighted to give the best least squares prediction of the factor. Unfortunately, the authors did not use the factor scores to predict the achievement variables. They plan to do this in another study.

Conclusion

The main pedagogical purpose for presenting multivariate methods other than multiple regression has been to deepen understanding of the generality and applicability of multiple regression ideas and methods. We are now able to see that discriminant analysis, canonical correlation, and multivariate analysis of variance can be conceived as first cousins of multiple regression. Factor analysis can be called a second cousin. In addition to deepened understanding, we have perhaps achieved greater generality of methodological and research outlook. If we understand the common core or cores of methods, then, like the musician who has mastered all forms of harmony and counterpoint, we can do what we will with research problems and research data. Hopefully, such deepened understanding can help us better plan and execute research and analyze data. One of the major points of this book has been that methods influence selection of research problems and even the nature of research problems. In brief, generality and understanding go together: if general principles are grasped and mastered, greater depth of understanding is achieved, and research tools can be appropriately and flexibly used.

There is a curious mythology about understanding and mastery of the technical aspects of research. Statistics is often called "mere statistics," and many behavioral researchers say they will use a statistician and a computer expert to analyze their data. An artificial dichotomy between problem conception and data analysis is set up. While it would be foolish to expect all scientists to be highly sophisticated mathematically and statistically, we must understand that this does not mean relative ignorance. The researcher who does not understand multivariate methods, let alone simpler statistics like analysis of variance and chi square, is a scientific cripple. He simply will not be able to handle the kinds of complex problems that must be handled.

To illustrate to some small extent what we mean, let us take a published study of high quality and point out possibilities of deepening the analysis—and thus the conception of the problem and the enrichment of the results. Free and Cantril (1967) used an important theoretical psychological notion, inconsistency between ideological belief and operational action, as a main basis for exploring the political beliefs of Americans. They did this by conceiving and measuring two belief spectra: operational and ideological. "Operational" meant agreeing or disagreeing with actual assistance or regulation programs of the Federal Government. "Ideological" meant agreeing or disagreeing with more abstract political and economic ideas. Free and Cantril found a distinct difference between the results obtained with their two instruments to measure these spectra: Most Americans agreed with actual social welfare ideas and were thus "operational liberals," but disagreed with more abstract expressions of beliefs related to the operational beliefs. In other words, many Americans were operational liberals and, at the same time, ideological conservatives, according to the Free and Cantril study.

Free and Cantril's analyses were limited to crossbreaks with percentages, a perfectly legitimate although somewhat limited form of analysis. Evidently they did not calculate coefficients of correlation, even when such coefficients were clearly appropriate and readily calculable. We could easily go on to show the limited quality of the measurement of the operational and ideological spectra, the lack of multivariate analysis ideas, and the general disregard of indicating the strength of relations quantitatively. Our purpose is not critical, however. We merely want to point out some possibilities for enriching the research and developing sociological, social psychological, and political theory.

First, Free and Cantril measured only economic, political, and welfare aspects of liberalism and conservatism. They also conceived liberalism and conservatism as a one-dimensional phenomenon with liberalism at one end and conservatism at the other end of a continuum. Suppose they had used a broader range of items to include religious, ethnic, educational, and other aspects of social beliefs and attitudes. They might have found, as has been found in at least some research (Kerlinger, 1970, 1972), that liberalism and conservatism are two different dimensions. This would of course have required factor analysis of items. They might then have used whatever factors emerged from the factor analysis to calculate factor scores and use these in studying the relations between political beliefs, which are now conceived to be multidimensional and not unidimensional, and education, sex, political party, international outlook, prejudice, and so on.

The idea of the distinction between the operational spectrum and the ideological spectrum is strikingly good, and Free and Cantril used it effectively in conceptualizing their research problem and in analyzing their data. We suggest, however, that more powerful analyses and broader generalizations are possible by conceiving the problem as multidimensional, which it almost certainly is, and by using the spectra as independent variables in multiple regression analysis and perhaps even canonical correlation analysis. One can write, at a minimum, a regression equation at the most elementary level as follows: $Y = a + b_1 X_1 + b_2 X_2$, where a = the usual regression constant, X_1 = the operational spectrum, X_2 = the ideological spectrum, b_1 and b_2 the regression coefficients, and Y = a number of dependent variables, like Democrat and Republican (which would yield a discriminant function), education, prejudice, and so on. Naturally, one can add other independent variables to the equation, for example, sex, education, income, and religion.

We hurriedly repeat that Free and Cantril's study is good, even excellent. We picked it deliberately because it *is* good and because its ideas are rich in theoretical implications that can be studied in considerably greater depth and complexity, the latter mirroring the complexity of political and social beliefs and the actions springing from such beliefs.

The greater flexibility and generality of multivariate conceptions and methods entail risks. In general, the more complex an analysis, the more difficult the interpretation of the results of the analysis and the more danger there is of judgmental errors. Where a simple and comparatively clear-cut analysis may be interpreted with perhaps less ambiguity, it may also not reflect the essential complexity of the research problem. If, for example, one is testing a hypothesis of the interaction of two independent variables affecting a dependent variable, simple t tests can be misleading; a more complex test is imperative. When Berkowitz (1959) studied the effect on displaced aggression of hostility arousal and anti-Semitism, he found that more hostility was aroused in subjects high in anti-Semitism than in those low in anti-Semitism. Such a result could not have been found without an analytic method at least as complex as factorial analysis of variance.

As usual, we are faced with a dilemma to which there can never be a clear resolution. It does not solve a problem, however, to decide that one will always use simple analysis because the interpretation of the results is easier and clearer. Nor does it solve it to decide that one will always use a complex analysis because, say, that's the thing to do, or because computer programs are readily available. To be repetitious and perhaps tedious, the kind of analysis depends on the research problem, the theory behind the problem, and the limitations of the data. Our point, again, is that multivariate analysis, when understood and mastered, offers the greater probability of obtaining adequate answers to research questions and developing and testing theory in many areas and concerns of scientific behavioral research.

Study Suggestions

1. The student who is able to handle more than elementary statistics and who knows a little matrix algebra will profit greatly from careful study of Rulon and Brooks' (1968) fine chapter on univariate and multivariate tests of statistical significance. These authors start with a t test of the significance of

the difference between two means. They next discuss other tests to accomplish the same purpose and then extend the discussion to the significance of the differences among several means. The discussion broadens to two groups with any number of dependent variables, and, after describing further tests, finally ends with tests of any number of groups and any number of dependent variables. If one grasps Rulon and Brooks' chapter, one realizes that a single test can embrace virtually all the statistical tests, and that multivariate analysis of variance significance tests embraces all the other tests. We strongly urge the student to study the chapter, particularly the almost magical properties of Wilks' lambda, Λ. Then supplement your study with Tatsuoka's (1971b) monograph.

2. Unlike the literature of multivariate analysis, where truly elementary explanations with simple examples are rare, the factor analysis literature includes elementary treatments. One of the simplest is Kerlinger's (1964, Chapter 36). Cronbach's (1960, Chapter 9) lucid discussion, which concentrates on the factor analysis of tests, is excellent. Nunnally's (1967, Chapter 9) discussion, while more technical and thus more difficult, is very good indeed. The best all-round text on factor analysis is probably Harman's (1967). But the student will need help with it; it is not easy. Thurstone's (1947) book, which is one of the classics of the century, is still highly relevant today despite its age. Like Harman's book, however, it is not easy.

3. Reread the description of the fictitious experiment in the early part of this chapter (data of Table 13.2). Interpret the results of the multivariate analysis of variance. Use Figure 13.1 to help you. Are these results related to the results of a significant interaction in univariate analysis of variance?

4. Explain the basic purpose of factor analysis and use an example to flesh up the explanation.

5. Suppose that you are interested, like the authors of *Equality of Educational Opportunity* (Coleman et al., 1966), in the presumed determinants of verbal ability. You have 42 measures, many of which can be assumed to influence verbal ability. Will you do a multiple regression analysis with 42 independent variables? How can factor analysis be used to help you study the problem?

6. Explain how factor scores and multiple regression analysis can be profitably used together to solve certain research problems. Use an example in your explanation.

14

Multivariate Regression Analysis

Methods of multivariate analysis were introduced and illustrated in Chapters 12 and 13. In the course of the presentation it was noted that multivariate analyses may be viewed as extensions and generalizations of the multiple regression approach. In the present chapter some of these ideas are amplified and illustrated in relation to multivariate analysis of variance. Specifically, it is shown how one can do multivariate analysis of variance with multivariate regression analysis. The latter can be viewed as an extension of multiple regression analysis to accommodate multiple dependent as well as multiple independent variables.

The basic idea of using multivariate regression analysis to do multivariate analysis of variance is quite simple and is a direct extension of the analytic approach presented in Part II. With one dependent variable and multiple categorical independent variables (for example, group membership, treatments), the categorical variables were coded and multiple regression analysis was done, regressing the dependent variable on the coded vectors. With multiple dependent variables and categorical independent variables, the latter are coded in the same manner as in the univariate case. Instead of a multiple regression analysis, however, a canonical correlation analysis is done. While the method is illustrated for a one-way multivariate analysis of variance, extensions to analyses with more than one factor are straightforward.

In order to demonstrate the identify of the two methods, the fictitious data presented in Chapter 12 and 13 are also used in the present chapter. The reader is urged to make frequent references to Chapter 12 for the basic ideas underlying canonical correlation and to Chapter 13 for the ideas and the calculation of multivariate analysis of variance.

The Case of Two Groups

The calculation of multivariate analysis of variance for the case of two groups is relatively simple, regardless of the number of the dependent variables. Group membership is represented by a coded vector (any coding method will do). For the purpose of the analysis, the roles of the independent and dependent variables are reversed. In other words, the coded vector representing group membership is treated as the dependent variable, while the actual dependent variables are treated as independent variables—but only for analytic purposes. A multiple regression analysis is then done. The resulting R^2 is tested for significance in the usual manner. The F ratio associated with the R^2 is identical to the F ratio that would be obtained if the same data were subjected to a multivariate analysis of variance. In fact, $1 - R^2$ is identical to the Λ that is obtained in the multivariate analysis of variance.

It will be noted that the method outlined here was used in Chapter 12 to do a discriminant analysis for two groups. This is not surprising since in both approaches, groups are compared on multiple dependent variables. Although in Chapter 12 the emphasis was on the development of a discriminant function to classify individuals into one or the other group, the same analysis may be viewed as an attempt to determine whether the two groups differ significantly on a set of dependent variables taken simultaneously.

Multivariate Analysis of Variance: Numerical Example

In Chapter 12 a discriminant function analysis was introduced and illustrated for 10 high school sophomore students. Five were experiencing considerable academic difficulty and 5 were not having difficulty, in the judgment of their teachers. Two measures were obtained for each subject: verbal ability, X_1, and school motivation, X_2. The data for this example (Table 12.1) are repeated in Table 14.1, along with the preliminary calculations necessary for the multivariate analysis of variance.

Recall that in multivariate analysis of variance we calculate Wilks' Λ in the following manner:

$$\Lambda = \frac{|\mathbf{W}|}{|\mathbf{T}|} \tag{14.1}$$

where Λ = lambda; $|\mathbf{W}|$ = determinant of the matrix of within groups sums of squares and cross products; $|\mathbf{T}|$ = determinant of the matrix of total sums of squares and cross products. From the calculations in Table 14.1 the matrices for the present problem are

$$\mathbf{W} = \begin{pmatrix} ss_{w_1} & scp_w \\ scp_w & ss_{w_2} \end{pmatrix} = \begin{pmatrix} 23.6 & -1.2 \\ -1.2 & 14.4 \end{pmatrix}$$

$$\mathbf{T} = \begin{pmatrix} ss_{t_1} & scp_t \\ scp_t & ss_{t_2} \end{pmatrix} = \begin{pmatrix} 38.0 & 6.0 \\ 6.0 & 18.0 \end{pmatrix}$$

The calculation of determinants is explained in Appendix A. For the two

TABLE **14.1**   VERBAL ABILITY AND SCHOOL MOTIVATION FOR TWO
GROUPS OF HIGH SCHOOL SOPHOMORES[a]

	A_1		A_2			
	1	2	1	2		
	8	3	4	2		
	7	4	3	1		
	5	5	3	2		
	3	4	2	2		
	3	2	2	5		
Σ:	26	18	14	12	$\Sigma_{t_1} = 40$	$\Sigma_{t_2} = 30$
Σ^2:	156	70	42	38	$\Sigma_{t_1}^2 = 198$	$\Sigma_{t_2}^2 = 108$
M:	5.2	3.6	2.8	2.4		
CP:	95		31			

$$ss_{t_1} = 198 - \frac{(40)^2}{10} = 38.0$$

$$ss_{w_1} = \left[156 - \frac{(26)^2}{5}\right] + \left[42 - \frac{(14)^2}{5}\right] = 23.6$$

$$ss_{t_2} = 108 - \frac{(30)^2}{10} = 18.0$$

$$ss_{w_2} = \left[70 - \frac{(18)^2}{5}\right] + \left[38 - \frac{(12)^2}{5}\right] = 14.4$$

$$scp_w = \left[95 - \frac{(26)(18)}{5}\right] + \left[31 - \frac{(14)(12)}{5}\right] = -1.2$$

$$scp_t = (95 + 31) - \frac{(40)(30)}{10} = 6.0$$

[a]Fictitious data originally given in Table 12.1. A_1 = students who are achieving adequately; A_2 = students who are not achieving adequately; 1 = verbal ability; 2 = school motivation; CP = cross products sums; ss_{t_1}, ss_{t_2}, ss_{w_1}, and ss_{w_2} = total and within groups sums of squares for variables 1 and 2; scp_w and scp_t = within groups and total sums of cross products.

matrices the determinants are

$$|\mathbf{W}| = (23.6)(14.4) - (-1.2)(-1.2) = 338.4$$
$$|\mathbf{T}| = (38.0)(18.0) - (6.0)(6.0) = 648.0$$

Applying Equation (14.1), we obtain

$$\Lambda = \frac{|\mathbf{W}|}{|\mathbf{T}|} = \frac{338.4}{648.0} = .52222$$

By formula (13.8) the F ratio for the present case is

$$F = \frac{1 - \Lambda}{\Lambda} \cdot \frac{N - t - 1}{t} \qquad (14.2)$$

where N = total number of subjects; t = number of dependent variables. The F ratio of formula (14.2) has t and $N - t - 1$ degrees of freedom for the numerator and the denominator respectively. Applying formula (14.2) to the present data,

$$F = \frac{1 - .52222}{.52222} \cdot \frac{10 - 2 - 1}{2} = \frac{.47778}{.52222} \cdot \frac{7}{2} = 3.20$$

with 2 and 7 degrees of freedom, $p > .05$.

Regression Analysis: Numerical Example

We now demonstrate how the results obtained in the multivariate analysis of variance are obtained by regression analysis. The data of Table 14.1 are displayed in Table 14.2 for regression analysis. Note that Y is a coded vector representing group membership. Subjects in group A_1 are assigned 1's, while subjects in group A_2 are assigned -1's (compare Table 14.2 with Table 12.1, where dummy coding was used for group membership). Vectors 1 and 2 consist of measures of verbal ability and school motivation respectively.

As noted above, the roles of the independent and the dependent variables are reversed for the purpose of calculation. For the data of Table 14.2, Y is the independent variable, group membership, while 1 and 2 are the dependent variables, verbal ability and school motivation respectively. Accordingly, we

TABLE **14.2** VERBAL ABILITY AND SCHOOL ACHIEVEMENT FOR TWO GROUPS OF HIGH SCHOOL SOPHOMORES. DATA DISPLAYED FOR REGRESSION ANALYSIS[a]

Group	Y	1	2
	1	8	3
	1	7	4
A_1	1	5	5
	1	3	4
	1	3	2
	-1	4	2
	-1	3	1
A_2	-1	3	2
	-1	2	2
	-1	2	5
Σ:	0	40	30
M:	0	4	3
ss:	10	38	18
$\Sigma_{y1} = 12$	$\Sigma_{y2} = 6$	$\Sigma_{12} = 6$	
$r_{y1} = .61559$	$r_{y2} = .44721$	$r_{12} = .22942$	

[a]Y = coded vector for group membership, where A_1 = adequate achievement, A_2 = inadequate achievement; 1 = verbal ability; 2 = school motivation. All sums of squares and sums of products are in deviation form.

calculate $R^2_{y.12}$. Using the zero-order correlations from Table 14.2 and applying formula (6.5), we obtain

$$R^2_{y.12} = \frac{r^2_{y1} + r^2_{y2} - 2r_{y1}r_{y2}r_{12}}{1 - r^2_{12}}$$

$$= \frac{(.61559)^2 + (.44721)^2 - 2(.61559)(.44721)(.22942)}{1 - (.22942)^2}$$

$$= \frac{.57895 - .12632}{1 - .05263} = \frac{.45263}{.94737} = .47778$$

As usual, the R^2 can be tested for significance:

$$F = \frac{.47778/2}{(1 - .47778)/(10 - 2 - 1)} = \frac{.23889}{.07460} = 3.20$$

with 2 and 7 degrees of freedom, $p > .05$. The same F ratio was obtained above using Λ. Furthermore, with two groups, $\Lambda = 1 - R^2$. For the present data, $\Lambda = 1 - .47778 = .52222$, the value obtained in the calculations of the multivariate analysis of variance.

In sum, then, with two groups and any number of dependent variables, a coded vector for group membership is generated. This vector is regressed on the dependent variables, which for the purpose of calculations are treated as independent variables. The resulting R^2 indicates the proportion of variance shared by group membership and the dependent variables. Λ equals $1 - R^2$, and the F ratio for R^2 is the same as that calculated with Λ.

Multiple Groups

In Chapter 13 a multivariate analysis of variance for three groups was presented and discussed in detail. An example of an experiment on changing attitudes was given, in which three kinds of appeals, A_1, A_2, and A_3, were used with prejudiced individuals. A_1 was a democratic appeal or argument in which prejudice was said to be incommensurate with democracy. A_2 was a fair play appeal: the American notion of fair play demands equal treatment for all people. A_3 was a religious appeal: prejudice and discrimination are violations of the ethics of the major religions. Two dependent variables were used, namely attitudes toward blacks and toward Jews (higher scores indicate greater acceptance). The fictitious data used in connection with this experiment are now used to demonstrate the method of multivariate regression analysis for multiple groups. The reader is urged to make frequent comparisons between the analysis presented here and the multivariate analysis of variance of these data in Chapter 13. This will enhance the understanding of the two approaches, as well as the relations between them.

The data of Table 13.2 are repeated in Table 14.3, but this time they are displayed in a form suited for multivariate regression analysis. Note that there are two vectors, Y_1 and Y_2, for the dependent variables, and two dummy vectors, X_1 and X_2, for the independent variable. Since the coding is used to identify group membership, it is no different from the coding methods used in the univariate analysis (see Part II). It is obvious that the representation of group membership is not affected by the number of measures taken on each individual. Because there is more than one dependent variable (in the present case there are two) and more than one dummy vector, it is not possible to do a multiple regression analysis. It is, however, possible to do a canonical correlation analysis, using the dependent variables as one set and the dummy vectors as the other set.

Canonical correlation was introduced and discussed in Chapter 12, where it was noted that the calculations can become quite complex and that it is therefore best to use a computer to do the analysis. Since, however, we are dealing

TABLE **14.3** DATA FROM HYPOTHETICAL EXPERIMENT ON
ATTITUDE CHANGE DISPLAYED FOR MULTIVARIATE
REGRESSION ANALYSIS[a]

Group	X_1	X_2	Y_1	$\dot{Y}_2$
	1	0	3	7
	1	0	4	7
A_1	1	0	5	8
	1	0	5	9
	1	0	6	10
	0	1	4	5
	0	1	4	6
A_2	0	1	5	7
	0	1	6	7
	0	1	6	8
	0	0	5	5
	0	0	6	5
A_3	0	0	6	6
	0	0	7	7
	0	0	7	8

Correlation Matrix

	X_1	X_2	Y_1	Y_2
X_1	1.000	−.500	−.420	.600
X_2	−.500	1.000	−.168	−.200
Y_1	−.420	−.168	1.000	.252
Y_2	.600	−.200	.252	1.000

[a]The data for the dependent variables are taken from Table 13.2. See Chapter 13 for a multivariate analysis of variance of these data. X_1 and X_2 = dummy coding for group membership; Y_1 = attitudes toward blacks; Y_2 = attitudes toward Jews. The zero-order correlations among the four vectors are given in the correlation matrix.

here with the simplest form of canonical analysis — two variables in each set — it seems worthwhile to go through the basic calculations. It is hoped that the presentation of the calculations will shed more light on the analysis and the interpretation of the results.

In Chapter 12, the basic information necessary for canonical correlation analysis was displayed in the following supermatrix:

$$\mathbf{R} = \left[\begin{array}{c|c} \mathbf{R}_{xx} & \mathbf{R}_{xy} \\ \hline \mathbf{R}_{yx} & \mathbf{R}_{yy} \end{array}\right]$$

where $\mathbf{R}$ = the whole correlation matrix of $p+q$ variables; $\mathbf{R}_{xx}$ = the correlation among the p independent variables; $\mathbf{R}_{yy}$ = the correlations among the q dependent variables; $\mathbf{R}_{xy}$ = the correlations between the independent and the dependent variables; $\mathbf{R}_{yx}$ = the transpose of $\mathbf{R}_{xy}$.

From Table 14.3 we obtain

$$\mathbf{R} = \left[\begin{array}{cc|cc} \multicolumn{2}{c}{\mathbf{R}_{xx}} & \multicolumn{2}{c}{\mathbf{R}_{xy}} \\ 1.000 & -.500 & -.420 & .600 \\ -.500 & 1.000 & -.168 & -.200 \\ \hline \multicolumn{2}{c}{\mathbf{R}_{yx}} & \multicolumn{2}{c}{\mathbf{R}_{yy}} \\ -.420 & -.168 & 1.000 & .252 \\ .600 & -.200 & .252 & 1.000 \end{array}\right]$$

We calculate: $\mathbf{R}_{yy}^{-1}\mathbf{R}_{yx}\mathbf{R}_{xx}^{-1}\mathbf{R}_{xy}$ where $\mathbf{R}_{yy}^{-1}$ is the inverse of the matrix $\mathbf{R}_{yy}$, and $\mathbf{R}_{xx}^{-1}$ is the inverse of $\mathbf{R}_{xx}$.[1]

---

[1]See Appendix A for a discussion of the inverse of a matrix. While the calculation of an inverse is generally quite laborious and is therefore done by a computer, the calculation of the inverse of a 2×2 matrix is simple. Let

$$\mathbf{R} = \begin{bmatrix} a & b \\ c & d \end{bmatrix}$$

Then

$$\mathbf{R}^{-1} = \left[\begin{array}{cc} \dfrac{d}{ad-bc} & \dfrac{-b}{ad-bc} \\ \dfrac{-c}{ad-bc} & \dfrac{a}{ad-bc} \end{array}\right]$$

For example,

$$\mathbf{R}_{yy} = \begin{bmatrix} 1.000 & .252 \\ .252 & 1.000 \end{bmatrix}$$

$$\mathbf{R}_{yy}^{-1} = \left[\begin{array}{cc} \dfrac{1.000}{(1.000)(1.000)-(.252)(.252)} & \dfrac{-.252}{(1.000)(1.000)-(.252)(.252)} \\ \dfrac{-.252}{(1.000)(1.000)-(.252)(.252)} & \dfrac{1.000}{(1.000)(1.000)-(.252)(.252)} \end{array}\right]$$

$$\mathbf{R}_{yy}^{-1} = \begin{bmatrix} 1.068 & -.269 \\ -.269 & 1.068 \end{bmatrix}$$

As a check, multiply $\mathbf{R}_{yy}^{-1}$ by $\mathbf{R}_{yy}$ to obtain an identity matrix.

For the present data,

$$
\begin{bmatrix} \mathbf{R}_{yy}^{-1} \\ 1.068 & -.269 \\ -.269 & 1.068 \end{bmatrix}
\begin{bmatrix} \mathbf{R}_{yx} \\ -.420 & -.168 \\ .600 & -.200 \end{bmatrix}
\begin{bmatrix} \mathbf{R}_{xx}^{-1} \\ 1.333 & .667 \\ .667 & 1.333 \end{bmatrix}
\begin{bmatrix} \mathbf{R}_{xy} \\ -.420 & .600 \\ -.168 & -.200 \end{bmatrix}
$$

$$
\mathbf{R}_{yy}^{-1}\mathbf{R}_{yx} = \begin{bmatrix} 1.068 & -.269 \\ -.269 & 1.068 \end{bmatrix} \begin{bmatrix} -.420 & -.168 \\ .600 & -.200 \end{bmatrix} = \begin{bmatrix} -.610 & -.126 \\ .754 & -.168 \end{bmatrix}
$$

$$
\mathbf{R}_{yy}^{-1}\mathbf{R}_{yx}\mathbf{R}_{xx}^{-1} = \begin{bmatrix} -.610 & -.126 \\ .754 & -.168 \end{bmatrix} \begin{bmatrix} 1.333 & .667 \\ .667 & 1.333 \end{bmatrix} = \begin{bmatrix} -.897 & -.575 \\ .893 & .279 \end{bmatrix}
$$

$$
\mathbf{R}_{yy}^{-1}\mathbf{R}_{yx}\mathbf{R}_{xx}^{-1}\mathbf{R}_{xy} = \begin{bmatrix} -.897 & -.575 \\ .893 & .279 \end{bmatrix} \begin{bmatrix} -.420 & .600 \\ -.168 & -.200 \end{bmatrix} = \begin{bmatrix} .473 & -.423 \\ -.422 & .480 \end{bmatrix}
$$

It is now necessary to solve the following:

$$
\begin{vmatrix} .473 - \lambda & -.423 \\ -.422 & .480 - \lambda \end{vmatrix} = 0
$$

We need to find the values of λ so that the determinant of the matrix will be equal to zero.[2] Therefore,

$$
(.473 - \lambda)(.480 - \lambda) - (-.423)(-.422) = 0
$$
$$
.2270 - .480\lambda - .473\lambda + \lambda^2 - .1785 = 0
$$
$$
\lambda^2 - .953\lambda + .0485 = 0
$$

Upon solving the above quadratic equation one obtains

$$
\lambda_1 = .8979 \qquad \lambda_2 = .0540
$$

λ_1 and λ_2 are called the roots of the matrix whose determinant is set to equal zero. Each root (λ) is equal to a squared canonical correlation. Accordingly,

$$
R_{c1} = \sqrt{\lambda_1} = \sqrt{.8979} = .9476
$$
$$
R_{c2} = \sqrt{\lambda_2} = \sqrt{.0540} = .2324
$$

where R_{c1} and R_{c2} are the first and second canonical correlations respectively. If p is the number of dummy vectors (variables on the left) and q is the number of dependent measures (variables on the right), then the number of nonzero roots (squared canonical correlations) that can be extracted is equal to the smaller of these two values. Since the number of dummy vectors is equal to the number of groups minus one, the number of nonzero roots is equal to the number of groups minus one or the number of dependent variables, whichever is smaller. In other words, if there are, for example, only two dependent variable measures the number of roots will be two, regardless of the number of groups involved. If, on the other hand, there are, for example, 12 dependent variables and four groups, the number of roots will be three (number of groups minus one).

[2]See Appendix A for a discussion of determinants and their calculation.

After obtaining the roots it is possible to calculate Λ:

$$\Lambda = (1-\lambda_1)(1-\lambda_2) \cdots (1-\lambda_q) \tag{14.3}$$

where Λ = Wilks' lambda (see Chapter 13 for a detailed discussion of Λ); q = number of roots. Or equivalently,

$$\Lambda = (1-R_{c1}^2)(1-R_{c2}^2) \cdots (1-R_{cq}^2) \tag{14.4}$$

where R_c^2 = squared canonical correlation. For the present example,

$$\Lambda = (1-.8979)(1-.0540) = (.1021)(.9460) = .0966$$

The same value of Λ was obtained in Chapter 13 in the conventional multivariate analysis of variance. The formula for the F ratio associated with Λ was given in Chapter 13 [formula (13.8)] and is not repeated here. Nor is the calculation of the F ratio repeated. Instead, another test of significance for Λ is introduced and illustrated.

Bartlett (1947) offered the following test for Λ:

$$\chi^2 = -[N-1-.5(p+q+1)] \log_e \Lambda \tag{14.5}$$

where N = number of subjects; p = number of variables on the left; q = number of variables on the right; $\log_e$ = natural logarithm. The degrees of freedom associated with this χ^2 are pq. For the present example, $\log_e .0966 = -2.3372$.

$$\chi^2 = -[15-1-.5(2+2+1)](-2.3372)$$
$$= -(14-2.5)(-2.3372) = (-11.5)(-2.3372) = 26.88$$

with 4 (pq) degrees of freedom, $p < .001$.

The test just performed refers to all the roots extracted. In the present example it refers to the two roots. It is desirable, however, to test each of the roots individually, thereby being in a position to determine which of them is significant. Formula (14.5) can be used for this purpose. The Λ associated with each root is tested separately. The degrees of freedom for the first root are $p+q-1$; for the second root $p+q-3$; for the third root $p+q-5$, and so on.

In the present example,

$$\Lambda_1 = 1-.8979 = .1021$$
$$\log_e .1021 = -2.2818$$
$$\chi_1^2 = -[15-1-.5(2+2+1)](-2.2818)$$
$$= (-11.5)(-2.2818) = 26.24$$

with 3 degrees of freedom $(p+q-1)$, $p < .001$.

$$\Lambda_2 = (1-.0540) = .9460$$
$$\log_e .9460 = -.0555$$

$$\chi_2^2 = -[15-1-.5(2+2+1)](-.0555)$$
$$= (-11.5)(-.0555) = .64$$

with 1 degree of freedom $(p+q-3)$, not significant.

Note that the chi squares are additive, as are the degrees of freedom.

$$\chi_1^2 + \chi_2^2 = 26.24 + .64 = 26.88$$

This is the value of the overall χ^2 obtained above, with 4 degrees of freedom. By decomposing the overall χ^2 it becomes evident that only the first root is significant. This means that only one dimension is necessary to describe the separation among the groups. In other words, only the first canonical vector is necessary.[3]

Proportion of Variance

In the univariate case η^2 (or R^2 with the coded vectors) was used to assess the proportion of variance accounted for by the categorical variables. In an analogous manner $1-\Lambda$ may be used to assess the proportion of variance accounted for in the multivariate case. The overall Λ for the present example is .0966. Therefore $1-.0966 = .9034$. About 90 percent of the variance is accounted for by the two roots. Note, however, that Λ_1 is .1021, so that the first root accounts for almost all of this variance $(1-.1021 = .8979)$. In fact, after allowing for the first root, the second root accounts for .0055 of the variance $(.9034 - .8979)$. As noted earlier, the first root is sufficient to account for the separation among the groups.

Other Test Criteria

Unlike univariate analysis of variance, more than one criterion is currently used by researchers for the purpose of performing tests of significance in multivariate analyses. While the various criteria used generally yield similar tests of significance results, it is possible for the different test criteria not to agree. The purpose of this section is not to review and discuss the merits and demerits of available criteria for tests of significance in multivariate analysis, but rather to

---

[3]As pointed out in Chapter 13, there are weights associated with each root. In the present example the weights for the first root are

| Left side (dummy vectors): | .903 | .429 |
| Right side (dependent variables): | −.705 | .709 |

The weights for the second root are

| Left side: | .033 | .999 |
| Right side: | −.711 | −.703 |

The weights associated with the dummy vectors are affected by the specific coding system used, and will change accordingly. On the other hand, the weights associated with the dependent variables will remain unchanged regardless of the coding system used for the categorical variables. Because only the first root was found to be significant, the weights for this root only are necessary for the purpose of prediction or classification of individuals.

introduce two such criteria because they are obtainable from the canonical analysis without further calculation.

The first criterion was developed by Roy (1957) and is referred to as Roy's largest root criterion, or the largest characteristic root. It is the largest root obtained from the canonical analysis, or the largest R_c^2. In the above example, the largest root was .8979, and it is this root that is tested for significance. Heck (1960) has provided charts for the significance of the largest characteristic root.[4] Pillai (1960) has provided tables for the significance of the largest root. The charts as well as the tables are entered with three values: s, m, and n. For the canonical analysis $s =$ number of nonzero roots; $m = .5(q-p-1)$, where $q \geqq p$; $n = .5(N-p-q-2)$.

For the example analyzed above,

$$s = 2$$

$$m = .5(2-2-1) = -.5$$

$$n = .5(15-2-2-2) = 4.5$$

It is with these values that one enters Heck's charts, for example. If the value of the largest root exceeds the value found in the chart, the result is significant at the level indicated in that chart.

A second criterion for the multivariate analysis is the sum of the roots. The sum of the roots is equal to the trace (the sum of the elements in the principal diagonal) of the matrix used to solve for λ. In other words, it is the trace of the matrix whose determinant was set equal to zero. Look back at this matrix and note that the two elements in its principal diagonal are .473 and .480. Their sum (.953) is equal, within rounding errors, to the sum of the two roots extracted: $.8979 + .0540 = .952$. Pillai (1960) has provided tables for testing the sum of the roots. The tables are entered with values of s, m, and n, where these are as defined above.

The Use of Orthogonal Coding

It was said earlier that it makes no difference which coding method is used for the purpose of representing the categorical variable (group membership, treatments, and so on). Nevertheless, the orthogonal coding method has some interesting properties. In order to demonstrate these properties we use orthogonal coding in an analysis of the above example.

In Table 14.4 we present the original data for the three groups and three sets of orthogonal coding. In the first set, vector 1 contrasts group A_1 with A_2, while vector 2 contrasts the means of groups A_1 and A_2 with the means of group A_3. In the second set, composed of vectors 3 and 4, vector 3 contrasts group A_2 with A_3, while vector 4 contrasts the means of groups A_2 and A_3 with the means of group A_1. In the third set, vector 5 contrasts groups A_1 with A_3, while vector

---

[4]These charts are reproduced in Morrison (1967) and in Press (1972).

TABLE **14.4** DATA FROM HYPOTHETICAL EXPERIMENT ON ATTITUDE
CHANGE; THREE SETS OF ORTHOGONAL CODING, AND
TWO DEPENDENT VARIABLES[a]

Group	1	2	3	4	5	6	7	8
	1	-1	0	2	1	-1	3	7
	1	-1	0	2	1	-1	4	7
A_1	1	-1	0	2	1	-1	5	8
	1	-1	0	2	1	-1	5	9
	1	-1	0	2	1	-1	6	10
	-1	-1	1	-1	0	2	4	5
	-1	-1	1	-1	0	2	4	6
A_2	-1	-1	1	-1	0	2	5	7
	-1	-1	1	-1	0	2	6	7
	-1	-1	1	-1	0	2	6	8
	0	2	-1	-1	-1	-1	5	5
	0	2	-1	-1	-1	-1	6	5
A_3	0	2	-1	-1	-1	-1	6	6
	0	2	-1	-1	-1	-1	7	7
	0	2	-1	-1	-1	-1	7	8

[a]The three sets of orthogonal coding are: vectors 1 and 2; 3 and 4; 5 and
6. Vector 7 = attitudes toward blacks; 8 = attitudes toward Jews.

6 contrasts groups A_1 and A_3 with group A_2. (See Chapter 7 for a discussion of
orthogonal coding.)

Rather than perform a canonical analysis, which will take a similar form to
the one performed with the dummy coding and will result in the same roots,[5]
we take a different approach to the analysis. Let us direct our attention to the
first orthogonally coded set (vectors 1 and 2 of Table 14.4) and to the two dependent variables (vectors 7 and 8). We calculate two multiple regression
analyses. For the purpose of these calculations, however, we reverse the roles
of the independent and the dependent variables, so that in both analyses vectors 7 and 8 (the actual dependent measures in the study) are treated as the
independent variables, while vectors 1 and 2 (the coded vectors) act, in turn,
as dependent variables. In other words, we calculate $R^2_{1.78}$ and $R^2_{2.78}$. These
calculations are summarized in Table 14.5. $R^2_{1.78} = .28641$, $R^2_{2.78} = .66556$.

Since vectors 1 and 2 are orthogonal, their regressions on vectors 7 and 8
do not overlap. Stated differently, the variance of vectors 7 and 8 is sliced in
two nonoverlapping parts. When these parts are added (that is, $R^2_{1.78} + R^2_{2.78}$)
their sum $-.28641 + .66556 = .95197 -$ is equal to the sum of the roots obtained

[5]Incidentally, when orthogonal coding is used, the calculations are somewhat simplified. The
matrix $\mathbf{R}_{xx}$ (the correlation matrix of the coded vectors) will be an identity matrix. Therefore,
$\mathbf{R}_{xx}^{-1} = \mathbf{R}_{xx}$, and the multiplication of a matrix by an identity matrix leaves the original matrix
unchanged. For a definition of identity matrices, see Appendix A.

TABLE **14.5**   CALCULATION OF $R^2_{1.78}$ AND $R^2_{2.78}$,
ORIGINAL DATA OF TABLE 14.4

Correlation Matrix

Variable	1	2	7	8
1	1.00000	.00000	−.14535	.46188
2	.00000	1.00000	.58743	−.40000
7	−.14535	.58743	1.00000	.25175
8	.46188	−.40000	.25175	1.00000

$$R^2_{1.78} = \frac{(-.14535)^2 + (.46188)^2 - 2(-.14535)(.46188)(.25175)}{1 - (.25175)^2}$$

$$= \frac{.23446 + .03380}{1 - .06338} = \frac{.26826}{.93662} = .28641$$

$$R^2_{2.78} = \frac{(.58743)^2 + (-.40000)^2 - 2(.58743)(-.40000)(.25175)}{1 - (.25175)^2}$$

$$= \frac{.50507 + .11831}{1 - .06338} = \frac{.62338}{.93662} = .66556$$

earlier in the canonical correlation analysis. The sum of the roots can, of course, be tested for significance in the manner outlined above in the section dealing with other test criteria.

In addition to demonstrating that the sum of the roots can be obtained through the calculation of a set of multiple regression analyses with orthogonal vectors as the dependent variables, note another property of this approach. Let us assume that the researcher had made the a priori hypotheses reflected by the two orthogonal contrasts (vectors 1 and 2). As a consequence of the multiple regression analyses it becomes evident that the second contrast (vector 2) contributes more than twice as much as does the first contrast (vector 1) to the sum of the roots. Stated differently, this means that contrasting the means of groups A_1 and A_2 with the means of group A_3 leads to a more pronounced separation than contrasting the means of group A_1 with those of group A_2.

In order to be better able to appreciate the insights yielded by an analysis with orthogonal coding, the means of the three groups on the dependent variables, along with the R^2's for the three sets of contrasts originally depicted in Table 14.4, are reported in Table 14.6. Note that for each set of contrasts the sum of the R^2's is equal to .952, which is the sum of the roots. The manner in which the variance is sliced, however, is quite different in the three sets. Note, for example, that in the third set (contrasts 5 and 6 of Table 14.6) the bulk of the variance is accounted for by contrasting group A_1 with group A_3 ($R^2_{5.78} = $ .89724). The second contrast in this set (contrast 6) accounts for relatively little variance ($R^2_{6.78} = .05474$). Study of the means of the groups in relation to these contrasts will reveal why this is so.

TABLE **14.6**   GROUP MEANS ON TWO DEPENDENT VARIABLES, THREE
SETS OF ORTHOGONAL CONTRASTS, AND R^2'S,
ORIGINAL DATA OF TABLE 14.4[a]

	Group Means		Contrasts					
	7	8	1	2	3	4	5	6
A_1	4.6	8.2	1	−1	0	2	1	−1
A_2	5.0	6.6	−1	−1	1	−1	0	2
A_3	6.2	6.2	0	2	−1	−1	−1	−1

$$R^2_{1.78} = .28641 \qquad R^2_{3.78} = .24431 \qquad R^2_{5.78} = .89724$$
$$R^2_{2.78} = .66556 \qquad R^2_{4.78} = .70767 \qquad R^2_{6.78} = .05474$$
$$\Sigma R^2\text{'s:}\ .952 \qquad\qquad .952 \qquad\qquad .952$$

[a] 7 = attitudes toward blacks; 8 = attitudes toward Jews. 1 and 2; 3 and 4; 5 and 6 are three sets
of orthogonal contrasts.

The presentation of the three sets of orthogonal coding should not lead one
to the erroneous conclusion that it is permissible to perform analyses with all
the possible orthogonal contrasts and then pick the result that one likes most.
On the contrary, orthogonal contrasts must be stated a priori as a consequence
of theoretical and practical considerations. The three sets of orthogonal contrasts were presented to demonstrate how they differ in appropriating different
proportions of the variance. In addition, the three contrasts have shown that
regardless of the specific orthogonal contrasts chosen by the researcher, the
sum of the R^2's is equal to the sum of the roots.

Needless to say, orthogonal coding may be used as a purely computational
device. Let us assume that one does not have at his disposal a computer
program for canonical correlation, but that he does have a computer program
for multiple regression analysis. By using orthogonal coding and regressing, in
turn, each coded vector on the dependent variables, one will obtain a number of
R^2's equal to the number of the orthogonal vectors. The sum of the R^2's will be
equal to the sum of the roots, which can be tested for significance in the manner
described above.

In the present example there were three treatments and therefore two
coded vectors were needed. The same approach, of course, can be used with
any number of treatments or groups. As usual, the number of coded vectors is
equal to the number of treatments minus one. When orthogonal coding is used,
one can either do a canonical correlation as shown in the first part of this chapter, or regress, in turn, each coded vector on the dependent measures, as shown
in the latter part of the chapter. For five groups, for example, four orthogonal
vectors are generated. Four multiple regression analyses are then done, in
each case using one of the coded vectors as the dependent variable and the
dependent measures as the independent variables. (Various computer programs
for multiple regression enable one to do multiple analyses in a single run.) The
sum of the four R^2's thus obtained equals the sum of the roots.

As in univariate analysis, it is possible in multivariate analysis to perform post hoc multiple comparisons between means of the dependent variables. This topic is beyond the scope of this chapter. The reader is referred to Morrison (1967) for a good introduction to multiple comparisons among means.

Summary

One of the purposes of multivariate regression analysis is the same as univariate analysis: to explain variance. The ideas and methods of this chapter were shown to be extensions of those of multiple regression analysis. Multivariate regression analysis is a powerful technique for studying multiple dependent variables simultaneously.

When the independent variables are categorical one can either do a multivariate analysis of variance or a canonical correlation analysis in which one set of variables consists of the dependent variables, while the second set consists of coded vectors representing the categorical variables. In the latter case, the use of orthogonal coding will enhance the interpretation of the results, when the researcher formulates a priori hypotheses about differences between groups. As shown in this chapter, however, the overall results are the same whether one does a multivariate analysis of variance or a canonical correlation analysis with any coding method.

As in univariate analysis, one should not categorize continuous variables in multivariate analysis. Consequently, when the independent as well as the dependent variables are continuous, canonical correlation analysis is the most appropriate analytic method. In sum, then, canonical correlation analysis is the most general of the analytic methods presented in this book.

Study Suggestions

1. A canonical correlation analysis is done in a study with four groups and five dependent variable measures.
    (a) How many coded vectors are necessary to represent group membership?
    (b) How many nonzero roots are there in the solution?
    (*Answers*: (a) 3; (b) 3.)
2. In a canonical correlation analysis with three coded vectors and four dependent variable measures, the following squared canonical correlations were obtained: $R_{c1}^2 = .7652$, $R_{c2}^2 = .2532$, $R_{c3}^2 = .1456$.
    (a) What is Λ?
    (b) What is the sum of the roots?
    (*Answers*: (a) .1498; (b) 1.1640.)
3. A researcher hypothesized that middle-class adolescents perceive themselves as more in control of their destiny and have higher career aspirations than lower-class adolescents. He administered a locus of control scale and a career aspirations scale to samples of middle- and lower-class adolescents. The following are fictitious data for two groups, each consisting of 10 subjects. Higher scores indicate greater feelings of control and higher career aspirations.

| Lower Class | | Middle Class | |
Locus of Control	Career Aspirations	Locus of Control	Career Aspirations
2	3	5	4
3	3	6	5
4	4	6	5
4	4	7	7
6	5	7	7
6	6	9	7
7	8	10	9
7	8	10	9
8	9	11	10
8	10	12	10

Do a multivariate regression analysis of the above data.

(a) What is the proportion of variance accounted for by group membership in the two measures?

(b) What is the F ratio for the difference between the groups on the two measures?

(c) What is Λ?

Interpret the results.

(*Answers*: (a) .5957; (b) $F = 12.53$, 2 and 17 *df*; (c) $\Lambda = .4043$.)

PART

4

Research Applications

The Use of Multiple Regression in Behavioral Research: I

Uses of multiple regression in actual research have been cited in previous chapters, usually to illustrate particular points being made. In this chapter and in Chapter 16, we describe and comment upon a number of research studies that have used multiple regression and related methods of analysis. Three major purposes guided choice and discussion of the studies. One, we wanted to give the reader a clear feeling for the research uses of multiple regression and its almost protean manifestations in different fields and different kinds of research. Two, we thought it wise to illustrate and reinforce certain points about multiple regression made in earlier chapters. And three, certain other points either not mentioned or, if mentioned, not elaborated to any degree, needed illustration and clarification. Two such points, for example, are the use of residual scores and factor scores. We start with fairly simple predictive studies and progress to rather complex predictive and explanatory uses of multiple regression. In addition to a mixture of simple and complex studies, studies in different fields have been cited.

Some of the multiple regression analyses we present in Chapters 15 and 16 were done by authors of the studies; some of them, however, were done by us from data published by the authors. Because we usually analyzed correlation matrices, or parts of such matrices, our analyses will lack certain statistics, for example, *b* weights. (Why?) All such analyses were done with the computer program MULR given in Appendix C. When an analysis was done by us, we will say so in the text. If we say nothing, the multiple regression analysis was done by the original author(s) of the studies.

Predictive Studies

Scannell: Prediction of College Success

High school grade-point average (GPA) has been found to be a good, perhaps the best, predictor of success in college. Scannell (1960), for example, found correlations of .67 between high school GPA and freshman college GPA and .59 between high school GPA and 4-year college GPA. This means that college success can be partially predicted from knowledge of high school achievement as reflected in high school grades. High school GPA, in Scannell's study, accounted for approximately 35 percent of the variance of 4-year college grades: $r^2 = .59^2 = .35$.

An important question that can be asked, however, is: Can the prediction of college success be improved by additional information? Scannell also had test scores of educational growth for his subjects. The correlation between this measure and 4-year GPA was .52, accounting for $.52^2 = 27$ percent of the variance of college grades. In addition, Scannell had the high school academic ranks of his subjects. The correlation between rank in class and college GPA was .39. There were, then, three independent variable measures: high school GPA, educational growth in high school, and rank in class in high school. The correlations between these three independent variables and the dependent variable, college GPA, were: .59, .52, and .39. Combining these in a multiple regression analysis yielded a multiple correlation coefficient of only .63, not much of an increase over the .59 obtained with high school GPA alone, about 5 percent: $.63^2 - .59^2 = .40 - .35 = .05$. Are such increases always small?

In the Holtzman and Brown (1965) study summarized in Chapter 1, a substantial increase was obtained. Holtzman and Brown, in studying the prediction of high school GPA, used study habits and attitudes (SHA) and scholastic aptitude (SA) as independent variables. Had they used SHA alone, they would have obtained $r^2 = .55^2 = .30$. Using SA alone they would have obtained $.61^2 = .37$. By adding scholastic aptitude to study habits and attitudes, however, they obtained an R^2 of .52, a substantial increase from .30 to .52. While such a substantial increase is not common, it is clearly possible. The difference in the predictions of the two studies was of course due to the correlations among the independent variables. (The correlations between the independent variables and the dependent variable were roughly similar.) In the Scannell study, they were substantial: mostly in the .60's. In the Holtzman and Brown study, on the other hand, the correlation between the independent variables was .32.

Worell: Level of Aspiration and Academic Success

Worell (1959) tested the notion that the more realistic an individual's level of aspiration, the greater the probability that he will be successful in college. Worell measured level of aspiration by asking students questions about their study habits and grades. Four such measures were used. Two other independent variables were scholastic aptitude and high school achievement.

Worell used two dependent variables, but we use only one of them, total college GPA. We ask: Does adding the level of aspiration measures to scholastic aptitude and high school achievement measures improve the prediction?

The increase in prediction was even more striking than the increase in the Holtzman and Brown study. Among 99 college sophomores, the regression of GPA on scholastic aptitude and high school achievement was expressed by $R = .43$. When Worell added his four levels of aspiration measures, R leaped to .85! In short, the addition of the noncognitive measures to the more conventional cognitive measures increased predictive efficiency dramatically: $.85^2 - .43^2$ $= .72 - .18 = .54$. This seems to be one of the largest reported increases in R^2 obtained by adding independent variables to other independent variables. Obviously, cross-validation is in order before faith can be put in such a large increase.

Layton and Swanson: Differential Aptitude Test Prediction

Lest the reader become too enthusiastic about increasing prediction by adding independent variables, we hastily mention, rather briefly, another predictive study in which the addition to prediction by adding independent variables was much more modest. We also want to show the effect of changing the order of independent variables. Layton and Swanson (1958) published the correlations among the six tests of the Differential Aptitude Test (DAT) and high school percentile rank. Using the published correlations for boys, $N = 628$ (*ibid.*, p. 154, Table 1), we did three multiple regression analyses. The analyses differed only in the order in which the variables, the DAT subtests, entered the regression equation. The R^2 and the betas were of course the same in all three analyses. But the squared semipartial correlations changed — as usual. The results are given in Table 15.1. In the last line of the table, we also report the beta weights.

In studying the table, ignore the last two entries on the right: we did not change their order. Now note different values of the squared semipartial correlations, which are percentages of the total variance. The differences are pronounced. VR, for instance, which accounts for 31 percent of the total

TABLE **15.1**   SQUARED SEMIPARTIAL CORRELATIONS AND BETA WEIGHTS, LAYTON AND SWANSON STUDY[a]

Order 1:	.3136 (VR)	.0735 (NA)	.0027 (AR)	.0000 (SR)	.0005 (MR)	.0147 (CSA)
Order 2:	.2025 (AR)	.1330 (VR)	.0542 (NA)	.0000 (SR)	.0005 (MR)	.0147 (CSA)
Order 3:	.1296 (SR)	.0922 (AR)	.1223 (NA)	.0456 (VR)	.0005 (MR)	.0147 (CSA)
Betas:	.2963 (VR)	.2953 (NA)	.0653 (AR)	.0111 (SR)	−.0164 (MR)	.1290 (CSA)

[a]The names of the six tests are: Verbal Reasoning (VR), Numerical Ability (NA), Arithmetic Reasoning (AR), Space Relations (SR), Mechanical Reasoning (MR), Clerical Speed and Accuracy (CSA). $R^2 = .41$.

variance in the first order of entry, accounts for only 5 percent in the third order when it enters the equation fourth. AR, which accounts for almost none of the variance (.0027) in the first order when it is the third independent variable, jumps to .20 in the second order when it is the first independent variable. Obviously the order of entry of independent variables in the regression equation is highly important. The reader should note other differences, for example, SR.

The beta weights, interpreted in conjunction with the squared semipartial correlations and with circumspection, also throw light on the relative contributions of the different independent variables to the variance of the dependent variable. Taken at face value, VR and NA are the most important — which is probably correct — and AR, SR, and MR less important. CSA is surprising. Judging from the squared semipartials, it is not important. Had it been entered in the regression equation first, however, its value would have been .10, not as high as VR, AR, and SR, but still not negligible.

The analytic and interpretative problem raised by this example — and, indeed, by all examples of multiple regression analysis — is so important that we pause to discuss it a little more. If the reader feels baffled, let him realize that he has company. The companion problems of the order of entry of variables in the regression equation and the relative contributions of the independent variables to the variance of the dependent variable are difficult and slippery ones, as we have said more than once. Actually, there is no "correct" method for determining the order of variables, unless the investigator has clear-cut theoretical presuppositions to guide him. Practical considerations and experience may be guides, too. In the DAT example, it makes good sense, for example, to enter Verbal Reasoning and Numerical Ability first; they are basic to school achievement.

A researcher may let the computer choose the order of variables with, say, a stepwise multiple regression program. For some problems this may be satisfactory, but for others it may not be satisfactory. As always, there is no substitute for extent and depth of knowledge of the research problem and concomitant knowledge of the theory behind the problem. We repeat: The research problem and the theory behind the problem should determine the order of entry of variables in multiple regression equations.

In one problem, for instance, intelligence may be a variable that is conceived to act in concert with other variables, compensatory methods and social class, say, to produce changes in verbal achievement. Intelligence would then enter the equation after compensatory methods and before (or after) social class. A researcher doing this would probably be influenced by the idea of interaction: the compensatory methods work differently at different levels of intelligence. In such a case, one would need a product vector or vectors to assess the interaction. Suppose, however, that the researcher wants only to control intelligence, to eliminate its influence on verbal achievement so that the influence of the compensatory methods, if any, can be seen without being muddied by intelligence. In this case he would treat intelligence as a covariate and enter it as the first variable in the regression equation. He can then remove its influence and study the influence of the compensatory methods and, perhaps, social class without being concerned about the confounding influence of intelligence.

Astin: Academic Achievement and Institutional Excellence

The studies discussed to this point have been relatively simple. Similar and few variables were used as predictors, and interpretation was fairly straightforward. We now examine an excellent, perhaps classic, but more complex predictive study in higher education in which much of the full power of multiple regression analysis was used to throw light on the relative contributions to academic achievement of student and institutional characteristics. The study also illustrates other features of multiple regression analysis that will justify discussing it in some detail.

Astin (1968) addressed himself to the complex question: Are students' learning and intellectual development enhanced by attendance at "high quality" institutions? Using a sample of 669 students in 248 colleges and universities, Astin sought an answer to this question by controlling student input—characteristics that might affect students' performance in college—and then studying the relations between characteristics of the institutions and measures of intellectual achievement. The basic thrust of the study, then, was to assess the effects of institutional quality on student intellectual achievement. The results were surprising.

One of the most interesting features of the study and the analysis was the use of student input measures as controls. That is, Astin's basic interest was in the relations between environmental measures, measures of institutional quality such as selectivity, per-student expenditures for educational purposes, faculty-student ratio, affluence, and so on, as independent variables, and student intellectual achievement, measured by the Graduate Record Examination (GRE) as dependent variables. A number of the institutional characteristics were scored using dummy coding (1's and 0's), for example, type of control (public, private, Protestant, Catholic) and type of institution (university, liberal arts college, and so on). Other institutional characteristics were treated as continuous variables, for example, total undergraduate enrollment, curricular emphases (percentage of degrees in six fields: scientific, conventional, artistic, and so on), and many others. Finally, Astin used interaction measures because, as he says, some versions of the folklore of institutional excellence say that it is the interaction of student characteristics and institutional quality that produces desirable effects on students. The interaction terms were: the product of the students' academic ability, measured by a composite test score, and the average ability of undergraduate students at the institutions, and the product of the students' academic ability and the institutions's per-student expenditures for educational and general purposes.

In order to study the relations between institutional characteristics and student intellectual achievement, that part of the GRE measures due to student input characteristics had to be removed or controlled. If the institutional characteristics measures were correlated with the GRE tests in their original form, the resulting correlations would be affected by student-input characteristics. For example, some substantial portion of the GRE variance was due to intelligence or high school achievement. To obtain GRE scores purged, at least to some extent, of such input influences, it was necessary to determine the relations of such measures to the GRE scores — and then remove the influences. To do this, Astin used a stepwise regression method in which the student-input variables were selectively entered in the regression analyses in order to obtain those input variables that contributed significantly to the regression of GRE on the input measures.

After identifying these input variables, the regression of GRE on all of them was calculated. Residual scores were then calculated. In this case, these scores reflected whatever influences there were on the GRE scores after the composite effect of the student input characteristics had been removed. If these residual scores are correlated with the measures of institutional characteristics, they should estimate the effect of the institutional characteristics on the GRE scores with student-input characteristics controlled. When Astin did this, some of the latter correlations, which had at least been statistically significant, dropped to near zero. The same was true of the interaction or product variables. The startling conclusion, taking the evidence at face value, is that quality of institution evidently made little difference. In fact, the most important influence on level of achievement was academic ability as measured in high school.

Astin did other regression analyses only one of which we discuss because of its relation to the above method and findings and its pertinence to the themes of this book. The three GRE area tests were Social Science, Humanities, and Natural Science. In addition to running multiple regression analyses by entering student input variables first — which was one of the important analyses Astin did — and then assessing the additional contributions using R^2 increments, Astin calculated the R^2 for the regression of the GRE on the institutional variables alone. The R^2's for all three GRE area tests were similar. That for the Social Science GRE, for instance, was .198, compared to the student input alone R^2 of .482. The joint contribution of both student input and college environment variables was .515. Astin also estimated the student-input effect independent of college environment and the college-environment effect independent of student input. The two R^2's were .317 and .033. These results showed that although there was some influence of institutional environment, it was quite small compared to the student-input influence.

We have taken a good deal of space to describe the Astin study to show the virtue and power of multiple regression in helping to answer complex research questions. Although Astin's results might have been approximated by calculating and interpreting individual correlations and partial correlations, nothing like the thoroughness and richness of his study could have been achieved without multiple regression analysis. We rate this research high, and urge students to read and carefully study the original report. It is a convincing analysis of institutional quality and effectiveness. We might also say that it is a discouraging one.

Miscellaneous Studies

In this section we describe five studies that illustrate certain points made in previous chapters. They also illustrate research in different fields: sociology, political science, psychological measurement, and education.

Cutright: Ecological Variables and High Correlations

In an unusual application of multiple regression analysis, Cutright (1963) studied the political development of 77 nations. He constructed a rather complex measure of political development by giving points for relative developments in the legislative and executive branches of government, for example, one point for each year a nation had a chief executive who had been elected by direct vote in an open competitive election. This measure was the dependent variable. The independent variables were also complex measures of communication, urbanization, education, and agriculture. The correlations of the independent variables with the dependent variable were all high: .81, .69, .74, and −.72. But the intercorrelations of the independent variables were even higher: .74 to .88. We have here, then, a difficult problem. It is called the problem of multicollinearity (Gordon, 1968). The reader will remember earlier discussions of the difficulties of multiple regression analysis, particularly in interpretation, when the correlations among the independent variables are high. Cutright was quite aware of these difficulties (see *ibid.*, footnote 13).

R was .82 and R^2 was .67. Cutright noted, however, that the communications variable (X_1) r^2 was .65. The main point of the example is that because of the high intercorrelations of the independent variables — communications and education, for instance, were correlated .88 — one would have great difficulty in interpreting the solved regression equation. It seems clear, nevertheless, that political development can be predicted very well from one variable alone: communications.

Cutright was not content with the basic regression statistics. He calculated a predicted T score ($T = 50 + 10z$, where $z =$ the standard score for a nation) for each nation and subtracted them from the observed T scores. (He had converted all his measures to T scores.) This procedure, of course, yielded for each nation residual scores on political development. Cutright then used regression reasoning and the residual scores to interpret the political development in individual nations. For example, he grouped nations geographically — North America, South America, Europe, and Asia — and, since positive and negative residuals indicate greater than and less than expected political development, the preponderance of positive or negative residuals or the means of the residuals indicate the state of political development in different parts of the world. (The results for South America were surprising and upset the stereotypical notion of lack of political stability there.) In addition, Cutright said that when a nation has departed from its predicted value there will be pressure to move toward the predicted value. Be this as it may, the method is suggestive and points toward fertile uses of regression thinking in such research.

Garms: Ecological Variables in Education[1]

Garms (1968) has done a study similar to Cutright's but in education: he used ecological, or environmental, variables in 78 countries. The hypothesis tested was that public effort for education, as measured by public educational expenditures as a percentage of national income, depends on ability to support education (measured by kilowatt hours of electricity per capita), the people's expectations for education, and the manner in which a government uses the ability and expectations in providing support for education.

Actually, Garms used more independent variables than indicated in the hypothesis, but, with one exception, we need not discuss them here. The exception is kinds of government: representative of the people, nonrepresentative but have the country's resources mobilized toward the achievement of national goals, and nonrepresentative and nonmobilizational, or preservative of the status quo of the ruling class. Two dummy variables using 1's and 0's made it possible to include these variables (or variable) in the analysis.

The multiple correlation coefficient for all 78 nations was .63 and its square .40. The four variables that seemed to contribute most to the regression were kilowatt hours of electricity, enrollment ratio—this is part of the expectations for education mentioned above and is the number of students attending all first and second level education as a percentage of the population in the age group eligible for such education—mobilization positively, and status quo negatively (judged by beta weights).

Wolf: The Measurement of Environmental Variables in Education

Wolf (1966) sought to measure those aspects of the environment that affect human characteristics. Rather than view the environment as a single entity, he believed that a certain physical environment consisted of a number of sub-environments that influence the development of specific characteristics. In his study he focused on those environmental aspects that influence intelligence and academic achievement. Three so-called environmental process variables were defined: Achievement Motivation, Language Development, and Provision for General Learning. Thirteen environmental process characteristics or variables, measured by responses of mothers to interview questions and by ratings, were subsumed under the three headings. Examples from each of the categories are: nature of intellectual expectations of the child, emphasis on use of language in a variety of situations, and opportunities for learning provided in the home.

Although Wolf's measurement work and the reasoning behind it are interesting and potentially of great importance in behavioral and educational

---

[1]The Cutright and Garms studies can be found in Eckstein and Noah (1969, pp. 367–383 and 410–428). It is possible to reanalyze most of Cutright's data from the correlation matrix he publishes (Eckstein & Noah, 1969, p. 377) and all of Garms' data from the raw data (*ibid.*, pp. 416–419).

research, we are only concerned here with his use of multiple regression to test the validity of the measurement of the environmental variables. The multiple correlation coefficient between the measures of the intellectual environment, as independent variables, and measured general intelligence, as dependent variable, was .69; between the measures of environment and a total achievement test battery score it was .80. The two R^2's are .48 and .64. The R^2's are quite high, and are indices of the validity of Wolf's reasoning and measurement.

An additional feature of Wolf's work was his use of regression for cross-validation, a topic discussed in Chapter 11. He divided his total sample into two subsamples at random, calculated the regression weights for each subsample separately, and applied the weights calculated for one subsample to the other subsample. The multiple R's for the relation of the intellectual environment and measured intelligence, using this procedure, were .66 and .66. These results supported the R of .69 reported above.

Powers, Sumner, and Kearl: Testing Readability Formulas

Powers, Sumner, and Kearl (1958) ingeniously used multiple regression to compare the relative efficacies of four reading difficulty formulas. They used, as independent variables in regression equations predicting reading difficulty, the indices of the four formulas. That is, the indices—for example, sentence length and syllables per 100 words—were applied to prose passages of certain reading tests, and four multiple regression analyses calculated. In sum, there were two independent variables for each formula: sentence length, X_1, which was the same in the four formulas, and a different variable, X_2, for each formula. These two independent variables were used to predict to reading difficulty, measured by the average score of pupils answering half the questions correctly. The relative efficacies of the readability formulas were judged from the four R^2's.

To make clear what Powers and his colleagues did, we write out two of the regression equations:

$$Y' = -2.2029 + .0778 \text{ (sentence length)} + .0455 \text{ (syllables per 100 words)}$$
$$Y' = 3.0680 + .0877 \text{ (sentence length)} + .0984 \text{ (percent polysyllables)}$$

where Y = average grade score of pupils answering half the questions on the McCall-Crabbs reading tests correctly. These equations were obtained by applying the measures to prose passages of the McCall-Crabbs tests. The R^2's were calculated with each of the four sets of data.

The names of the four readability formulas and the R^2's associated with them are: Flesch = .40; Dale-Chall = .51; Farr-Jenkins-Patterson = .34; Gunning = .34. Evidently the Dale-Chall formula is the best. The regression equation obtained with this formula predicts reading difficulty best; the formula accounts for more of the variance of reading difficulty than any of the formulas.

Lee: Attitudes and the Computer

The substance of our last example is quite different from the other examples. In a nationwide survey of 3000 persons 18 years of age and older, Lee (1970) had people respond to a 20-item scale that measured attitudes toward the computer. Factor analysis of the correlations among the items yielded two factors: I. "Beneficial Tool of Man's Perspective," and II. "Awesome Thinking Machine Perspective." Lee used the latter factor as a dependent variable in multiple regression analysis. His purpose was to throw light on, to "explain," this factor, which expressed a science fiction view of the computer as awesome and as inspiring inferiority. Here are two of the Factor II items: "They can think like a human being thinks," and "There is no limit to what these machines can do."

To achieve his purpose, Lee used the following variables as independent variables in a multiple regression analysis to "explain" Factor II: intolerance of ambiguity, alienation, education, and four others. His R was .50, and R^2 was .25. He found, however, that intolerance of ambiguity and alienation, without the other variables, had an R of .48 with Factor II, a potentially important finding theoretically. Again we can see the benefit of using factor analysis and multiple regression analysis in conjunction and the virtue of attempting to explain a complex phenomenon, in this case attitudes toward computers, with multivariate thinking.

Conclusion

Our purpose in this book is nowhere better illustrated than in the examples of this chapter and the next chapter. In this chapter, we have seen the predictive and explanatory purposes of multiple regression analysis, although the emphasis has been on the predictive purpose. Just as important from a practical standpoint are the variety of studies and uses of multiple regression we have seen: from the relative simplicity of Scannell's study to the complexity of Astin's; from psychology and sociology to education; from results with low R's to those with high R's. The flexibility, applicability, and power of multiple regression have shown themselves rather well. They will continue to show themselves in the next chapter.

Study Suggestions

1. What is the difference between predictive studies and "explanatory" studies? Does it make a difference in using multiple regression analysis and in interpreting data obtained from such analysis which kind of study one is doing? Why?
2. Suppose that the results of a multiple regression analysis with the dependent variable, science achievement, and the three independent variables, social class, race, and intelligence, are as follows. The correlations between the independent variables and the dependent variable are: .40, .44, and .67. $R^2 = .48$, significant at the .01 level. The regression equation (with betas) is
$$z'_y = .10z_1 + .20z_2 + .55z_3$$
z_1 = race, z_2 = social class, and z_3 = intelligence. The squared semipartial correlations are: .19, .14, .19.

Interpret the results of the analysis. In your interpretation, think of the problems of the order of variables and the "importance" of variables.

3. In the Worell study summarized in the chapter, the addition of four non-cognitive measures to the first two independent variables increased R^2 dramatically: from $R = .43$ to $R = .85$, an increase of 54 percent. Are large increases like this common in research, do you think? Give reasons for your answer.

4. In Wolf's (1966) study of environmental variables in education, summarized in the chapter, it was said that the multiple regression analysis was used, in effect, to test the validity of the measurement of the environmental variables. Explain how this was accomplished in the study. Focus on how multiple regression analysis can be used to test the validity of the measurement of variables.

The Use of Multiple Regression in Behavioral Research: II

The main difference between the studies and methods of Chapter 15 and those of this chapter is complexity of conceptualization and technique. While one or two of the studies of Chapter 15 were complex, the studies of this chapter are in general even more complex (except those in the beginning of the chapter). Thus our task is more difficult. We will have to provide more analytic and substantive details if we hope to make matters clear. The specific techniques to be illustrated include two or three already discussed, like factor scores and residual scores. Some of them, however, were not illustrated in Chapter 15, though they were discussed and illustrated in earlier chapters. We also focus again on unusual and particularly fruitful uses of multiple regression. In most of the studies to be reported, the authors did the multiple regression analysis. In certain others, however, we have ourselves done the analyses using data published in the articles. For one study, an experiment of the factorial kind, we did the analysis with the original data supplied by the senior author of the study.

Miscellaneous and Unusual Uses of Multiple Regression

Koslin et al.: Sociometric Status

Koslin et al. (1968), in an unusual study of determinants of sociometric status, used social judgment theory as the basis of their measurement procedures—which was the unusual part of the study. Social judgment theory (Sherif & Hovland, 1961) says, among other things, that individuals in groups develop reference scales for judging and evaluating the behavior of others in the group.

Sherif and Hovland (*ibid.*, p. 12) say, "Once a psychological scale is formed, subsequent judgment of a similar stimulus is greatly affected by the position of that stimulus relative to the prevailing reference scale." In other words, once a reference scale is formed based on psychological and social realities, it serves as a basis for comparison and appraisal of similar stimuli on subsequent occasions (*ibid.*, p. 13).

Koslin and his colleagues measured certain tasks or group variables of central concern to a group of boy campers: Rifle and Canoe tasks. In the Rifle task, each group member shot four bullets, and the other group members watched him. Each time a bullet hit the target the target disappeared and the group members recorded the score they believed the individual had made. The individual's actual scores were subtracted from these subjective estimates, after a suitable transformation of the scores. These difference scores, averaged, were measures of the over- and under-estimations of the performance of each of the campers.

We will not describe the remaining independent variables in this detail. The next variable, Canoe, was a similar individual task that was judged by campers. The third independent variable was a measure of sociometric status in the group (loosely, group prestige of individuals), and the fourth independent variable was an ingenious height perception test in which group members estimated the heights of the other members from stick figures. (This test correlates positively with group status.) In short, there were four independent variables all of which measured the sociometric status of individuals in groups. The dependent variable was an objective measure of group status. The researchers observed individual groups in interaction, and members were assigned scores on the basis of offering suggestions and having them translated into action (+2), receiving suggestions (+1), and showing deference (−2).

Multiple regression analysis yielded a multiple correlation of .79, with an F ratio of 9.68, significant at the .001 level. Clearly the subjective sociometric measures predicted objective group status successfully. The authors reported beta weights but did not interpret them — perhaps wisely in view of the N of 29. Our analysis, which included squared semipartial correlations, showed that the Rifle and Canoe measures were sufficient to predict the objective measure of group status; the sociometric and height measures added nothing statistically significant to them.

This study is satisfying to read, except for the much too small N of 29. It is based on social psychological theory, it used measurement procedures closely tied to the theory, and its analysis was well-suited to the data.

Heise: Sentence Dynamics

In two theoretically important studies, Heise (1969a, 1970) imaginatively explored the dynamics of simple sentences. In the first of these studies, he investigated affective dynamics and in the second potency dynamics. The technical details of the studies are much too complicated to present in a summary. We will try, however, to give Heise's basic idea and his use of multiple regression to test different models.

The basic notion under test is expressed by the question: Do attitudes toward individual words predict attitudes toward sentences in which the words are embedded?[1] That is, from knowledge of individuals' feelings about words, can we predict their feelings of sentences that contain these words? Heise (1969a) tested four prediction models expressed by three kinds of regression equations. There were three kinds of variables, labeled S, V, and Q. S and V mean the subject and verb of a sentence, and Q the predicate or object. For example, "The man kicked the dog," or S V the Q. An original attitude toward a subject is S, and similarly with V and Q. A resultant attitude, when S is in a sentence, is S', or predicted S.

The simplest predictive model can be written: $S' = K_s + a_s S$, which merely means that the predicted attitude, S', is a function of some constant, K, plus the original attitude S. Heise constructed four kinds of equation. One was like that just given. A second was: $S' = K_s + a_s S + b_s V + c_s Q$, which says, in effect, that the attitude, embedded in a sentence context, is some function of the subject, verb, and predicate. The third and fourth models included interaction forms like $d_s(V \times Q)$, where d_s is a weight. Equations for V' and Q' were also constructed.

Heise, using the semantic differential, measured individuals' attitudes toward words alone and in sentence contexts. These measures were inserted in the regression equations of the four models. Multiple regression analysis then yielded R^2's and regression weights, which were used to compare the prediction models. For example, the regression equation obtained for Model III was: $S' = -.15 + .37S + .55V + .07Q + .25VQ$. R^2 for this model was .70, quite high indeed. For this particular dependent variable, S', the four models yielded R^2's of .20, .56, .70, and .71. Clearly, Model III, the model expressed by the equation given above, predicts the dependent variable very well. Model IV, the most complicated model with all interactions of S, V, and Q added, adds little to Model III. These are the results for the evaluative dimension of attitudes. Heise also reported results for the potency and activity measures, but we omit their consideration here. We also omit consideration of the regression weights and Heise's findings on the relative contributions to the attitude resultants of S, V, and Q.

Anderson: Product Terms and Interaction

In Anderson's (1970) study of the effects of classroom social climate on learning, he calculated product terms of the form $X_i X_j$, where X_i and X_j are different independent variables, to study the interaction or joint influence of independent variables on a dependent variable. The method is discussed in detail by Ezekiel and Fox (1959, Chapter 21). (Difficulties connected with its use are discussed in a later section of this chapter.) The values of two independent variables are simply multiplied over all cases to create a third variable. For example, if we have X_1 and X_2, a third "variable" $X_1 X_2$ is created. This new variable is entered

---

[1]Heise's use of the word "attitude" is different from social psychological use. Perhaps a more accurate word would be "feeling," or simply "reaction to." Yet the theoretical ideas and the methods can certainly be used in attitude theory and research.

in the regression equation as another variable. If there is a statistically significant interaction between the variables in their effect on the dependent variable, the t test of the significance of the regression coefficient of the new variable will reveal it. Anderson also investigated possible curvilinear relations, but we omit this part of his method (mainly because the trends were not statistically significant).

Classroom climate was measured by having students respond to the 14 subscales of the *Learning Environment Inventory* (LEI), an instrument constructed by Anderson. Some of these subscales are: Intimacy (members of the class are personal friends), Difficulty (students are constantly challenged), and Disorganization (the class is disorganized). These were the independent variables. There were four dependent variables, only one of which we consider here, Understanding Science. In other words, Anderson predicted Understanding Science (and the other three dependent variables) from each of the 14 LEI measures. In addition, he included intelligence (IQ) and the interaction term, $IQ \times LEI$.

With four of the subscales, the interactions were statistically significant. The beta weights in the three-variable equation predicting Understanding Science, with girls, when LEI-Intimacy (or cohesiveness) is one of the independent variables, were .27, .08, and .28, for IQ, LEI, and $IQ \times LEI$, respectively. The successive R's were: .30, .30, and .42. The increment added by $IQ \times LEI$ was significant at the .01 level. Evidently there is an interaction between intelligence and the intimacy of a classroom in its effect on understanding science. Intimacy appears to be positively related to gains of girls of high ability in understanding science, but it has a negative relation to understanding science with girls of low ability.

Although we have chosen to highlight only the product-term analysis, Anderson's work both in the measurement of classroom climate and in the use of multiple regression analysis deserves more careful attention. It is part of the enrichment, through more adequate conceptualization and analysis, that is more and more becoming characteristic of educational research.

Cronbach: Reanalysis of Wallach-Kogan Data

In what he called a parsimonious interpretation, Cronbach (1968) reanalyzed some of the data and questioned some of the conclusions of Wallach and Kogan's (1965) *Modes of Thinking in Young Children*. In that portion of the reanalysis that we focus upon, the independent variables are intelligence, called A by Cronbach, and creativity, called F. He called dependent variables Z. The dependent variable of this example, the one Cronbach used to illustrate part of the method, is *deprecates*, which is a single rating scale that presumably measures achievement orientation. In their report, Wallach and Kogan (1965, p. 85) did analyses of variance for the sexes separately. Cronbach combined the sexes for greater power and simplicity. (He also demonstrated that there were few interactions involving sex.) In any case, Wallach and Kogan found a significant effect of A on Z in both sexes. Neither F nor the AF interaction was significant in either analysis. A major purpose of Wallach and Kogan was to study the relations between creativity as a mode of thinking and a wide variety of dependent variables, of which *deprecates* is only one. On the basis of his extensive analysis, Cronbach (1968, pp. 508–509) concluded that creativity, or the F index, explains very little of the variance of the dependent variables and has "disappointingly limited psychological significance." Let's see why he said this.

Cronbach's basic analysis used the R^2 incremental method that we have discussed in earlier chapters. The correlation between A and Z is $-.552$; $R^2_{Z.A}$ is therefore .305, which is significant at the .05 level. Adding F (creativity) to this yields $R^2_{Z.A,F} = .306$. The increment is not significant. (Wallach and Kogan, too, in their analysis of variance, found F not to be significant.) Adding the interaction of A and F yields $R^2_{Z.A,F,AF} = .338$. In this case the increment is significant at the .05 level. As Cronbach points out and we have said earlier, splitting A and F at the median discards variance. Multiple regression analysis uses all the information.

Cronbach, in reanalyzing Wallach and Kogan's data in this manner, found out that of about 100 tests 33 reached the .05 level, in contrast to Wallach and Kogan's findings of 27 significant out of about 200 tests. Obviously the multiple regression method is superior at detecting significant relations when they exist. Cronbach found that A was related to the classroom behaviors, but F accounted for very little of the variance in the dependent variables. In his Table 2, we counted five cases in which the increments in R^2 due to F were statistically significant — out of 33 tests. The total number of statistically significant increments added by the interaction of A and F was seven. Cronbach's conclusion, mentioned above, is apparently correct.

We emphasize that we are not derogating Wallach and Kogan's study. (Nor was Cronbach derogating it.) Indeed, we believe it to be one of the most significant researches in an exceedingly difficult and elusive area. We simply wished to illustrate the power of regression analysis where it is appropriate. And it is certainly appropriate for these kinds of data. (It must again be pointed out, however, that there are difficulties connected with the use of product vectors to assess the "interaction" of continuous variables, as Cronbach did. See later section in this chapter, "Product Variables and Interactions.")

Comparison of Regressions[2]

Suppose you had done a study in New York and had repeated it in North Carolina. The two main variables of the study, let us say, were agreement between one's own and others' perceptions of one's competence, or consistency for short, and observed ratings as successful in teaching, or success. That is, you are predicting from consistency, or agreement in perception of self, to

[2]The comparison of regressions was discussed in Chapter 10. There, however, the emphasis was mainly on the technical and calculation aspects of such comparisons. Here we emphasize the substantive aspects, but inevitably have to mention technical matters.

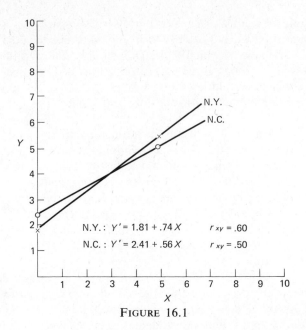

N.Y. : $Y' = 1.81 + .74\,X$ $r_{xy} = .60$

N.C. : $Y' = 2.41 + .56\,X$ $r_{xy} = .50$

FIGURE 16.1

teaching success. Further, suppose that the correlation for the New York sample was .60, while that for the North Carolina sample was .50. The two regression equations were

$$\text{N.Y.:} \qquad Y' = 1.81 + .74\,X$$

$$\text{N.C.:} \qquad Y' = 2.41 + .56\,X$$

Are these regression the "same"? Do the b's differ significantly from each other? In other words, are the slopes the same? Do the intercepts differ significantly from each other? In short, do the regressions differ?

These questions are not too troublesome to answer. First, test the difference between the two, or more, regression coefficients or slopes. The regression lines of the two regression equations just given have been drawn in Figure 16.1. Note that the two regressions are similar, even though one might not be able to judge whether the regression coefficients of .74 and .56 are significantly different. It seems clear, too, that the intercept constants do not differ much. A visual impression would lead us to believe that the regressions are much the same. This conclusion happens to be correct by actual test. If so, then we can say that the two b's and the two a's differ only by chance, and the regressions in New York and North Carolina of success in teaching on consistency are the same.

In Figure 16.2 we have drawn four regression lines. We need not bother with actual numbers. Regression lines A and B are like those of Figure 16.1. They do not differ much from each other. Now regard the other regression lines. Compare A and D. The lines are rather widely separated. The slopes or

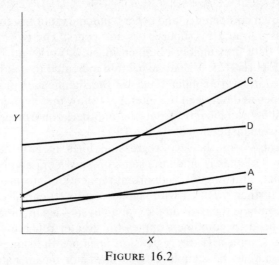

FIGURE 16.2

regression coefficients of A and D are much the same, but the intercepts are different. Next, compare A and C. Although the intercepts are close together, the slopes of the two lines are quite different. Finally, study C and D. These two regression lines are similar to A and B: they cross each other. But in A and B the slopes and intercepts are close; in C and D both the slopes and the intercepts are different. C and D illustrate the interaction of independent variables in their effect on a dependent variable. (So do A and C.) In the consistency and teaching success in New York and North Carolina example, if C and D were the New York and North Carolina regression lines, then we would have an interaction between consistency and geographic location (or group) in their relation to teaching success.

There are, then, four possibilities of comparison and sources of differences between regression lines: AB, where both slopes and intercepts do not differ; AD, where the slopes are the same or similar and the intercepts are different; AC, where the slopes are different and the interaction is ordinal; CD, where the slopes are different and the interaction is disordinal. (Recall that when the slopes are different one does not speak of the differences between intercepts.)

These different regression situations of course have different research meanings. We hesitate to try to catalog all such meanings because we would probably fail, and, maybe more important, we might give the reader a spurious feeling of completeness. About all we can do is again to give an intuitive sense and understanding of the differences by outlining three or four fictitious and actual research examples. We will not discuss how differences in regression are tested statistically. That was done in Chapter 10.

Fictitious Examples

Take the example given earlier of the relation between consistency and teaching success, consistency meaning agreement between an individual's feeling or sense of his own competence and others' judgments of his competence, and teaching success as rated by supervisors and peers. The two regression lines for two sets of data presumably obtained in New York and North Carolina were drawn in Figure 16.1. Visual inspection indicated that although the lines cross the regressions were similar. Tests of the significance of the differences between the slopes and between the intercepts show that neither the slopes nor the intercepts differ significantly from one another, confirming the visual impression gotten from inspection of Figure 16.1.

The "sameness" of the two regression lines means that the relation between the two variables is much the same in New York and North Carolina, other things equal. This might be important replication information, especially if supported by further research with similar results. It can also mean that the researcher can combine the two samples for analysis — again, other things equal. Frequently we need to combine samples for greater reliability, for example, male and female samples. If the regression lines are the same or similar, we can probably combine the samples. (This does not mean, naturally, that all relations between other variables will be the same for males and females.)

Suppose the situation had been different, say like that of A and D in Figure 16.2. Note the substantial separation between the A and D regression lines but the similarity of slopes. Statistical tests of the differences will probably indicate that the difference in slopes is not significant but the difference in intercepts is significant. This means that the relation between consistency and teaching success is the same in both states, but that the levels are different. For example, the individuals in the New York sample (the lower of the two intercepts) might have been generally less successful in teaching than the individuals of the North Carolina sample. Of course, a t test of the Y group means would also show this.

Now suppose the situation had been like that of A and C in Figure 16.2. Here the statistical tests will probably show the slopes to be significantly different. This kind of research result is more complex and harder to deal with, as is the CD result to be discussed below. It means that the relations between consistency of perception of competence and teaching success are different in the two states. If we make up regression equations that roughly depict the two regression lines, they might be

$$\text{A, N. Y.:} \quad Y' = 1.25 + .10X$$

$$\text{C, N. C.:} \quad Y' = 1.50 + .50X$$

Clearly the regression coefficients are sharply different: the regression coefficient in A indicates that with a one-unit change in X, there is only a tenth of a unit change in Y. With C, on the other hand, a one-unit change in X means a half-unit change in Y.

Notice the radically different way of thinking here: we are talking about differences in relations, a considerably more complex and interesting problem than differences in means or frequencies. Such relational differences can mean a great deal, although they are admittedly hard to cope with. It is easy to see the meaning of such a difference between boys and girls. Suppose the regression coefficient of achievement on intelligence for girls in high school was .60, but that obtained with boys was .20. The difference between these two regressions may be very important: it may be that girls are in general highly motivated and that boys are not. Thus boys may not have worked up to their intellectual capacity. This would tend to attenuate the correlation between intelligence and achievement.

Regression Differences in the Coleman Report

Take two or three examples of differences in relations of variables among whites and among blacks in the Coleman Report, *Equality of Educational Opportunity* (Coleman et al., 1966, Supplement, pp. 61–69, 75–83). The most important dependent variable was verbal ability or achievement. There were a large number of independent variables, including an interesting measure of self-concept which we discuss later in this chapter. For the total white sample, the correlation between self-concept and verbal achievement was .46; it was .28 in the total black sample (r^2's of .21 and .08). Supporting this difference were the correlations between self-concept and reading comprehension: .39 and .26 (r^2's of .15 and .07). The differences are substantial. With the huge samples involved, we need no tests of statistical significance.

In the regression of verbal ability on self-concept the slopes are different. This is roughly depicted, in Figure 16.2, by regression lines A and C, except that the intercept constants would be farther apart. The interpretation of these differences is not simple. With whites, the higher the self-concept the higher the verbal achievement. With blacks, the same tendency is present, but it is considerably weaker. Again, we have interaction of independent variables. If we assume that adequate self-concept is an important ingredient in school achievement, is it that the difficulties and disadvantages of Negro life attenuate the relation? That is, even when a black child has an adequate perception of himself, he has less of a chance of achieving verbally at a high level because of disadvantages and barriers that white children do not have — other things equal of course. Our purpose here is not to labor the social psychological educational problem, however. It is to show the reader the high importance of differences in the slopes of regression equations.

Astin: Comparison of Expected and Actual Ph.D. Outputs of Colleges

To assess the Ph.D. outputs of different kinds of colleges and universities, Astin (1962) compared actual slopes with expected slopes. The basic problem being investigated was the role played by the undergraduate college in stimulating its students to go on to the Ph.D degree. To study the problem properly, Astin reasoned, required the control of student input. A college may be considered highly "productive," defined as Ph.D. output, when in fact its Ph.D. output may be due merely to brighter or "Ph.D.-prone" students selecting the college. As a measure of Ph.D. productivity, Astin used the ratio between a college's actual output and its expected output. Actual output was measured as follows: the total number of Ph.D.'s obtained by an institution's graduates during 1957–1959 was divided by the total number of bachelor's degrees awarded in 1951, 1952, and 1953. These rates ranged from .000 to .229. The calculation of expected Ph.D. output was more complex and we need not go into it here.

The correlation between expected and actual outputs were calculated for all institutions and for different types of institutions (men's colleges, coeducational colleges, and so on) and in three broadly different fields (natural sciences and arts, humanities, and social science). The method for doing this took account of the student-input variables mentioned above. For example, the correlation between expected and actual output in the natural sciences in 97 coeducational universities was .72. Astin graphed the relation between expected output, as the independent variable, and actual output, as the dependent variable. He reasoned, rather nicely we must say, that if an institution's plotted point was above the regression line it was "overproductive," and if it was below the regression line it was "underproductive."

The important part of the analysis for our purpose was the comparison of the slopes of different kinds of institutions, using a method like the one described in Chapter 10 but with an original twist. Astin calculated the mean regression coefficient or slope; it was .81. He then tested each type of institution's slope against the mean slope of .81. He found that technological institutions and men's colleges had slopes that differed significantly from .81. The technological institution slope was steeper, 1.13, indicating overproductivity, and the men's college slope was .37, indicating underproductivity. Astin, using these regression slopes, further analyzed the data by categorizing the individual institutions as overproductive, ratios greater than 1.00, or underproductive, ratios less than 1.00. He came to the conclusion that men's colleges and universities were underproductive, institutions in the Northeast were overproductive, and public institutions seemed to be overproductive. Astin's use of slopes to compare institutions of higher learning in their productivity of Ph.D.'s was clearly fruitful.

Warr and Smith: Traits Inferences, Set Theory, Slopes, and Intercepts

What method of combining traits is best to draw inferences about personal traits? Warr and Smith (1970) used six models to predict response items — confident, dull, imaginative, sincere, and so on — from cue traits — ambitious, conceited, humorous, inconsiderate, intelligent. The six predictive models ranged from simple averaging through complex set theoretical probability models. For instance, the simple averaging model was

$$p(C/A \ \& \ B) = 1/2[p(C/A) + p(C/B)]$$

which is read: the probability of C, given A and B, equals one-half the probability of C, given A, plus one-half the probability of C, given B, where A, B, and C are traits, as above. An example is to predict the response trait *original* from the cue traits *humorous* and *intelligent*, each of these cue trait pairs being weighted in six different ways (one of them, for instance, weighting the two cue traits equally according to the above model).

Through an independent procedure, Warr and Smith determined the observed inferences by asking 105 subjects, in one task, the following question: "Consider a person who is [cue trait or pair of traits]. How likely is it that this same person would have the following characteristics [the response traits]?" There was more to their procedure than this, but the details do not concern us. The measures obtained were the dependent variables.

The six prediction models were used to calculate *predicted* measures using measures obtained from averaging the subjects' inferences. Correlations and regression statistics of the observed and predicted inferences were calculated. The correlations for five of the six models were .93 and greater. The important statistics to test the relative predictive powers of the models, however, were the regression coefficients and the intercepts. Warr and Smith reasoned that a perfect predicting model should have a regression coefficient of 1.00 and an intercept of 0. These values would indicate perfect prediction since a regression coefficient of 1.00 means that a unit change in X is accompanied by a unit change in Y, and an intercept of 0 means perfect correspondence of X and Y, or $Y = X$. While all the models did rather well, one of them, a model using the set-theoretic idea of union, or $A \cup B$, satisfied the postulated criteria almost perfectly: the regression coefficient was 1.01 and the intercept constant .00.

This is a remarkable example of the fruitfulness of theoretical and technical thinking. Few examples in the literature are better. The flexibility and power of set theory were put to excellent use, and the use of regression analysis to test theoretical predictions was particularly satisfying. The student is advised to study and compare Warr and Smith's use of regression ideas and statistics with those of Heise described earlier. Both are unusual and original, yet quite different.

A Practical Application of Slope and Intercept Differences[3]

There has been much talk about the bias of teachers in grading working-class and black students. Regression analysis can provide a means of objectively testing for the possible existence of such bias. The method is based on the question: What grades are expected from knowledge of ability? Suppose all students in an integrated high school have been tested with a reliable and valid ability test, say the verbal and numerical subtests of the Differential Aptitude Test. The combined verbal and numerical score is a predictor of success in high school. Although there may be mean differences between groups, we can

---

[3]This example was inspired by Cleary's (1968) study of bias in Scholastic Aptitude Test scores in integrated colleges. We have borrowed Cleary's fruitful idea but have changed and extended it a bit.

assume that for any particular individual, white or black, if we wish to concentrate on this source of possible bias, a predicted grade on the basis of DAT score can be calculated. (Of course, we can use the verbal and numerical scores as separate independent variables in a multiple regression analysis. The basic idea is the same.) In other words, we have DAT scores as the independent variable and grades, as reflected in grade-point average (GPA), as the dependent variable.

For a combined white and black sample we calculate the regression of Y, the grades, on X, the DAT scores. We then calculate the predicted or *expected* scores of all individuals. These scores are expected on the basis of ability and, of course, not on the basis of race. If the relation between DAT scores and grades is substantial, say $r = .70$, and we plot the obtained, or Y, scores against the expected, or Y', scores — and there is no bias in grading — then white and black students should appear about equally above and below expectation, or about equally often above and below the regression line. If, on the other hand, there is bias in grading, positive or negative, then black students will appear more often than white students either above or below the regression line. Again, we must say "other things equal." For example, we must assume that motivation is about the same for whites and blacks, a somewhat questionable assumption. It would be better, therefore, if we had a measure of academic motivation to put into the regression equation. We could then say that the predicted or expected scores reflect both ability and motivation.

The argument is important. On the basis of reliable and valid information we say what we expect of an individual. In this case, we say what grade we expect for the individual on the basis of his measured ability. Theoretically, an individual's race or social class — or anything else — should make no difference in the sense that other characteristics are not relevant, or should not be relevant, to the relation between ability and grades. (We must, of course, assume a substantial relation between ability and grades. If the system is such as to lower this relation, or to use another basis of prediction, say personality, then the argument breaks down.) If, after calculating the regression of grades on ability and predicting the scores of all individuals, white and black, we find that systematically more blacks fall below or above the regression line, then bias apparently has operated.

There is another way to study the same problem. Instead of calculating the regression for the combined white and black groups, calculate the regressions for the separate groups, as was done in Chapter 10. If there is no bias in grading, the regressions should be approximately the same. A statistical test of the significance of the difference between slopes should not be significant. And, if there is no bias, there should be no significant difference between the intercepts.

To illustrate what is meant, study Figure 16.3. The three plots of DAT scores and grade-point average (GPA) represent three likely regressions among the many regressions possible. While contrived, they suffice to demonstrate bias as we relate it to regression analysis. Bias is here defined as significant group departure from expectation. If we take the common regression line — the

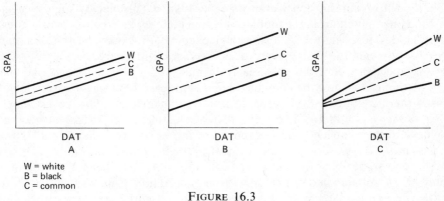

W = white
B = black
C = common

FIGURE 16.3

dashed lines between the two regression lines in the figure — as expectation, then bias can be defined as significant departure from expectation. That is, on the basis of ability, as measured by the DAT, we predict a student's GPA. The prediction is based on the common regression line. We say what we expect from him from knowledge of his DAT test scores. If his GPA departs considerably from this expectation, then, other things equal, we may suspect grading bias.

In Figure 16.3, the regression lines of white and black students were calculated separately. Study the A regression. As can be seen, the slopes are virtually the same and the lines are close together. That is; tests of the significance of the differences between the b's and the a's of the whites and the blacks show no significant differences. Another way to say this is: the common regression line, C, expresses the regression of GPA on DAT scores without distortion.

B and C, however, depict possible bias. In B, the W and B regression lines are relatively far apart. Suppose a test of the difference between the b's is not significant, as it would not be in B. Now suppose a test of the two intercepts, a_W and a_B, is significant, as it would be in B. Other things equal, it appears that there is bias. Blacks are systematically and regularly graded lower than whites, *on the basis of the common expectation.* This is the same idea, fundamentally, as that outlined earlier except that the two regression lines emphasize the systematic *group* discrepancy. A significant difference between means does not in and of itself indicate bias. It is systematic discrepancy between expectation and outcome that does. For example, suppose a white student who has a DAT of 40 gets a GPA of 2.0, but a black student with a DAT of 40 gets a GPA of 1.5. If this situation is repeated for many white and black students, there is bias in grading — again, other things equal.

Data set C is the most interesting — and perhaps the most likely — example of possible grading bias. Suppose the b's are tested and found to be significantly different. Then there is a difference between the slopes of whites and blacks.

The slope for whites is considerably steeper than that for blacks. The change in Y for a one-unit change in X for blacks is considerably less than that for whites. The grades for blacks are lower than expectation (based on the common regression), but the discrepancy also increases with increasing ability: the greater the ability of blacks, the greater the bias.[4]

Although bias may be demonstrated—or at least there may be suspicion of bias—there are of course plausible alternative explanations. The most compelling, perhaps, is cultural. It may be that black students of high ability do not achieve to their capacity because they have little need for achievement.[5] They have not been rewarded for achievement of the kind found in school. There is no need to argue about the sources of differences because the method operates only on expectation and discrepancy from expectation. Conclusions that there is bias in grading, however, have to be based on more than discrepancies in regression slopes and intercepts. The statistical picture is, so to speak, a necessary but not sufficient condition for such conclusions.

Product Variables and Interactions

Certain studies we have summarized (Anderson, 1970; Astin, 1968; Cronbach, 1968; Heise, 1969a) have used product variables—or "products," "product vectors," or "cross products"—to represent interactions of independent variables in their effects on dependent variables. A product variable is the product of the vectors of two (or more) independent variables. For example, if one has three independent variables in a regression analysis, X_1, X_2, and X_3, there are four possible cross products: X_1X_2, X_1X_3, X_2X_3, and $X_1X_2X_3$. X_1X_2 is formed by multiplying the values of the X_1 vector by the values of the X_2 vector.[6]

In earlier chapters, but especially in Chapter 8, we showed how the interaction term is calculated by multiplying the coded vectors of the main effects to produce new "interaction" vectors, one for each degree of freedom. The product terms now being discussed are produced in the same way except that the actual values of continuous and categorical variables are multiplied.

While similar in conception, however, the two procedures are different in outcome. The main effects in factorial analysis of variance are orthogonal to

[4]One suspects that some such bias may operate with both blacks and women. A confusing factor in the black case, however, is the tendency for some professors to exercise a positive bias, especially at the lower ability levels. If the latter were the case, then the lines would cross and we would have disordinal interaction.

[5]Judging from evidence in the Coleman report (Coleman et al., 1966, Supplement, pp. 59, 73), this statement is probably not true. The means for blacks for one of the important measures in the report, Interest in School and Reading, one of whose questions, at the twelfth-grade level, was "How good a student do you want to be in school?" was higher for blacks than for whites.

[6]The reader who wishes to try out cross-product analysis should note that MULR, the computer program given in Appendix C, provides a way to generate the cross-product vectors and include them in the complete regression analysis. It also provides an option of raising vectors to powers and including the powered vectors in the analysis. The same options are provided by the BMD programs in their transgeneration features (Dixon, 1970, pp. 15–21). One might use BMD03R, for example, and use the transgeneration feature to obtain the cross-product vectors. MULR is somewhat easier to use than BMD03R, but the latter has greater flexibility.

each other (when the cell n's are equal or proportional). They spring from an experimental conception in which independent variables are consciously and systematically kept independent by random assignment and orthogonality. The correlations between the main effects, then, are zero. Continuous variable vectors, however, are usually correlated, sometimes substantially so. They spring from an ex post facto situation in which random assignment and orthogonality of independent variables are not possible. It is the orthogonal condition of experimental factorial designs that permits unambiguous statements about the contributions of independent variables and their interactions. With continuous variables, as we have seen, unambiguous statements are hard to come by because of the correlations among the variables.

There are other technical difficulties with product variables. Their correlations with the variables from which they were formed are affected by the means and variances of the original variables and the correlations between them (Althauser, 1971; Glass, 1968). It is clear that they do not reflect interaction in the analysis of variance experimental sense of the word. We counsel great caution if they are used. They seem to have been used effectively, however, by Cronbach (1968) in his reanalysis of Wallach and Kogan's (1965) data on creativity and intelligence, by Anderson (1970) in his study of classroom social climate, and by others. Moreover, their legitimate use with categorical and continuous variables was shown in Chapter 10.

Residuals and Control

We have more than once in earlier pages discussed the control of variables through regression analysis and the use of residuals as "controlled" variables. In Chapter 15 we gave examples of the use of residuals in actual research. Because of the importance and complexity of the subject we need to give more examples and to explain some of the problems involved in the use of residuals. Before such discussion, let us look at two research uses of residuals as "controlled" or "purged" variables and as variables in their own right.

Hiller, Fisher, and Kaess: Residuals as Effectiveness Criterion

Although we have selected a highly interesting, even creative, study by Hiller, Fisher, and Kaess (1969) to illustrate an unusual use of residuals, we want also to say that the study is a good example of the use of modern research and analytic thinking and technology to study an old and difficult problem: effective classroom lecturing.

Social studies teachers of senior high school classes delivered two 15-minute lectures on Yugoslavia and Thailand on successive days. The classes were tested for comprehension immediately after each lecture. The mean comprehension scores of classes were called the basic effectiveness scores. In addition, Hiller and his colleagues wanted to control the variables of ability and interest. To do this, they had all classes listen to a tape-recorded lecture on Israel and then take a test on the material presented in the lecture. One might say that this was a base measure, the variance of which should reflect *ability and interest*. Before the original Yugoslavia and Thailand lectures, the teachers had been sent half the items of the two tests used. They therefore knew half the content of the tests. The lectures they gave were rated for coverage of the test items of the Yugoslavia and Thailand comprehension tests. These ratings were measures of *relevance*. Two measures, class mean scores on the Israel test and the relevance ratings were used as independent variables and class scores on the Yugoslavia and Thailand tests as the dependent variable in a multiple regression analysis. The residual scores obtained from this analysis constituted the effectiveness criterion, the dependent variable of the whole study.

We must understand Hiller et al.'s reasoning and method. Their conclusions depend on them. The teachers of course differed in their ability to lecture and in their interest. Thus these sources of variance in the class mean comprehension scores had to be controlled or removed. In addition, because the teachers had prior knowledge of some of the test content and because they may have differed in the knowledge pertinent to the tests they conveyed in their lectures, the lectures were rated for relevance, another possible source of variance in the class comprehension scores. If we now write a regression equation with two independent variables, one for teacher ability and interest and the other for pertinence or relevance, we can then calculate the regression of the class comprehension scores on the ability-interest and relevance measures.

After obtaining the regression equation, predicted scores are calculated. These predicted scores of course reflect ability-interest and relevance in the dependent variable, comprehension. If they are subtracted from the obtained class comprehension scores, the residuals should reflect sources of variance other than ability-interest and relevance—and, of course, error. They can then be used as measures "purged" of the unwanted influence of ability-interest and relevance. (See footnote 8.) Judging from the results, the method was successful.

Although our main point in presenting this study was the apparently successful use of residual scores, we want to go a little further. Hiller et al. had 35 measures of characteristics of lecturers and lectures. Perhaps the most important feature of their study was the content analysis by computer, too detailed to explain here, of the lectures. Hiller et al. used five a priori categories under which to subsume the 35 measures.[7] They then obtained the correlations among the five "factors" as independent variables and the correlations between the five "factors" and the dependent variable, comprehension (the residual scores). Multiple regression analyses of the Yugoslavia and Thailand lectures were done. The squared multiple correlation coefficients were .46 and .60, quite substantial portions of the variance of the comprehension scores.

[7]Note that factor analysis would have been desirable. Perhaps the authors did not use factor analysis because of the small N's. Unfortunately, the study is marred by these N's, a function of the method of using class means as scores. On the other hand, the reliability of mean scores is always higher than the reliability of the scores from which they are calculated, thus balancing the smallness of the N's, at least to some extent.

(Hiller et al. report .53 rather than .46, but they are evidently in error: $.675^2 =$ .456.)

The five a priori lecture variables were Verbal Fluency, Optimal Information Amount, Knowledge Structure Cues, Interest, and Vagueness. From Hiller et al.'s published correlation matrix for the Thailand data, we calculated the regression equation, as well as other regression statistics. The equation, with beta weights, is

$$Y' = .1439X_1 + .3749X_2 - .1005X_3 + .4168X_4 - .4480X_5$$

The squared semipartial correlations were .14, .22, .01, .11, and .18. All the variables, except Knowledge Structure, contributed significantly to the regression. Hiller et al. reasoned that Vagueness (X_5) was the strongest influence. We would hardly go this far, especially with an N of only 23 and considering the beta weights and squared semipartial correlations just reported, but there seems to be little doubt that Vagueness is an important negative predictor of lecturer effectiveness, as measured by comprehension.

Although we are getting tired of our homiletic advice, we cannot forbear urging the reader to examine this study in the original. He can learn a good deal about fruitful ways to approach an old, interesting, and difficult research problem from study of Hiller et al.'s excellent reasoning and methodology.

Thistlethwaite and Wheeler: Student Aspirations and Peer and Teacher Subcultures

In their highly sophisticated study of the dispositions of students to seek graduate study and degrees, Thistlethwaite and Wheeler (1966) used multiple regression analysis to control a number of independent variables that contribute to graduate aspirations in order to study, in as uncontaminated a manner as possible, the effects of the demands, expectations, and activities of teachers and students on the graduate aspirations of students. Thistlethwaite and Wheeler were interested in *changes* in aspirations to graduate study. This is the core of their use of residual scores.

There were 33 so-called college press scales, which measured the demands, expectations, and activities of teachers and fellow students mentioned above. Factor analysis of the intercorrelations of these items yielded six factors, which were then correlated with the residual scores. Thistlethwaite and Wheeler wanted to know how the college press variables affected changes in aspirations of students. They also obtained measures of aspirations of students just before they were to be graduated from college. One of these simply asked the students whether or not they were enrolling in graduate or professional schools. (We omit consideration of the other measures.) They then developed multiple regression equations on random samples of 475 men and 412 women drawn from the panel of 1772 students who had completed college by regressing the intention or aspiration measure on eight independent variables thought (and known) to be related to aspiration: sex, degree aspiration at the beginning of college, National Merit Test scores, father's educational level, and so on.

Note especially that one of these variables was degree aspiration at the beginning of college. These were the variables to be controlled.

The obtained regression equations (two of them, one for each of the entry dependent variables) were then applied to the remaining students, 461 men and 424 women, who constituted the calibration sample. From these regression equations the aspiration variables were predicted and residuals calculated. That is, the regression equation, as we know, yields a composite, Y', that is maximally correlated with the dependent variable, Y. Through subtraction, $Y - Y' = d$, the residuals are obtained, and they contain sources of variance in Y other than those of the independent variables — plus, of course, error variance. Thistlethwaite and Wheeler reasoned that since the predicted variable represented variance in the dependent variable due to the precollege characteristics, the independent variables, the residual scores represented the variance in the aspiration measures, the dependent variables, not due to the eight independent variables. They further reasoned that this remaining or residual variance must reflect change in student intentions or motivation. Remember that one of the measures was degree of aspiration at the beginning of college. Therefore this source of variance, plus sources due to sex, the National Merit Scholarship Test, father's educational level, and so on, were removed from the scores.

If the residuals are correlated with the aspiration or intention measures at the end of college, the correlations should indicate the relations between the college press and other variables — which indicate faculty and peer influence — and change in aspiration. The results, a bit disappointing, indicated that the college press and other variables had little influence on change in aspiration. Part of the reason for the low correlations, according to Thistlethwaite and Wheeler, was the necessarily restricted range of scores and the unreliability of the residual scores.

The reasoning used in this study and in the Hiller et al. study is very nice indeed. One wonders, however, whether the residual scores in the Thistlethwaite and Wheeler study really measure change in aspirations. There is little doubt of the ingenuity and depth of Thistlethwaite and Wheeler's thinking; but there can be some doubt about the validity of the residuals. After all, there must have been sources of variance in the original dependent variable that the eight independent variables did not themselves have. Perhaps these "hidden" variables took up a disproportionate share of the variance in the residual scores.

Residual scores also have errors of measurement, like any other scores. But they probably have more measurement error than ordinary scores. In short, the approach of using residual scores should be used — but with more than ordinary care and circumspection. To some extent at least, one does not quite know what such scores measure, and one can assume that they are not as reliable as the measures from which they are calculated.[8]

[8]For an excellent discussion of difference scores, residuals, measurement errors, and reliability, see Thorndike (1963). See, also, Cronbach and Furby (1970). The use of residuals is virtually the same as analysis of covariance: one gets rid of the influence of a variable or variables by extracting, so to speak, its variance from the dependent variable and then seeing how much of the dependent variable's variance is accounted for by the subsequent variables.

Analysis of Variance and Multiple Regression: The Jones et al. Study[9]

We are in a peculiar position when we wish to use an actual research study to illustrate the use of multiple regression to do analysis of variance. Actually, a researcher would not say it this way. He would probably use either analysis of variance, even when inappropriate, or use multiple regression analysis and report such use accordingly. So, admitting the slight peculiarity of the procedure, we take a study in which the data were analyzed using factorial analysis of variance. Had the original investigators had equal numbers of cases in the cells of their analysis, the different analyses would be an academic exercise of no great importance, although perhaps mildly interesting as a trick. In any case, in the study we now report, the numbers of the cases in the cells were not equal, and, while factorial analysis of variance yielded satisfactory results, multiple regression analysis would have been more appropriate.

When there are unequal numbers of cases in the cells of a factorial analysis of variance, the essential simplicity and elegance of the ideas behind analysis of variance break down because the independent variables are no longer orthogonal. With small inequalities and large numbers of cases, it does not matter too much because the correlations between the conditions will be fairly close to zero. Sometimes, however, they become rather substantial — and the more substantial they are, the more inappropriate the use of analysis of variance of the conventional sort. (See Overall & Spiegel, 1969, and the related discussion in Chapter 8.)

In an excellent set of studies of high theoretical and technical sophistication, Jones and his colleagues (Jones et al., 1968) sought answers to questions about inferences that people make about other people on the basis of observation of performance. One of their important variables was called Ascending-Descending. This referred to an experimental manipulation in which observer subjects were led to perceive other subjects, called stimulus persons (SP), as increasingly (Ascending), decreasingly (Descending), or sporadically correct on a series of rather difficult problem-solving tasks. How would the observer predict future success or failure for Ascending and Descending subjects? Would more ability be attributed to those individuals who were initially successful or those who were successful later in the series?

In one of their experiments, Jones et al. included three independent variables: Sex, Predict-Solve, and Ascending-Descending. The dependent variable was prediction of success of the SP's on a second set of tasks (after the first set of experimental tasks). Predict-Solve means that some subjects were

[9]We are grateful to Professor Edward Jones for making the original data of one of his studies available to us for multiple regression analysis. The analysis was done with the program MULR, Appendix C. Overall and Spiegel (1969), in their detailed article on the regression analysis of factorial analysis of variance data when n's are unequal, discuss three models or methods of least-squares solutions. MULR uses their model III. Strictly speaking, as shown in Chapter 8, their model II is probably more appropriate. The practical difference between models II and III, however, is ordinarily slight. Overall and Spiegel's article is an important contribution. Researchers who are going to use least squares methods to analyze factorial design data should study their essay carefully.

asked to solve the first set of experimental problems themselves while others did not have this experience. We are interested mainly in the Ascending-Descending effect. Similar factorial designs were illustrated and discussed earlier in this book. The Jones et al. design was an ordinary $2 \times 2 \times 2$.

We analyzed the original data with the multiple regression method outlined earlier. A summary paradigm of how this was done is given in Table 16.1. There were unequal n's in the cells, precluding vector orthogonality, at least by easy means. The numbers in the Y column are the scores for the first subject of each group. Each degree of freedom of the factorial analysis has a coded vector of 1's and -1's. This system of coding gives a fairly close approximation to orthogonality of vectors with these data where the n's are not too different. (The greater the difference in n's the greater the correlations between vectors will be.)

The analysis of variance results yielded by our analysis are reported in Table 16.2. This table is the same in form as the table reported by Jones et al.

TABLE **16.1**   CODING SCHEME OF FACTORIAL DESIGN, JONES et al. STUDY[a]

Y	$X_1 = A$	$X_2 = B$	$X_3 = C$	$X_4 = AB$	$X_5 = AC$	$X_6 = BC$	$X_7 = ABC$
15	1	1	1	1	1	1	1
9	1	1	-1	1	-1	-1	-1
20	1	-1	1	-1	1	-1	-1
7	1	-1	-1	-1	-1	1	1
20	-1	1	1	-1	-1	1	-1
20	-1	1	-1	-1	1	-1	1
14	-1	-1	1	1	-1	-1	1
18	-1	-1	-1	1	1	1	-1

[a]The values given in the Y column are the first values for each cell of the design. $N = 141$. The n's for each cell were $A_1B_1C_1 = 18$; $A_1B_1C_2 = 18$; $A_1B_2C_1 = 14$; $A_1B_2C_2 = 20$; $A_2B_1C_1 = 20$; $A_2B_1C_2 = 18$; $A_2B_2C_1 = 18$; $A_2B_2C_2 = 15$.

(*ibid.*, p. 333, Table 7), except that their table reports only mean squares and F ratios. Our table reports degrees of freedom, sums of squares, mean squares, F ratios, probability levels of the effects (Jaspen, 1965), and the proportions of the variance accounted for by the effects (squared semipartial correlations). The values in Table 16.2 are slightly different from those reported in the original article, but the results are essentially the same. They differ because of the difference in the methods or models used (see Overall & Spiegel, 1969, and footnote 9).

$R^2 = .1366$, indicating that all the main effects and interactions accounted for only a modest proportion of the variance of the dependent variable, predicted success or failure. In an experiment of this kind, however, one does not expect the proportion to be large. The three main effects F ratios were statistically significant; none of the interactions was significant. The theoretically most important main effect, Ascending-Descending, was significant, indicating that observers expect greater success from those subjects who succeeded earlier in the problem-solving series. We may attribute greater ability to those individuals who immediately impress us with their ability (success) than to those who achieve success later. Perhaps slow learners have a built-in psychological handicap.

The proportions of the variance accounted for, reported in the extreme right column of Table 16.2, give us estimates of the contributions of the main effects: .07, .03, and .03. Since the correlations among the effects are low — they range from −.08 to .05 — the order in which the three main effects were entered in the regression equation does not matter too much. Interpreted as absolute values, these estimates are not impressive, especially for Ascending-Descending, the most important one (.03). But two important points must be borne in mind: the R^2 of the total regression is only .14, and the nature of the research and the variables, as indicated above, are such that low values are to be expected. After all, Jones et al. were working with a dependent variable that

TABLE **16.2**   ANALYSIS OF VARIANCE OF JONES et al. DATA: $2 \times 2 \times 2$ FACTORIAL DESIGN DONE WITH MULTIPLE REGRESSION ANALYSIS

Source[a]	df	ss	ms	F	p	Prop. of Variance
A	1	114.2298	114.2298	10.8626	.002	.071
B	1	50.5486	50.5486	4.8069	.028	.031
C	1	45.6395	45.6395	4.3401	.037	.028
A × B	1	.0149	.0149	.0014	.969	.000
A × C	1	3.6905	3.6905	.3510	.562	.002
B × C	1	1.2133	1.2133	.1154	.734	.001
A × B × C	1	5.8576	5.8576	.5570	.537	004
Residual	133	1398.6071	10.5158			
Total	140	1619.8014				$R^2 = .137$

[a]*A*: Sex; *B*: Predict-Solve; *C*: Ascending-Descending.

required from subjects subjective impressions of the future success of other people on the basis of little real information.

Complex Explanatory Studies

We conclude this chapter and our review of studies with two large and important researches. The first of these is Fredericksen, Jensen, and Beaton's (1968) sophisticated and imaginative experiment that competently used a variety of design, measurement, and statistical techniques to study the effects of organizational climates on the administrative performance of executives. The second study is the massive and well-known Coleman (1966) Report, *Equality of Educational Opportunity*, in which multiple regression was the basic technique used to analyze the data. Both studies can be called explanatory because the authors' apparent intent was to "explain" dependent variables rather than merely to predict them. It will be necessary to describe these studies and their methods in considerable detail if the reader is to understand them and their findings. Despite the detail, we can only talk about a relatively small portion of their analyses, especially with the *Equality* study.

Frederiksen, Jensen, and Beaton: Organizational Climate and Administrative Performance

This is an extraordinary study on four main counts. One, it is substantively strong. That is, its problem and the thinking behind it are solid, backed up by theory and earlier research. Two, it is methodologically sophisticated, and the design and the methodology nicely fit the problem. Three, the analysis of the data shows depth of understanding and high competence. Finally, it is important because it may lead to theoretical development, further research, and, perhaps, practical consequences. We should probably add that it is also very interesting. Although the results are far from dramatic, the ideas behind the study may help to open an important area of research.

Frederiksen et al. (1968) asked the basic question: What are the effects of the climates of organizations on the administrative performance of executives? To answer the question, at least in part, they set up an experiment using a 2×2 factorial design. The independent variables were two dichotomized aspects of organizational climate. The first dichotomy was *innovation and originality* versus *rules and standard procedures*. The second was supervisory practices: *global supervision* versus *detailed supervision*. In the global supervision condition, work was assigned and the subordinate given the freedom to get the work done as he saw fit. In the detailed supervision condition, the supervisor monitored the subordinate's work in detail. The basic design is given in Figure 16.4. Each executive subject "worked" in a simulated organization that reflected the above independent variables, which were manipulated by presenting information about the organization. Appropriate documents with different information were given to the subjects in their in-baskets.

	Innovation and Originality	Rules and Standard Procedures
Global Supervision	Dependent Variables:	
Detailed Supervision	Administrative Performance	

FIGURE 16.4

The subjects, simulating executive behavior, were required to take appropriate administrative action on the documents they received. The In-Basket Test (Frederiksen, 1962; Frederiksen, Saunders, & Ward, 1957; Hemphill, Griffiths, & Frederiksen, 1962) was the basic instrument used. This test is a set of elaborate administrative situations and problems that a subject works his way through. It has a number of measures—Requires Further Information, Delays or Postpones Decision, Takes Terminal Action, and so on—on which it is scored. Its "items" have been factor analyzed. Eight first-order and two second-order factors have been found. Two of the first-order factors were Exchanging Information and Analyzing the Situation. Since the first-order factors were correlated—the two factors just named had a correlation of .37— it was possible to extract second-order factors, factors of the factors. They were Preparation for Decision versus Final Action and Amount of Work Expended in Handling the Item. In short, an executive who had worked through the items of the In-Basket Test could be scored on the first- and second-order factors, the factor scores used in the analysis.

In the present research, 55 in-basket scores were factor analyzed and ten factors found; for example, Productivity, Thoughtful Analysis of Problems, Accepts Administrative Responsibility. These ten measures were the dependent variables. An additional dependent variable was used: the average of the in-basket scorers' ratings of overall quality of performance.

One general finding was obtained by calculating factor scores for each subject, calculating the variance-covariance matrices of the factor scores for each cell of the experimental conditions separately, and testing the significance of the differences of these matrices. This is a multivariate analysis of variance test as outlined in Chapter 13. Since this is an important and potentially useful method, we will explain it a bit more. Each person has a factor score on each of the ten factors. To test the hypothesis that different climates have different effects, it is of course possible to do factorial analysis of variance on each dependent variable at a time. But Frederiksen et al. wanted to know about the combined effects of each of the organizational climate dichotomies and their interaction on the in-basket factors together. Look back at Figure 16.4. Using the factor scores, they calculated variance-covariance matrices and tested their differences according to the dictates of the design of Figure 16.4. They found that the organizational climates did influence the interrelations of the factor scores.

The most important of the in-basket factors, the dependent variables, is Thoughtful Analysis of Problems, which was negatively correlated with Interacts with Superiors and Accepts Administrative Responsibility and positively correlated with Interacts with Peers. The authors say (*ibid.*, p. 342):

> It would appear that the climates that provide more freedom of thought and action to employees—innovation and global supervision—tend to send the more thoughtful subjects out to deal directly with their peers (heads of other divisions, who happen to be the source of many of the in-basket problems). In the more restrictive and controlled climates—rules and detailed supervision—the thoughtful people are, on the other hand, constrained to work through their superiors and through that part of the organization for which they are responsible.

Another interesting interaction hypothesis was that the organizational climates affect people differently depending on their personal characteristics. This hypothesis was tested by regressing the in-basket factors on measures of personal characteristics derived from scores on various tests and scales, and then comparing regression slopes. Eleven slope comparisons for each of the contrasting experimental conditions were made. For example, the regression of Productivity, one of the in-basket factors, on a weighted combination of personal characteristics for subjects in the innovation climate was compared to the same regression for subjects in the rules climate. If the slopes are significantly different, the authors say, the effect of an experimental treatment depends on one or more of these characteristics, and examination of the regression weights will show which characteristics are involved. The way the analysis was done seems to mean that a significant difference in slopes indicates that the relation between personal characteristics and productivity is different in the two organizational climates, or that the organizational climates affected the relation between personal characteristics and productivity. This is an intriguing way to study complex organizational phenomena. Unfortunately, only one difference out of 33 tested was significant.

In another kind of approach, Frederiksen and his colleagues used multiple regression analysis in which the regression of the in-basket factors on a variety of cognitive, personality, and biographical measures was studied. The three most predictable in-basket factors were: Productivity, Thoughtful Analysis of Problems, and Defers Judgment and Action. Surprisingly, the Hidden Patterns Test, a measure of field dependence, turned out to be the best predictor of Productivity and a significant predictor of the other two in-basket variables. Can it be that better administrators are more field independent—whatever field independence means?[10]

---

[10]Unfortunately, the report of this study is difficult to obtain. We hope that it is soon published in more readily available form.

Coleman et al.: Equality and Inequality in American Education

Equality of Educational Opportunity (Coleman et al., 1966) is perhaps the most important single study of American education in the last three decades. While it has defects, probably due to the hurry in which it was done to meet a Congressional deadline, it stands as a landmark of educational, sociological, and psychological research in its breadth and understanding of education, its use of modern techniques of data collection, and its analysis of huge amounts of data. It serves a very useful purpose for the present book because it illustrates how multiple regression analysis, its basic analytic tool, can be used for explanatory as well as predictive purposes.

One of the basic purposes of the study was to explain school achievement, or, more accurately, inequality in school achievement. The most important dependent variable was verbal ability or achievement (VA), as measured by various tests. Some 60 independent variables believed to be directly or indirectly related to achievement were correlated with VA. We will give one or two examples taken from the many reported in *Equality*, and then we will do an analysis of our own with selected variables and the correlations among them reported in the Appendix of *Equality*.

Much of the data of the study were obtained from the sixth, ninth, and twelfth grades of schools all over the country. In one table (*ibid.*, Table 3.221.1, p. 299), Coleman et al. reported for the three grade levels the percentage of variance in verbal achievement among whites, blacks, Puerto Ricans, and other groups accounted for by school-to-school differences (A), objective background factors (B), subjective background factors (C: parents' interests and educational desires), attitudes toward school of child (D: interest in school, self-concept, and sense of control of environment). The table reported the successive increments of the variance added, that is, A, A + B, A + B + C, A + B + C + D. This is the R^2 increment method we have met repeatedly. For Northern twelfth-grade black students the successive percentages were: 11.19, 15.34, 18.85, 31.04. Note the substantial increase added by D, attitudes. By contrast, the percentages for Northern twelfth-grade white students were: 8.25, 17.24, 27.12, 40.09. Evidently these factors account for more of the variance of verbal achievement with white students than with black students: 40.09 versus 31.04. The contribution of attitudes (D) is about the same and substantial in both groups: 14 percent and 12 percent.

One of the most controversial points made in the Coleman Report was that the differences between schools had little relation to verbal achievement compared to the relations between verbal achievement and the child's own background and attitudes. In technical language, this means that the variance in verbal achievement accounted for by background factors and the attitudes of the child is much greater than the variance accounted for by differences among schools. This does *not* means, as some have taken it to mean, that schools make no difference. They do. But background factors, like an encyclopedia in the home—in the total Negro sample, the correlation between VA and an

TABLE **16.3**   CORRELATIONS AMONG FIVE INDEPENDENT VARIABLES
AND VERBAL ACHIEVEMENT, *Equality of Educational Opportunity* DATA,
NORTHERN WHITE AND NEGRO TWELFTH-GRADE SAMPLES[a]

	1	2	3	4	5	6
1	1.0000	−.0676	.0094	.0689	−.0204	.0321
2	.1671	1.0000	.0244	−.0663	.0291	.0824
3	−.0222	.0280	1.0000	−.0273	.3455	.4645
4	.3669	−.0376	−.1059	1.0000	.0335	.0622
5	−.0101	.0183	.3018	−.0596	1.0000	.3383
6	.1265	.0006	.3289	.1463	.3464	1.0000

[a]1: Verbal Ability, Teacher; 2: Per Pupil Expenditure; 3: Self-Concept; 4: Proportion White;
5: Control of Environment; 6: Verbal Achievement. Data from Coleman et al. (1966, Supplemental Appendix, pp. 89ff. and 117ff.). Correlations of the white sample are above the diagonal and those of the black sample below the diagonal.

encyclopedia in the home was .38, and in the total white sample it was .19 — structural integrity of the home, siblings, parents' education, parents' interest, and students' attitudes — self-concept, interest in school, and sense of control of environment — accounted for much of the variance. Moreover, school facilities contributed relatively little to the variance. For twelfth-grade Negroes in both the North and the South, the variance in VA contributed by school facilities is .02 percent compared to 6.77 percent for student body quality (*ibid.*, Table 3.23.1, p. 303). The authors say, pithily: "For equality of educational opportunity through the schools must imply a strong effect of schools that is independent of the child's immediate social environment, and that strong independent effect is not present in American schools (*ibid.*, p. 325)."

The example to be given now was calculated by us from the published correlation tables of the Coleman Report (*ibid.*, Supplemental Appendix, pp. 89ff. and 117ff.). We selected five independent variables for their importance and interest. The example is included here because it illustrates what can be done with simple correlations and multiple regression analysis. Two sets of correlations were taken from the Northern white twelfth-grade sample data and the Northern black twelfth-grade sample data. The dependent variable was verbal achievement. The identification of the five independent variables is given in the footnote of Table 16.3, which contains the correlation matrices of the two samples. The correlations above the diagonal are from the white sample, while those below the diagonal are from the Negro sample. The R^2's, the beta weights, and the squared semipartial correlations (SP^2) of the regression analysis of the two samples are given in the body of Table 16.4. In addition to the comparisons between white and Negro sample results, the results gotten from three different orders of entry of the variables in the regression equation are also given in the table. The R^2's and the beta weights for the three orders of entry, of course, are the same, since changing the order does not change R^2 or the beta weights. The SP^2's are different.

TABLE **16.4** MULTIPLE REGRESSION ANALYSIS: R^2's, BETA
WEIGHTS, AND SQUARED SEMIPARTIAL CORRELATIONS, SELECTED
VARIABLES FROM *Equality of Educational Opportunity*

	VAT[a]	PPE	SC	PW	CE	R^2
White:						
β	.033	.074	.396	.069	.198	
SP^2	.001	.007	.214	.006	.034	.262
Negro:						
β	.079	−.019	.265	.161	.277	
SP^2	.016	.000	.111	.020	.070	.217
	CE	PW	SC	PPE	VAT	
White:						
β	.198	.069	.396	.074	.033	
SP^2	.114	.003	.139	.005	.001	.262
Negro:						
β	.277	.161	.265	−.019	.079	
SP^2	.120	.028	.063	.000	.005	.217
	PW	SC	CE	PPE	VAT	
White:						
β	.069	.396	.198	.074	.033	
SP^2	.004	.218	.035	.005	.001	.262
Negro:						
β	.161	.265	.277	−.019	.079	
SP^2	.021	.120	.070	.000	.005	.217

[a]VAT: Verbal Ability, Teacher; PPE: Per Pupil Expenditure; SC: Self-Concept; PW: Proportion White; CE: Control of Environment.

Most of the variance of verbal ability or verbal achievement seems to be due to Self-Concept, a measure constructed from three questions the answers to which reveal how a pupil perceives himself (for example, "I sometimes feel that I just can't learn"). The SP^2's show that this is so in all three orders of entry for whites (.214, .139, .218). It is less true for black students (.111, .063, .120). The only other variable that accounts for a substantial amount of verbal achievement variance ($\geq .10$) is Control of Environment, CE, which is another variable involving the concept of self but which adds the notion of control over one's fate. Here whites and blacks are similar except that CE appears to be somewhat weightier for blacks.

One of the most interesting comparisons is that between kinds of variables. SC and CE are both "subjective" variables: the student projects his own image. The other variables are "objective": they are external to the student; they are part of the objective environment, so to speak. Per Pupil Expenditure, PPE, is one such variable. An important finding of the study, mentioned to some extent above, was that things like tracking (homogeneous grouping), per pupil expenditure (PPE), and school facilities accounted for little of the variance in achievement compared to certain other measures. (Note the SP^2 values for PPE in Table 16.4.) The so-called attitude variables, two of which are SC and CE in the table, accounted for more variance than any other variables in the study (*ibid.*, pp. 319–325).

Study of the beta weights of Table 16.4 is profitable. Since they were calculated from data of very large samples, they are probably stable. They seem to reflect accurately the relative importance of the five variables. For whites SC has the largest value, .396. Evidently self-concept is a most important variable in explaining the verbal achievement of white students. And this interpretation is supported by the SP^2's in all three orders of entry. For black students CE is largest, followed by SC. And this is the pattern of the findings of *Equality*. Evidently self-perception or self-attitudes are of great importance in achievement—and perhaps in all of education. The major American faith in school facilities, equipment, expenditure per pupil, and other material supports of education is a bit shaken, if we are to believe the data of the Coleman Report. At the very least, the report has called two or three assumptions about education into serious question. And the multivariate approach and the use of multiple regression analysis made it possible for Coleman and his colleagues to accomplish the feat of challenging basic and probably false assumptions and to show rather clearly that educational equality is a myth.[11]

Conclusion

From our review of studies in these two chapters, it should be obvious that multiple regression can be applied in different ways to a variety of research problems and data. It is well-suited to predictive studies, as we have seen, especially in the early part of Chapter 15. But it is also well-suited to the more

[11]The Coleman Report should not be accepted as dogma. While it is probably the best and certainly the largest study of its kind, it has defects, as we said before. One of these is inadequate responses: about 30 percent of the schools selected for the study did not participate. Information on schools was obtained from teachers and administrators. Information on the pupils' backgrounds was obtained from the pupils. Thus there are two important sources of unknown bias in the data. Moreover, the presentation of some of the statistics leaves something to be desired. Even a careful and knowledgeable student would have considerable difficulty tracking down the validity of some of the conclusions. Basic regression statistics, like R^2, are not reported. And the sheer mass of tables and statistics, with insufficient help from the authors, hardly makes for clear grasp and understanding. Evidently squared semipartial correlations were used in certain tables; yet this is not made clear. (Professor Coleman told one of us how the proportions of variance statistics were calculated; from his description we inferred that they are squared semipartial correlations.)

There have been good reviews of the study, its methodology, and its findings. One of the best is Nichols' (1966). This review also includes a convenient summary of the major findings. Nichols somewhat cynically but accurately points out that educational practice in the United States is not usually based on research and that the results of the Coleman report will probably have little influence on educational policy. He does not think, however, that this is too bad since "the findings are too astonishing to be accepted on the basis of one imperfect study (Nichols, 1966, p. 1314)." Since this chapter was written, a large book of reanalysis, commentary, and criticism has been published: *On Equality of Educational Opportunity* (Mosteller & Moynihan, 1972). This book may become the definitive critique of the Coleman Report.

difficult explanatory function of science and research. Prediction is never enough. To understand phenomena, we must be able to "explain" them by showing, as precisely as possible and in as controlled a way as possible, their relations to other phenomena. No method seems so well-suited to doing this as multiple regression. Our study of Worell's work with level of aspiration, although perhaps basically predictive, illustrates such explanation: level of aspiration helped considerably to "explain" academic success. Astin's study of the effect of colleges on undergraduate achievement, although again basically predictive, "explained" academic success by showing that, compared to student input, institutional characteristics do not influence academic success to any great extent. The Coleman study is a splendid example of the explanatory use of multiple regression to throw light on the verbal achievement of majority and minority group children.

The research studies of these two chapters were, with two exceptions, nonexperimental or ex post facto studies. We want to make two points. One, multiple regression can be used with both experimental and nonexperimental studies and data; we hope our examples have dispelled any misunderstanding about this. Two, despite the ability of multiple regression to handle both experimental and nonexperimental data, it seems especially suited to studies in which experimental and nonexperimental variables are used together. Analysis of variance can handle ordinary experimental data very well. But it cannot handle, at least easily and naturally, both kinds of variables. Multiple regression can. The Frederiksen, Jensen, and Beaton study of administrative performance showed this. When the numbers of cases in the cells of an analysis of variance are unequal, analysis of variance, although it can of course be used, labors under difficulties. Multiple regression takes the situation in stride, as we hope our analysis of the Jones et al. study data showed.

Multiple regression has a certain flexibility, adaptability, and generality about it that helps. We tried to demonstrate some part of these qualities by citing rather unusual uses. The Heise study in which multiple regression was used to test different prediction models is a good example. Our analysis of Koslin et al.'s sociometric data is another. Astin's comparison of slopes to study the productivity of undergraduate institutions is still another. Hiller et al.'s use of residual scores in their computer content analysis investigation of classroom lecturing illustrates another facet of multiple regression. So does Thistlethwaite and Wheeler's use of residual scores to study student aspirations.

To learn about phenomena and fields of study–like attitudes and risk taking in social psychology, achievement and anxiety in educational psychology, or social class and role prescriptions in sociology—a good textbook can help because it imposes structure and, hopefully, clarity on the fields and the relations among the phenomena. But for depth of learning and understanding the intensive reading of original research reports is indispensable. We can summarize Jones et al.'s study of attribution of ability from perception of success, but our summary necessarily sacrifices both the guts and nuances of Jones' interesting and important work.

Similarly, to understand multiple regression analysis in its actual use, intensive reading and study of original research reports are necessary. While it may be a bit tedious to read and study the Coleman report, or the similar Wilson (1967) study, there is no other way to integrate the substance of research and its methodology and analysis. Understanding and mastery of research require the integration of all three. Scientific research is not something we read just for enjoyment. We read to understand what the authors did, how they did it, and why they did it. In short, there can be no divorce of substance and technique. We believe that this may be the main advantage of reading and studying original research reports.

Study Suggestions

1. From his reanalysis of the Wallach and Kogan (1965) data (summarized in the chapter), Cronbach (1968) concluded that creativity (the F index) explained little of the variance of the dependent variables. Lay out the method he used. How was he able to come to his conclusion? Why is multiple regression a "better" method than analysis of variance for analyzing the Wallach and Kogan data?
2. When a researcher compares slopes, what is he doing in essence? If the regression lines are about the "same" in two samples, what does this mean? If they are different, however, what may it mean?
3. Suppose you were asked to help plan a study of bias in grading women students in college. What might your advice be? How can regression analysis be used to study grading bias? What is the basic idea behind a regression approach to the problem?
4. Residuals, or residual scores, are being used more and more in behavioral research. Explain what residual scores are and what they can accomplish in research. Give an example to show what you mean.

PART

S

Scientific Research and Multiple
Regression Analysis

Theory, Application, and Multiple Regression Analysis in Behavioral Research

It is fitting that we end a book on the regression analysis of scientific data with an extended consideration of behavioral scientific theory and the use of multiple regression analysis to help test the theory. Rather than discuss theory itself to any great extent, however, we will manufacture a rather complex example of particular pertinence to educational research and hope that it will make some of our main points. Although educational in content, the example will be based primarily on social psychological theory and research but with adequate consideration of sociological variables. Our major concern, of course, will be to apply multiple regression to the analysis of the synthetic data of the example and to show how such analysis can be used to test theoretical propositions and predictions.

Although theory will be the main preoccupation of the chapter, three or four other topics will be discussed, though briefly. The first of these, research design, is closely related to the discussion of theory. The second topic is the strengths and weaknesses of multiple regression analysis. Here and there in the book we have discussed the strengths and weaknesses of the method, but particularly the strengths. We want now to recapitulate some of the points discussed earlier and to try to put the whole subject in perspective. A third topic is the reliability of regression statistics. In this discussion, the desirability and necessity of replication will be stressed.

Finally, we will attempt to pull our central arguments together. Again, we want perspective and, hopefully, a balanced view. Perhaps we can achieve perspective and a balanced view and also epitomize the central purpose and function of multiple regression analysis in scientific behavioral research.

A Synthetic Theory of Achievement

A theory, as we said earlier in this volume, is an interrelated set ". . . of constructs (concepts), definitions, and propositions that presents a systematic view of phenomena by specifying relations among variables, with the purpose of explaining and predicting the phenomena (Kerlinger, 1964, p. 11)." In this section, we want to give a final idea of how multiple regression analysis can be used to test theoretical notions. The basic ideas to be used are borrowed mainly from two sources: the Coleman Report (Coleman et al., 1966) and Jones and Gerard's (1967, Chapter 9) theoretical exposition of social comparison processes. We wish to "explain" achievement theoretically, set up a paradigm for empirical testing of aspects of the theory, and use multiple regression to analyze the "data." Since we want only to sketch the outline of the whole procedure we will discuss measurement and other technical matters only peripherally.

Social Comparison Theory

One of the most important concepts in social psychology is "social reality," which means, in brief, other people (Festinger, 1950, 1954). Beliefs about physical objects and factual matters can be validated rather directly. If a person believes that Republican presidents have favored big business, he can check his belief by an extensive study of the economic actions of presidents. It may be difficult, but it can be done. Beliefs about matters that have little or no basis in physical reality or fact, however, cannot be checked directly. But such beliefs also have to be validated or they will probably die. If a person believes that Jews are clannish, or that the Irish are pugnacious, or that blacks are more musical than whites, he has little adequate way to check the validity of the beliefs. It is next to impossible to run an objective test of the clannishness of Jews or the superior musical ability of blacks. Even when it is possible to test the correctness of a belief, as the musical ability of blacks' belief, it is unlikely the person will do so. Besides, he does not need to; there is a readily available test of such beliefs: other people and their beliefs.

"Social reality," then, is other people and particularly the beliefs, attitudes, and values of other people. When beliefs cannot be validated directly, or when such validation is difficult, people will "test" or "check" their beliefs against those of other people. And the other people are those who are significant to us, the people of the groups to which we belong or to which we refer our beliefs and attitudes (reference groups). Beliefs are usually not checked directly and consciously. Rather, their "validation" is picked up, as it were, indirectly and unconsciously.

Jones and Gerard have proposed two related notions that we can use to help build a theory of achievement. One of these, dependence, has two aspects: information dependence and effect dependence. *Information dependence* is the dependence a child has on his parents because early in his life parents virtually control information flow. Jones and Gerard (1967, p. 127) say, "Because of his lack of ready access to nonsocial sources of information, the child is peculiarly vulnerable to those social sources appearing and reappearing in the immediate environment." The power of the parents or peers in information dependence is due of course to the child's need to know things in order to live and to a presumed tendency to seek information and reduce uncertainty.

Effect dependence is more general and probably underlies information dependence. It is the dependence a child has on parents for achieving ends. In other words, he depends on the parents – and later on peers and teachers – to help him get what he needs and wants. Information dependence, which is our concern and which seems to be an aspect of effect dependence, arises because of the child's need for clarification and structure (*ibid.*, p. 79). One can also say that adults develop information dependence on others because of their need to validate beliefs, attitudes, and values, their need to check social reality.

Let the impatient reader know that we *are* working our way to school achievement. Part of social reality testing is our testing of ourselves. We have to know how we are doing, particularly how well we are doing. While we of course look to ourselves, we look even more to other people for appraisals of how we do and how we think, and we look especially to those presumably in a position to tell us. *Reflected appraisal* is the evaluation, the estimate, of ourselves bounced back to us from other people. It is particularly applicable to self-ability estimates, and is part of effect and information dependence. Teachers are important sources of self-appraisal for their pupils. They are second only to parents and peers in this respect. Children, who are effect dependent and even more information dependent on teachers, must continually appraise themselves and their work. Teachers are the official sources of the reflected appraisal of children in school work and achievement. The pupil's perception of himself is in part a reflection of the teacher's appraisal of him, and the more effect and information dependent the child, the more reflective appraisal power the teacher has. This is of course similar to the general socialization process of "learning the self." We "know ourselves" mainly through the eyes of others. This is Cooley's "social self," or "looking glass self," and Mead's self-other idea (see Newcomb, 1950, pp. 312ff.). In short, a youngster's self-image in relation to school work is strongly influenced by the social reality of the teacher. His general self-image is anchored in the social reality of parents, peers, and teachers. He sees himself "as others see him."

We have explained these ideas in some detail because they are crucial to our theory of achievement. We are saying that a child's perception and judgment of himself and how he does, his achievement motivation, his level of aspiration, his attitudes toward school, teachers, and authority figures, and sense of worth and control of the environment are tied to social reality and social comparison processes. We hypothesize that this interrelated set of variables accounts for a substantial portion of the variance of school achievement; indeed, a more substantial portion of the variance than so-called school variables. Background variables are of course also important because the kind of home and community a child lives in – the social reality of his parents' and his peers' beliefs, attitudes, values, and expectations — also have a profound influence (Berkowitz, 1969; Zigler & Child, 1969).

Testing the Theory

The variables of the theory are given in Figure 17.1. Ten variables pertinent to achievement have been selected for study. The basic notion under test is that the variables discussed in the preceding section, henceforth called "subjective" variables, account for more of the variance of school achievement than so-called school variables, race, social class, and teacher characteristics, henceforth called "objective" variables. Only one other variable, intelligence, will account for more achievement variance than the subjective variables. Further theoretical details will be given in subsequent discussion.

	Independent Variables		Dependent Variable
I	**II**	**III**	
X_1: Intelligence	X_3: Race	X_7: Self-Concept	
X_2: Social Class	X_4: Home Background	X_8: Need for Achievement	Y: Verbal Achievement
	X_5: School Quality	X_9: Level of Aspiration	
	X_6: Teacher Characteristics	X_{10}: Reflected Appraisal	

FIGURE 17.1

In the figure, there are three levels of variables. The first level, I, can be called control variables. While there may be interesting relations between intelligence and social class and the other independent variables in their impact on verbal achievement (VA), the dependent variable, we are here concerned primarily with the variables of Levels II and III and their impact on VA.

Level II consists of background variables, each of which should have some influence on VA. Three of these variables, home background, school quality, and teacher characteristics, are conceived to be single variables distilled from a number of other variables. For example, home background is some sort of composite of several indices of the social, cultural, and educational level of the home. We believe that all three of these variables have some influence on VA; we are most interested in school quality, however. Is it possible, for example, that school quality has less influence than the subjective variables? The Coleman Report suggests this, but we are not sure. The variables of Levels I and II are, with the exception of intelligence, the "objective" variables discussed earlier. They are "objective" in the sense that their measurement is based on relatively objective indices. The remaining background variable, race, may be quite important in the theory. We could have conceived of race as a control variable, but we thought it better to treat it as a background variable. We will return to this point later.

The variables of Level III are the most interesting and pertinent in this particular formulation. They are the "subjective" variables, the psychological variables that are presumed to have the most influence on the dependent variable, VA. We are particularly interested in self-concept and reflected appraisal because, in the theory, they offer a key explanation of the pupil's perception and judgment of himself and his performance. This was explained earlier.

On the basis of knowledge, the theory outlined above, and conjecture, we "manufactured" the correlation matrix of Table 17.1. For example, we knew that intelligence almost always correlates highly with verbal achievement. So we "assigned" a correlation of .60 between intelligence and VA. But how about the correlations between intelligence and the other nine variables? We know that intelligence is correlated with social class and race, but not too substantially. We assigned an r of .30 to both relations. On the other hand, we thought that there would be hardly any correlation between intelligence and teacher characteristics (or quality). One can of course reason differently than we did and say that there *is* a positive correlation between the intelligence of the pupils and the characteristics or quality of teachers. We chose not to. Further, we assigned r's of .04 and .07 to intelligence and self-concept and intelligence and reflected appraisal. Having little or no basis for knowing what such correlations might be, we assigned values close to zero. In some cases, we simply guessed.

In a matrix of 55 correlations our estimates must sometimes, perhaps often, be in error. Nevertheless, we tried to be realistic and still load the R matrix in favor of the hypothesis. That we missed to some extent is attested by a low negative beta weight attached to social class (see below). Social class may have a low beta weight but it should be positive. The results of the multiple regression analysis to be reported below, however, are surprising: in general we hit the mark and are able to illustrate what we want to illustrate with this completely synthetic R matrix.

We did a multiple regression analysis of the data of Table 17.1, entering the variables as they are designated in the table, that is, X_1, X_2, and so on. The variables were so entered to test the theory as outlined above. Since the variables of Level I have a control function, they were entered first. The other objective variables, race and the variables associated with schools, were entered next. Do they add anything to the variance of Y, verbal achievement, after intelligence and social class? The subjective or psychological variables were entered last. What effect do they have on VA after the control and objective variables have been entered? One can of course study the regression of VA on the subjective variables alone, but we felt it more realistic—and more interesting—to embed these variables in a social and school context, as they are usually embedded in real school situations. Perhaps more important, one wants to know the effect of these variables collectively and individually *after* the other variables have entered the regression equation because logically and psychologically they operate after the other variables.

The multiple correlation coefficient was .8164 and its square was .6665.

TABLE **17.1**   SYNTHETIC CORRELATION MATRIX: TEST OF A THEORY OF
VERBAL ACHIEVEMENT[a]

		I		II				III				
		1	2	3	4	5	6	7	8	9	10	11
I	1	1.00	.30	.30	.15	.12	.05	.04	.15	.10	.07	.60
	2	.30	1.00	.40	.30	.20	.14	.20	.22	.24	.30	.30
II	3	.30	.40	1.00	.25	.18	.05	.15	.30	.29	.30	.35
	4	.15	.30	.25	1.00	.21	.07	.09	.14	.21	.08	.18
	5	.12	.20	.18	.21	1.00	.22	.01	.05	.11	.07	.15
	6	.05	.14	.05	.07	.22	1.00	.06	.12	.10	.07	.15
III	7	.04	.20	.15	.09	.01	.06	1.00	.36	.34	.12	.46
	8	.15	.22	.30	.14	.05	.12	.36	1.00	.37	.35	.36
	9	.10	.24	.29	.21	.11	.10	.34	.37	1.00	.25	.35
	10	.07	.30	.30	.08	.07	.07	.12	.35	.25	1.00	.40
	11	.60	.30	.35	.18	.15	.15	.46	.36	.35	.40	1.00

[a]I: control variables: X_1 = intelligence, X_2 = social class; II: objective variables: X_3 = race, X_4 = home background, X_5 = school quality, X_6 = teacher characteristics; III: subjective variables: X_7 = self-concept, X_8 = need for achievement, X_9 = level of aspiration, X_{10} = reflected appraisal; dependent variable: $X_{11} = Y$ = verbal achievement. $N = 1200$.

The F ratio was 237.616, which, at 10 and 1189 degrees of freedom (we set the sample size at 1200), is highly significant. Approximately 67 percent of the variance of verbal achievement is accounted for by the ten variables, a respectable portion of the dependent variable variance. If these were real research results, they would be gratifying indeed.

We need to go deeper into the results to test the hypothesis and to examine the contributions of the different independent variables. Bear in mind, however, that the analysis must necessarily be limited because we do not have all the results possible with analysis that starts with raw data. The full regression equation with betas as the regression coefficients is

$$Y' = .5622X_1 - .0935X_2 + .0286X_3 + .0267X_4 + .0405X_5 + .0703X_6 + .3796X_7 - .0014X_8 + .0864X_9 + .3036X_{10}$$

If we take this equation and the beta weights at face value, the hypothesis is supported. It is wise, however, to set down and study the squared semipartial correlations as we did earlier with other data. This is probably best done in a table so that we can see the betas and the SP^2's together. They are given in the first two data rows of Table 17.2. The third row of indices are the partial correlations. That is, each such correlation is an estimate of the correlation between that variable and the dependent variable with the other nine independent variables controlled

First, we should know which variables' contributions were statistically significant. Successive F tests (or t tests) were calculated. These amounted to testing the statistical significance of the increment added to the variance by any variable *in the given order of variables*. For example, the F ratio to test the

TABLE **17.2** BETA WEIGHTS, SQUARED SEMIPARTIAL CORRELATIONS,
AND PARTIAL CORRELATIONS, DATA OF TABLE **17.1**[a]

	1	2	3	4	5	6	7	8	9	10
β	.56	−.09	.03	.03	.04	.07	.38	−.00	.09	.30
SP^2	.36	.02	.02	.00	.00	.01	.17	.01	.01	.07
PC	.67	−.13	.04	.04	.07	.12	.51	−.00	.13	.42

[a]SP^2: Squared Semipartial Correlations; PC: Partial Correlations. $R^2 = .67$.

statistical significance of the variance increment added by variable 3, .02, was 41.078, which, at 1 and 1196 degrees of freedom, is highly significant. These F tests, of course, amount to testing the statistical significance of each of the squared semipartial correlations. All the increments are statistically significant except variables 4 and 5, home background and school quality. Evidently these two variables contribute nothing to VA after intelligence, social class, and race are taken into account.[1]

First, study the beta weights. Variables 1, 7, and 10 stand out: .56, .38, and .30. Moreover, they are supported by the SP^2's: .36, .17, and .07. The substantial contribution of variable 1, intelligence, was expected. The substantial contributions of variables 7 and 10, self-concept and reflected appraisal, were hypothesized. But we also thought that variables 8 and 9 would contribute more than they did. The SP^2's of 8 and 9, however, are only .01 and .01. The betas of −.00 (actually −.0014) and .09 are congruent with the SP^2's. Taking the first six independent variables into account, then, the subjective variables, self-concept and reflected appraisal, are evidently important in the prediction and explanation of verbal achievement. The presumably related subjective variables, need for achievement (X_8) and level of aspiration (X_9), although correlated with verbal achievement .36 and .35, had little or no effect: beta weights of .00 and .09 and SP^2's of .01 and .01. An estimate of the total effect of the subjective variables (7, 8, 9, and 10) on verbal achievement can be obtained by adding the four increments, or SP^2's: .26.

If we now examine the objective variables (3, 4, 5, and 6) — we can also include variable 2, social class, although it was conceived as a "control" variable in this analysis — we find low betas of .03, .03, .04, and .07 and SP^2's of .02, .00, .00, and .01. The total of the SP^2's is only .03.[2]

The hypothesis that the subjective variables account for more of the variance of verbal achievement than the objective variables is supported. Moreover, what we earlier called the key variables of the theory, self-concept and

---

[1]These results are a bit unfortunate, just as the negative beta weight for variable 2, social class, is unfortunate. We wanted all the objective variables to contribute a small but statistically significant amount to the variance of Y. This shows how difficult it is to synthesize data adequately. Although there are certain other defects in our data, the results are in general satisfactory and we will interpret them as though there were no such inadequacies.

[2]To supplement what was said before, especially in footnote 1, such results are unlikely. While we would expect that these variables, on the basis of the results of the Coleman Report, would have lower values than the subjective variables, they would hardly be this low. In other words, in constructing the R matrix to support the hypothesis we overdid it.

reflected appraisal, account for more of the variance of verbal achievement (24 percent) than any of the other variables or any combination of them, except intelligence. These conclusions are also supported by the partial correlations reported in the third data line of Table 17.2. (Of course, the beta weights discussed above are also partial indices.) The partial correlation of intelligence (X_1) with verbal achievement, controlling all the other independent variables, is .67. The partial correlations of self-concept and reflected appraisal $(X_7$ and $X_{10})$ with verbal achievement, controlling the other independent variables including intelligence, are .51 and .42. In this case, then, the three kinds of indices are consistent and point to the same conclusions.

Even though our synthetic data may be here and there unrealistic, they are adequate to illustrate how multiple regression analysis can be used to test theoretical notions. We are not of course saying that such an analysis would be a definitive test of the theory. One might well use path analysis, as was done in Chapter 11, to trace the influence of the variables. One would like to explore the interactions of certain of the independent variables in their influence on the dependent variable. One would want to do regression analyses of different social classes and races. Nevertheless, the approach to testing theoretical ideas should be clear. In this example, the "correct" order of entry of the independent variables was fairly obvious. There is little doubt that intelligence should be the first to enter the regression equation. If the theory spoke of an interaction between intelligence and other variables, however, then one might need to change the order and to insert product vectors of intelligence and other variables.[3] The background variables (the objective variables) should enter the equation before the subjective variables because one assumes that they lie behind, so to speak, the subjective variables. Again, however, one or more interactions might be important. Does the effect of reflected appraisal on verbal achievement, for instance, differ depending on race?

Ideally, an experimental approach should also be used. In our theory, reflected appraisal might be manipulated. We might give different groups of pupils who had been randomly assigned to experimental groups differing feedback on how teachers view them and their achievement. One can conceive of a factorial experimental design in which the manipulated independent variables might be reflected appraisal and need for achievement (Kolb, 1965; McClelland, 1965). In such an experiment, or better, series of experiments, one can of course include only the experimental variables trusting to random assignment to control other variables. But one can also include other variables with the experimental variables and get the benefit of both random assignment and the control and information attainable by adding control, background, and subjective measured variables. Such considerations lead us to other possibilities.

Another Example

Suppose that an educational-psychological researcher contemplates an experiment in which reflected appraisal (RA) and need for achievement (NA) will

---

[3]But note our admonitions on the use of product vectors in Chapter 16.

be manipulated variables, as described above. He can of course do an experiment to test the effects of these variables. Now let us suppose that his theory dictates different effects of RA depending on race: white students receiving RA will exhibit greater VA (verbal achievement) than white students not receiving RA, and black students receiving RA will exhibit greater VA than black students not receiving RA, but black students receiving RA will exhibit relatively greater VA than white students. Suppose, also, that the same kind of interaction is expected with the manipulated variable NA (need for achievement). The researcher can of course do a $2 \times 2 \times 2$ factorial experiment. If the experiment is properly done, he will have adequately tested his expectations.

To continue the suppositions, however, suppose the researcher wants to include two other independent variables: intelligence and self-concept (SC); the first as a control variable (as before) and the second because of its presumed relations to RA and VA. We now have five independent variables: two manipulated variables (RA and NA), one categorical variable (race), and two continuous variables (intelligence and SC). One can construct, say, some sort of factorial-type paradigm by dichotomizing intelligence and SC and adding them to the basic $2 \times 2 \times 2$ factorial design, but the paradigm becomes awkward. One can conceivably use certain modified designs and analysis of covariance; intelligence being the covariate. Such devices are not really necessary, however. Instead, multiple regression analysis can be used and a single equation written to express the expectations of the researcher.

In such situations, analysis of variance paradigms can be used in connection with regression equations. The paradigms have value as conceptual devices. One can see the relations, as it were. One sees the experimental variables in a way that one does not when one uses only multiple regression equations. Moreover, the analysis and study of the means in the cells is invaluable.[4] With experiments it can even be said that the study of the means is the fundamental mode of analysis (Glass & Hakstian, 1969). The actual model for basic conceptualization and analysis, however, is the multiple regression equation.

To facilitate the grasp of the whole problem and how it can be handled we lay out the regression equation and the variables in Table 17.3. Intelligence, as a control variable, can be handled as it was in the previous example: simply entered first in the equation and the effects of the other variables assessed after the effect of intelligence has been assessed. Another method is to calculate the regression of verbal achievement on intelligence alone and then to calculate the residual Y scores. In other words, these residuals presumably contain all sources of variance other than intelligence; they are dependent variable measures purged of the effect of intelligence. Whichever way is chosen, the remainder of the analysis proceeds as usual. Regression weights, R^2's, SP^2's, and t and F ratios are calculated and interpreted as usual but with special attention given to the tests of the hypotheses. To study the possible interaction between race and reflected appraisal mentioned earlier, a cross-product term, in this case X_2X_5, can be used.

---

[4]The detailed study of cell means after regression analysis was discussed in Part II.

TABLE **17.3**   VARIABLES AND REGRESSION TERMS AND EQUATION
IN FICTITIOUS RESEARCH PROBLEM

X_1 = Intelligence (Continuous)
X_2 = Race (Dummy: coded 1,0)
X_3 = Self-Concept (Continuous)
X_4 = Need for Achievement (Experimental)
X_5 = Reflected Appraisal (Experimental)
X_6 = Interaction: Race and Reflected Appraisal (X_2X_5)
Y = Verbal Achievement (Dependent Variable)

$$Y' = a + b_1X_1 + b_2X_2 + b_3X_3 + b_4X_4 + b_5X_5 + b_6X_6$$

A severe danger of using multiple regression analysis and planning is that it is so easy to include variables that one is inclined to do so without sufficient thought and care. We are most concerned, therefore, that in what we think is our justifiable enthusiasm for multiple regression we do not give the impression that basic ideas of research design and careful hypothesis-testing in a theoretical framework are in the least derogated. To the contrary, they are even more important because of the ease and facility of operating in the multiple regression framework. Perhaps a bit monotonously and sententiously, we again say that the research problem and its theoretical framework are the most important considerations in research planning and execution. Their demands are even more pressing in an ordinary analysis of variance or other experimental framework where one can say that healthy constraints and parsimony are embedded in the intertwined design paradigms and analysis systems.

Strengths and Weaknesses of Multiple Regression Analysis

In our discussions throughout this book we have certainly brought out the strengths of multiple regression analysis. And occasionally we have mentioned weaknesses. A systematic recapitulation of the points previously made and the presentation of one or two new points are in order, as are advice and admonition. Since our purpose is summation more than exposition, and since some of the points have been made in considerable detail earlier, we only mention most of them without much elaboration.

Weaknesses of Multiple Regression

There are five or six weaknesses of multiple regression analysis that make its use difficult. The first two of these are the tendency of researchers to throw variables indiscriminately into the multiple regression pot and thus let the method and the computer do one's thinking and to obscure research design paradigms by depending completely on multiple regression equations and related statistics. We want to comment on the former tendency. The practice of throwing many variables into the research pot is still with us, although not as much as it used to be. This is, in effect, a shotgun approach: shoot enough shot often enough and you are bound to hit something. Give many tests and scales to a group of individuals and you are bound to get some significant, maybe even substantial, correlations. Such an approach is rarely justified. It is based on naive and false assumptions on what research is and should be. The student may ask: How about the Coleman (1966) study in which 60 and more measures were correlated with measures of achievement? To be sure, the study had a bit of a shotgun flavor about it, but most of the measures were chosen for good reason. It was certainly not a blatant shotgun approach.

In general, before using many variables in multiple regression analysis, some attempt should be made to reduce the number through theory and factor analysis. How much more compelling and convincing it is to have 5 instead of 15 or 50 variables in a multiple regression analysis! One thinks of the wisdom of the study of organizational climates and administrative performance of Frederiksen et al. (1966) in which the number of the In-Basket Test variables was reduced from 55 to 11 through factor analysis. (We even wonder if there might be fewer than 11 factors. The method of factor rotation used tends to spread factor variance excessively.)

A serious weakness of multiple regression analysis is what can be called the unreliability of regression weights. With large samples and relatively few independent variables, the problem is not severe. For example, if a beta weight was .40, R^2 was .60, and the R^2 between the variable in question and the other independent variables were .30, the standard error of the beta weight would be about .08 in a sample of 100 with five independent variables. With a sample of only 20, however, the standard error would be about .19. Clearly there can be considerable fluctuation in beta weights even in samples of 100 and certainly in small samples. Moreover, when variables are added to or subtracted from a problem, all the regression weights change. There is nothing absolute or fixed, then, about regression weights.

What can be done either to make regression weights more dependable or to have some fairly clear notion of how much they are likely to fluctuate? We have already given one piece of advice by implication: use large samples. While multiple regression may not need the very large samples that factor analysis does, it needs samples of sufficient size to keep standard errors small. A sample size of 100 will yield a standard error of beta weights of about .08 (with the figures given above). A sample size of 200 cuts this down to about .05. This means that the beta weights obtained from samples of 200 will probably not fluctuate more than .10. A sample size of 40, however, will yield fluctuations as high as .25. If the independent variables are highly correlated among themselves, then the beta weights are less reliable. The rule, then, is to use large samples — over 100 and preferably 200 or more — and independent variables whose intercorrelations are as low as possible.

This rule makes it clear that the practice of throwing variables into an analysis may be expensive in reliability of regression weights. If variables are redundant, that is, if they tend to measure the same things, then their intercorrelations will be high. This increases the standard errors of the regression weights. (Note that in experiments, where orthogonality of independent variables is possible, regression weights are more reliable.) The more important consideration, however, is size of sample. Lest the reader conclude that regression coefficients are untrustworthy, that they should not be used, or at least not interpreted, and that multiple regression analysis is therefore suspect, he should know that there is more to the story. We will discuss strengths of regression coefficients later.

Another weakness of multiple regression is the changing nature of squared semipartial correlations with different orders of entry of independent variables in the regression equation. It would be nice if there was a foolproof way to calculate the contributions of the independent variables to the variance of the dependent variable. But there is no such foolproof method (Darlington, 1968). The virtue of the squared semipartial correlations is that their meaning is unambiguous: they are the differences between "adjacent" R^2's. Nevertheless, it is annoying to a researcher to realize that a variable added to a regression analysis may yield a substantial SP^2 with one order of entry, but that it can drop sharply with another order. We give the advice we gave earlier: Always try to enter variables according to the dictates of the theory and the research problem. If the problem and the theory behind the problem dictate the order of entry or even the approximate order of entry, then there is little difficulty.

The last weakness of multiple regression analysis to be discussed was just mentioned in connection with the changing nature of SP^2's with different orders of entry of independent variables. This is that there is no one and only way to estimate the presumed "importance" of independent variables to the variance of the dependent variable. Contrast the usual multiple regression situation with an experimental analysis of variance of the factorial kind. Since the various effects are orthogonal—provided, of course, that subjects have been assigned to cells at random—interpretation of the results is straightforward and relatively unambiguous: the SP^2's, like the beta weights, will be the same no matter what the order of entry of the variables. One can have some confidence in the estimates of variance accounted for.[5] In the usual multiple regression analysis, on the other hand, the independent variables are correlated causing the complications we have been struggling with. But these are the complications of the real psychological, sociological, economic, political scientific, and educational worlds. In real life, independent variables, the p's of our If p, then q propositions, are correlated. And much, perhaps most, research in the behavioral sciences has to be ex post facto in nature, as we said before, and the independent and dependent variables consequently have a messiness

---

[5]Estimates of variance accounted for do not necessarily give information on "importance" and "significance" of variables, although they may do so. While the proportion of variance is probably the best index in most research, it can be misleading in some research. See, for example, the Jones study in Chapter 16, Table 16.3, where the theoretically most important independent variable, Ascending-Descending, accounted for less variance than sex.

about them that controlled experimentation does not have. Unfortunately, such messiness is, in essence, inescapable, although there are ways to clean it up to some extent by judicious use of methods discussed earlier.

Strengths of Multiple Regression Analysis

Now look at the other side of the coin. That multiple regression has great strengths should by now be monotonously obvious. Let us try to be clear about them in this final attempt at balanced appraisal. Perhaps the most important strength of the multiple regression approach is that it is closely related to the basic purpose of science, the explanation of natural phenomena. In most basic research at least, the major effort is directed toward explaining a single phenomenon, although the phenomenon may be complex and have various facets (for example, aggression, authoritarianism, or achievement). Thus the single dependent variable aspect of multiple regression fits this scientific preoccupation. In applied research, where the emphasis is more on prediction, it is still single phenomena that are predicted, although one can of course predict successively different dependent variables. The feature of multiple regression that ties it closely to the explanation of phenomena, however, is its multiple independent variables. This strength is so obvious by now we need say no more about it.

The nature of multiple regression equations also reflects the close relation between the method and scientific research. The scientific enterprise is expressed succinctly, parsimoniously, and elegantly by multiple regression equations. Moreover, such equations express the logic of scientific inquiry in the sense that human logical reasoning and inference are based on conditional statements of the If p, then q kind. Multiple regression equations in effect say: If $X_1, X_2, \ldots$, then Y.

Another strength of multiple regression is its ability to handle any number and kind of independent variables. While we have extensively discussed and illustrated this strength by discussing kinds of variables and how to handle them, the point can stand repetition because it is still not generally appreciated by researchers in the behavioral sciences.

No elaboration is needed, either, of another strength: multiple regression analysis can do everything that analysis of variance can do—and more. This strength, too, like its ability to handle different kinds of independent variables, is either not known or is insufficiently appreciated.

The next strength has also not been appreciated as much as it should be by behavioral researchers: multiple regression is often the best method of analysis of nonexperimental data. Although this statement can be challenged, especially if we make it without qualification, we think it is generally accurate. Some investigators may say that much behavioral data can and should be analyzed with frequency and percentage crossbreak analysis, and we agree with them. Unfortunately, such analysis is limited. The most it can intelligibly do is to present relations among three variables at a time, and such three-way crossbreaks are difficult to grasp and interpret. In short, while a viable and useful, even indispensable, form of analysis, it can miss the mark in much research because it cannot answer certain important questions except in a roundabout and often clumsy way. A good example of what we mean has already been cited: the excellent Free and Cantril (1967) study of the political beliefs of Americans (see Chapter 13). One wonders what might have been done in several of the larger and better studies of important theoretical and practical problems had a thorough-going multivariate attitude and approach been used.

To get back to the point, multiple regression analysis is suited to almost any nonexperimental research in which there are several independent variables and one dependent variable (or one dependent variable at a time). No matter what the scales of measurement or what the kind of variable, useful analysis can be done and interpretations made. Of course, if scales of measurement differ widely — for example, if continuous measures are mixed with dichotomous measures, or if interval measures are mixed with ordinal and nominal measures, then there will be difficulty in interpretation because of the different meanings of the different measures and the mixing of kinds of correlation coefficients. Nevertheless, with adequate knowledge and care, valuable results that are unobtainable with crossbreak and univariate analysis can often be gotten with multiple regression analysis of nonexperimental data. Again, we are not advocating mindless throwing together of widely disparate variables nor casting aside conventional kinds of analysis. We *are* advocating expanding such analysis in a multiple regression fashion, with care and circumspection, to obtain better answers to research questions.

Multiple regression opens up research possibilities not available, or at least not generally and readily available, in the past. This strength is of course related to the strengths already discussed. For example, take the inclusion of variables by coding. In Astin's (1968) study of institutional quality and undergraduate achievement, a number of variables — type of institution, type of control, and intended field of study — were used in the analysis by creating dummy variables with 1 and 0 scoring. Astin also tested interaction hypotheses by multiplying variable vectors and treating the products as separate variables. Such possibilities, still relatively new and untried, should help to enrich behavioral research.

A final strength of multiple regression is its rich yield of various statistics to be used in the interpretation of data. First, measures of the overall relation between the independent variables and the dependent variable, R^2, an estimate of the proportion of variance accounted for by all the variables or any subset of them, and F tests of the statistical significance of different R^2's are routinely produced. Second, regression coefficients, both b's and β's, are calculated, with t or F tests of their statistical significance. Third, auxiliary measures to aid interpretation are calculated: squared semipartial correlations and partial correlations. That interpretation is not easy, as we have repeatedly pointed out, does not alter this rather remarkable wealth of statistical resources.

The Reliability of Multiple Regression
Results: Replication

The reliability of the results of multiple regression analysis is clearly a major problem of the method. The tendency of regression coefficients to change with different samples, to have rather large standard errors, to change with different numbers of independent variables are brute facts of multiple regression and its use. Reliability of results, then, is a central difficulty. We want to know how much regression weights will fluctuate with different samples at different times. How much dependence can we put on the set of regression weights of this study? Similar questions must of course be asked of other regression statistics, like R^2's and squared semipartial correlations, but the question is probably most important and difficult with regression coefficients. Much of the discussion of this section is tentative: to our knowledge the answers to these questions are not fixed or definite. While a good deal is "known" theoretically about regression statistics, there appears to have been few systematic studies of their reliability in actual research. We focus what we say, therefore, on what we think researchers should do, when possible, to improve their estimates and assess the reliability of regression statistics.

Regression and Replication

Regression coefficients are perhaps the nearest that scientists get to causal indices. Blalock (1964, pp. 51, 87) even says that it is regression coefficients that give us the laws of science and that they are to be used when attention is focused on causal laws. If we could take them at face value a regression coefficient indicates the change in the dependent variable with a change of one unit in the independent variable. With standard scores, this means that with a change of one standard deviation in the independent variable the dependent variable will change β standard deviations. In multiple regression analysis, as we know, it is not this simple. Errors of measurement and correlations between the independent variables cloud the picture. Still, the basic idea is that regression weights help "explain" the dependent variable. Other things being equal, the larger a regression weight the greater is its variable's contribution to the dependent variable. This "greater" contribution does not always mean "most important." It is conceivable that a variable that has a moderate or even relatively small weight may be the "most important." In the Jones et al. study, the theoretically most important main effect, Ascending-Descending, had the lowest of the three main effect beta weights. In general and other things equal, however, magnitude of regression weights, especially beta weights which are generally comparable, usually indicates "importance," "significance," or contribution to the dependent variable.

The best advice we have to ensure the reliability of regression statistics has been given before: use as large and as representative samples as possible and replicate research studies. Any multiple regression analysis, and especially those with many independent variables, should have at least 100 subjects, preferably 200 or more. This does not mean that thousands are needed, it can even be said to be undesirable to have huge samples because very low correlations and very slight differences have a greater probability of statistical significance since many or most statistical tests of significance are in part a function of the sample size (Hays, 1963, p. 333). In any case, the larger the sample size the more precise the statistical estimate. The two most important regression statistics, R^2 and β (or b), are in most cases biased estimates: R^2 is overestimated and β is either over- or underestimated. The larger the sample the less the bias of both statistics.

Perhaps as important as size of sample is replication. A good rule to follow in both experimental and nonexperimental research is: Always replicate research studies with different samples in different places. If it can be arranged, do your study over in another state, or better, two other states. At least try to get over 100 miles away. If the study calls for elementary school teachers, use three samples of such teachers in, say, New York, South Carolina, and Michigan. If the nature of the research is such that specific kinds of people are not needed—almost any category of person will do—then use different kinds of people in the replicated samples. Relations and statistics that hold up under replication, especially with different kinds of subjects, can be trusted much more than the relations and statistics of only one study. With six independent variables, say, the probability of obtaining the same pattern of relations among regression coefficients in three different samples in three different places is quite low. (Naturally, if one does not obtain the same or similar pattern in the replications one is in trouble.)

Replication is a broader word than repetition. While it can mean repetition with an attempt to duplicate the study as closely as possible, it always can mean "duplication" of a study with changes of minor details. Using different kinds of subjects is such a change. Another change is the addition or deletion of variables. Of course, the more changes the less likely that a repeated study is truly a replication. Replication is associated with what has been called external validity and generalizability (Campbell, 1957; Campbell & Stanley, 1963). Are the multiple regression results obtained general? Do they apply to other similar populations or samples? Will this pattern of regression coefficients obtained from a sample in California be the same or similar to that obtained with a sample in Louisiana? If three factors to be used in regression analysis have been found with such-and-such attitude items, will another set of similar attitude items yield virtually the "same" factors?

Replication also bears on internal validity. Internal validity has been defined for experiments (Campbell, 1957): Did the experimental treatments in fact make a difference in this particular case? This definition is a bit restricted. Actually, internal validity means the adequacy of a study's design and execution to estimate the relations of a study accurately and without spuriousness. So viewed, internal validity, like external validity, applies to both experimental and ex post facto studies. Research seeks empirical evidence and brings it to bear on conditional statements of the If p, then q kind. If we find that when p

varies q also varies, as predicted, and the criteria of research design have been satisfied, then we can say that the study is internally valid. Even if there is a zero relation between p and q, we can, if the design criteria have been met, say that the study is internally valid. In other words, internal validity refers to the controlled conditions of the study and not to its results as such. If a study is internally valid, the relations found can be trusted. We "know," or rather, have considerable reason to believe, that this q varies because of this p and not because of other p's.

This definition of internal validity makes two or three things clear. One, the criteria of scientific research must be applied as much as possible to nonexperimental as well as to experimental studies. Indeed, internal validity questions, particularly ones about alternative hypotheses or alternative independent variables, have to be asked even more consciously and systematically with nonexperimental studies than with experimental studies in which random assignment has been used.

Two, estimates of regression weights and other statistics of multiple regression are not simple. It is not enough to talk about the accuracy of estimates; we must talk about their "validity." Suppose we find in a nonexperimental study with three independent variables the following regression equation (with beta weights):

$$z'_y = .30z_1 + .60z_2 - .10z_3$$

We have to ask such questions as: Will the relative sizes of the beta weights be about the same, within sampling fluctuations, if we add one or two more variables? Will testing alternative hypotheses upset the pattern and the relations? These questions and others like them are internal validity questions. Replication of studies bears on such questions because it is through replication (in the broad sense) that we can obtain answers to the questions.

Conclusion: Multiple Regression, Theory, and Scientific Research

The various aspects of research go together. One is unthinkable without the others. When the researcher makes observations and gathers data, he must have in mind an overall design, the kind of measurement to use, and, most important for our purpose, the analysis of the data. We said earlier that design is data discipline. We now want to add to this: analysis is data reduction and clarification. The technical aspects of research are inextricably linked together because they all have the same purpose: to bring controlled evidence to bear on the relations of the research. It should be clear that the technical aspects of research strongly influence each other as well as research problems. Investigators literally do not think of problems whose variables cannot be measured. The scientific study of repression simply has not progressed because we have had no valid and reliable way to measure repression, at least repression in the Freudian sense. Similarly, without the availability of multiple regression analysis, investigators did not really think seriously of several independent variables mutually influencing a dependent variable.

We have said again and again that the substantive aspects of research are more important than the technical aspects. And we reaffirm this seemingly obvious point. But we hasten to add that without adequate technical means of observation, measurement, and analysis the substantive aspects of research remain belief and mythology. Unless the hypothesis that aggression, frustration, and anti-Semitism are related in an interactive manner can be tested empirically, the notion that anti-Semites, under conditions of frustration (hostility arousal), will show more displaced aggression than less anti-Semitic subjects remains a belief that may or may not be true. Berkowitz (1959), to test this hypothesis, had to have the technical means at his disposal to measure anti-Semitism and displaced aggression, as well as to manipulate hostility arousal. Just as important, he had to have the technical means of analysis, in this case factorial analysis of variance, at his disposal to be able to test what is essentially an interaction hypothesis. In short, the theoretical reasoning that led to the hypothesis (Dollard et al., 1939) is scientifically empty without the technical means of testing the implications of the theory.

Analysis breaks down large, complex, and even incomprehensible sets of data into units, patterns, and indices that are comprehensible and capable of being applied to the research problems under study. Most important, unlike the original raw data, the products of analysis are interpretable. To examine the scores of 100 subjects on five variables is, to say the least, bewildering. To examine the means, standard deviations, and correlations among the variables is much more comprehensible. And, if one of the five variables is a dependent variable that we seek to explain, then multiple regression analysis results are still more comprehensible. In other words, analysis is used to bare the underlying relations in masses of data and, in so doing, to obtain answers to research questions.

The main point of all this argument is that we seek to draw inferences about relations among variables from the data. We seek, in other words, to interpret the empirical evidence. The main question, then, is how best to analyze the data so that inferences can be reliably and validly made. Interpretation means to make inferences pertinent to the research relations studied and to draw conclusions about the relations on the basis of the results of analysis. It can be said that interpretation is the purpose of analysis; all analysis leads to interpretation, to inferences and conclusions about the relations under study. We want, for example, to be able to say, with a hopefully high degree of confidence, that the statement If $p_1, p_2, \ldots, p_k$, then q is empirically valid.

We have tried to show that multiple regression analysis is particularly suited to much, perhaps most, research of a nonexperimental nature. In studies such as *Equality of Educational Opportunity* (Coleman et al., 1966), *Organizational Climates and Administrative Performance* (Frederiksen et al., 1968), and Astin's (1968) exploration of institutional excellence and undergraduate achievement, there can be no doubt whatever of the demand and need for multivariate techniques, and particularly for multiple regression analysis. There is no adequate way to interpret the complex data and to infer what has to be inferred without multivariate analysis. But we hope we have also showed, beyond caviling doubt, that experimental data can also be handled with multiple regression analysis, although it is often not necessary to use such analysis. It is desirable, however, to use multiple regression analysis either when the variables of a study are all nonexperimental or when they are a mixture of experimental and nonexperimental.

The basic question is not whether this method or that method should be used. Such questions, while important, are essentially trivial compared to the larger questions of the development and testing of theory. Thus a more important question is: Which methods — of observation, measurement, analysis — help the development and testing of theory? The answer to this question gives the most compelling — we are almost inclined to say "overwhelmingly compelling" — reason for the intelligent use of multiple regression and related analyses. There is a great need in the behavioral sciences for the development of theory and the precise statement of theory (Kemeny, 1959, Chapter 15). The physical sciences progressed from verbal statements and manipulations to mathematical statements and manipulations (*ibid.*, p. 257). They thus progressed enormously. Just so, the behavioral sciences will progress from their present largely verbal level to more abstract and mathematical levels in the statements of their theories.

Multiple regression analysis is of course part of mathematics. Already we have ample evidence of the power of its formulations for solving behavioral research problems of almost frightening complexity. In educational research, for instance, the study of the behavior of pupils, teachers, and administrators of schools is itself unbelievably complex. The plethora of possible variables bewilders one. Yet techniques such as factor analysis and multiple regression analysis are helping us attack the complexity in order to understand educational processes; they have already made substantial inroads into this forbidding territory. If we have not succeeded in showing this, we have failed in what we set out to do. The next steps are the continuation of the inroads, the exploitation of the power of the methods, and, above all, the development in more precise form of theoretical explanations of psychological, sociological, and educational processes. The multiple regression approach, as well as multiple regression analysis itself, is a definite and clear way to formulate research problems and to help develop and test theory.

We can do nothing better, we think, than to conclude our discussion with some words from a distinguished philosopher of science, Braithwaite (1953):

Man proposes a system of hypotheses: Nature disposes of its truth or falsity. Man invents a scientific system, and then discovers whether or not it accords with observed fact . . . . The function of mathematics in science has been shown to be . . . that of providing a variety of methods for arranging hypotheses in a system; knowledge of new branches of mathematics opens up new possibilities for the construction of such systems (p. 368).

Study Suggestions

1. What is a theory? How can multiple regression analysis be used to help test theory?
2. Discuss the strengths and weaknesses of multiple regression analysis. Use examples of actual research in your discussion.
3. Why are researchers advised to use large samples when using multiple regression analysis (and factor analysis and other forms of multivariate analysis)? Be as precise as possible in your answer. Give an example. Show specifically what happens in multiple regression analysis with small and large samples.
4. Consider order of entry of variables in regression equations. Which regression statistics stay the same no matter what the order of entry? Which statistics change with different orders of entry? What implications do these changes (or lack of changes) have for research and multiple regression?
5. In the text, it was said that multiple regression is closely related to the basic purpose of science, the explanation of natural phenomena. What does this statement mean? Give an example.
6. It was also said in the text that multiple regression analysis can do all the analysis of variance can do—and more. Defend this statement. Be explicit in your defense.
7. Why is replication important in scientific research? Couch your answer in the context of multiple regression analysis. Does "replication" mean "repetition" of research studies?

Appendixes

Matrix Algebra in Multiple Regression Analysis

Matrix algebra is one of the most useful and powerful branches of mathematics for conceptualizing and analyzing psychological, sociological, and educational research data. As research become more and more multivariate, the need for a compact method of expressing data becomes greater. Certain problems require that sets of equations and subscripted variables be written. In many cases the use of matrix algebra simplifies and, when familiar, clarifies the mathematics and statistics. In addition, matrix algebra notation and thinking fit in nicely with the conceptualization of computer programming and use.

This chapter provides a brief introduction to matrix algebra. The emphasis is on those aspects of the subject that can be used in multiple regression analysis. Thus many matrix algebra techniques, important and useful in other contexts, are omitted. In addition, certain important derivations and proofs are neglected.

Basic Definitions

A *matrix* is an n-by-k rectangle of numbers or symbols that stand for numbers. The order of the matrix is n by k. It is customary to designate the rows first and the columns second. That is, n is the number of rows of the matrix and k the number of columns. A 2-by-3 matrix called **A** might be

$$\mathbf{A} = \begin{matrix} & 1 & 2 & 3 \\ 1 & \\ 2 & \end{matrix} \begin{pmatrix} 4 & 7 & 5 \\ 6 & 6 & 3 \end{pmatrix}$$

(We use parentheses to indicate matrices.) This matrix may be symbolized a_{ij}, where i refers to rows and j to columns, a common way to designate rows and columns. **A** can also be written

$$\mathbf{A} = \begin{pmatrix} a_{11} & a_{12} & a_{13} \\ a_{21} & a_{22} & a_{23} \end{pmatrix}$$

The *transpose* of a matrix is obtained simply by exchanging rows and columns. In the present case, the transpose of **A**, written **A**′, is

$$\mathbf{A}' = \begin{pmatrix} 4 & 6 \\ 7 & 6 \\ 5 & 3 \end{pmatrix}$$

If $n = k$, the matrix is square. A square matrix can be symmetric or asymmetric. A *symmetric* matrix has the same elements below the diagonal as above the diagonal except that they are transposed. The diagonal is the set of elements from the upper left corner to the lower right corner. The following correlation matrix is symmetric:

$$\mathbf{R} = \begin{pmatrix} 1.00 & .70 & .30 \\ .70 & 1.00 & .40 \\ .30 & .40 & 1.00 \end{pmatrix}$$

A *vector* is an n-by-1 or 1-by-n matrix. The first row vector of **A** is

$$\mathbf{a}_1' = (4 \quad 7 \quad 5).$$

(We designate most vectors by lower-case boldface letters.) This vector can of course be expressed as a column vector:

$$\mathbf{a}_1 = \begin{pmatrix} 4 \\ 7 \\ 5 \end{pmatrix}$$

a is used for column vectors; **a**′, or **a**-prime, is used for row vectors (because **a**′ is the transpose of **a**).

A *diagonal* matrix is frequently encountered in statistical work. It is simply a matrix in which some values other than zero are in the diagonal of the matrix, from upper left to lower right, and all the remaining cells of the matrix have zeros in them. Here is a diagonal matrix:

$$\begin{pmatrix} 2.759 & 0 & 0 \\ 0 & 1.643 & 0 \\ 0 & 0 & .879 \end{pmatrix}$$

A particularly important form of a diagonal matrix is an *identity* matrix, **I**, which has all 1's in the diagonal and 0's elsewhere:

$$\mathbf{I} = \begin{pmatrix} 1 & 0 & 0 \\ 0 & 1 & 0 \\ 0 & 0 & 1 \end{pmatrix}$$

Any matrix pre- or post-multiplied by the identity matrix remains the same:

$$IA = AI = A$$

Matrix Operations

The power of matrix algebra becomes apparent when we explore the operations that are possible. The major operations are addition, subtraction, multiplication, and inversion. A large number of statistical operations can be done by knowing the basic rules of matrix algebra. We define and illustrate some but not all of these operations.

Addition and Subtraction

Two or more vectors can be added or subtracted provided they have the same number of elements. The laws of algebra are applicable. We add two vectors:

$$\begin{pmatrix} 4 \\ 3 \\ 5 \end{pmatrix} + \begin{pmatrix} 7 \\ 7 \\ 4 \end{pmatrix} = \begin{pmatrix} 11 \\ 10 \\ 9 \end{pmatrix}$$

$$\textbf{a} \qquad \textbf{b} \qquad \textbf{c}$$

Now we add two 3-by-2 matrices:

$$\begin{pmatrix} 6 & 4 \\ 5 & 6 \\ 9 & 5 \end{pmatrix} + \begin{pmatrix} 7 & 4 \\ 7 & 4 \\ 1 & 3 \end{pmatrix} = \begin{pmatrix} 13 & 8 \\ 12 & 10 \\ 10 & 8 \end{pmatrix}$$

$$\textbf{A} \qquad\quad \textbf{B} \qquad\quad \textbf{C}$$

Subtraction is equally simple. We subtract **B** from **A**:

$$\begin{pmatrix} 6 & 4 \\ 5 & 6 \\ 9 & 5 \end{pmatrix} - \begin{pmatrix} 7 & 4 \\ 7 & 4 \\ 1 & 3 \end{pmatrix} = \begin{pmatrix} -1 & 0 \\ -2 & 2 \\ 8 & 2 \end{pmatrix}$$

$$\textbf{A} \qquad\quad \textbf{B} \qquad\quad \textbf{C}$$

Multiplication

For statistical purposes multiplication of matrices is the most important operation. The basic rule is: Multiply rows by columns. An illustration is easier than verbal explanation. We want to multiply two matrices, **A** and **B**, to produce a product matrix, **C**:

$$\begin{pmatrix} 3 & 1 \\ 5 & 1 \\ 2 & 4 \end{pmatrix} \times \downarrow \begin{pmatrix} 4 & 1 & 4 \\ 5 & 6 & 2 \end{pmatrix} = \begin{pmatrix} 17 & 9 & 14 \\ 25 & 11 & 22 \\ 28 & 26 & 16 \end{pmatrix}$$

$$\textbf{A} \qquad\qquad \textbf{B} \qquad\qquad \textbf{C}$$

Following the row-by-column rule, we multiply and add as follows (follow the arrows):

$$(3)(4) + (1)(5) = 17 \qquad (3)(1) + (1)(6) = 9 \qquad (3)(4) + (1)(2) = 14$$

$$(5)(4) + (1)(5) = 25 \qquad (5)(1) + (1)(6) = 11 \qquad (5)(4) + (1)(2) = 22$$

$$(2)(4) + (4)(5) = 28 \qquad (2)(1) + (4)(6) = 26 \qquad (2)(4) + (4)(2) = 16$$

The rule, of course, is: Multiply each row element of the first matrix by each column element of the second matrix, adding the resulting products within each row and column. The only restriction is that the rows and columns to be multiplied must have the same number of elements. The number of columns of the first matrix must equal the number of rows of the second matrix. There is no restriction on the other dimensions of the matrices. Symbolically, we can multiply an n-by-k matrix and a k-by-m matrix

$$n\begin{pmatrix} \checkmark & \checkmark \\ \checkmark & \checkmark \\ \checkmark & \checkmark \\ \checkmark & \checkmark \end{pmatrix} \times k\begin{pmatrix} \checkmark & \checkmark & \checkmark \\ \checkmark & \checkmark & \checkmark \end{pmatrix} = n\begin{pmatrix} \checkmark & \checkmark & \checkmark \\ \checkmark & \checkmark & \checkmark \\ \checkmark & \checkmark & \checkmark \\ \checkmark & \checkmark & \checkmark \end{pmatrix}$$

to obtain an n-by-m matrix. Note that k, the number of columns of the first matrix, must equal k, the number of rows of the second matrix.[1] Most of the matrix calculations with which we will be concerned involve square matrices. Thus, the rule and calculations are straightforward.

Three more operations, which of course follow the matrix multiplication rule, need to be clarified. Vectors can be multiplied. For example, the multiplication of $(a_1 \ a_2 \ a_3)$ by $(b_1 \ b_2 \ b_3)$ is accomplished by

$$(a_1 \quad a_2 \quad a_3)\begin{pmatrix} b_1 \\ b_2 \\ b_3 \end{pmatrix} = a_1 b_1 + a_2 b_2 + a_3 b_3$$

Using actual numbers,

$$(4 \quad 1 \quad 3)\begin{pmatrix} 1 \\ 2 \\ 5 \end{pmatrix} = (4)(1) + (1)(2) + (3)(5) = 21$$

The product of a column vector of k elements and a row vector of k elements is

---

[1] It is useful to keep the rule in mind: The "outside" dimensions of the two matrices being multiplied become the dimensions of the product matrix. For example, if we multiply a 3-by-2 matrix and a 2-by-5 matrix, we obtain

$$(3\text{-by-}2) \times (2\text{-by-}5) = (3\text{-by-}5)$$

In symbols, as above,

$$(n\text{-by-}k) \times (k\text{-by-}m) = (n\text{-by-}m)$$

a $k \times k$ matrix:

$$\begin{pmatrix} 4 \\ 1 \\ 3 \end{pmatrix}(1 \quad 2 \quad 5) = \begin{pmatrix} 4 & 8 & 20 \\ 1 & 2 & 5 \\ 3 & 6 & 15 \end{pmatrix}$$

A matrix can be multiplied by a single number called a *scalar*. Suppose, for example, we want to calculate the mean of each of the elements of a matrix of sums. Let $N = 10$. The operation is

$$\frac{1}{10}\begin{pmatrix} 20 & 48 \\ 30 & 40 \\ 35 & 39 \end{pmatrix} = \begin{pmatrix} 2.0 & 4.8 \\ 3.0 & 4.0 \\ 3.5 & 3.9 \end{pmatrix}$$

The scalar is $1/10$. A matrix can be multiplied by a vector. The first example given below is pre-multiplication by a vector, the second is post-multiplication:

$$(6 \quad 5 \quad 2) \times \begin{pmatrix} 7 & 3 \\ 7 & 2 \\ 4 & 1 \end{pmatrix} = (85 \quad 30)$$

$$\begin{pmatrix} 7 & 7 & 4 \\ 3 & 2 & 1 \end{pmatrix} \times \begin{pmatrix} 6 \\ 5 \\ 2 \end{pmatrix} = \begin{pmatrix} 85 \\ 30 \end{pmatrix}$$

[*Note the rule:* In the latter example, (2-by-3) × (3-by-1) becomes (2-by-1).] This sort of multiplication of a matrix by a vector is done frequently in multiple regression analysis.

Matrix Inversion and the Matrix Inverse

The reader may have noted that nothing has been said about matrix division. Recall that the division of one number into another number amounts to multiplying the dividend by the reciprocal of the divisor:

$$\frac{a}{b} = \frac{1}{b}a$$

For example, $12/4 = (12)(1/4) = (12)(.25) = 3$. Analogously, in matrix algebra, instead of dividing a matrix **A** by another matrix **B** to obtain matrix **C**, we multiply **A** by the *inverse* of **B** to obtain **C**. The inverse of **B** is written $\mathbf{B}^{-1}$. Suppose, in ordinary algebra, we had $ab = c$, and wanted to find a. We would write

$$ab = c$$
$$a = c/b$$

In matrix algebra, we write

$$\mathbf{AB} = \mathbf{C},$$
$$\mathbf{A} = \mathbf{B}^{-1}\mathbf{C}$$

(Note that **C** is pre-multiplied by $\mathbf{B}^{-1}$ and not post-multiplied. In general, $\mathbf{B}^{-1}\mathbf{C} \neq \mathbf{C}\mathbf{B}^{-1}$.)

The formal definition of the inverse of a square matrix is: Given $\mathbf{A}$ and $\mathbf{B}$, two square matrices, if $\mathbf{AB} = \mathbf{I}$, then $\mathbf{A}$ is the inverse of $\mathbf{B}$. The inverse of the correlation matrix

$$\mathbf{A} = \begin{pmatrix} 1.00 & .14 & .35 \\ .14 & 1.00 & .02 \\ .35 & .02 & 1.00 \end{pmatrix}$$

is

$$\mathbf{A}^{-1} = \begin{pmatrix} 1.17 & -.16 & -.41 \\ -.16 & 1.02 & .03 \\ -.41 & .03 & 1.14 \end{pmatrix}$$

Then $\mathbf{A}^{-1}\mathbf{A} = \mathbf{I}$, or

$$\underbrace{\begin{pmatrix} 1.17 & -.16 & -.41 \\ -.16 & 1.02 & .03 \\ -.41 & .03 & 1.14 \end{pmatrix}}_{\mathbf{A}^{-1}} \underbrace{\begin{pmatrix} 1.00 & .14 & .35 \\ .14 & 1.00 & .02 \\ .35 & .02 & 1.00 \end{pmatrix}}_{\mathbf{A}} = \underbrace{\begin{pmatrix} 1.00 & 0 & 0 \\ 0 & 1.00 & 0 \\ 0 & 0 & 1.00 \end{pmatrix}}_{\mathbf{I}}$$

A difficulty that occasionally gives trouble in the actual analysis of data is that some matrices have no inverses. A so-called singular matrix, for example, has no inverse. Note the following matrix:

$$\begin{pmatrix} .70 & .30 \\ .35 & .15 \end{pmatrix}$$

Row 2 is half of row 1. If any rows or columns of a matrix can be produced from any other row or column, or combination of rows or columns (like row $1 + \text{row } 4 = \text{row } 7$), the matrix is singular. If we had a matrix of correlations among the items of a scale and, in addition, had a vector that represented the correlations between each item and the total score, the matrix would be singular. Fortunately, few actual data matrices are singular.

Another definition of the singularity of matrices uses determinants, which are defined later. For now, the determinant of the above matrix is $(.70)(.15) - (.35)(.30) = 0$. When the determinant of a matrix is zero, the matrix is singular. A singular matrix has no inverse. That is, if $\mathbf{A}$ is a square matrix and is singular, then $\mathbf{A}^{-1}$ does not exist.

Although it is possible to calculate matrix inverses with a desk calculator, it is tedious and prone to error. Besides, computer programs for calculating inverses are readily available. The subroutine, INVERT, in the multiple regression program, MULR, given in Appendix C at the end of the book, calculates inverses and determinants of matrices.

To show the usefulness of the matrix inverse in the solution of certain difficult analytic problems, we outline, algebraically, the operations involved in solving a set of simultaneous linear equations for the unknown beta weights.[2]

---

[2]The example to be used now is explained in greater detail in Chapter 4. The student should not be too concerned if he does not understand all of the discussion. Its purpose is to focus on matrix operations and not to elucidate regression theory.

Suppose we have three independent variables. The basic regression equation is

$$Y' = a + b_1 X_1 + b_2 X_2 + b_3 X_3$$

The a and the b's must be found. We find those values of the b's that minimize the sum of squares of the deviations from prediction (the residuals). The calculus is used to do this. A set of simultaneous linear equations called *normal equations* results.

A set of such normal equations, using coefficients of correlation and beta weights, with three independent variables, is as follows:

$$r_{11}\beta_1 + r_{12}\beta_2 + r_{13}\beta_3 = r_{y1}$$
$$r_{21}\beta_1 + r_{22}\beta_2 + r_{23}\beta_3 = r_{y2}$$
$$r_{31}\beta_1 + r_{32}\beta_2 + r_{33}\beta_3 = r_{y3}$$

It is easy to write the set of equations in matrices:

$$\begin{pmatrix} r_{11} & r_{12} & r_{13} \\ r_{21} & r_{22} & r_{23} \\ r_{31} & r_{32} & r_{33} \end{pmatrix} \begin{pmatrix} \beta_1 \\ \beta_2 \\ \beta_3 \end{pmatrix} = \begin{pmatrix} r_{y1} \\ r_{y2} \\ r_{y3} \end{pmatrix}$$

It is much more compact, however, to write

$$\mathbf{R}_{ij}\boldsymbol{\beta}_j = \mathbf{r}_{yj}$$

Since we know the correlations, we need only to determine the β_j. This is done by using matrix algebra:

$$\boldsymbol{\beta}_j = \mathbf{R}_{ij}^{-1}\mathbf{r}_{yj}$$

Thus, to find the betas, we must first find the inverse of $\mathbf{R}_{ij}$ and then postmultiply this inverse by the correlations between each of the independent variables with the dependent variable, $\mathbf{r}_{yj}$.

Determinants

A most important idea is that of the determinant of a matrix. A *determinant* is a certain numerical value associated with a square matrix. We indicate determinants by vertical straight lines instead of by parentheses. For example, the determinant of the matrix $\mathbf{B}$, $\det \mathbf{B}$, is written

$$\det \mathbf{B} = \det \begin{pmatrix} 4 & 2 \\ 1 & 5 \end{pmatrix} = \begin{vmatrix} 4 & 2 \\ 1 & 5 \end{vmatrix}$$

Since the calculation of determinants is complex, and since discussions are readily available (Aitken, 1956, Harman, 1967; Kemeny, Snell, & Thompson, 1966; Thurstone, 1947), they will not be discussed here except to show the reader the simplest form of a determinant. The determinant of $\mathbf{B}$, above, is calculated:

$$\det \mathbf{B} = (4)(5) - (2)(1) = 20 - 2 = 18$$

In letter symbols and subscripts, the calculation is

$$\det \mathbf{B} = b_{11}b_{22} - b_{12}b_{21}$$

where the cells of the square matrix are

$$\mathbf{B} = \begin{pmatrix} b_{11} & b_{12} \\ b_{21} & b_{22} \end{pmatrix}$$

The calculations are more complicated when k is greater than 2 (see Aitken, 1956, p. 40).

An Application of Determinants

To give the flavor of the place and usefulness of determinants, know that the square of the multiple correlation coefficient, R^2, can be calculated with determinants. The problem is to calculate the determinants. A bit more concrete taste of determinants may be given by two correlation examples. Suppose we have two correlation coefficients, r_{y1} and r_{y2}, calculated between a dependent variable, Y, and two variables, 1 and 2. The correlations are $r_{y1} = .80$ and $r_{y2} = .20$. We set up two matrices that express the two relations, but we express the matrices immediately as determinants and calculate their numerical values:

$$\begin{matrix} & 1 & y \\ 1 & \\ y \end{matrix} \begin{vmatrix} 1.00 & .80 \\ .80 & 1.00 \end{vmatrix} = (1.00)(1.00) - (.80)(.80) = .36$$

and

$$\begin{matrix} & 2 & y \\ 1 & \\ y \end{matrix} \begin{vmatrix} 1.00 & .20 \\ .20 & 1.00 \end{vmatrix} = (1.00)(1.00) - (.20)(.20) = .96$$

The two determinants are .36 and .96.

Now, let us do the usual thing to determine the percentage of variance shared by y and 1 and by y and 2; square the r's:

$$r_{y1}^2 = (.80)^2 = .64$$

$$r_{y2}^2 = (.20)^2 = .04$$

If we subtract each of these from 1.00, we obtain $1 - .64 = .36$, and $1.00 - .04 = .96$. These values are the determinants just calculated. They are $1 - r^2$, or the proportions of the variance not accounted for.

This rather simple demonstration becomes less simple and more meaningful when we have more than one independent variable. In such cases, the squared multiple correlation coefficient, R^2, which is analogous to the zero-order r^2, can be calculated using certain determinants of the correlation matrix. We will give examples later. (See Study Suggestions 2, 3, and 4 at the end of this Appendix.)

Linear Dependence and Independence and Orthogonality

Linear dependence means that one or more vectors of a matrix, rows or columns, are a linear combination of other vectors of the matrix. The vectors $\mathbf{a}' = (3 \quad 1 \quad 4)$ and $\mathbf{b}' = (6 \quad 2 \quad 8)$ are dependent since $2\mathbf{a}' = \mathbf{b}'$. If one vector is a function of another in this manner, the coefficient of correlation between them is 1.00. If two vectors are independent, then we cannot write a functional equation to express the relation between them, and the coefficient of correlation cannot be 1.00. For instance, $\mathbf{a}' = (2 \quad 1 \quad 1)$ and $\mathbf{b}' = (3 \quad 1 \quad 4)$ are independent vectors and their correlation is less than 1.00. Dependence in a matrix can also be defined by determinants. If $\det \mathbf{A} = 0$, $\mathbf{A}$ is singular and there is linear dependence in the matrix. Take the following matrix in which the values of the second row are twice the values of the first row — and thus there is linear dependence in the matrix (the matrix is singular):

$$\begin{pmatrix} 3 & 1 \\ 6 & 2 \end{pmatrix}$$

The determinant of the matrix is 0:

$$\begin{vmatrix} 3 & 1 \\ 6 & 2 \end{vmatrix} = (3)(2) - (1)(6) = 0$$

The notions of dependence and independence of vectors and singular and nonsingular matrices are sometimes very important in multiple regression analysis. If a matrix is singular, this means that at least two vectors are dependent and the matrix cannot be inverted, as said above. Thus multiple regression computer programs that depend on inverting matrices (some do not), like MULR in Appendix C, will not work. As we said earlier, however, most correlation matrices are nonsingular and are amenable to the basic analyses presented in this book.

Orthogonality

Orthogonal means right angled. The usual axes, x and y, on which the values of two variables are plotted, are at right angles. They are orthogonal to each other. The correlation between two orthogonal vectors is zero. The sum of the cross products is zero. These three vectors are orthogonal to each other:

1.	0	0	-1	1
2.	1	-1	0	0
3.	$\frac{1}{2}$	$\frac{1}{2}$	$-\frac{1}{2}$	$-\frac{1}{2}$

Note the sum of the cross products:

1×2: $(0)(1) + (0)(-1) + (-1)(0) + (1)(0) = 0$

1×3: $(0)(\frac{1}{2}) + (0)(\frac{1}{2}) + (-1)(-\frac{1}{2}) + (1)(-\frac{1}{2}) = 0$

2×3: $(1)(\frac{1}{2}) + (-1)(\frac{1}{2}) + (0)(-\frac{1}{2}) + (0)(-\frac{1}{2}) = 0$

This is, of course, vector multiplication. If vector 1, above, is **a**, vector 2 is **b**, and their product is **c**, then we simply write

$$\mathbf{a'b = c}$$

or

$$(0 \quad 0 \quad -1 \quad 1)\begin{pmatrix} 1 \\ -1 \\ 0 \\ 0 \end{pmatrix} = 0$$

We study vector orthogonality in this way because it is a condition of great importance in multiple regression analysis. As several discussions in Part II of this book show, coded vectors can be used to represent experimental treatments and categorical variables. Orthogonal vectors "create" the desirable condition of factorial and other designs: independence of factors or conditions. They can also be used to help in the comparisons of the means of experimental treatments. *Coding* is the assignment of numerals —{0,1}, {−1,0,1}, {1,−1}, and so on—to the individuals of different experimental treatments or subgroups to denote group membership. When coded vectors are orthogonal, the analysis and interpretation of multiple regression data and results are simplified and clarified.

Statistical and Multiple Regression Applications

The purpose of the present section is to help the student think vectors and matrices using sums and sums of squares and cross products. Ability to think in this way will facilitate later study and work. The need for such calculations occurs repeatedly in multivariate analysis.

To calculate the simple sum of a vector, we simply multiply a row vector of 1's by a column vector of X's:

$$\sum X: \quad (1 \quad 1 \quad 1 \quad 1 \quad 1)\begin{pmatrix} 1 \\ 4 \\ 1 \\ 3 \\ 7 \end{pmatrix} = 16$$

In practical work, there is little need to do this, however. Much more useful is the calculation of the sums of squares and cross products in one matrix operation, $\sum X_i X_j$, or, in matrix notation, $\mathbf{X'X}$:

$$k\begin{pmatrix} 1 & 4 & 1 & 3 & 7 \\ 2 & 3 & 3 & 4 & 6 \\ 2 & 5 & 1 & 3 & 5 \end{pmatrix} n\begin{pmatrix} 1 & 2 & 2 \\ 4 & 3 & 5 \\ 1 & 3 & 1 \\ 3 & 4 & 3 \\ 7 & 6 & 5 \end{pmatrix} = \begin{pmatrix} 76 & 71 & 67 \\ 71 & 74 & 64 \\ 67 & 64 & 64 \end{pmatrix}$$

$$\mathbf{X'} \qquad\qquad \mathbf{X} \qquad\qquad \mathbf{X'X}$$

In statistical symbols, $\mathbf{X'X}$ is

$$\sum X_i X_j = \begin{pmatrix} \Sigma X_1^2 & \Sigma X_1 X_2 & \Sigma X_1 X_3 \\ \Sigma X_2 X_1 & \Sigma X_2^2 & \Sigma X_2 X_3 \\ \Sigma X_3 X_1 & \Sigma X_3 X_2 & \Sigma X_3^2 \end{pmatrix}$$

Using the usual formulas for the sums of squares of deviations from the mean, $\Sigma x^2 = \Sigma X^2 - (\Sigma X)^2/N$, and the deviation cross products, $\Sigma x_i x_j = \Sigma X_i X_j - (\Sigma X_i)(\Sigma X_j)/N$, we obtain the useful deviation sums of squares and cross products matrix, $\mathbf{C}$:

$$\sum x_i x_j = \begin{pmatrix} \Sigma x_1^2 & \Sigma x_1 x_2 & \Sigma x_1 x_3 \\ \Sigma x_2 x_1 & \Sigma x_2^2 & \Sigma x_2 x_3 \\ \Sigma x_3 x_1 & \Sigma x_3 x_2 & \Sigma x_3^2 \end{pmatrix} = \mathbf{C}$$

If we now divide all the terms by N, we obtain the variance and covariance matrix. We can easily obtain the correlation matrix by using the formula

$$r_{x_i x_j} = \frac{\Sigma x_i x_j}{\sqrt{\Sigma x_i^2 \Sigma x_j^2}}$$

A mathematician might write this formula compactly in matrix symbols. If we change x_j to y,

$$\mathbf{R} = (\mathbf{x'y})[(\mathbf{x'x})(\mathbf{y'y})]^{-1/2}$$

(The reciprocal of a number, x, is $1/x$. This is indicated, as we showed earlier with the inverse of a matrix, by a superscript -1, $x^{-1} = 1/x$. The 1/2 superscript indicates the second root, or square root, $x^{-1/2} = 1/\sqrt{x}$. Similarly, $x^{1/3} = \sqrt[3]{x}$ and $x^{-1/3} = 1/\sqrt[3]{x}$.)

Study Suggestions

1. The student will find it useful to work through some of the rules of matrix algebra. Use of the rules occurs again and again in multiple regression, factor analysis, discriminant analysis, canonical correlation, and multi-variate analysis of variance. The most important of the rules are as follows:

    (1) $\mathbf{ABC} = (\mathbf{AB})\mathbf{C} = \mathbf{A}(\mathbf{BC})$

    This is the *associative rule* of matrix multiplication. It simply indicates that the multiplication of three (or more) matrices can be done by pairing and multiplying the first two matrices and then multiplying the product by the remaining matrix, or by pairing and multiplying the second two and then multiplying the product by the first matrix. Or we can regard the rule in the following way:

    $$\mathbf{AB} = \mathbf{D}, \text{ then } \mathbf{DC}$$
    $$\mathbf{BC} = \mathbf{E}, \text{ then } \mathbf{AE}$$

    (2) $\mathbf{A} + \mathbf{B} = \mathbf{B} + \mathbf{A}$

    That is, the order of addition makes no difference. And the associative rule applies:

$$A+B+C = (A+B)+C = A+(B+C)$$

(3) $A(B+C) = AB+AC$

This is the *distributive rule* of ordinary algebra.

(4) $(AB)' = B'A'$

The transpose of the product of two matrices is equal to the transpose of their product in reverse order.

(5) $(AB)^{-1} = B^{-1}A^{-1}$

This rule is the same as that in (4), above, except that it is applied to matrix inverses.

(6) $AA^{-1} = A^{-1}A = I$

This rule can be used as a proof that the calculation of the inverse of a matrix is correct.

(7) $AB \neq BA$

This is actually not a rule. It is included to emphasize that the order of the multiplication of matrices is important.

Here are three matrices, A, B, and C.

$$\begin{pmatrix} 2 & 3 \\ 1 & 2 \end{pmatrix} \quad \begin{pmatrix} 3 & 4 \\ 0 & 1 \end{pmatrix} \quad \begin{pmatrix} 0 & 2 \\ 5 & 3 \end{pmatrix}$$

$$\quad A \qquad\qquad B \qquad\qquad C$$

(a) Demonstrate the associative rule by multiplying:

$$A \times B; \text{ then } AB \times C$$

$$B \times C; \text{ then } A \times BC$$

(b) Demonstrate the distributive rule using A, B, and C of (a), above.
(c) Using B and C, above, show that $BC \neq CB$.

$$\left(Answers: \quad \text{(a) } ABC = \begin{pmatrix} 55 & 45 \\ 30 & 24 \end{pmatrix} \quad \text{(b) } A(B+C) = \begin{pmatrix} 21 & 24 \\ 13 & 14 \end{pmatrix} \right)$$

2. Calculate the determinant of the following correlation matrix:

$$R = \begin{pmatrix} 1.00 & .70 \\ .70 & 1.00 \end{pmatrix}$$

Now calculate r_{12}^2. What is the relation between the results of the two calculations?
(*Answer*: det $R = .51$; $r_{12}^2 = .49$, and det $R + r_{12}^2 = 1.00$.)

3. In a study of Holtzman and Brown (1968), the correlations among measures of study habits and attitudes, scholastic aptitude, and grade-point averages were reported. The correlations are

	SHA	SA	GPA
SHA	1.00	.32	.55
SA	.32	1.00	.61
GPA	.55	.61	1.00

The determinant of the 2-by-2 matrix of the correlations of the independent variables, SHA and SA, is $(1.00)(1.00) - (.32)(.32) = 1.0000 - .1024 = .8976$. The determinant of the whole 3-by-3 matrix is $.4377$. If we divide the first of these determinants into the second, $.4377/.8976$, we get $.4876$. The determinant of a 2-by-2 correlation matrix represents the variance of the second variable not accounted for by the first variable. Actually, the variance not accounted for in a dependent variable is the ratio of the determinant of the matrix of all the variables to the determinant of the matrix of the independent variables. Therefore, $.4377/.8976 = .4876$ represents the proportion of variance in grade-point average not accounted for by study habits and attitudes and scholastic aptitude. The variance accounted for must then be $1.0000 - .4876 = .5124$, or 51 percent. This is, of course, the squared coefficient of multiple correlation, R^2.

4. Liddle (1958) reported the following correlations among intellectual ability, leadership ability, and withdrawn maladjustment:

$$
\begin{array}{c@{\quad}ccc}
 & \text{IA} & \text{LA} & \text{WM} \\
\text{IA} & 1.00 & .37 & -.28 \\
\text{LA} & .37 & 1.00 & -.61 \\
\text{WM} & -.28 & -.61 & 1.00
\end{array}
$$

The determinant of the whole matrix is $.5390$. Calculate the determinant of the matrix of independent variables. Then calculate the proportion of the variance of the dependent variable not accounted for by the independent variables. Finally, calculate R^2, the proportion of variance accounted for by the independent variables.
(*Answer*: $R^2 = .3755$; $1 - R^2 = 1 - .3755 = .6245$.)

5. Solve the following regression equation for the β_j using matrix algebra. In addition, write out the matrices and vectors that correspond to the matrix algebra.

$$Y' = \beta_1 X_1 + \beta_2 X_2$$

(*Answer*: $\beta = \mathbf{X}^{-1}\mathbf{Y}$.)

6. The student should study one or more of the following references: Bush, Abelson, and Hyman (1956). Part III of this unusual work summarizes many uses of matrices and matrix algebra in the psychological literature. Examples M35, M49, M60, M61, M62, M63, and M64 are pertinent to multiple regression analysis.
    Harman (1967, Chapter 3). Although geared basically to the use of matrix algebra in factor analysis, this is a fine chapter, much of which is pertinent to multiple regression analysis.
    Horst (1965, Chapter 2). A good reference for the beginner; it is particularly thorough on matrix multiplication.
    Kemeny, Snell, and Thompson (1966, Chapter V). Although little attention is paid to the needs of statistics, the exposition of this chapter repays study.
    Searle (1966). This solid and useful book has a chapter on the application of matrix algebra to regression analysis (Chapter 9).
    Thurstone (1947, "Mathematical Introduction"). This was a pioneering and classic chapter when it was written. It is still a classic.

7. It was said in the beginning of this appendix that matrix algebra is useful for conceptualizing, as well as analyzing, psychological, sociological, and educational research data. Explain this statement. Outline the advantages of matrix algebra. Give an example. Compare the matrix algebra statement of the example with a statistical statement of it.

The Use of the Computer in Data Analysis

Multivariate analysis and the computer are like husband and wife: they are inextricably linked for better or for worse. This should have become quite apparent to anyone who has read even one-third of this book. Appendix A on matrix algebra makes this especially clear because matrix operations require many calculations that are not only difficult to do by hand; they can generate considerable error. To be sure, we have used many examples in this book that can satisfactorily be done with a desk calculator so that the student could follow them without being overwhelmed by complex, lengthy, and tedious calculations. Most real problems in multiple regression, however, almost have to be done by computer. Once an investigator uses more than two independent variables and, say, 30 cases, the desk calculator going can be quite rough – and prone to error.

In this appendix, we emphasize certain points that we have found valuable when using computers. Many students have to flounder until they strike the solutions to their computer problems. We hope we can help to cut down some of the ignorance and reduce the haphazard use of the computer. Our own experiences and observations of many professors and students at two or three universities compel us to the uncomfortable conclusion that more than half the individuals who use or try to use computers are ill-informed. One cannot make a computer work with ignorance. One cannot make it work with faith or hope. It only responds to knowledge and a deft strong hand. This appendix's main goals are to help cut down the ignorance, to give some hopefully sensible guidance not so much to computers and computing but rather to understanding computers and how to go about achieving a fair degree of mastery. Of one thing we are fairly sure: the behavioral researcher who does not tool himself to use the computer with some ease, who does not know how to program, even in an elementary way, who has to depend on others for his computer work, or, worse, who avoids and evades the computer and rationalizes his avoidance and evasion, is obsolescent if not obsolete.

Computer Characteristics[1]

The most important characteristics of the modern electronic computer are its speed, tireless but finite accuracy, flexibility, and ductility. The computer user, particularly the user who must do multiple regression, discriminant analysis, canonical correlation, multivariate analysis of variance, and factor analysis, must constantly keep these characteristics, but especially the last, in mind. As we will try to show, their understanding helps him to master the machine, at least sufficiently to do his work sensibly. We cannot overemphasize this. The weakness of computer use is usually the human user and not the machine.

The speed of the modern large computer is well-known, although perhaps a bit unbelievable. A factor analytic program the writers use on the CDC-6600 computer calculates all the basic statistics and correlations and extracts the factors and rotates them successively — first two, then three, and so on. A 35-variable problem takes about 20 seconds, a 50-variable problem less than a minute and a half, and a 100-variable problem less than 2 minutes! We ran ten multiple regression problems, both small and large, at one pass with MULR, the program given in Appendix C. The total computing time (excluding input and output time) for complete and extensive analyses was 14 seconds!

Phenomenal as it is, speed in and of itself is not really important. The important thing is that such high speed changes the nature of research because it makes the analysis of large quantities of data with many variables possible. It also makes possible repeated analyses of multiple regression problems such as some of those discussed in this volume. In short, the speed of the modern computer makes flexible analysis of multivariate problems almost routine. Both conception of problems and their analysis can hardly help but be affected.

The computer calculates with high but finite accuracy. Earlier we brought one source of error to the attention of the student: rounding errors. While the computer for the most part avoids the errors that were common when desk calculators were used almost exclusively, there can still be errors. The computer user can help avoid them by knowing that the possibility of error is always present, even with eight decimal places.

The computer's accuracy inheres in its memory, which is limited to numbers of finite accuracy within a wide range of magnitude. This is especially so in multivariate analysis. In calculating sums of squares or cross products of large sets of numbers, for example, if the number of significant digits exceeds the machine's finite accuracy, the resulting sums will not be correct. Computer

[1]The discussion in this section and in certain other parts of this appendix is based in part on the discussion in Kerlinger (1964, Appendix C).

output *can* be meaningless. The important thing is for the researcher to be able to tell when results are meaningless or questionable. Researchers have to be constantly alert to the possibility of inaccurate results. It is always necessary to match the computer output results with the original data to see if the results "look right" and "make sense." A small error in input like a misplaced decimal point or a number punched in the wrong column of a card can create large errors in output.[2]

The flexibility of computers is really a characteristic of the use of the machine. There are always several ways to tell a computer how to do its operations. The beauty of the flexibility feature is that the machine will produce identical results with different sets of instructions. This means that even the inexpert programmer can achieve satisfactory results. Elegance may be sacrificed — professional programmers are usually proud of the elegance of their work — but accurate results can be obtained even with what experts would call clumsy programming. The program MULR in Appendix C is an example. Although much of its programming is inelegant, it achieves its objectives quite well. In other words, the computer and program language permit flexibility of programming. This is a distinct advantage to the researcher who is not and cannot be expected to be an expert programmer.

Another aspect of flexibility is the modern computer's adaptability to many different kinds of problems and operations. The computer can be used effectively and efficiently by physicists, chemists, biologists, meteorologists, sociologist, economists, psychologists, and educators. It can be programmed to handle mathematical operations of all kinds: symbolic logic, statistics, analysis of verbal materials. This generality and flexibility distinctly aid the behavioral science researcher. Take one example. A computer installation, for instance, may not have just the program a researcher needs. He may have to develop and write his own program, perhaps with expert help. We have seen earlier that it is frequently necessary to invert matrices and to calculate the determinants of matrices. That is, the researcher may want to solve the equation $\beta_j = R_{ij}^{-1} r_{yj}$ [see Chapter 4, Equations (4.2), (4.3), (4.4), and (4.5)], which of course calls for the inverse of a correlation matrix. Research workers in the natural sciences also have to invert matrices, and all computer installations have computer routines to do so. In other words, with a little programming skill, the researcher can write his own program to invert the R matrix and solve the above equation.

The final computer characteristics, ductility, can be loosely defined as stupidity. We insist on the importance of this characteristic for good reasons. Man has a tendency to anthropomorphize animals and natural and manufactured nonhuman things. Ships take on life — female life of course — dogs acquire personality, mountains become forbidding, even malevolent, and computers have all these characteristics — and more. Much of the difficulty that

[2]Unfortunately, some computer programs do not provide the option of printing the input data. This can be a serious omission. All computer programs should have an option for printing the original data and the results of certain intermediate calculations like correlation matrices, matrix inverses and their proofs ($A^{-1}A = I$), variances as well as standard deviations, and so on.

intelligent people have in understanding computers is the esoteric, magical, and even mystical properties attributed to the giants. The computer is a complete idiot, though, to be sure, a remarkable idiot. It does exactly what a programmer tells it to do, no more and no less. (This is sometimes hard to believe, however.) If the programmer programs intelligently, the machine performs intelligently. If the programmer errs, the machine faithfully does as it has been told to do: it errs. In other words, the modern computer is highly reliable: it performs faithfully and obediently. It even makes the programmer's mistakes faithfully and obediently. When things go wrong, one can usually assume that it is one's own fault or the fault of the program. It is rarely the computer's fault (except in the early shakedown period of its installation and when it occasionally develops difficulties). Computers do not often make mistakes. Thus when we say computers are stupid, we mean they are reliable and virtually errorless; they do precisely and stupidly what we tell them to do.[3]

Programming Language

One of the great achievements of this century is the invention of intermediary languages to communicate with computers. The heart and soul of the researcher's use of the computer is the programming language he uses. In the early days of computers, programmers had to work in the operational language of the computer. This was tedious, difficult, and error-prone. Now, the programmer can use FORTRAN, say, which is more or less an English computer language. The computer translates the FORTRAN into machine language and operations.

A computer program is a set of instructions in some sort of machine intermediary language, like FORTRAN, COBOL, ALGOL, or PL/I, that tells the machine what to do and how to do it. The commonest language in the behavioral sciences at present is FORTRAN (FORmula TRANslation). There have been different versions of FORTRAN; the version in use at this writing is FORTRAN-IV (see McCracken, 1965; Mullish, 1968), a highly deveoped, powerful, efficient, and flexible means of telling computers what to do.[4] It consists,

---

[3]This does not mean that a computer's performance cannot seem magical. Sometimes we do not know what can happen if we carry out an operation many times because we ourselves are unable to do so, or even to imagine what can happen after a long series of operations. The computer can do so, however, and the results are sometimes surprising. An example is the use of computer-generated random numbers and computer calculations with the numbers to help solve otherwise virtually insoluble problems. Humphreys and Ilgen (1969), for example, generated random factors in order to help solve the difficult problem of how many factors to extract and rotate in factor analysis. The computer results were gratifying, even though the combined use of random numbers and the repetitive operations of the computer made the results seem almost magical. For an excellent introduction to the use of random numbers and the computer use of such numbers, see Lohnes and Cooley (1968).

[4]It appears doubtful that FORTRAN will continue to be preeminent. We have been informed by computer experts that PL/I (Programming Language I) (see Bates & Douglas, 1967; Pollack & Sterling, 1969) will probably supersede FORTRAN. PL/I can handle verbal materials, as well as numerical operations, easily and flexibly, whereas FORTRAN is best adapted to numerical operations. The need for a more general language is clear, and within about five or ten years, many or most scientific computer installations will probably have changed to some language like PL/I.

basically, of simple statements such as DO, READ, PRINT, CALL, IF, and GO TO. These instructions mean what they say: they tell the computer to do this, do that, go here, go there, read this instruction, print that output, and so on. The power of this language cannot be exaggerated. There is almost no numerical or logical operation that cannot be done with it.

"Package" Programs

The availability of statistical and mathematical programs all over the world, including multiple regression and other multivariate analysis programs, is truly remarkable. To be realistic, most researchers doing multivariate analysis will have to rely on computer programs written by others, and most reliance will have to be put on so-called "package" programs. Package programs are generalized programs of considerable complexity written to do a variety of analyses of a certain kind. BMD03R, called "Multiple Regression With Case Combinations," which was used to solve many of the problems in this book, is a versatile package multiple regression program that can be applied to a number of multiple regression problems. In fact, all the BMD programs, the Cooley and Lohnes programs, and the Veldman programs mentioned in Appendix A are package programs.

Our purpose here is to give the student some practical guidance to widely available programs that he can use in the immediate future. Because these programs have been mentioned and discussed in Appendix A, and because it is highly probable that some of them will have become obsolescent, if not obsolete, a few years after this book is published, we will confine ourselves to discussing one widely available set of programs and certain other important programs and approaches. There is little doubt that even if computers adopt a new language like PL/I, the standard sets will be rather quickly translated to the new language.[5]

Although many multiple regression and other multivariate analysis and factor analysis programs now exist, the programs of the so-called BMD series (Dixon, 1970) are perhaps the most widespread and available. There are six regression programs; BMD03R is perhaps the most useful. The set also includes other multivariate analysis programs. The BMD programs are highly sophisticated, accurate, and dependable. They have, from our point of view, two major drawbacks. One, their input and output features can be considerably improved. We advise the researcher to alter the input to read identification numbers of cases in the first columns of cards and to improve the format of the output. It is advisable, too, to make it possible to print the input data if the user desires. Two, the BMD programs are sometimes difficult to use. This is a characteristic of many so-called generalized programs. A generalized program

---

[5]Such translation can be done by the computer. It will not really be necessary for a programmer to laboriously rewrite programs in a new language. Instead, a translation program can be written and a program in one language can be converted to the new language by the computer itself. Because most programs are very complex, however, the human programmer will usually have to intervene.

is written so that it can be used with many different kinds of problems with different numbers of variables and different kinds of solutions. Such a program does a lot and provides the user with a variety of choices. In other words, a price in difficulty of use is paid for the generalizability. Taken as a whole and considering their reliability and wide availability the BMD programs are probably a good set for a researcher to concentrate upon, especially if he does not write programs himself.

Completely Generalized Programs

Another significant approach to programming and data analysis must be mentioned because it is possible that, as statistical and computer sophistication increases, it will supersede some of the present practices. This approach is a generalized one that relies upon and uses fundamental subroutines (a subroutine is a relatively autonomous part of a larger program that ordinarily accomplishes some specific purpose, like inverting a matrix, and that can be called into action at any point in a main program) as a basic core that can be used to accomplish a variety of purposes. Such an approach is particularly appropriate in multivariate analysis and, of course, in regression analysis and factor analysis. We briefly describe two of these "programs," one because it is a giant all-purpose package and the other because it is a relatively small, compact, and efficient unit that may revolutionize programming.

Buhler's PSTAT.[6] Buhler's PSTAT is a large package of over 60 programs that can be called by using a few IBM cards with special punched instructions. The entire package is recorded on a tape that is mounted by a machine operator. The cards just mentioned call PSTAT into operation and activate that part of the program that is needed. Suppose, for instance, that one wishes to do a factor analysis. One uses entry cards with the name of the program and other pertinent information such as what one wants the program to do. The factor analysis itself then uses not only its own program but a number of subsidiary and complementary programs. One of the advantages of PSTAT is thus relative simplicity and ease of input. Another is uniformity of input and output. The output, for example, is always completely labeled and in such a form — we might say almost elegant, unlike most package outputs — that it can be readily and easily used by researchers.

PSTAT and programs like it have, then, the advantage of encyclopedic use. Buhler's aim has been to supply a package that can do most statistical tasks from calculating a mean to all the complex calculations of factor analysis. At the present writing, changes in computer design appear to be causing difficulties because its original tape feature consumes too much computer time. No doubt PSTAT will be adapted to the computer changes and be the excellent set of programs it was.

Beaton's Matrix Operators. Beaton's (1964) approach, worked out in

---

[6]Unfortunately, there is no published source for PSTAT although there is a manual of instructions (produced by the computer).

his doctoral thesis and later improved, is quite different. Beaton emphasizes mathematical models and deemphasizes special techniques for statistical calculations. The approach is based on six special matrix operators called SCP (sum cross products), SWP (sweep), TCM (transform cross-products matrix), and so on. Using these operators statistical techniques are themselves redefined. In any case, the range of statistics and multivariate analysis, including multiple regression, can be calculated with the operators.

Highly important, the researcher who knows the elements of programming can in effect write his own programs using the operators to help accomplish a variety of calculations. The operators make programming a flexible, more efficient, and simpler procedure, and enable the researcher to be his own programmer with much less effort in actual programming.

We believe Beaton's use of matrix operators in statistical calculus to be a unique, original, and perhaps outstanding achievement. There are three rather large difficulties. One, the system is not widely available; it still has not been published. Two, its use appears to require considerable mathematical and statistical knowledge, more indeed, than other programs. And three, Beaton's discussion of the system is at a level beyond that of most researchers. At present, we cannot recommend that researchers seize the system and use it. We believe, however, that either Beaton or someone else will translate it in such a way that researchers can study it and use it. We also believe that, when so translated, it may revolutionize programming in the sense that it can put the researcher and the computer into a close and profitable relationship, a relationship that is not now prevalent.

Testing Programs

Package programs have to be approached and used with care and circumspection. If the user is not alert he can obtain incorrect results. He must know what a program can do and cannot do. He may sometimes even have to "go into" a package program to find out just how it does a certain calculation or set of calculations.

An example or two may reinforce the argument. Take first a simple example. Almost all statistical package programs calculate and print standard deviations. Does the program calculate the standard deviations with N or $N-1$ in the denominator? The BMD manual gives almost all the formulas used; so there is no problem if one takes the trouble to check the manual. Certain other programs do not tell how the standard deviation is calculated. One has to examine a listing of the program. There are more complicated and difficult cases. In multiple regression analysis, for example, there are various ways that have been recommended to estimate the variance of the dependent variable that an independent variable contributes. BMD03R, as part of its output, prints the squared semipartial correlations, but labels them "PROP. VAR. CUM.," which means, according to the manual (Dixon, 1970, p. 268), "PROPORTION OF TOTAL VARIANCE ADDED." The unwary user, however, may not even be aware of squared semipartial correlations. He may believe that these values were calculated another way. (The manual is clear on this point, but it has to be read carefully. It does not use the expression "squared semipartial correlations.")

Computer programs can and do have incorrect procedures. Even the best routines sometimes have errors and omissions. BMD03R, for instance, omits beta weights in its output. An easy error to make is to use incorrect numbers of degrees of freedom. We have encountered one such case in a packaged regression program. We have also had difficulty with formulas for the standard error of regression coefficients. Another problem is when a computer program does not calculate or at least print statistics needed by an analyst. The BMD03R example just mentioned is an example. The user who needs beta weights — and one would think that any user of multiple regression would need them — will have to go into the BMD03R program and insert the calculation and printing of beta weights, or have a computer programmer do it. (The alternative, of course, is to calculate them by hand from the output b weights.)

Our main point, again, is: Know what you're doing. We suggest, moreover, that researchers routinely test package and other programs with one or more sets of "model" data, the complete solutions of which are available. This is sometimes easier said than done because original sets of data used in published analyses are often not given. There is no problem with multiple regression: standard texts publish the original data. To test factor analysis and multivariate analysis programs, however, is another matter because the original data are seldom given. We recommend three things. One, if at all possible work a small problem with a desk calculator and use this to test a computer program. Since errors are easy to make, this may not always work. It has the virtue, however, of forcing you to know what you're doing and what the computer program does and does not do. Two, take two or even more examples from two or more different texts. For multiple regression we recommend Snedecor and Cochran (1967, Chapter 13) and this book. For factor analysis, Harman (1967, Table 2.1, p. 14) gives a small but good example. The only published raw data for canonical correlation analysis we know of are given in Veldman (1967, pp. 290–291). Most discussions give only the correlations. Strangely enough, on the other hand, BMD03R, for all its complexity and sophistication, will accept only raw scores as input. As we said earlier, the researcher who wants to use multiple regression to analyze a correlation matrix will have to go into BMD03R and alter it — or use a program like MULR (Appendix C). For discriminant analysis, Veldman (*ibid.*, p. 277) gives an example. So does BMD04M and BMD05M (Dixon, 1970, pp. 192 and 204). Multivariate analysis of variance is a problem: published examples of raw data are as rare as raw data for factor analysis. Morrison (1967, p. 179), however, gives data that are manageable.

Concluding Remarks, Advice, and Admonition

The individual who first uses the computer, and even the person who has already used it a good deal, can have frustrating experiences, waste a good deal of time, and perhaps even be turned off by their experiences. We hope that what we have said in this appendix will help to ameliorate the more difficult aspects of the student's experience. We now want to concentrate on certain problems of computer use that cause problems. Being aware of the problems the researcher can perhaps learn to cope with them.

The greatest problem in computer usage stems from ignorance, misconceptions, and incorrect assumptions. There is a widespread and erroneous belief that when one has a problem for the computer that one goes to a computer expert who solves the problem and turns over a finished computer answer to the researcher. This belief leads only to grief, unless one is very lucky. It is similar to the belief that one goes to a statistician when one has a statistical problem, and the statistician will either do the problem or tell one precisely how it should be done. Both beliefs are based on assumptions that are often not warranted. One of the most prevalent of such assumptions is that the computer expert or statistician understands behavioral science problems, data, and methods. Another is that computer and statistical methods are uniform, applicable to all substantive and analytic problems. It is unrealistic, even unreasonable to expect computer experts and statisticians to understand and know the substance and methods of, say, psychology or sociology. Although they are, for the most part, highly competent people, many of them cannot be expected to know the special requirements of a particular field.

One of the most difficult problems of computer work is communication between researcher and programmer. Since it is highly unrealistic to expect professional programmers to understand the substance and methodology of the behavioral sciences, the best solution of the problem is clear: the researcher must learn at least enough programming to be able to talk knowledgeably and intelligently to the programmer. Such programming knowledge can be acquired in a matter of months, whereas it would take the programmer years to learn enough about behavioral science and behavioral scientific methods and analysis to communicate with the researcher at the researcher's level. We are not advocating that the researcher become an expert programmer, although this would be highly desirable since our hunch is that all researchers will, within the next decade, have to become fairly good programmers. We *are* advocating that researchers become sufficiently expert so that they will not be overly dependent on programmers. We go so far as to say that graduate programs in the behavioral sciences and education that do not include computer know-how and programming in their curricula are and will increasingly be woefully deficient. The researcher, in other words, must be able to communicate with the programmer to some reasonable extent at the programmer's level. Fortunately, programming, once learned to even a fair degree, is a fascinating business. It is so fascinating, in fact, that there is a danger of spending too much time at it.

Computers are extremely useful, reliable, and obedient servants. One must always remember, however, that they are utterly stupid and that their facile output can never substitute for competence, knowledge, and understanding. Our final word is that we urge the student to learn how to use this fascinating phenomenon of our times, and to learn it well. The work involved and the frustrations encountered are more than balanced by the power acquired over research data and analysis and by the sheer interest, even wonder, of the subject. The student who writes his first program for real data and makes it work is on the road to being hooked, as we have been for some years.

Study Suggestion

One or more of the following references on computer programming will be helpful to the student. They are all good. All have valuable features. Veldman's Chapter 7 integrates matrix algebra and computer programming. Lohnes and Cooley skilfully combine computer programming with statistics. Both books are recommended to students who intend to learn programming. McCracken is an excellent standard general text. Mullish's book, in addition to its pedagogical clarity, has many useful short routines. The Cooley and Lohnes book contains multivariate analysis programs. [Cooley and Lohnes (1971); Lohnes and Cooley (1968); McCracken (1965); Mullish (1968); Veldman (1967, Chapter 7).]

MULR : Multiple Regression Program

The examples in this book were run with either or both of two computer programs, BMD03R or MULR.[1] We suggest that the reader use one or the other of these unless, of course, he already has a good program. The choice depends on one's purposes. If one wishes to vary the order of the independent variables, as well as to vary their number, *in one pass*, BMD03R is the choice. BMD03R also has other advantages that will become evident after careful study of its manual (Dixon, 1970, pp. 258–275c; 14–21). On the other hand, MULR, while not as flexible a program, has several virtues: ease of use, variety of analyses available, acceptance of a correlation matrix as input, and so on.

The purpose of this appendix is to make MULR available to the reader. It is written in FORTRAN-IV and should be readily adaptable to most computers. Although long and complex, one need prepare only three to seven program-control cards, all of them simple.

We suggest that the reader who decides to use MULR do so with some of the problems in the text. They have simple numbers and their answers are known. After successfully working through, say, one or two of the examples of Tables 3.1, 4.2, 4.6, and 5.7 (an *R* matrix), try the harder examples of the middle of the book. Run the data of one or more examples in different ways. For example, delete each of the independent variables of the Table 4.2 data in turn. Then vary the order of the variables. These runs will be valuable because they will familiarize the user with several of the features of MULR; for

---

[1] There is one exception to this statement: In Chapter 11 we used BMD02R for stepwise regression. In examples with more than one dependent variable, naturally, neither BMD03R nor MULR can be used. Instead, a canonical correlation, multivariate analysis of variance, or multiple discriminant analysis program must be used.

instance, the variable format feature (see instructions, paragraph 6) and the variable rearrangement feature (see instructions, paragraph 7).

Next, use MULR to do analysis of variance. Note that only columns 1 to 40 of the main program-control card (no. 4, under the instructions) need be used for regular regression analysis. Columns 45 to 60 govern analysis of variance. One weakness of MULR is that the coded vectors have to be punched by the user. (It would have made the instructions and input considerably more complex to have the computer generate the coded vectors.[2]) This is no burdensome job, however, since it means punching only a few more columns on each IBM card. In doing factorial analysis of variance, one need not punch the interaction vectors (the cross products); the computer generates them from the coding of the main effects.

Finally, try running problems with added power vectors, for example, X^2 and X^3, and interaction (cross-product) vectors, X_1X_2, X_1X_3, and so on. Columns 65 and 70 govern these maneuvers. Also, see paragraphs 8 and 9 of the instructions.[3]

MULR Instructions

MULR is a general multiple regression program that can be used to solve most ordinary multiple regression problems and some analysis of variance problems. It will read in either raw data or a correlation matrix. If desired, the variables can be rearranged. Some variable or variables can be dropped by appropriate use of variable format. Variables can be raised to powers (for trend analysis, for example) and their interactions (cross products) can be generated and studied. The capacity of the program is 30 variables and 2000 cases. (If there are more than 2000 cases, the program can be easily altered by a programmer.)

The calculations and printed output of the program are in two parts. The first part includes the usual multiple regression calculations. Here is how the results of the calculations are printed. First, the title of the project and the parameters that were fed in on the program-control card are given. The data (variable) format is also given. Second, if the user has called for printing of the raw data, they will be printed in the order the user specified, provided, of course, he specified some order of the variables other than the order on the data cards. The specified order is also printed. Third, the means, variances, and standard deviations will be printed, followed by the correlation matrix. Immediately after the R matrix, the matrix of sums of squares of deviations, labeled SSD(I,J), is printed. The inverse of the independent variable R matrix and the proof that the inverse matrix is correct (an identify matrix) are also

[2]Note that one can generate coded vectors, powered vectors, and interactions by using the so-called transgeneration and other features of the BMD set of programs. See Dixon (1970, pp. 15–21).

[3]There are no restrictions on the use of MULR. That is, permission need not be asked to copy and use MULR or any part of it for scientific research purposes.

printed. The determinant of the independent variable R matrix is also calculated and printed.

On the next page, the complete regression analysis is given: the correlations of each predictor variable with Y, labeled $RY(J)$; the betas, the betas squared, and the b coefficients; multiple R, multiple R squared, the F ratio, the degrees of freedom, and the probability estimates of the analysis of variance; the regression sum of squares, the mean square, the deviation sum of squares and mean square, and the intercept constant.

The following page of output gives the regression coefficients, the standard errors of the regression coefficients, the t ratios, and the probabilities associated with each of the t ratios. Next, the standard errors of estimate for z scores and for raw scores are given. Finally, the observed Y scores, the predicted Y scores, and the deviations from prediction (residuals) are printed on the following pages. These are the predicted scores obtained from the regression equation.

While some of the above statistics tell the user about the relative importance of the different independent variables in contributing to the variance of the criterion or Y variable, the second large set of calculations and printouts was specifically included to study the relative importance of independent variables. The user is cautioned, however, that the correlations that usually exist among independent variables make interpretations of these "relative statistics" very tricky.

The rationale of the first set of these "special" analyses was given in the text (see Chapter 4, for example) and originally came from Snedecor and Cochran (1967, pp. 385–389, 398–399). This is a set of analyses of "separate" sums of squares. An example will perhaps help to clarify what the analyses accomplish. Suppose we have three independent variables, X_1, X_2, and X_3, and, of course, the Y or criterion variable. After finishing the usual analysis of the regression of Y on X_1, X_2, and X_3, MULR calculates the regression of Y on X_1 *alone*. It then immediately calculates the regression of Y on the *remaining* variables, X_2 and X_3. In so doing, it takes account of the influence of X_1, that is, it shows the increment added by X_2 and X_3. The sums of squares, the R^2's and R's, and the F ratios are calculated (with appropriate degrees of freedom and probability values). After completing this first analysis, the program calculates the regression of Y on X_2 *alone*, after which it calculates the regression statistics of Y on the *other* variables, X_1 and X_3, taking account of the influence of X_2. Finally, it does the same analysis with Y and X_3 and Y and the other variables, X_1 and X_2.

The final set of analyses is also aimed at understanding the relative contributions of the various independent variables to the dependent variable. It is a sequential analysis that starts with variable 1 and calculates certain statistics. Then it adds variables 2, 3, and so on and calculates the same statistics. These statistics are the determinants of each of the successive R matrices: those of the correlation of variable 1 and Y, the correlations of variables 1 and 2 and Y, the correlations of variables 1, 2, and 3 and Y, and so on. Using the determinantal formula, the successive R^2's are calculated and printed. The determinantal formula is $R^2 = 1 - \det_l/\det_s$, where $\det_l =$ the determinant of the larger matrix, or the matrix of independent variables *and* the dependent variable, and $\det_s =$ the determinant of the smaller matrix, or the matrix of independent variables only. (See Appendix A, Study Suggestions 2, 3, and 4.) Next, the regression weights, betas and b's, for each successive analysis are given. The partial r's are also given. These are the partial correlations between the successive variables and Y, partialing out the influence of the other independent variables. Then, the differences between the successive R^2's are given on the last page of the output, that is, $R^2_{y.1}$; $R^2_{y.12} - R^2_{y.1}$; $R^2_{y.123} - R^2_{y.12}$. These are actually the squared semipartial correlations. (See Chapters 4 and 5 of the text.) Accompanying these squared semipartial r's are F and t ratios of the successive contributions, with appropriate degrees of freedom and probability estimates.

The user can vary the order of the variables of a problem simply by putting the data through the computer again, using the rearrangement choice card mentioned earlier and described below. For instance, if one were doing a forward or backward solution, one would have to do this. Variables can also be easily deleted by the use of the variable format card.

The user can elect to raise any or all of his independent variables to powers of the variables. For example, X_1, X_3, and X_4, say, of five independent variables can be raised to the second and third powers: X_1^2, X_1^3, X_3^2, and X_4^2. This feature of MULR is useful for trend analysis (see Chapter 9). These powered variables become new vectors that are added to the data matrix in the order specified by the user. If the user specifies, for instance, raising the second and third variables of four independent variables to the second power, the new vectors will be the fifth and sixth columns of the data matrix. The dependent variable, Y, is, as usual, put last: in this example it will be the seventh vector. The whole data matrix, then, will be X_1, X_2, X_3, X_4, X_5 $(= X_2^2)$, X_6 $(= X_3^2)$, X_7 $(= Y)$. Similarly, interactions or cross products of selected independent variables can be specified and included in the regression analysis. Suppose the user wants the interactions (cross products) of variable 1, 2, and 4 of four independent variables. Properly instructed (see paragraph 9, below), the program will create new vectors ("variables") $X_1 X_2$, $X_1 X_4$, and $X_2 X_4$, and insert them as X_5, X_6, and X_7. Again, the dependent variable vector, Y, will be placed last.

It is suggested that the user study the appropriate discussions of powers and interactions in the text before using either or both the above features of MULR. Indiscriminate use can lead only to peculiar results, to say the least.

Instructions for Use

Prepare program-control and data cards as follows — and in the indicated order:
1. *System cards.*   Identification, and so on.
2. *Program cards.*   FORTRAN deck.
3. *Project description.*   Any description desired in Cols. 1–80 (alphanumeric).

4. *Program-control card.*

Cols. 1–5	N = No. of subjects
Cols. 6–10	KT = Total no. of variables (including the criterion or Y variable)
Cols. 11–15	K = No. of predictor or independent variables $(KT - 1)$
Col. 20	$NFVC$ = No. of variable format cards (no more than five cards)
Col. 25	1 = print raw scores
(IPRX)	0 = do not print raw scores
Col. 30	1 = read in correlation matrix
(IRMAT)	0 = do not read in correlation matrix
Col. 35	1 = calculate and print predicted Y scores
(IPRY)	0 = do not calculate and print predicted Y scores
	[*Note*: If IRMAT = 1, punch 0.]
Col. 40	1 = want data to be rearranged
(ICHVAR)	0 = do not want data to be rearranged
	(If ICHVAR = 1, then a rearrange card must be inserted. See below.)
Col. 45	1 = analysis of variance wanted
(INOVA)	0 = no analysis of variance
Col. 50	leave blank
(IFACAN)	
Col. 55	1 = factorial ANOVA, 2 effects
(IFACNO)	0 = not factorial ANOVA, 2 effects
Col. 60	1 = factorial ANOVA, 3 effects
(IFAC3)	0 = not factorial ANOVA, 3 effects
Col. 65	1 = want powers of vectors
(IIPOW)	0 = do not want powers of vectors
Col. 70	1 = want interaction (cross-product) vectors generated
(IINT)	0 = do not want interaction vectors generated

5. *Factorial analysis of variance parameter card.* If 1 has been punched in Col. 55 (IFACNO) or 60 (IFAC3) of the preceding card, this card must be inserted. It contains the number of partitions of the A and B variables of a factorial analysis of variance. Punch the smaller of these in Col. 5 and the larger in Col. 10. For example, if the factorial analysis is a 2×3 design, punch 2 in Col. 5 and 3 in Col. 10. If it is a $2 \times 3 \times 3$ (three effects) factorial design, then punch 2 in Col. 5, 3 in Col. 10, and 3 in Col. 15. [*Note*: See "Additional Notes on Factorial Analysis of Variance," below, for complete instructions for factorial analysis of variance. If not doing factorial analysis of variance, omit this card.]

6. *Variable format card(s).* Specify the format of the data cards, whether raw scores or R matrix. Format must have an I specification before the data specification, which must be F specification. Note that by proper use of the variable format card and the rearrange choice (ICHVAR = 1, Col. 40, of Card 4, above), together with the rearrange card, No. 7, below, a set of data can be run in different ways, including the omission of one or more of the independent variables. (Up to 5 variable format cards can be used.)

7. *Rearrange card.*   If Col. 40 of the program-control card (No. 4, above) has had a 1 punched in it, this card must be inserted. If Col. 40 = 0, omit this card entirely.

Cols. 8–10	Punch number of variables to be read from data card, including the Y variable.
Cols. 11–12, 13–14,.. etc. (IVAR(J))	Punch variable order desired. The criterion or Y variable must always come last. If, for example, the Y variable is the first variable on the cards and there are 4 variables, punch 4 in Col. 10(the number of variables), and, in Cols. 12, 14, 16, and 18, punch 4 1 2 3. In other words, the Y variable must always have the number equal to the number of variables: this will make it come last. The user can choose any number of variables from the data cards and can put them in any order desired. This is useful usually after an initial run, and the user wants different analyses of the same data.

8. *Powers card.*   If 1 has been punched in Col. 65 of card No. 4, above (IIPOW = 1), insert this card. If IIPOW = 0, or is left blank, omit the card. Punch the number of variables whose powers will be calculated (KPOW) in Col. 10. Punch the variables to be raised to powers and the powers desired in fields of 5, as follows. Punch the first variable number in Col. 13 and the power to which the first variable is to be raised in Col. 15. Punch the second variable and its power in Col. 18 and Col. 20. Continue similarly with succeeding variables. Note that the same variable can be raised to more than one power. For example, suppose that 3 has been punched in Col. 10, and the first variable is to be raised to the second and third powers and the third variable to the second power. The numbers punched in the card, then, will be: 3 in Col. 10; 1 in Col. 13 and 2 in Col. 15; 1 in Col. 18 and 3 in Col. 20; 3 in Col. 23 and 2 in Col. 25. In other words, the first variable, X_1, will be raised to the second and third powers, and the third variable, X_3, will be raised to the second power. These powered "variables" will be new vectors added to the data matrix. If there are four independent variables, then the added vectors will be $X_5 = X_1^2$, $X_6 = X_1^3$, $X_7 = X_3^2$. The dependent variable, Y, will be placed last, as usual: $X_8 = Y$.

9. *Interactions (cross products) card.* If 1 has been punched in Col. 70 of card No. 4, above (IINT = 1), insert this card. If IINT = 0, or is left blank, omit the card. Punch the number of variables whose interactions (cross products) are desired in Col. 10. In Cols. 12, 14, 16, and so on, punch the numbers of the variables whose interactions are desired. For example, suppose a user has four independent variables and he wants the interactions of three of them: X_1, X_2, and X_4. He would punch 3 in Col. 10 and 1 2 4 in Cols. 12, 14, and 16. This will produce the new vectors added as independent variables: $X_5 = X_1 X_2$, $X_6 = X_1 X_4$, $X_7 = X_2 X_4$. The dependent variable, Y, will, as usual, be last, $Y = X_8$.

10. *Data cards.*   The input to the program may be either raw data or the correlations of an R matrix. Precede the actual data of each subject, or variable, in the case of an R matrix, with an I specification for identification. Punch the

data with or without decimals. If the input to MULR is a correlation matrix, certain of the regression and other statistics mentioned in the general description above will not be calculated. For example, the means, variances, and standard deviations will obviously not be calculated. Most of the values essential to the usual multiple regression analysis, however, will be calculated and printed.

The storage allocation for MULR is 135,000 (octal) words, CDC-6600.

Additional Notes on Factorial Analysis of Variance

The factorial analysis of variance specified by a 1 in Col. 55 or Col. 60 of the program-control card, above, will do any factorial analysis of variance of two and three (and no more) main effects. Call the analysis SPEC.

In order to use SPEC, 1 must have been punched in Col. 55 (IFACNO = 1) or in Col. 60 (IFAC3 = 1). [If 1 has been punched in Col. 55 or 60, then 1 must be punched in Col. 45 (IANOVA = 1).] That is, only one of these choices can be made, and note that a 1 in Col. 55 means two main effects, A and B, and a 1 in Col. 60 means three main effects, A, B, and C. The independent variables in a SPEC analysis are the coded variables, the coding, of course, indicating group membership. The coding can be any of the kinds described in Chapter 7. *It is only necessary to punch the main effects' coding.* The computer will calculate the interaction vector coding and will also recalculate the number of variables parameters. Further instructions are given below.

On the program-choice card, KT and K (Cols. 9–10 and 14–15) should be the number of coded vectors plus the Y vector for KT and K $= KT - 1$. In a 3×5 factorial analysis of variance problem, for instance, the degrees of freedom for the main effects are 2 and 4. KT is thus $2 + 4 + 1 = 7$, and $K = 7 - 1 = 6$. In a $2 \times 3 \times 5$ (three main effects) problem, the degrees of freedom and thus the number of coded vectors are 1, 2, 4. Thus $KT = 1 + 2 + 4 + 1 = 8$, and $K = 8 - 1 = 7$. (The computer will, as indicated earlier, calculate the interaction vectors and new values of KT and K. In the 3×5 problem, for example, the user will feed in KT as 7 and K as 6. The computer will calculate the complete KT as 15 and K as 14.)

SPEC expects each main effect degree of freedom to be coded. It first calculates the interaction vectors from the coded main effects vectors. In a 3×4 problem, for example, the user should have punched $2 + 3 = 5$ main effects vectors. The computer will then calculate the six interaction coded vectors: A_1B_1, A_1B_2, A_1B_3, A_2B_1, A_2B_2, A_2B_3. SPEC then calculates the sums of squares of each of the coded vectors. For example, the 3×4 analysis just mentioned is being done and each degree of freedom of the main effects has been coded and punched on the data cards — with one of the kinds of coding discussed in Chapter 7. The sums of squares are calculated with the formula: $ss_i = (\Sigma y_i^2)(rs_i^2)$, where $ss_i =$ sum of squares of vector i: $\Sigma y_i^2 =$ total sum of squares (deviation sum of squares) of Y, the criterion variable, and $rs_i^2 = $ semi-partial (part) correlations squared. These are the sums of squares that *each vector* contributes to Y. Since an effect may have more than one vector, SPEC adds the sums of squares of the vectors that belong to a particular effect (in the analysis of variance sense) to form the sum of squares due to the effect. For instance, in a 3×5 design, there will be two vectors for A and four vectors for B and eight vectors for the interaction between A and B. (Since there are 15 cells, there are 14 degrees of freedom, and $2 + 4 + 8 = 14$.) These sums of vector sums of squares are then used for factorial analysis of variance, SPEC calculating the appropriate degrees of freedom, the mean squares, the F ratios, and the probabilities associated with each of the F ratios.

Summary Instructions. To be quite clear about how to handle factorial analysis of variance, we repeat the instructions in different words. To obtain factorial analysis of variance, a 1 is punched in Col. 45 (IANOVA = 1) on card 4, calling for analysis of variance in general, and a 1 is punched in Col. 55 (IFACNO) *or* in Col. 60 (IFAC3). The 1 in Col. 55 is for a two-effect ANOVA; the 1 in Col. 60 is for a three-effect ANOVA. (Col. 50 should always be left blank.) In short, only one of the choices can be made. Code only the main effects. (See Chapter 8.) The program will generate the interaction vectors. The sums of squares of the separate coded variables are printed followed by the usual ANOVA table. Note that when factorial ANOVA is done with IFACNO = 1 or IFAC3 = 1, much of the output described earlier is not printed.

```
      PROGRAM MULR (INPUT,OUTPUT,TAPE5=INPUT)
C
C     PROGRAM MULR.  MULTIPLE REGRESSION--1969, 1970, FEB., 1972.
C
C     KERLINGER PROGRAM.  (CDC-6600)
C
C     USES SUBROUTINES MATOUT AND LEQ.  ALSO JASPEN, FOR P OF F-RATIO.
C     CALCULATES ALL THE USUAL MULTIPLE REGRESSION STATISTICS,
C     INCLUDING F TEST. ALTHOUGH DOES NOT HAVE STEPWISE FEATURE,
C     PERMITS USER TO PLACE ANY VARIABLE AS THE Y, OR CRITERION
C     VARIABLE. I.E.,PERMITS REARRANGEMENT OF DATA (RAW SCORES)
C     IN ANY ORDER.  THIS INCLUDES THE OPTION OF DROPPING VARIABLES.
C     ALSO CALCULATES SUPPLEMENTARY REGRESSION STATISTICS--
C     PARTIAL CORRELATIONS, SEMIPARTIAL (PART) CORRELATIONS SQUARED,
C     WHICH INDICATE  THE PROPORTIONS OF VARIANCE ACCOUNTED FOR
C     BY THE SUCCESSIVE CUMULATED INDEPENDENT VARIABLES,  T TESTS
C     OF THE SIGNIFICANCE OF THE REGRESSION COEFFICIENTS, AS WELL
C     AS F TESTS OF VARIOUS STATISTICS.
C     THE PREDICTED CRITERION MEASURES ARE ALSO CALCULATED AND
C     PRINTED, IF DESIRED.
C     THE USER CAN GENERATE CROSS PRODUCTS (FOR INTERACTIONS)
C     AND POWERS OF VARIABLES (FOR TREND ANALYSIS), IF NEEDED.
C     THE PROGRAM ALSO DOES 1- AND 3-VARIABLE FACTORIAL ANOVA.
C
C     ****************************************************************
C     STORAGE ALLOCATION FOR MULR   IS 135000 (OCTAL) WORDS OF CENTRAL M
C     ****************************************************************
C
C     PREPARE PROGRAM AND DATA CARDS AS FOLLOWS--
C
C     1.  DESCRIPTION CARD
C         COLS. 1-80   ANY DESCRIPTION DESIRED (ALPHANUMERIC)
C
C     2.  CHOICE-CONTROL CARD
C
C         COLS. 1 - 5    N=NO. OF SUBJECTS
C         COLS. 6 -10    KT=TOTAL NO. OF VARIABLES (INCLUDING THE
C                            CRITERION OR Y-VARIABLE)
C         COLS. 11-15    K=NO. OF PREDICTOR VARIABLES (KT-1)
C         COL.    20     NFVC = NO. OF VARIABLE FORMAT CARDS
C         COL.    25     1= PRINT RAW SCORES
C           (IPRX)       0= NO PRINT SCORES
C         COL.    30     1= READ IN CORRELATION MATRIX
C          (IRMAT)       0= READ IN RAW SCORES
C         COL.    35      1 = CALCULATE AND PRINT PREDICTED Y SCORES
C          (IPRY)         0 = DO NOT CALCULATE AND PRINT PREDICTED Y
C              (NOTE--IF IRMAT=1, PUNCH 0 HERE.)
C         COL.    40     1= WANT DATA TO BE REARRANGED.
C          (ICHVAR)      0= WANT DATA TO BE READ IN AS ON DATA CARDS.
C                         (DOES NOT APPLY IF R-MATRIX READ IN.
C                          THUS, PUNCH 0  )
C         COL.    45     1=ANALYSIS OF VARIANCE WANTED
C          (IANOVA)      0=NO ANOVA
C         COL.    50     ---LEAVE BLANK---
C         COL.    55     1 = FACTORIAL ANOVA, WITH DF
```

```
C        (IFACNO)              CODING, 2 EFFECTS OR FACTORS.
C                              0 = NOT FACTORIAL ANOVA, 2 EFFECTS
C        COL.    60            1 = FACTORIAL ANOVA, 3 EFFECTS OR FACTORS.
C                              0 = NOT 3 EFFECTS
C        COL. 65               1=WANT SOME VARIABLES RAISED TO POWERS
C        (IIPOW)               0=DO NOT WANT POWERS
C        COL. 70               1=WANT INTERACTIONS OF SOME VARIABLES
C        (IINT)                1=DO NOT WANT INTERACTIONS
C
C    IF IFACNO=1 (COL. 55) OR IFAC3=1 ( COL. 60), THEN IANOVA, COL.
C    45, MUST BE 1.  ONLY ONE OF THESE CHOICES CAN BE MADE.
C    IF IFACNO=1 OR IFAC3=1, I.E., 2-WAY OR 3-WAY ANOVA,
C    CODE ONLY THE MAIN EFFECTS.  THE COMPUTER WILL CALCULATE THE
C    INTERACTION CODING.  K WILL BE THE NUMBER OF CODED VECTORS
C    FOR THE MAIN EFFECTS AND KT = K + 1.  E.G., IF A 3 X 4
C    FACTORIAL DESIGN, THEN DF = 2 AND 3 AND K = 2 + 3 = 5
C    AND KT = 5 + 1 = 6.
C
C
C    3. FACTORIAL ANALYSIS PARAMETER CARD.  IF HAVE SELECTED
C       FACTORIAL ANOVA CHOICE (1 IN COL. 50, 55, OR 60,
C       MUST INSERT THIS CARD.  PUNCH THE NUMBER OF PARTITIONS
C       OF THE A-VARIABLE IN COL. 5 AND THE NUMBER OF PARTITIONS
C       OF THE B-VARIABLE IN COL. 10.  THAT IS, IF THE FACTORIAL
C       ANOVA IS A 2 BY 3 DESIGN, PUNCH 2 IN COL. 5 AND 3 IN COL. 10.
C    ENTER THE VARIABLE WITH THE SMALLER NUMBER OF PARTITIONS FIRST
C    AND CALL IT THE A-VARIABLE.  THE VARIABLE WITH THE LARGER
C    NUMBER OF PARTITIONS, OF COURSE, WILL BE THE B-VARIABLE.
C    IF IFAC3=1 (COL. 60) IN 2,ABOVE, THEN PUNCH THE NUMBERS OF
C    PARTITIONS IN COLS. 5, 10, AND 15, PUNCHING THE PARTITIONS
C    IN RANK ORDER FROM LOW TO HIGH, E.G., IN A 2 X 2 X 4 DESIGN,
C    PUNCH 2  2  4.    CALL THESE EFFECTS A, B, AND C.  THE CODED
C    VECTORS MUST BE IN THIS ORDER.
C       IF NOT DOING FACTORIAL ANOVA, OMIT THIS CARD.
C
C    4.  VARIABLE FORMAT CARD(S). SPECIFY THE FORMAT OF THE DATA CARDS
C            WHETHER RAW SCORES OR R-MATRIX. FORMAT MUST HAVE AN I-SPECI
C            FICATION BEFORE THE DATA SPECIFICATION, WHICH MUST BE
C            F-SPECIFICATION
C
C    NOTE--IF WANT SUCCESSIVE RUNS WITH CERTAIN VARIABLES DELETED
C    AND PERHAPS REARRANGED, CHANGE THE VARIABLE FORMAT CARD TO READ
C    IN THOSE VARIABLES WANTED IN THE ANALYSIS AND PUNCH THE REARRANGE
C
C    5.  REARRANGE CARD.  IF COL. 40 OF THE CHOICE-CONTROL CARD HAS
C    CARD (SEE NO. 5, BELOW) ACCORDINGLY.  (CARDS 1 AND 2 CAN BE
C
C    THE SAME.)
C            BEEN PUNCHED 1, THIS CARD MUST BE INSERTED. IF COL. 40=0,
C            OMIT THIS CARD.
C
C        COLS. 8-10     PUNCH NUMBER OF VARIABLES TO BE READ FROM
C        (KKT)          DATA CARDS--INCLUDING THE Y VARIABLE
C        COLS. 11-12,13-14,...ETC. PUNCH ORDER DESIRED.
C        (IVAR(J))      IF, FOR EXAMPLE, THE Y VARIABLE IS THE FIRST
C                       VARIABLE ON THE CARDS AND THERE ARE FOUR
C                       VARIABLES, PUNCH 4 IN COL. 10 (THE NUMBER
C                       OF VARIABLES), AND, IN COLS. 12, 14, 16,
C                       AND 18, PUNCH 4  1  2  3.  THE Y VARIABLE
C                       MUST ALWAYS HAVE THE NUMBER EQUAL TO THE
```

```
C              NUMBER OF VARIABLES--I.E., IT MUST COME LAST.
C              THE OTHER VARIABLES CAN HAVE ANY
C              NUMBERS DESIRED.
C              (USER CAN THUS CHOOSE ANY NUMBER OF VARIABLES
C              FROM THE DATA CARDS, AND CAN PUT THEM IN ANY
C              ORDER DESIRED. THIS IS USEFUL USUALLY AFTER
C              AN INITIAL RUN, AND THE USER WANTS DIFFERENT
C              ANALYSES. (REORDERING AND SELECTION OF
C              VARIABLES IS NOT POSSIBLE WITH AN R-MATRIX.)
C              NOTE--THE NUMBERS PUNCHED IN COLS. 11-12,
C                13-14, ETC. WILL BE ATTACHED TO THE VARIABLES
C                AS THEY APPEAR ON THE DATA CARDS. IF MORE
C                THAN ONE CARD IS NEEDED, CONTINUE PUNCHING
C                SECOND CARD STARTING FROM COLS. 11-12.
C
C      6.   POWERS CARD. IF IIPOW=1 (ABOVE), INSERT THIS CARD.
C              OTHERWISE, OMIT. PUNCH THE NO. OF VARIABLES TO BE RAISED
C              TO POWERS IN COL. 10. THEN, PUNCHTHE NUMBERS OF THE VARIABLE
C              WHOSE POWERS ARE DESIRED AND THE POWERS TO WHICH TO RAISE
C              THE VARIABLES IN FIELDS OF 5, AS FOLLOWS--THE FIRST
C              VARIABLE NUMBER IN COL. 13 AND THE POWER TO WHICH TO RAISE
C              IT IN COL. 15, THE SECOND VARIABLE NUMBER IN COL. 18 AND ITS
C              POWER IN COL. 20, AND SO ON.
C
C      7.   INTERACTIONS (CROSS PRODUCTS) CARD. IF IINT=1, INSERT
C              THIS CARD. OTHERWISE, OMIT. PUNCH THE NUMBER OF VARIABLES
C              WHOSE INTERACTIONS ARE WANTED IN COL. 10. IN COLS.
C              12, 14, 16, ETC., PUNCH THE NUMBERS OF THE VARIABLES
C              WHOSE INTERACTIONS ARE WANTED.
C
C      (NOTE--IF BOTH POWERS AND INTERACTIONS ARE WANTED, NEW VECTORS FO
C      BOTH ARE GENERATED. THE USER IS CAUTIONED, HOWEVER,
C      AGAINST INDISCRIMINATE USE OF THIS FEATURE OF THE PROGRAM.)
C
C      8.   DATA CARDS.     MAY BE EITHER RAW SCORES OR THE CORRELATIONS
C                           OF AN R-MATRIX. PRECEDE THE ACTUAL DATA OF
C                           EACH SUBJECT, OR VARIABLE, WITH AN I-SPECI-
C                           FICATION FOR IDENTIFICATION.
C
C         NOTE--MORE THAN ONE SET OF DATA CAN BE RUN ON ONE PASS.
C               INSERT BEFORE SUCH SETS OF DATA CARDS  PROGRAM
C               CARDS 1, 2, 3, 4, 5, 6, AND 7, AS NEEDED.
C
C      NOTE--IF THE INPUT TO MULR IS A CORRELATION MATRIX,
C      CERTAIN REGRESSION STATISTICS CALCULATED WHEN THE INPUT IS IN
C      RAW SCORE FORM CANNOT BE CALCULATED.
C
C      ALL OUTPUT IS APPROPRIATELY LABELED. NOTE THAT AFTER THE MAIN
C      REGRESSION ANALYSIS THE REGRESSION OF THE Y VARIABLE ON THE
C      SEPARATE X VARIABLES IS DONE.
C      IMMEDIATELY AFTER THIS ANALYSIS, THE REGRESSION OF Y
C      ON THE REMAINING VARIABLES IS DONE. THE USER CAN THUS COMPARE
C      THE SEPARATE AND COMBINED EFFECTS OF THE INDEPENDENT VARIABLES.
C         SEE INSTRUCTIONS FOR NOTES ON THE SUPPLEMENTARY
C         REGRESSION ANALYSES THE PROGRAM DOES.
C
C      THE PRESENT CAPACITY OF THE PROGRAM IS 30
C      INDEPENDENT VARIABLES. THE CAPACITY CAN BE INCREASED, IF NEEDED,
C      BY CHANGING THE DIMENSION STATEMENTS OF THE MAIN PROGRAM
C      AND CERTAIN CALL STATEMENTS.
```

```
C
C
      DIMENSION RR(30,30  ),R(30,30  ),ID(2000),AM(100),V(100),SD(100)
      DIMENSION  X(30),  SX(100),SY(100),SSX(30,30  ),SSD(30,30  )
      DIMENSION B(100),BSQ(100),BW(100),BWT(100),IVAR(100),RY(100)
      DIMENSION FMT(32),DESCR(8),XORD(30),  DIAG(100),BORD(100)
      DIMENSION IORD(100),SSSEP(100),L(100)
      DIMENSION SSDIF(30),AMSQDF(30),AMSQSP(30)
      DIMENSION XX(500,30),DD(500),YPR(500),YY(500)
      DIMENSION U(30,30),PI(30,30),QQ(30,30),BETA(30),LABB(20)
      DIMENSION INTVEC(10),IPVEC(10),IPOW(10)
      COMMON R,SD,SSD,C
C
   10 FORMAT (8A10)
   11 FORMAT (1H0)
   13 FORMAT (1H1)
  100 PRINT 13
      PRINT 11
      PRINT 104
  104 FORMAT (2X,*PROGRAM MULR (FNK)*///)
  105 READ 10,(DESCR(J),J=1,8)
      IF (EOF,5)  1003,1005
 1003 CALL EXIT
 1005 CONTINUE
      PRINT 108,(DESCR(J),J=1,8)
  108 FORMAT (2X,8A10)
      PRINT 11
      PRINT 11
  110 READ 114,N,KT,K,NVFC,IPRX,IRMAT,IPRY,ICHVAR,IANOVA,IFACAN,IFACNO,
     1IFAC3,IIPOW,IINT
  114 FORMAT (14I5)
      KTSAVE = KT
      KSAVE = K
      PRINT 116,N,KT,K,NVFC,IPRX,IRMAT,IPRY,ICHVAR
  116 FORMAT (3X,4HN = I5,2X,5HKT = I4,2X,4HK = I4,2X,7HNVFC = I3,2X,
     17HIPRX = I3,2X,8HIRMAT = I3,2X,7HIPRY = I3,2X,9HICHVAR = I3//)
      PRINT 118,IANOVA,IFACAN,IFACNO,IFAC3,IIPOW,IINT
  118 FORMAT (3X,9HIANOVA = I3,2X,9HIFACAN = I3,2X,9HIFACNO = I3,2X,8HI
     1AC3 = I3,15X,8HIIPOW = I2,2X,7HIINT = I3//)
      IF (IANOVA.EQ.0)  GO TO 120
 8005 IF (IFACAN.EQ.0)  GO TO 8050
 8010 READ 8014,KA,KB
 8014 FORMAT (2I5)
      PRINT 8018,KA,KB
 8018 FORMAT (3X,*THIS WILL BE A* I2,* BY*I2,* FACTORIAL ANALYSIS OF VA
     1IANCE - REGRESSION ANALYSIS*//)
      KGROUP=KA*KB
      AKTM1=KGROUP-1
      ANMKT=N-KGROUP
      ANDF1=AKTM1
      ANDF2=ANMKT
      NDF1=ANDF1
      NDF2=ANDF2
 8050 CONTINUE
      IF (IFACNO.EQ.0)  GO TO 8055
      READ 8014,KA,KB
      PRINT 8018, KA,KB
 8055 CONTINUE
      IF (IFAC3.EQ.0)  GO TO 120
      READ 8024,KA,KB,KC
```

```
8024 FORMAT  (3I5)
     PRINT 8028,KA,KB,KC
8028 FORMAT (3X,*THIS WILL BE A*I2,* BY*I2,* BY*I2,* FACTORIAL ANALYSI
    1 OF VARIANCE - REGRESSION ANALYSIS*//)
 120 NFC = NVFC*8
     READ 124,(FMT(J),J=1,NFC)
 124 FORMAT (32A10)
     PRINT 128,NFC,NVFC,(FMT(J),J=1,NFC)
 128 FORMAT (3X,6HNFC = I3,2X,7HNVFC = I3,6X,*DATA FORMAT  *32A10)
     PRINT 13
     PRINT 11
     II=1
     IF (IRMAT) 135,135,155
 135 IF (IPRX) 199,199,140
 140 PRINT 144
 144 FORMAT (2X,*ORIGINAL DATA*//)
     GO TO 199
 155 IF  (ICHVAR.EQ.1)  GO TO 200
 157 DO 160 I=1,KT
 160 READ FMT, I, (R(I,J),J=1,KT)
     PRINT 164
 164 FORMAT (2X,*R-MATRIX*//)
C
     CALL MATOUT (R,KT,30)
C
     GO TO 599
 199 CONTINUE
C      **********  PROGRAM CHANGED BELOW KT AS UPPER LIMIT TO 30  *****
 300 DO 305  J=1,30
     SX(J)=0.0
     DO 305  I=1,30
 305 SSX(I,J)=0.0
C
C    READ IN NUMBER AND ORDER OF VARIABLES DESIRED IN ANALYSIS.
C    (ICHVAR=1).
C
 200 IF (ICHVAR) 229,229,205
 205 READ 218,KKT,(IVAR(J),J=1,KKT)
 218 FORMAT (I10,35I2)
     KT=0
     KT=KKT
     K =KT-1
     PRINT 224,KKT,(IVAR(J),J=1,KKT)
 224 FORMAT (2X,*NUMBER OF VARIABLES IN ANALYSIS    *I5/2X,*DESIRED OR
    1ER OF VARIABLES    *30I3)
     PRINT 11
 229 CONTINUE
     IF (IINT.EQ.0.AND.IIPOW.EQ.0)  GO TO 7551
5301 IF (IIPOW)  5310,5310,5305
5305 READ 5308,KPOW, (IPVEC(J),IPOW(J),J=1,KPOW)
5308 FORMAT (8X,I2,10(I3,I2))
5310 IF (IINT)  5319,5319,5311
5311 READ 5314,KIVEC, (INTVEC(J),J=1,KIVEC)
5314 FORMAT (8X,I2,10I2)
5319 CONTINUE
     IF (IIPOW)  5330,5330,5321
5321 PRINT 5324,KPOW
5324 FORMAT (3X,*NO. OF VARIABLES POWERED = *I3//)
     PRINT 5328, (IPVEC(J),IPOW(J), J=1,KPOW)
5328 FORMAT (3X, *VARIABLES POWERED AND POWERS--*//7X,10(2I3,3X))
```

```
      PRINT 11
 5330 IF (IINT)  5339,5339,5331
 5331 PRINT 5334,KIVEC, (INTVEC(J),J=1,KIVEC)
 5334 FORMAT (3X, *NO. OF INTERACTIONS (CROSS PRODUCTS) = *I4,4X,*VARS.
     1SELECTED--*10I2//)
 5339 CONTINUE
      PRINT 11
      PRINT 11
 7551 CONTINUE
      KT = KTSAVE
      K = KSAVE
      IF (ICHVAR)  310,310,230
  230 IF (IRMAT.EQ.0)  GO TO 237
C
      CALL REARR  (KT,K,FMT,IVAR)
C
      GO TO 599
C
C     READ DATA IN REARRANGED ORDER.
C
  237 CONTINUE
      KT = KTSAVE
      K = KSAVE
      READ FMT, ID(II), (XORD(J),J=1,KT)
      DO 239  J=1,KT
      NUM = IVAR(J)
      X(NUM)=XORD(J)
  239 CONTINUE
      IF (IIPOW.EQ.1.OR.IINT.EQ.1)  GO TO 9812
 9800 DO 9810  J=1,KT
 9810 XX(II,J)=X(J)
 9812 CONTINUE
      IF (IFACNO.EQ.0.AND.IFAC3.EQ.0)  GO TO 351
C
      CALL VECINT  (KT,XX,II,X,KA,KB,K,IFACNO,IFAC3,KC)
C
      IF (IPRX)  260,260,250
  250 PRINT 254,ID(II),(X(J),J=1,KT)
  254 FORMAT (4X,I5,3X,10F11.3/ (12X,10F11.3))
  260 CONTINUE
      GO TO 400
  310 CONTINUE
      KT = KTSAVE
      K = KSAVE
  311 READ FMT,ID(II), (X(J),J=1,KT)
      IF (IIPOW.EQ.1.OR.IINT.EQ.1)  GO TO 9839
 9820 DO 9830  J=1,KT
 9830 XX(II,J)=X(J)
 9839 CONTINUE
      IF (IFACNO.EQ.0.AND.IFAC3.EQ.0)  GO TO 351
C
      CALL VECINT  (KT,XX,II,X,KA,KB,K,IFACNO,IFAC3,KC)
C
  351 CONTINUE
      IF (IIPOW)  5520,5520,5501
C
 5501 CALL POW (X,IPVEC,KT,K,KTOT,KPOW,IPOW)
      KT=KTOT
      K=KTOT-1
 5601 DO 5610  J=1,KT
```

```
 5610 XX(II,J)=X(J)
C
 5520 CONTINUE
      IF (IINT)    5771,5771,5510
 5510 CALL SELCP (X,KIVEC,INTVEC,K,KT,KTOT)
      KT=KTOT
      K=KTOT-1
 5701 DO 5715   J=1,KT
 5715 XX(II,J) = X(J)
 5771 CONTINUE
 5551 CONTINUE
      IF(IPRX) 359,359,355
  355 IF (KT.LE.7)   GO TO 7739
      PRINT 6254, ID(II), (X(J),J=1,KT)
 6254 FORMAT (4X,I5,3X,7F16.2/(12X,7F16.2))
      GO TO 359
 7739 PRINT 7744,ID(II), (X(J),J=1,KT)
 7744 FORMAT (4X,I5,3X,7F16.2)
  359 CONTINUE
  400 DO 410   J=1,KT
  410 SX(J)=SX(J)+X(J)
      DO 420   I=1,KT
      DO 420   J=1,KT
  420 SSX(I,J)=SSX(I,J)+X(I)*X(J)
      IF(II.EQ.N) GO TO 1999
      II=II+1
      GO TO 237
 1999 CONTINUE
      IF (IFACNO.EQ.0.AND.IFAC3.EQ.0)   GO TO 6731
      PRINT 11
      PRINT 11
      PRINT 6724,KT,K
 6724 FORMAT   (3X, *CHECK ON KT AND K      KT = * I4,4X, *K = * I4//)
 6731 CONTINUE
C
C     DEVIATION SUMS OF SQUARES AND CROSS-PRODUCTS, MEANS, VARIANCES,
C      STANDARD DEVIATIONS, R-MATRIX CALCULATIONS.
C
      AN=N
  500 DO 510  I=1,KT
      DO 510  J=1,KT
  510 SSD(I,J)=SSX(I,J)-(SX(I)*SX(J))/AN
  515 DO 530  J=1,KT
  520 AM(J)=SX(J)/AN
  525 V(J)=SSD(J,J)/(AN-1.0)
  530 SD(J)=SQRT(V(J))
  540 DO 550  I=1,KT
      DO 550  J=1,KT
  550 R(I,J)=SSD(I,J)/SQRT(SSD(I,I)*SSD(J,J))
C
C     PRINT SSD(I,J),AM(J),V(J),SD(J),R(I,J)
C
  560 PRINT 13
      PRINT 11
      PRINT 564
  564 FORMAT (2X,*MEANS, VARIANCES,  STANDARD DEVIATIONS*//)
      PRINT 574, (J,AM(J), J=1,KT)
  574 FORMAT (6X, *MEANS*,14X, 4(I2,F14.4,5X) / (25X, 4(I2,F14.4,5X))
      PRINT 11
  578 FORMAT  (6X, *VARIANCES*,10X,4(I2,F14.4,5X)/(25X,4(I2,F14.4,5X))
```

```
      PRINT 578, (J, V(J), J=1,KT)
      PRINT 11
      PRINT 584, (J, SD(J), J=1,KT)
  584 FORMAT (6X,*STAN. DEVS.*,8X,4(I2,F14.4,5X)/(25X,4(I2,F14.4,5X)))
      PRINT 11
      PRINT 11
      PRINT 11
      PRINT 588
  588 FORMAT (3X,*R-MATRIX*//)
C
      CALL MATOUT (R,KT,30)
C
      PRINT 13
      PRINT 592
  592 FORMAT (3X,*SSD(I,J)*/ )
      DO 1120  J=1,KT
 1120 L(J)=J
      IJ=KT/6
      IF (KT-6*IJ)  1127,1127,1125
 1125 IJ=IJ+1
 1127 DO 1139  IK=1,IJ
      KK=IK*6
      II=KK-5
      IF (KK-KT)  1131,1131,1129
 1129 KK=KT
 1131 PRINT 1134,(L(J),J=II,KK)
 1134 FORMAT  (2X,6I18)
      PRINT 11
      DO 1135  I=1,KT
 1135 PRINT 1136,I,(SSD(I,J),J=II,KK)
 1136 FORMAT (I6,6F18.4)
      PRINT 11
 1139 CONTINUE
  599 CONTINUE
C
C     PARTITION R-MATRIX--R INTO RR, K=KT-1, R(KT) TO RY, RY(I) TO B(I)
C
  600 DO 610  I=1,K
      DO 610  J=1,K
  610 RR(I,J)=R(I,J)
      DO 620  I=1,K
  620 RY(I)=R(I,K+1)
      DO 625  I=1,K
  625 B(I)=RY(I)
C
C     SOLVE SET OF EQUATIONS,(RR)B=R(I,KT),B=(RR)-1R(I,KT), USING LEQ.
C
      DO 6310  I=1,K
      DO 6310  J=1,K
 6310 U(I,J) = R(I,J)
C
      CALL INVERT  (K,DET,30,30,U,PI)
C
      PRINT 11
      PRINT 6324
 6324 FORMAT (3X, *R-INVERSE PI--OF INDEPENDENT VARIABLES */ )
C
      CALL MATOUT (PI,K,30)
C
C     PROOF--R-INVERSE*R = I
```

```
C
      DO 6320   I=1,K
      DO 6320   J=1,K
 6320 QQ(I,J) = 0.0
      DO 6352   I=1,K
      DO 6351   M=1,K
      DO 6350   J=1,K
      QQ(I,M) = QQ(I,M) + PI(I,J) * R(J,M)
 6350 CONTINUE
 6351 CONTINUE
 6352 CONTINUE
      PRINT 11
      PRINT 6354
 6354 FORMAT (3X,*PROOF OF INVERSE MATRIX--PI R=QQ=I*/ )
C
      CALL MATOUT (QQ,K,30)
C
      PRINT 6358,DET
 6358 FORMAT (3X,*DETERMINANT OF R(I,J) = * F13.8//)
C
C     CALCULATE BETAS--
C     BETA(J) = PI(I,J)*RY(J)
C
      DO 6410   J=1,K
 6410 BETA(J) = 0.0
      DO 6420   I=1,K
      DO 6420   J=1,K
 6420 BETA(I) = BETA(I) + PI(I,J)*RY(J)
      PRINT 11
C
      CALL LEQ (RR,B,K,K,30,30,DET)
C
C     CALCULATE AND PRINT R,RSQ,DF,F-RATIO, BSQ,B WEIGHTS,
C     AND INTERCEPT CONSTANT (SUM OF PRODUCTS OF B WEIGHTS AND MEANS).
C
  700 RSQ=0.0
      DO 710   I=1,K
  710 RSQ=RSQ+B(I)*RY(I)
      RMULT=SQRT(RSQ)
  800 DO 810   I=1,K
  810 BSQ(I)=B(I)**2
      IF (IRMAT)  815,815,7010
  815 DO 820   I=1,K
  820 BW(I)=SD(K+1)/SD(I)
      DO 830   I=1,K
  830 BWT(I)=BW(I)*B(I)
 7100 SSREG=0.0
      DO 7110   I=1,K
 7110 SSREG=SSREG + BWT(I)*SSD(I,KT)
      SSDEV = SSD(KT,KT) - SSREG
      DO 9060   I=1,K
 9060 SSSEP(I)=SSD(I,KT)**2/SSD(I,I)
 7010 IF (IFACNO.EQ.1.OR.IFAC3.EQ.1)   GO TO 8170
      PRINT 13
      PRINT 714
  714 FORMAT (10X,*MULTIPLE REGRESSION ANALYSIS*////)
  720 PRINT 724, (J,RY(J), J=1,K)
  724 FORMAT  (3X,*RY(J)=R(I,KT)*,9X,6(I2,F10.4,5X) /(25X,6(I2,F10.4,5X
     1))
      PRINT 11
```

```
      PRINT 668, (J,B(J), J=1,K)
  668 FORMAT  (3X,*BETAS*, 17X, 6(I2,F10.4,5X) / (25X, 6(I2,F10.4,5X)))
      PRINT 11
      PRINT 864, (J,BSQ(J), J=1,K)
  864 FORMAT  (3X,*BETAS SQUARED*,9X, 6(I2,F10.4,5X) /(25X, 6(I2,F10.4,
     1X)))
      PRINT 11
      IF (IRMAT)  7020,7020,7025
 7020 PRINT 854, (J,BWT(J), J=1,K)
  854 FORMAT (3X, *B COEFFICIENTS*, 8X, 6(I2,F10.4,5X) / (25X,6(I2,F10.
     1,5X)))
 7025 PRINT 11
      PRINT 11
  730 PRINT 734,RMULT,RSQ
  734 FORMAT (6X,*MULTIPLE R = *F8.4,8X,*MULTIPLE R SQUARED = *F8.4/)
 8170 IF (IFACAN.EQ.1)  GO TO 755
  750 NDF1=KT-1
      NDF2=N-KT
      ANDF1=NDF1
      ANDF2=NDF2
  755 F=(RSQ*ANDF2)/((1.0-RSQ)*ANDF1)
C
      CALL JASPEN (F,NDF1,NDF2,P)
C
      SVE=1.0-RSQ
      SEM=SQRT(SVE)
      IF (IFACNO.EQ.1.OR.IFAC3.EQ.1)  GO TO 7765
  760 PRINT 764,F,NDF1,NDF2,P
  764 FORMAT(6X,4HF = F10.4,6X,6HDF1 = I4,3X,6HDF2 = I4,3X,4HP = F12.7/
      PRINT 11
 7765 CONTINUE
      IF (IFACAN)  8175,8175,8180
 8175 AKTM1=KT-1
      ANMKT=N-KT
 8180 IF (IRMAT)  9070,9070,4060
 9070 AMSQRG=SSREG/AKTM1
      AMSQDV=SSDEV/ANMKT
      IF (IFACNO.EQ.1.OR.IFAC3.EQ.1)  GO TO 7030
      PRINT 9064,SSREG,SSDEV
 9064 FORMAT (6X,*REGRESSION SUM OF SQUARES = *F14.4,6X,*DEVIATION SUM
     1F SQUARES = *F12.4//)
      PRINT 9068,AMSQRG,AMSQDV
 9068 FORMAT (6X,*REGRESSION MEAN SQUARE = *F10.4,6X,*DEVIATION MEAN SQ
     1ARE = *F10.4//)
      PRINT 9074, (J,SSSEP(J),J=1,K)
 9074 FORMAT (6X,*SEPARATE SUMS OF SQUARES   * 5(I2,F13.4,3X) / (34X,
     1(I2,F13.4,3X)))
      PRINT 11
C
C     INTERCEPT CONSTANT
 7030 CC=0.0
      DO 840  I=1,K
  840 CC=CC+BWT(I)*AM(I)
      C=AM(K+1)-CC
      SERAW=SD(KT)*SQRT(1.0-RSQ)
      IF  (IFACNO.EQ.1.OR.IFAC3.EQ.1)  GO TO 7769
      PRINT 858,C
  858 FORMAT (10X,*INTERCEPT CONSTANT = *F10.4//)
      PRINT 13
C
```

```
      CALL REGSTAT  (K,KT,SSD,BWT,PI,N,SEEST,SSDEV,SEM,SVE)
C
      PRINT 11
      PRINT 768, SEEST, SEM
  768 FORMAT   (6X, *STANDARD ERROR OF ESTIMATE = * F10.4,5X, *SE ESTIMA
     1E. Z-SCORES = *  F10.4////)
      PRINT 5004
 5004 FORMAT  (6X, *REGRESSION EQUATION*//)
      LABB(1) = 4HX(1)
      LABB(2) = 4HX(2)
      LABB(3) = 4HX(3)
      LABB(4) = 4HX(4)
      LABB(5) = 4HX(5)
      LABB(6) = 4HX(6)
      LABB(7) = 4HX(7)
      LABB(8) = 4HX(8)
      LABB(9) = 4HX(9)
      LABB(10) = 5HX(10)
      LABB(11) = 5HX(11)
      LABB(12) = 5HX(12)
      LABB(13) = 5HX(13)
      LABB(14) = 5HX(14)
      LABB(15) = 5HX(15)
      LABB(16) = 5HX(16)
      LABB(17) = 5HX(17)
      LABB(18) = 5HX(18)
      LABB(19) = 5HX(19)
      LABB(20) = 5HX(20)
      PRINT 5008,C, (BWT(J), LABB(J), J=1,K)
 5008 FORMAT  (10X,*YP = *F10.4, 6(2X,*+*,F10.4,A5) // (15X, 6(2X,*+*,
     110.4,A5)))
 7769 CONTINUE
      IF (IPRY.EQ.0)  GO TO 9898
      DO 9855  I=1,N
 9855 YPR(I)=0.0
 9860 DO 9875  I=1,N
      DO 9870  J=1,K
 9870 YPR(I) = YPR(I) + BWT(J)*XX(I,J)
      YPR(I) = YPR(I) + C
 9875 CONTINUE
      PRINT 13
      PRINT 11
      PRINT 9874
 9874 FORMAT (1X,*OBTAINED Y-SCORES, PREDICTED Y-SCORES, AND DEVIATION
     1CORES*///)
      PRINT 9878
 9878 FORMAT (5X,*SUBJECT            Y        PREDICTED Y        DEVIATI
     1N*//)
 9880 DO 9885  I=1,N
 9885 YY(I)=XX(I,KT)
      DO 9890  I=1,N
 9890 DD(I) = YY(I) - YPR(I)
      DO 9891  I=1,N
 9891 PRINT 9894,I,YY(I),YPR(I),DD(I)
 9894 FORMAT (5X,I5,10X,F10.4,5X,F10.4,5X,F10.4)
 9898 CONTINUE
      PRINT 11
      PRINT 11
      PRINT 11
      PRINT 6674
```

```
6674 FORMAT  (25X, *END OF REGULAR REGRESSION ANALYSIS*//)
C
C      SEPARATE SUMS OF SQUARES AND F RATIOS--X2, . . . XK AFTER X1, ETC
C
       IF (K.EQ.1)  GO TO 4065
       IF (IFACNO.EQ.1.OR.IFAC3.EQ.1)  GO TO 4060
       PRINT 13
       PRINT 4004
 4004 FORMAT (1X,*SUPPLEMENTARY ANALYSIS OF SEPARATE SUMS OF SQUARES--*
       PRINT 4006
 4006 FORMAT(1X,*SIGNIFICANCE OF REMAINING VARIABLES AFTER REMOVING,*)
       PRINT 4007
 4007 FORMAT (1X,*ONE AFTER ANOTHER, VARIABLES 1, 2,  . . ., K*)
       PRINT 9018
 9018 FORMAT (1X,*SEE SNEDECOR AND COCHRAN, PP. 398-399 AND P. 407, FOR
      1DISCUSSION OF THIS FORM OF REGRESSION ANALYSIS.*//)
       PRINT 4014
 4014 FORMAT (2X,*SEPARATE SUMS OF SQUARES AND F RATIOS.    SSDIF(I) = S
      1REG-SSSEP(I),  FMI = SSDIF(I)/AMSQD *//)
C
       KKK=1
       AKKK=KKK
       MDF=K-1
       AMDF=MDF
       AK=K
C
       DO 4050  I=1,K
 8200 IF (IFACAN)  4020,4020,8220
 8220 AMSQDV= SSDEV/ANMKT
       GO TO (8230,8240,8250), I
 8230 KKK=KA-1
       AKKK=KKK
       MDF=KGROUP-1-KKK
       AMDF=MDF
       GO TO 8260
 8240 KKK=KB-1
       AKKK=KKK
       MDF=KGROUP-1-KKK
       AMDF=MDF
       GO TO 8260
 8250 KKK=(KA-1)*(KB-1)
       AKKK=KKK
       MDF=KGROUP-1-KKK
       AMDF=MDF
 8260 CONTINUE
 4020 SSDIF(I)=SSREG-SSSEP(I)
       AMSQDF(I)=SSDIF(I)/AMDF
       AMSQSP(I)=SSSEP(I)/AKKK
       FI = AMSQSP(I)/AMSQDV
       FMI = AMSQDF(I)/AMSQDV
C
       CALL JASPEN (FI,KKK,NDF2,P)
       CALL JASPEN  (FMI,MDF,NDF2,P2)
C
       PRINT 4038,I,SSSEP(I),AMSQSP(I)
 4038 FORMAT (4X,*SEPARATE VARIABLE NO.*I3,10X, *SUM OF SQUARES = *F11.
      1,7X,*MEAN SQUARE = *F11.4/)
       PRINT 764,FI,KKK,NDF2,P
       RSQSEP = SSSEP(I) / SSD(KT,KT)
       RSEP = SQRT(RSQSEP)
```

```
      PRINT 4044,RSEP,RSQSEP
 4044 FORMAT (8X,4HR = F8.4,6X,6HRSQ = F8.4//)
      PRINT 4034,SSDIF(I),AMSQDF(I)
 4034 FORMAT (6X,*XK AFTER XI (I=1,2,. . .,K)        SUM OF SQUARES = *F11
     1 4.7X,*MEAN SQUARE = *F11.4/)
      PRINT 764,FMI,MDF,NDF2,P2
      RSQDIF = SSDIF(I) / SSD(KT,KT)
      RDIF = SQRT(RSQDIF)
      PRINT 4044,RDIF,RSQDIF
 4050 CONTINUE
 4060 CONTINUE
C
C     CALL RZSEMP TO CALCULATE R-SQUARES VIA DETERMINANTS AND
C     SEMIPARTIAL R-SQUARES.
      CALL RZSEMP(K,KT,N,IRMAT,IFACNO,IFAC3,KA,KB,KC,SSDEV)
C
 4065 CONTINUE
  899 CONTINUE
      GO TO 100
      END
C
C

      SUBROUTINE MATOUT(A,K,IA)
C
      DIMENSION  A(IA,IA), L(100)
   11 FORMAT(1H0)
      DO 1120 J=1,K
 1120 L(J)=J
      IJ=K/12
      IF(K-12*IJ)1127,1127,1125
 1125 IJ=IJ+1
 1127 DO 1139 IK=1,IJ
      KK=IK*12
      II=KK-11
      IF(KK-K)1131,1131,1129
 1129 KK=K
 1131 PRINT 1134, (L(J),J=II,KK)
 1134 FORMAT (7X,12I10)
      PRINT 11
      DO 1135 I=1,K
 1135 PRINT 1136,I,(A(I,J),J=II,KK)
 1136 FORMAT(4X,I2,4X,12F10.4)
      PRINT 11
 1139 CONTINUE
      RETURN
      END
C
C

      SUBROUTINE LEQ(A,B,NEQS,NSOLNS,IA,IB,DET)
C
C     THIS SUBROUTINE IS FROM THE MATHEMATICS SUBROUTINE
C     LIBRARY OF THE COMPUTING CENTER OF THE COURANT
C     INSTITUTE OF MATHEMATICAL SCIENCES, NEW YORK UNIVERSITY.
C     IT IS REPRODUCED WITH THE PERMISSION OF THE CENTER.
C
C     SOLVE A SYSTEM OF LINEAR EQUATIONS OF THE FORM AX=B BY A MODIFIED
C     GAUSS ELIMINATION SCHEME
C
C     NEQS = NUMBER OF EQUATIONS AND UNKNOWNS
C     NSOLNS = NUMBER OF VECTOR SOLUTIONS DESIRED
```

```
C         IA = NUMBER OF ROWS OF A AS DEFINED BY DIMENSION STATEMENT ENTRY
C         IB = NUMBER OF ROWS OF B AS DEFINED BY DIMENSION STATEMENT ENTRY
C         ADET = DETERMINANT OF A, AFTER EXIT FROM LEQ
C
          DIMENSION A(IA,IA),B(IB,IB)
          NSIZ = NEQS
          NBSIZ = NSOLNS
C         NORMALIZE EACH ROW BY ITS LARGEST ELEMENT, FORM PARTIAL DETERNT
          DET=1.0
          DO 1 I=1,NSIZ
          BIG=A(I,1)
          IF(NSIZ-1)50,50,51
     51   DO 2 J=2,NSIZ
          IF(ABS (BIG)-ABS (A(I,J))) 3,2,2
   3      BIG=A(I,J)
   2      CONTINUE
          DO 4 J=1,NSIZ
   4      A(I,J)=A(I,J)/BIG
          DO 41 J=1,NBSIZ
   41     B(I,J)=B(I,J)/BIG
          DET=DET*BIG
   1      CONTINUE
C         START SYSTEM REDUCTION
          NUMSYS=NSIZ-1
          DO 14 I=1,NUMSYS
C         SCAN FIRST COLUMN OF CURRENT SYSTEM FOR LARGEST ELEMENT
C         CALL THE ROW CONTAINING THIS ELEMENT, ROW NBGRW
          NN=I+1
          BIG=A(I,I)
          NBGRW=I
          DO 5 J=NN,NSIZ
          IF(ABS (BIG)-ABS (A(J,I))) 6,5,5
   6      BIG=A(J,I)
          NBGRW=J
   5      CONTINUE
C         SWAP ROW I WITH ROW NBGRW UNLESS I=NBGRW
          IF(NBGRW-I) 7,10,7
C         SWAP A-MATRIX ROWS
   7      DO 8 J=I,NSIZ
          TEMP=A(NBGRW,J)
          A(NBGRW,J)=A(I,J)
   8      A(I,J)=TEMP
          DET = -DET
C         SWAP B-MATRIX ROWS
          DO 9 J=1,NBSIZ
          TEMP=B(NBGRW,J)
          B(NBGRW,J)=B(I,J)
   9      B(I,J)=TEMP
C         ELIMINATE UNKNOWNS FROM FIRST COLUMN OF CURRENT SYSTEM
   10     DO 13 K=NN,NSIZ
C         COMPUTE PIVOTAL MULTIPLIER
          PMULT=-A(K,I)/A(I,I)
C         APPLY PMULT TO ALL COLUMNS OF THE CURRENT A-MATRIX ROW
          DO 11 J=NN,NSIZ
   11     A(K,J)=PMULT*A(I,J)+A(K,J)
C         APPLY PMULT TO ALL COLUMNS OF MATRIX B
          DO 12 L=1,NBSIZ
   12     B(K,L)=PMULT*B(I,L)+B(K,L)
   13     CONTINUE
   14     CONTINUE
```

```
C       DO BACK SUBSTITUTION
C          WITH B-MATRIX COLUMN = NCOLB
   50 DO 15 NCOLB=1,NBSIZ
C       DO FOR ROW = NROW
        DO 19 I=1,NSIZ
        NROW=NSIZ+1-I
        TEMP=0.0
C       NUMBER OF PREVIOUSLY COMPUTED UNKNOWNS = NXS
        NXS=NSIZ-NROW
C       ARE WE DOING THE BOTTOM ROW
        IF(NXS) 16,17,16
C       NO
   16   DO 18 K=1,NXS
        KK=NSIZ+1-K
   18   TEMP=TEMP+B(KK,NCOLB)*A(NROW,KK)
   17   B(NROW,NCOLB)=(B(NROW,NCOLB)-TEMP)/A(NROW,NROW)
C       HAVE WE FINISHED ALL ROWS FOR B-MATRIX COLUMN = NCOLB
   19   CONTINUE
C       YES
C       HAVE WE JUST FINISHED WITH B-MATRIX COLUMN NCOLB=NSIZ
   15   CONTINUE
C       YES
C       NOW FINISH COMPUTING THE DETERMINANT
        DO 20 I=1,NSIZ
   20   DET=DET*A(I,I)
        ADET = DET
C       WE ARE ALL DONE NOW
C       WHEW...
        RETURN
        END
C
C
        SUBROUTINE INVERT   (N,DET,IA,IB,A,B)
C
C       A IS MATRIX TO BE INVERTED.  N IS ORDER OF MATRIX A.
C       GENERATES I MATRIX IN B AND CALLS LEQ.  I IS PLACED IN B,
C       AND A IS DESTROYED.   U=A.
C
        DIMENSION  A(IA,IA),B(IB,IB)
        DO 4  I=1,N
        DO 4  J=1,I
        IF (J-I)  2,3,2
    2   B(I,J)=0.0
        B(J,I)=0.0
        GO TO 4
    3   B(I,J)=1.00
    4   CONTINUE
C
        CALL LEQ (A,B,N,N,30,30,DET)
C
        RETURN
        END
C
C
        SUBROUTINE JASPEN   (F,I,J,P)
C
   11   FORMAT(1H0)
        P=1.0
        IF(F) 100,100,10
   10   IF(I) 100,100,20
```

```
   20 IF(J) 100,100,30
   30 IF(F-1.) 40,50,50
   40 B=J
      W=I
      G=1./F
      GO TO 60
   50 B=I
      W=J
      G=F
   60 ALPHA=2./(9.*B)
      BETA=2./(9.*W)
      TOP=(1.-BETA)*G**(1./3.)-1.+ALPHA
      BOT=SQRT(BETA*G**(2./3.)+ALPHA)
      Z=ABS (TOP/BOT)
      IF (W-3.) 70,70,80
   70 Z=Z*(1.+.0800*Z**4/W**3)
   80 CA=.196854
      CB=.115194
      CC=.000344
      CD=.019527
      P=.5/(1.+Z*(CA+Z*(CB+Z*(CC+Z*CD))))**4
      IF(F-1.) 90,100,100
   90 P=1.-P
  100 CONTINUE
      RETURN
      END
C
C
      SUBROUTINE RZSEMP    (K,KT,N,IRMAT,IFACNO,IFAC3,KA,KB,KC,SSDEV)
C
C     RZSEMP CALCULATES R SQUARES VIA DETERMINANTS, AFTER PARTITIONING
C     THE R-MATRIX.  THE PARTITIONED R-MATRICES ARE RZ(I,J).
C     SEMIPARTIAL R-SQUARES ARE THEN CALCULATED USING THE DIFFERENCES
C     BETWEEN THE DIFFERENT ORDER R-SQUARES, E.G., RSQ Y.1234 - RSQ Y.1
C
C     RSQ = 1.0 - (DET. OF LARGER MATRIX / DET. OF SMALLER MATRIX).
C     SEMP = RZSQ(J+1) - RZSQ(J).
C
      COMMON  R,SD,SSD,C
      DIMENSION  R(30,30),RZ(30,30),DETT(30), RZSQ(30),SEMPSQ(30),
     * RZIN(30,30),RZZ(30,30), RZZIN(30,30)
      DIMENSION DFF1(30),DFF2(30),NDFF1(30),NDFF2(30),F(30),TT(30)
      DIMENSION  PARR(30), W(30), SD(100),BB(30),BWGHT(30)
      DIMENSION RZR(30), PQQ(30),DETIND(30),SSD(30,30),SSQV(30)
   11 FORMAT (1H0)
   13 FORMAT (1H1)
      AN=N
      L=2
      PRINT 13
      PRINT 7704
 7704 FORMAT ( / ,2X,*R-SQUARES VIA DETERMINANTS, SEMIPARTIAL R-SQUARES
     1-SEQUENTIAL ANALYSIS*/1H0)
      M=1
 7700 MM=M+1
 7705 DO 7710  J=1,MM
 7710 RZ(J,J) = R(J,J)
      IF (M -1)  7720,7720,7715
 7715 DO 7719   I=1,M
      DO 7719   J=1,M
 7719 RZ(I,J) = R(I,J)
```

```
 7720 DO 7725  J=1,M
 7725 RZ(J,MM) = R(J,KT)
      DO 7730  I=1,M
 7730 RZ(MM,I) = R(KT,I)
      DO 8710  I=1,M
      DO 8710  J=1,M
 8710 RZZ(I,J) = RZ(I,J)
C
      CALL INVERT (M,DETIND(M),30,30,RZZ,RZZIN)
      CALL INVERT (MM,DETT(MM),30,30,RZ,RZIN)
C
C        PARTIAL R.
      DO 7805  I=1,MM
 7805 PARR(I) =(-RZIN(I,MM)) / (SQRT (RZIN(I,I) * RZIN (MM,MM)))
C
C          CALCULATION OF REGRESSION WEIGHTS, W(J).
C
 7830 DO 7835  J=1,MM
 7835 W(J) = -RZIN(J,MM) / RZIN (MM,MM)
      PRINT 4204, M,MM
 4204 FORMAT   (2X,4HM = I5,4X,5HMM = I3,3X,*(M = 1, 2, ...,M NO. OF IND
     1PENDENT VARIABLES)*//)
      PRINT 7838
 7838 FORMAT (5X,*BETAS*//)
      PRINT 7844, (J,W(J), J=1,M)
 7844 FORMAT   (7X, 6(I2,F11.5,5X))
      PRINT 11
      IF (IRMAT.EQ.1)  GO TO 7809
  815 DO 820  J=1,M
  820 BWGHT(J) = SD(KT) / SD(J)
      DO 830  J=1,M
  830 BB(J) = BWGHT(J) * W(J)
      PRINT 834
  834 FORMAT  (5X, *B-WEIGHTS*//)
      PRINT 7844,  (J,BB(J), J=1,M)
      PRINT 11
 7809 CONTINUE
      PRINT 7818
 7818 FORMAT (5X,                        *PARTIAL CORRELATIONS*//)
      PRINT 7844, (I,PARR(I), I=1,M)
      AM=M
      ICHA=M/2
      CHA=AM/2.0
      CCHA=ICHA
      IF (CHA-CCHA)  8110,8111,8110
 8110 PRINT 11
      PRINT 11
      GO TO 8115
 8111 PRINT 13
      PRINT 11
 8115 CONTINUE
C
C     CALCULATION OF R-SQUARE (RZSQ(I)).
C
 7900 IF (M-1) 7910,7910,7919
 7910 AM=M
      AMM=MM
      DFF1(1) = AMM - AM
      DFF2(1) = AN - AM - 1.0
      NDFF1(1) = DFF1(1)
```

```
      NDFF2(1) = DFF2(1)
      RZSQ(1) = R (1,KT)**2
      SEMPSQ(1) = RZSQ(1)
      F(1) =  (( RZSQ(1) * DFF2(1)) / ( 1.0 - RZSQ(1)) * DFF1(1))
      TT(1) = SQRT (F(1))
C
      CALL JASPEN (F(1), NDFF1(1), NDFF2(1), PQQ(1))
C
 7915 CONTINUE
      M=M+1
      GO TO 7700
 7919 CONTINUE
      AM = M
      AMM = MM
      DFF1(L) = AMM-AM
      DFF2(L) = AN-AM -1.0
 7930 RZSQ(M) = 1.0 - (DETT(MM) / DETIND(M))
      RZR(M) = SQRT(RZSQ(M))
C
      L=L+1
      IF (M - K) 7940,7945,7945
 7940 M=M+1
      GO TO 7700
 7945 IF (K.EQ.1)  GO TO 7999
C
C     CALCULATE  SEMPSQ(I) VIA R-SQUARE DIFFERENCES.
C
      KM1=K-1
 7960 DO 7965  I = 1,KM1
      J=I+1
 7965 SEMPSQ(J) = RZSQ(J) - RZSQ(I)
C
C     PRINTING OF INTERMEDIATE CALCULATIONS.
C
      IF  (CHA-CCHA)   4223,4225,4223
 4223 PRINT 13
      PRINT 11
 4225 CONTINUE
      PRINT 4008
 4008 FORMAT (4X,*DETERMINANTS, INDEPENDENT VARIABLES (DETIND(J))*//)
      PRINT 4014, (J,DETIND(J), J=1,K)
 4014 FORMAT (6X, 6(I2,F12.7,3X))
      PRINT 11
      PRINT 4018
 4018 FORMAT (4X,*DETERMINANTS, WITH Y VARIABLE (DETT(J))*//)
      PRINT 4014, (J,DETT(J),J=2,KT)
 4079 CONTINUE
      PRINT 11
      PRINT 11
      PRINT 7982
 7982 FORMAT  (2X, *R-SQUARES--VARIANCE ACCOUNTED FOR* / 1H0)
      PRINT 7974, (J,  RZSQ(J),J=1,K)
      PRINT 11
      PRINT 7964
 7964 FORMAT  (2X, *SEMIPARTIAL (PART) CORRELATIONS SQUARED--PROPORTION
     1OF VARIANCE*/1H0)
      PRINT 7974, (I, SEMPSQ(I),I=1,K)
 7974 FORMAT  (8 (5X,I3,F8.4))
      PRINT 11
      IF  (IRMAT.EQ.1)  GO TO 7978
```

```
      DO 7975  I=1,K
 7975 SSQV(I) = SEMPSQ(I) * SSD(KT,KT)
      PRINT 7976
 7976 FORMAT  (2X,*SEPARATE SUMS OF SQUARES, EACH VECTOR*//)
      PRINT 7984, (J,SSQV(J),J=1,K)
 7984 FORMAT  (6X, 6(I2,F12.4,5X))
      PRINT 11
 7978 PRINT 11
      IF (IFACNO.EQ.1.OR.IFAC3.EQ.1)  GO TO 6000
C
C     F AND T TESTS.
C
      DO 7977  I=1,K
      NDFF1(I) = DFF1(I)
 7977 NDFF2(I) = DFF2(I)
 7980 DO 7985  I=1,KM1
      J=I+1
 7981 F(J) = ((RZSQ(J) - RZSQ(I))* DFF2(J)) / ((1.0 - RZSQ(J)) *DFF1(J))
      TT(J) = SQRT (F(J))
C
      CALL JASPEN (F(J), NDFF1(J), NDFF2(J), PQQ(J))
C
 7985 CONTINUE
      DO 7990  I=1,K
 7990 PRINT 7992,I,I,F(I),I, NDFF1(I),I,NDFF2(I), PQQ(I),I,TT(I)
 7992 FORMAT (5X,I5,5X, 2HF(,I2,4H) = F9.3,6X, 4HDF1(,I2,4H) = I4,6X,4H
     1F2(,I2,4H) = I4,6X,4HP =  F12.7,7X,2HT(,I2,4H) = F9.3//)
      PRINT 11
      IF (IRMAT.EQ.1)  GO TO 6019
 6000 PRINT 6008,C
 6008 FORMAT  (5X, *INTERCEPT CONSTANT = *F10.4////)
 6019 CONTINUE
      PRINT 6004
 6004 FORMAT  (20X, *END OF SEQUENTIAL ANALYSIS*/ )
 6007 CONTINUE
C
C     CALL SPEC FOR SPECIAL SUPPLEMENTARY ANALYSIS FOR FACTORIAL
C     ANALYSIS, 2 FACTORS OR EFFECTS, WHEN EACH DEGREE OF FREEDOM
C     HAS A CODED VECTOR.   CALL SPEC3 FOR 3 FACTORS OR EFFECTS.
C
      SSTY = SSD(KT,KT)
      IF  (IFACNO.EQ.0)  GO TO 7995
C
      CALL SPEC  (SSTY,K,KT,SEMPSQ, KA,KB, SSDEV,N)
      GO TO 7999
 7995 CONTINUE
      IF  (IFAC3.EQ.0)  GO TO 7999
C
      CALL SPEC3  (SSTY,K,KT,SEMPSQ,KA,KB,KC,SSDEV,N)
C
 7999 CONTINUE
      RETURN
      END
C
C
      SUBROUTINE REGSTAT  (M,MM,SSD, BWT, PI,N, SEEST,SSDEV,SEM,SVE)
C
C     STANDARD ERROR OF ESTIMATE, ST. ERRORS OF REGRESSION COEFFICIENTS
C     T-RATIOS OF REGRESSION COEFFICIENTS
C
```

```
C
      DIMENSION  TRAT(30),SERC(30), PI(30,30), BWT(30),SSD(30,30)
     * ,C(30),TRATSQ(30),P(30)
      AM=M
      AMM=MM
C     SE ESTIMATE.
      AN=N
  100 SEEST = SQRT (SSDEV / (AN-AM-1.0))
   11 FORMAT (1H0)
   13 FORMAT (1H1)
      PRINT 11
C
C     STANDARD ERRORS OF REGRESSION COEFFICIENTS,  SEEST/PI(II).
C
      DO 130  J=1,M
      C(J) = PI(J,J) / SSD(J,J)
  130 SERC(J) = SEEST * SQRT (C(J))
C
C     T-RATIOS OF REGRESSION COEFFICIENTS,  T(I) = BWT (I)/SERC(I)
C
      DO 140  J=1,M
  140 TRAT(J) = BWT(J) / SERC(J)
      NDFT = N-MM
      DO 143  J=1,M
  143 TRATSQ(J) = TRAT(J)**2
      DO 180  I=1,M
      CALL JASPEN (TRATSQ(I), 1, NDFT, P(I))
  180 CONTINUE
C
      PRINT 11
      PRINT  144
  144 FORMAT  (2X, *STANDARD ERRORS OF ESTIMATE AND REGRESSION COEFFICI
     1NTS, T RATIOS */ 1H0)
      PRINT 158
  158 FORMAT (23X,  *B              SERC           T-RATIO
     1     P*//)
      DO 160  I=1,M
  160 PRINT 164,I,BWT(I),SERC(I),TRAT(I),P(I)
  164 FORMAT  (5X,I5,5X,F12.5,6X,F12.5,6X,F12.5,9X,F9.6/)
      PRINT 11
      RETURN
      END
C
C
      SUBROUTINE SPEC  (SSTY,K,KT,SEMPSQ,KA,KB,SSDEV,N)
C
C     SPEC CALCULATES SUMS OF SQUARES OF INDIVIDUAL VECTORS, VIA
C     SSQR(I) = SEMPSQ(I) * SSD(KT,KT).  ALSO CALCULATES MEAN SQUARES
C     FOR MAIN EFFECTS AND INTERACTION OF FACTORIAL ANALYSIS,
C     AND PS.
C      THERE MUST BE A CODED VECTOR FOR EACH DEGREE OF FREEDOM,
C     ORTHOGONAL OR NON-ORTHOGONAL,  (1,0)  OR  (1,0,-1), ETC.
C
      DIMENSION  SEMPSQ(30), FRAT(30), SSQR(30), P(10), NDF(10)
      DIMENSION  SRSQ(10),  SR(10)
      DIMENSION DETIND(1)
      COMMON  R,SD,SSD,C
   11 FORMAT (1H0)
   13 FORMAT (1H1)
  100 DO 110  I=1,K
```

```
  110 SSQR(I) = SEMPSQ(I) * SSTY
      PRINT 13
      PRINT 11
      PRINT 114
  114 FORMAT  ( 5X,*FACTORIAL ANALYSIS OF VARIANCE AND VECTOR SUMS OF S
     1UARES  (VIA CODING OF EACH DEGREE OF FREEDOM*////)
      PRINT 118
  118 FORMAT  (4X, *SEPARATE SUMS OF SQUARES, EACH VECTOR*//)
      PRINT 124, (J,SSQR(J), J=1,K)
  124 FORMAT  (6X, 6(I2,F12.4,5X) / (6X, 6(I2,F12.4,5X)))
      PRINT 11
      PRINT 11
      PRINT 11
C     GROUPING SSQR(I) FOR EFFECTS AND F TESTS.
  120 KAB=KA*KB
      KALL= KAB-1
      NM1=N-1
      NRES=NM1-KALL
      KAM1=KA-1
      KBM1=KB-1
      KPR=KAM1*KBM1
      KBPA=KAM1 + KBM1
      KBPAP1 = KBPA + 1
      KEND = KAM1 + KBM1 + KPR
      ANRES = NRES
      AKAM1 = KAM1
      AKBM1 = KBM1
      SUM1 = 0.0
      SUM2 = 0.0
      SUM3 = 0.0
      DO 130  I=1,KAM1
  130 SUM1 = SUM1 + SSQR(I)
      DO 140  I=KA,KBPA
  140 SUM2 = SUM2 + SSQR(I)
      DO 150  I=KBPAP1,KEND
  150 SUM3= SUM3 + SSQR(I)
 1000 DO  1010  I=1,4
 1010 SRSQ(I) = 0.0
      DO 1030  I=1,KAM1
 1030 SRSQ(1) = SRSQ(1) + SEMPSQ(I)
      DO 1040  I=KA,KBPA
 1040 SRSQ(2) = SRSQ(2) + SEMPSQ(I)
      DO 1050  I=KBPAP1,KEND
 1050 SRSQ(3) = SRSQ(3) + SEMPSQ(I)
      DO 1060  I=1,KEND
 1060 SRSQ(4) = SRSQ(4) + SEMPSQ(I)
      DO 1070  J=1,4
 1070 SR(J) = SQRT(SRSQ(J))
      AKA = KA
      AKB =KB
      AKAB = KAB
  160 AMSQA = SUM1/(AKA-1.0)
      AMSQB = SUM2/(AKB-1.0)
      AMSQAB = SUM3/((AKA-1.0)*(AKB-1.0))
      AMSQER = SSDEV/ANRES
  170 NDF(1) = KA-1
      NDF(2) = KB-1
      NDF(3) = (KA-1)*(KB-1)
  180 FRAT(1) = AMSQA/AMSQER
      FRAT(2) = AMSQB/AMSQER
```

```
      FRAT(3) = AMSQAB/AMSQER
  200 DO 220 I=1,3
      CALL JASPEN  (FRAT(I), NDF(I),NRES, P(I))
  220 CONTINUE
      PRINT 244,KA,KB
  244 FORMAT(35X,*ANALYSIS OF VARIANCE     (*,I2,* BY*,I2,*)*////)
      PRINT 248
  248 FORMAT(14X,*SOURCE          DF         SS          MS          F
     1         P            PROP. VAR.*///)
      PRINT 254,NDF(1), SUM1, AMSQA, FRAT(1),P(1),SRSQ(1)
  254 FORMAT (15X,*A        * I3,3X,F13.4,1X,F11.4,2X,F11.4,2X,F10.7,
     1X,F11.4//)
      PRINT 258,NDF(2),SUM2,AMSQB, FRAT(2),P(2),SRSQ(2)
  258 FORMAT(15X,*B          *I3,3X,F13.4,1X,F11.4,2X,F11.4,2X,F10.7,5X
     1F11.4//)
      PRINT 264, NDF(3), SUM3, AMSQAB, FRAT(3), P(3),SRSQ(3)
  264 FORMAT(13X,*A X B       *I3,3X,F13.4,1X,F11.4,2X,F11.4,2X,F10.7,
     1X,F11.4//)
      PRINT 268,NRES,SSDEV,AMSQER
  268 FORMAT(11X,*RESIDUAL       *I3,3X,F13.4,1X,F11.4////)
      PRINT 274,NM1,SSTY,SRSQ(4)
  274 FORMAT(12X, *TOTAL       *I3,3X,F13.4,42X,F11.4//)
      PRINT 11
      PRINT 11
      PRINT 284
  284 FORMAT  (20X, *END OF FACTORIAL ANALYSIS OF VARIANCE (SPEC)*//)
      RETURN
      END
C
C

      SUBROUTINE SPEC3  (SSTY,K,KT,SEMPSQ,KA,KB,KC,SSDEV,N)
C
C     SAME AS SPEC BUT FOR 3 FACTORS OR EFFECTS.
C
      DIMENSION  SEMPSQ(30),FRAT(30),SSQR(30),P(10),NDF(10), SUM(10),
     * LAB(10)
      DIMENSION DF(10), AMSQ(10),SRSQ(10),SR(10)
      COMMON R,SD,SSD,C
   11 FORMAT (1H0)
   13 FORMAT (1H1)
  100 DO 110  I=1,K
  110 SSQR(I) = SEMPSQ(I) * SSTY
      PRINT 13
      PRINT 11
      PRINT 114
  114 FORMAT  (5X,*FACTORIAL ANALYSIS OF VARIANCE  (VIA CODING OF EACH
     1EGREE OF FREEDOM)*//// )
C
C     SUBSCRIPTS AND GROUPING SSQR(I) FOR EFFECTS AND F TESTS.
  120 KAB = KA * KB
      KABC = KA * KB * KC
      KALL = KABC - 1
      NM1 = N-1
      NRES = NM1 - KALL
  130 KAM1 = KA-1
      KBM1 = KB-1
      KCM1 = KC-1
  140 ANRES = NRES
      AKAM1 = KAM1
      AKBM1 = KBM1
```

```
        AKCM1 = KCM1
  145 AKA = KA
        AKB = KB
        AKC = KC
  150 AKAB = KAB
        AKAC = KAC
        AKBC = KBC
        AKABC = KABC
  155 KBPA = KAM1 + KBM1
        KBPAP1 = KBPA + 1
        KPRAB = KAM1 * KBM1
        KPRAC = KAM1 * KCM1
        KPRBC = KBM1 * KCM1
        KPRABC = KAM1 * KBM1 * KCM1
        KEND = KAM1 + KBM1 + KCM1
        KENDP1 = KEND + 1
  160 KENDAB = KEND + KPRAB
        KDABP1 = KENDAB + 1
        KENDAC = KENDAB + KPRAC
        KDACP1 = KENDAC + 1
        KENDBC = KENDAC + KPRBC
  165 KABCP1 = KENDBC + 1
        KDABC = KENDBC + KPRABC
C
  200 DO 210  I = 1,7
  210 SUM(I) = 0.0
        DO 220   I=1,KAM1
  220 SUM(1) = SUM(1) + SSQR(I)
        DO 225 I=KA,KBPA
  225 SUM(2) = SUM(2) + SSQR(I)
        DO 230  I=KBPAP1,KEND
  230 SUM(3) = SUM(3) + SSQR(I)
        DO 240   I=KENDP1,KENDAB
  240 SUM(4) = SUM(4) + SSQR(I)
        DO 250   I=KDABP1,KENDAC
  250 SUM(5) = SUM(5) + SSQR(I)
        DO 260  I=KDACP1,KENDBC
  260 SUM(6) = SUM(6) + SSQR(I)
        DO 270  I=KABCP1,KDABC
  270 SUM(7) = SUM(7) + SSQR(I)
C
 2000 DO 2010   I=1,8
 2010 SRSQ(I) = 0.0
        DO 2020   I=1,KAM1
 2020 SRSQ(1) = SRSQ(1) + SEMPSQ(I)
        DO 2025  I=KA,KBPA
 2025 SRSQ(2) = SRSQ(2) + SEMPSQ(I)
        DO 2030  I=KBPAP1,KEND
 2030 SRSQ(3) = SRSQ(3) + SEMPSQ(I)
        DO 2040   I=KENDP1,KENDAB
 2040 SRSQ(4) = SRSQ(4) + SEMPSQ(I)
        DO 2050  I=KDABP1,KENDAC
 2050 SRSQ(5) = SRSQ(5) + SEMPSQ(I)
        DO 2060   I=KDACP1,KENDBC
 2060 SRSQ(6) = SRSQ(6) + SEMPSQ(I)
        DO 2070  I=KABCP1,KDABC
 2070 SRSQ(7) = SRSQ(7) + SEMPSQ(I)
        DO 2080  I=1,KDABC
 2080 SRSQ(8) = SRSQ(8) + SEMPSQ(I)
        DO 2090   J=1,8
```

```
 2090 SR(J) = SQRT(SRSQ(J))
C
      PRINT 308,KA,KB,KC
  308 FORMAT (29X,*ANALYSIS OF VARIANCE   (*,I2,* BY*,I2,* BY*,I2,*)*/
     1//)
      PRINT 314
  314 FORMAT (12X,*SOURCE          DF          SS          MS
     1 F          P          PROP. VAR.*///)
C
C     CALCULATE DF.
C
  320 NDF(1) = KA-1
      NDF(2) = KB-1
      NDF(3) = KC-1
  325 NDF(4) = KAM1 * KBM1
      NDF(5) = KAM1 * KCM1
      NDF(6) = KBM1 * KCM1
  330 NDF(7) = KAM1 * KBM1 * KCM1
      NDFTOT = N - 1
  335 NDFALL = KALL
      NRES = NDFTOT - NDFALL
      DO 340  I=1,7
  340 DF(I) = NDF(I)
      DFTOT = NDFTOT
      DFALL = NDFALL
      DFRES = NRES
C
C     MEAN SQUARES
      DO 360 I=1,7
  360 AMSQ(I) = SUM(I)/ DF(I)
      AMSQER = SSDEV / DFRES
  400 DO 410  I=1,7
  410 FRAT(I) = AMSQ(I) / AMSQER
      DO 420  I = 1,7
C
      CALL JASPEN  (FRAT(I), NDF(I), NRES, P(I))
C
  420 CONTINUE
C
      LAB(1) =  9H     A
      LAB(2) =  9H     B
      LAB(3) =  9H     C
      LAB(4) =  9H   A X B
      LAB(5) =  9H   A X C
      LAB(6) =  9H   B X C
      LAB(7) =  9HA X B X C
C
      DO 450  I = 1,7
  450 PRINT 454, LAB(I), NDF(I), SUM(I), AMSQ(I), FRAT(I), P(I),SRSQ(I)
  454 FORMAT   (11X, A9,6X,I3,3X,F13.4,1X,F11.4,2X,F11.4,2X,F10.7,5X,F11
     14//)
      PRINT 458,NRES,SSDEV,AMSQER
  458 FORMAT  (/,11X,*RESIDUAL        *,I3,3X,F13.4,1X,F11.4////)
      PRINT 464, NM1,SSTY,SRSQ(8)
  464 FORMAT  (12X,*TOTAL         *,I3,3X,F13.4,42X,F11.4//)
      PRINT 11
      PRINT 11
      PRINT 468
  468 FORMAT  (20X,*END OF FACTORIAL ANALYSIS (SPEC3)*//)
      RETURN
```

```
      END
C
C
      SUBROUTINE VECINT  (KT,XX,II,X,KA,KB,K,IFACNO,IFAC3,KC)
C
C     CALCULATES INTERACTION VECTORS FOR FACTORIAL ANOVA
C        (IFACNO=1 OR IFAC3=1).
C
      DIMENSION  XX(500,30), X(30)
  100 IF  (IFAC3.EQ.1)  GO TO 200
  110 KAM1=KA-1
      KBM1=KB-1
      KINT=KAM1*KBM1
      KEND=KAM1+KBM1
      KENDP1=KEND+1
      KKALL=KEND+KINT
      KALL = KKALL + 1
      XX(II,KALL) = XX(II,KT)
      KT = KALL
      K = KT - 1
C     CALCULATE INTERACTION VECTORS,  2 EFFECTS, A AND B.
      L = KEND
      DO 120  J=1,KAM1
      DO 120  M=KA,KEND
      L = L+1
      XX(II,L) = XX(II,J) * XX(II,M)
  120 CONTINUE
      DO 140  J=1,KT
  140 X(J) = XX(II,J)
      GO TO 299
  200 CONTINUE
C
C     INTERACTION VECTORS,  3 EFFECTS, A, B, AND C.
C
  210 KAM1=KA-1
      KBM1=KB-1
      KCM1=KC-1
      KINTAB = KAM1 * KBM1
      KENDAB = KAM1 + KBM1
      KDABP1 = KENDAB+ 1
      KINTBC = KBM1 * KCM1
      KINTAC = KAM1 * KCM1
      KINABC = KAM1 * KBM1 * KCM1
      KEND = KAM1 + KBM1 + KCM1
      KAB = KEND + KINTAB
      KABP1 = KAB + 1
      KAC = KAB + KINTAC
      KACP1 = KAC + 1
      KKALL = KEND + KINTAB + KINTAC + KINTBC + KINABC
      KBC = KAC + KINTBC
      KALL = KKALL + 1
      XX(II,KALL) = XX(II,KT)
      KT = KALL
      K = KT - 1
C     A X B
      L=KEND
      DO 260 J=1,KAM1
      DO 260 M=KA,KENDAB
      L=L+1
      XX(II,L) = XX(II,J) * XX(II,M)
```

```
  260 CONTINUE
C
C     A X C
C
      LL = KAB
      DO 270  J=1,KAM1
      DO 270  M=KDABP1,KEND
      LL = LL+1
      XX(II,LL) = XX(II,J) * XX(II,M)
  270 CONTINUE
C
C     B X C.
C
      LLL = KAC
      DO 280  J=KA,KENDAB
      DO 280  M=KDABP1,KEND
      LLL = LLL + 1
      XX(II,LLL) = XX(II,J) * XX(II,M)
  280 CONTINUE
C
C     A X B X C
C
      LLLL = KBC
      DO 290  J=1,KAM1
      DO 290   M=KA,KENDAB
      DO 290   I=KDABP1,KEND
      LLLL = LLLL + 1
      XX(II,LLLL) = XX(II,J) * XX(II,M) *XX(II,I)
  290 CONTINUE
C
  291 DO 295  J=1,KT
  295 X(J) = XX(II,J)
  299 CONTINUE
      RETURN
      END
C
C
      SUBROUTINE  REARR  (KT,K,FMT,IVAR)
C
C     REARRANGES VARIABLES OF R-MATRIX ACCORDING TO REARRANGE CARD (IVA
C
      DIMENSION  FMT(32),R(30,30),RORD(30,30),RI(30,30),IVAR(100)
      COMMON  R,SD,SSD,C
   10 FORMAT  (8A10)
  100 PRINT 104
  104 FORMAT  (2X, *R-MATRIX*//)
      DO 110  I=1,KT
  110 READ FMT,I, (RORD(I,J), J=1,KT)
      DO 150   I=1,KT
      DO 150   J=1,KT
      TEMP = RORD(I,J)
      NUM = IVAR(J)
      RI(I,NUM) = TEMP
  150 CONTINUE
C
      DO 190   J=1,KT
      DO 190   I=1,KT
      TEMP = RI(I,J)
      NUM = IVAR(I)
      R(NUM,J) = TEMP
```

```
   190 CONTINUE
C
       CALL MATOUT  (R,KT,30)
C
       RETURN
       END
C
C
       SUBROUTINE POW (X,IPVEC,KT,K,KTOT,KPOW,IPOW)
C
       DIMENSION X(30),IPVEC(10),IPOW(10)
C
       KP1=K+1
       KK=K+KPOW
       KTOT=KK+1
C      CHANGE Y TO LAST VARIABLE.
       X(KTOT) = X(KT)
C
       DO 229  J=1,KPOW
       M=IPVEC(J)
       MM=IPOW(J)
C      GENERATE SELECTED POWERS OF X(J).
       X(KP1) = X(M)**MM
       KP1 = KP1+1
   229 CONTINUE
       RETURN
       END
C
C
       SUBROUTINE SELCP (X,KIVEC,INTVEC,K,KT,KTOT)
C
       DIMENSION X(30),CP(30),INTVEC(10),XX(30)
C
       M=INTVEC(1)
C      SELECT VARIABLES FOR CROSS-PRODUCTS.
       DO 55   JJ=1,KIVEC
       XX(JJ) = X(M)
       M = INTVEC(JJ+1)
    55 CONTINUE
       KM1 = KIVEC-1
       L=1
   100 DO 120   J=1,KM1
       JP1=J+1
   105 DO 115   JJ=JP1,KIVEC
       CP(L) = XX(J)*XX(JJ)
       L=L+1
   115 CONTINUE
   120 CONTINUE
C      REARRANGE VARIABLES--PUT Y LAST, MOVE CP INTO X.
       KINT= (KIVEC*(KIVEC-1))/2
       KTOT= KT + KINT
       X(KTOT) = X(KT)
       L=K+1
       DO 160   J=1,KINT
       X(L) = CP(J)
       L=L+1
   160 CONTINUE
       RETURN
       END
```

The 5 (Roman Type) and 1 (Boldface Type) Percent Points for the Distribution of F*

n_1 degrees of freedom (for greater mean square)

n_2	1	2	3	4	5	6	7	8	9	10	11	12	14	16	20	24	30	40	50	75	100	200	500	∞
1	161 **4,052**	200 **4,999**	216 **5,403**	225 **5,625**	230 **5,764**	234 **5,859**	237 **5,928**	239 **5,981**	241 **6,022**	242 **6,056**	243 **6,082**	244 **6,106**	245 **6,142**	246 **6,169**	248 **6,208**	249 **6,234**	250 **6,258**	251 **6,286**	252 **6,302**	253 **6,323**	253 **6,334**	254 **6,352**	254 **6,361**	254 **6,366**
2	18.51 **98.49**	19.00 **99.00**	19.16 **99.17**	19.25 **99.25**	19.30 **99.30**	19.33 **99.33**	19.36 **99.34**	19.37 **99.36**	19.38 **99.38**	19.39 **99.40**	19.40 **99.41**	19.41 **99.42**	19.42 **99.43**	19.43 **99.44**	19.44 **99.45**	19.45 **99.46**	19.46 **99.47**	19.47 **99.48**	19.47 **99.48**	19.48 **99.49**	19.49 **99.49**	19.49 **99.49**	19.50 **99.50**	19.50 **99.50**
3	10.13 **34.12**	9.55 **30.82**	9.28 **29.46**	9.12 **28.71**	9.01 **28.24**	8.94 **27.91**	8.88 **27.67**	8.84 **27.49**	8.81 **27.34**	8.78 **27.23**	8.76 **27.13**	8.74 **27.05**	8.71 **26.92**	8.69 **26.83**	8.66 **26.69**	8.64 **26.60**	8.62 **26.50**	8.60 **26.41**	8.58 **26.35**	8.57 **26.27**	8.56 **26.23**	8.54 **26.18**	8.54 **26.14**	8.53 **26.12**
4	7.71 **21.20**	6.94 **18.00**	6.59 **16.69**	6.39 **15.98**	6.26 **15.52**	6.16 **15.21**	6.09 **14.98**	6.04 **14.80**	6.00 **14.66**	5.96 **14.54**	5.93 **14.45**	5.91 **14.37**	5.87 **14.24**	5.84 **14.15**	5.80 **14.02**	5.77 **13.93**	5.74 **13.83**	5.71 **13.74**	5.70 **13.69**	5.68 **13.61**	5.66 **13.57**	5.65 **13.52**	5.64 **13.48**	5.63 **13.46**
5	6.61 **16.26**	5.79 **13.27**	5.41 **12.06**	5.19 **11.39**	5.05 **10.97**	4.95 **10.67**	4.88 **10.45**	4.82 **10.27**	4.78 **10.15**	4.74 **10.05**	4.70 **9.96**	4.68 **9.89**	4.64 **9.77**	4.60 **9.68**	4.56 **9.55**	4.53 **9.47**	4.50 **9.38**	4.46 **9.29**	4.44 **9.24**	4.42 **9.17**	4.40 **9.13**	4.38 **9.07**	4.37 **9.04**	4.36 **9.02**
6	5.99 **13.74**	5.14 **10.92**	4.76 **9.78**	4.53 **9.15**	4.39 **8.75**	4.28 **8.47**	4.21 **8.26**	4.15 **8.10**	4.10 **7.98**	4.06 **7.87**	4.03 **7.79**	4.00 **7.72**	3.96 **7.60**	3.92 **7.52**	3.87 **7.39**	3.84 **7.31**	3.81 **7.23**	3.77 **7.14**	3.75 **7.09**	3.72 **7.02**	3.71 **6.99**	3.69 **6.94**	3.68 **6.90**	3.67 **6.88**
7	5.59 **12.25**	4.74 **9.55**	4.35 **8.45**	4.12 **7.85**	3.97 **7.46**	3.87 **7.19**	3.79 **7.00**	3.73 **6.84**	3.68 **6.71**	3.63 **6.62**	3.60 **6.54**	3.57 **6.47**	3.52 **6.35**	3.49 **6.27**	3.44 **6.15**	3.41 **6.07**	3.38 **5.98**	3.34 **5.90**	3.32 **5.85**	3.29 **5.78**	3.28 **5.75**	3.25 **5.70**	3.24 **5.67**	3.23 **5.65**
8	5.32 **11.26**	4.46 **8.65**	4.07 **7.59**	3.84 **7.01**	3.69 **6.63**	3.58 **6.37**	3.50 **6.19**	3.44 **6.03**	3.39 **5.91**	3.34 **5.82**	3.31 **5.74**	3.28 **5.67**	3.23 **5.56**	3.20 **5.48**	3.15 **5.36**	3.12 **5.28**	3.08 **5.20**	3.05 **5.11**	3.03 **5.06**	3.00 **5.00**	2.98 **4.96**	2.96 **4.91**	2.94 **4.88**	2.93 **4.86**
9	5.12 **10.56**	4.26 **8.02**	3.86 **6.99**	3.63 **6.42**	3.48 **6.06**	3.37 **5.80**	3.29 **5.62**	3.23 **5.47**	3.18 **5.35**	3.13 **5.26**	3.10 **5.18**	3.07 **5.11**	3.02 **5.00**	2.98 **4.92**	2.93 **4.80**	2.90 **4.73**	2.86 **4.64**	2.82 **4.56**	2.80 **4.51**	2.77 **4.45**	2.76 **4.41**	2.73 **4.36**	2.72 **4.33**	2.71 **4.31**
10	4.96 **10.04**	4.10 **7.56**	3.71 **6.55**	3.48 **5.99**	3.33 **5.64**	3.22 **5.39**	3.14 **5.21**	3.07 **5.06**	3.02 **4.95**	2.97 **4.85**	2.94 **4.78**	2.91 **4.71**	2.86 **4.60**	2.82 **4.52**	2.77 **4.41**	2.74 **4.33**	2.70 **4.25**	2.67 **4.17**	2.64 **4.12**	2.61 **4.05**	2.59 **4.01**	2.56 **3.96**	2.55 **3.93**	2.54 **3.91**
11	4.84 **9.65**	3.98 **7.20**	3.59 **6.22**	3.36 **5.67**	3.20 **5.32**	3.09 **5.07**	3.01 **4.88**	2.95 **4.74**	2.90 **4.63**	2.86 **4.54**	2.82 **4.46**	2.79 **4.40**	2.74 **4.29**	2.70 **4.21**	2.65 **4.10**	2.61 **4.02**	2.57 **3.94**	2.53 **3.86**	2.50 **3.80**	2.47 **3.74**	2.45 **3.70**	2.42 **3.66**	2.41 **3.62**	2.40 **3.60**
12	4.75 **9.33**	3.88 **6.93**	3.49 **5.95**	3.26 **5.41**	3.11 **5.06**	3.00 **4.82**	2.92 **4.65**	2.85 **4.50**	2.80 **4.39**	2.76 **4.30**	2.72 **4.22**	2.69 **4.16**	2.64 **4.05**	2.60 **3.98**	2.54 **3.86**	2.50 **3.78**	2.46 **3.70**	2.42 **3.61**	2.40 **3.56**	2.36 **3.49**	2.35 **3.46**	2.32 **3.41**	2.31 **3.38**	2.30 **3.36**
13	4.67 **9.07**	3.80 **6.70**	3.41 **5.74**	3.18 **5.20**	3.02 **4.86**	2.92 **4.62**	2.84 **4.44**	2.77 **4.30**	2.72 **4.19**	2.67 **4.10**	2.63 **4.02**	2.60 **3.96**	2.55 **3.85**	2.51 **3.78**	2.46 **3.67**	2.42 **3.59**	2.38 **3.51**	2.34 **3.42**	2.32 **3.37**	2.28 **3.30**	2.26 **3.27**	2.24 **3.21**	2.22 **3.18**	2.21 **3.16**

Appendix D is reproduced from Snedecor: *Statistical Methods*, Iowa State College Press, Ames, Iowa, by permission of the author and publisher.

The 5 (Roman Type) and 1 (Boldface Type) Percent Points for the Distribution of F^*—Continued

n_1 degrees of freedom (for greater mean square)

n_2	1	2	3	4	5	6	7	8	9	10	11	12	14	16	20	24	30	40	50	75	100	200	500	∞
14	4.60 **8.86**	3.74 **6.51**	3.34 **5.56**	3.11 **5.03**	2.96 **4.69**	2.85 **4.46**	2.77 **4.28**	2.70 **4.14**	2.65 **4.03**	2.60 **3.94**	2.56 **3.86**	2.53 **3.80**	2.48 **3.70**	2.44 **3.62**	2.39 **3.51**	2.35 **3.43**	2.31 **3.34**	2.27 **3.26**	2.24 **3.21**	2.21 **3.14**	2.19 **3.11**	2.16 **3.06**	2.14 **3.02**	2.13 **3.00**
15	4.54 **8.68**	3.68 **6.36**	3.29 **5.42**	3.06 **4.89**	2.90 **4.56**	2.79 **4.32**	2.70 **4.14**	2.64 **4.00**	2.59 **3.89**	2.55 **3.80**	2.51 **3.73**	2.48 **3.67**	2.43 **3.56**	2.39 **3.48**	2.33 **3.36**	2.29 **3.29**	2.25 **3.20**	2.21 **3.12**	2.18 **3.07**	2.15 **3.00**	2.12 **2.97**	2.10 **2.92**	2.08 **2.89**	2.07 **2.87**
16	4.49 **8.53**	3.63 **6.23**	3.24 **5.29**	3.01 **4.77**	2.85 **4.44**	2.74 **4.20**	2.66 **4.03**	2.59 **3.89**	2.54 **3.78**	2.49 **3.69**	2.45 **3.61**	2.42 **3.55**	2.37 **3.45**	2.33 **3.37**	2.28 **3.25**	2.24 **3.18**	2.20 **3.10**	2.16 **3.01**	2.13 **2.96**	2.09 **2.89**	2.07 **2.86**	2.04 **2.80**	2.02 **2.77**	2.01 **2.75**
17	4.45 **8.40**	3.59 **6.11**	3.20 **5.18**	2.96 **4.67**	2.81 **4.34**	2.70 **4.10**	2.62 **3.93**	2.55 **3.79**	2.50 **3.68**	2.45 **3.59**	2.41 **3.52**	2.38 **3.45**	2.33 **3.35**	2.29 **3.27**	2.23 **3.16**	2.19 **3.08**	2.15 **3.00**	2.11 **2.92**	2.08 **2.86**	2.04 **2.79**	2.02 **2.76**	1.99 **2.70**	1.97 **2.67**	1.96 **2.65**
18	4.41 **8.28**	3.55 **6.01**	3.16 **5.09**	2.93 **4.58**	2.77 **4.25**	2.66 **4.01**	2.58 **3.85**	2.51 **3.71**	2.46 **3.60**	2.41 **3.51**	2.37 **3.44**	2.34 **3.37**	2.29 **3.27**	2.25 **3.19**	2.19 **3.07**	2.15 **3.00**	2.11 **2.91**	2.07 **2.83**	2.04 **2.78**	2.00 **2.71**	1.98 **2.68**	1.95 **2.62**	1.93 **2.59**	1.92 **2.57**
19	4.38 **8.18**	3.52 **5.93**	3.13 **5.01**	2.90 **4.50**	2.74 **4.17**	2.63 **3.94**	2.55 **3.77**	2.48 **3.63**	2.43 **3.52**	2.38 **3.43**	2.34 **3.36**	2.31 **3.30**	2.26 **3.19**	2.21 **3.12**	2.15 **3.00**	2.11 **2.92**	2.07 **2.84**	2.02 **2.76**	2.00 **2.70**	1.96 **2.63**	1.94 **2.60**	1.91 **2.54**	1.90 **2.51**	1.88 **2.49**
20	4.35 **8.10**	3.49 **5.85**	3.10 **4.94**	2.87 **4.43**	2.71 **4.10**	2.60 **3.87**	2.52 **3.71**	2.45 **3.56**	2.40 **3.45**	2.35 **3.37**	2.31 **3.30**	2.28 **3.23**	2.23 **3.13**	2.18 **3.05**	2.12 **2.94**	2.08 **2.86**	2.04 **2.77**	1.99 **2.69**	1.96 **2.63**	1.92 **2.56**	1.90 **2.53**	1.87 **2.47**	1.85 **2.44**	1.84 **2.42**
21	4.32 **8.02**	3.47 **5.78**	3.07 **4.87**	2.84 **4.37**	2.68 **4.04**	2.57 **3.81**	2.49 **3.65**	2.42 **3.51**	2.37 **3.40**	2.32 **3.31**	2.28 **3.24**	2.25 **3.17**	2.20 **3.07**	2.15 **2.99**	2.09 **2.88**	2.05 **2.80**	2.00 **2.72**	1.96 **2.63**	1.93 **2.58**	1.89 **2.51**	1.87 **2.47**	1.84 **2.42**	1.82 **2.38**	1.81 **2.36**
22	4.30 **7.94**	3.44 **5.72**	3.05 **4.82**	2.82 **4.31**	2.66 **3.99**	2.55 **3.76**	2.47 **3.59**	2.40 **3.45**	2.35 **3.35**	2.30 **3.26**	2.26 **3.18**	2.23 **3.12**	2.18 **3.02**	2.13 **2.94**	2.07 **2.83**	2.03 **2.75**	1.98 **2.67**	1.93 **2.58**	1.91 **2.53**	1.87 **2.46**	1.84 **2.42**	1.81 **2.37**	1.80 **2.33**	1.78 **2.31**
23	4.28 **7.88**	3.42 **5.66**	3.03 **4.76**	2.80 **4.26**	2.64 **3.94**	2.53 **3.71**	2.45 **3.54**	2.38 **3.41**	2.32 **3.30**	2.28 **3.21**	2.24 **3.14**	2.20 **3.07**	2.14 **2.97**	2.10 **2.89**	2.04 **2.78**	2.00 **2.70**	1.96 **2.62**	1.91 **2.53**	1.88 **2.48**	1.84 **2.41**	1.82 **2.37**	1.79 **2.32**	1.77 **2.28**	1.76 **2.26**
24	4.26 **7.82**	3.40 **5.61**	3.01 **4.72**	2.78 **4.22**	2.62 **3.90**	2.51 **3.67**	2.43 **3.50**	2.36 **3.36**	2.30 **3.25**	2.26 **3.17**	2.22 **3.09**	2.18 **3.03**	2.13 **2.93**	2.09 **2.85**	2.02 **2.74**	1.98 **2.66**	1.94 **2.58**	1.89 **2.49**	1.86 **2.44**	1.82 **2.36**	1.80 **2.33**	1.76 **2.27**	1.74 **2.23**	1.73 **2.21**
25	4.24 **7.77**	3.38 **5.57**	2.99 **4.68**	2.76 **4.18**	2.60 **3.86**	2.49 **3.63**	2.41 **3.46**	2.34 **3.32**	2.28 **3.21**	2.24 **3.13**	2.20 **3.05**	2.16 **2.99**	2.11 **2.89**	2.06 **2.81**	2.00 **2.70**	1.96 **2.62**	1.92 **2.54**	1.87 **2.45**	1.84 **2.40**	1.80 **2.32**	1.77 **2.29**	1.74 **2.23**	1.72 **2.19**	1.71 **2.17**
26	4.22 **7.72**	3.37 **5.53**	2.98 **4.64**	2.74 **4.14**	2.59 **3.82**	2.47 **3.59**	2.39 **3.42**	2.32 **3.29**	2.27 **3.17**	2.22 **3.09**	2.18 **3.02**	2.15 **2.96**	2.10 **2.86**	2.05 **2.77**	1.99 **2.66**	1.95 **2.58**	1.90 **2.50**	1.85 **2.41**	1.82 **2.36**	1.78 **2.28**	1.76 **2.25**	1.72 **2.19**	1.70 **2.15**	1.69 **2.13**

Appendix D is reproduced from Snedecor: *Statistical Methods*, Iowa State College Press, Ames, Iowa, by permission of the author and publisher.

The 5 (Roman Type) and 1 (Boldface Type) Percent Points for the Distribution of F^*—Continued

n_1 degrees of freedom (for greater mean square)

n_2	1	2	3	4	5	6	7	8	9	10	11	12	14	16	20	24	30	40	50	75	100	200	500	∞
27	4.21 / 7.68	3.35 / 5.49	2.96 / 4.60	2.73 / 4.11	2.57 / 3.79	2.46 / 3.56	2.37 / 3.39	2.30 / 3.26	2.25 / 3.14	2.20 / 3.06	2.16 / 2.98	2.13 / 2.93	2.08 / 2.83	2.03 / 2.74	1.97 / 2.63	1.93 / 2.55	1.88 / 2.47	1.84 / 2.38	1.80 / 2.33	1.76 / 2.25	1.74 / 2.21	1.71 / 2.16	1.68 / 2.12	1.67 / 2.10
28	4.20 / 7.64	3.34 / 5.45	2.95 / 4.57	2.71 / 4.07	2.56 / 3.76	2.44 / 3.53	2.36 / 3.36	2.29 / 3.23	2.24 / 3.11	2.19 / 3.03	2.15 / 2.95	2.12 / 2.90	2.06 / 2.80	2.02 / 2.71	1.96 / 2.60	1.91 / 2.52	1.87 / 2.44	1.81 / 2.35	1.78 / 2.30	1.75 / 2.22	1.72 / 2.18	1.69 / 2.13	1.67 / 2.09	1.65 / 2.06
29	4.18 / 7.60	3.33 / 5.42	2.93 / 4.54	2.70 / 4.04	2.54 / 3.73	2.43 / 3.50	2.35 / 3.33	2.28 / 3.20	2.22 / 3.08	2.18 / 3.00	2.14 / 2.92	2.10 / 2.87	2.05 / 2.77	2.00 / 2.68	1.94 / 2.57	1.90 / 2.49	1.85 / 2.41	1.80 / 2.32	1.77 / 2.27	1.73 / 2.19	1.71 / 2.15	1.68 / 2.10	1.65 / 2.06	1.64 / 2.03
30	4.17 / 7.56	3.32 / 5.39	2.92 / 4.51	2.69 / 4.02	2.53 / 3.70	2.42 / 3.47	2.34 / 3.30	2.27 / 3.17	2.21 / 3.06	2.16 / 2.98	2.12 / 2.90	2.09 / 2.84	2.04 / 2.74	1.99 / 2.66	1.93 / 2.55	1.89 / 2.47	1.84 / 2.38	1.79 / 2.29	1.76 / 2.24	1.72 / 2.16	1.69 / 2.13	1.66 / 2.07	1.64 / 2.03	1.62 / 2.01
32	4.15 / 7.50	3.30 / 5.34	2.90 / 4.46	2.67 / 3.97	2.51 / 3.66	2.40 / 3.42	2.32 / 3.25	2.25 / 3.12	2.19 / 3.01	2.14 / 2.94	2.10 / 2.86	2.07 / 2.80	2.02 / 2.70	1.97 / 2.62	1.91 / 2.51	1.86 / 2.42	1.82 / 2.34	1.76 / 2.25	1.74 / 2.20	1.69 / 2.12	1.67 / 2.08	1.64 / 2.02	1.61 / 1.98	1.59 / 1.96
34	4.13 / 7.44	3.28 / 5.29	2.88 / 4.42	2.65 / 3.93	2.49 / 3.61	2.38 / 3.38	2.30 / 3.21	2.23 / 3.08	2.17 / 2.97	2.12 / 2.89	2.08 / 2.82	2.05 / 2.76	2.00 / 2.66	1.95 / 2.58	1.89 / 2.47	1.84 / 2.38	1.80 / 2.30	1.74 / 2.21	1.71 / 2.15	1.67 / 2.08	1.64 / 2.04	1.61 / 1.98	1.59 / 1.94	1.57 / 1.91
36	4.11 / 7.39	3.26 / 5.25	2.86 / 4.38	2.63 / 3.89	2.48 / 3.58	2.36 / 3.35	2.28 / 3.18	2.21 / 3.04	2.15 / 2.94	2.10 / 2.86	2.06 / 2.78	2.03 / 2.72	1.98 / 2.62	1.93 / 2.54	1.87 / 2.43	1.82 / 2.35	1.78 / 2.26	1.72 / 2.17	1.69 / 2.12	1.65 / 2.04	1.62 / 2.00	1.59 / 1.94	1.56 / 1.90	1.55 / 1.87
38	4.10 / 7.35	3.25 / 5.21	2.85 / 4.34	2.62 / 3.86	2.46 / 3.54	2.35 / 3.32	2.26 / 3.15	2.19 / 3.02	2.14 / 2.91	2.09 / 2.82	2.05 / 2.75	2.02 / 2.69	1.96 / 2.59	1.92 / 2.51	1.85 / 2.40	1.80 / 2.32	1.76 / 2.22	1.71 / 2.14	1.67 / 2.08	1.63 / 2.00	1.60 / 1.97	1.57 / 1.90	1.54 / 1.86	1.53 / 1.84
40	4.08 / 7.31	3.23 / 5.18	2.84 / 4.31	2.61 / 3.83	2.45 / 3.51	2.34 / 3.29	2.25 / 3.12	2.18 / 2.99	2.12 / 2.88	2.07 / 2.80	2.04 / 2.73	2.00 / 2.66	1.95 / 2.56	1.90 / 2.49	1.84 / 2.37	1.79 / 2.29	1.74 / 2.20	1.69 / 2.11	1.66 / 2.05	1.61 / 1.97	1.59 / 1.94	1.55 / 1.88	1.53 / 1.84	1.51 / 1.81
42	4.07 / 7.27	3.22 / 5.15	2.83 / 4.29	2.59 / 3.80	2.44 / 3.49	2.32 / 3.26	2.24 / 3.10	2.17 / 2.96	2.11 / 2.86	2.06 / 2.77	2.02 / 2.70	1.99 / 2.64	1.94 / 2.54	1.89 / 2.46	1.82 / 2.35	1.78 / 2.26	1.73 / 2.17	1.68 / 2.08	1.64 / 2.02	1.60 / 1.94	1.57 / 1.91	1.54 / 1.85	1.51 / 1.80	1.49 / 1.78
44	4.06 / 7.24	3.21 / 5.12	2.82 / 4.26	2.58 / 3.78	2.43 / 3.46	2.31 / 3.24	2.23 / 3.07	2.16 / 2.94	2.10 / 2.84	2.05 / 2.75	2.01 / 2.68	1.98 / 2.62	1.92 / 2.52	1.88 / 2.44	1.81 / 2.32	1.76 / 2.24	1.72 / 2.15	1.66 / 2.06	1.63 / 2.00	1.58 / 1.92	1.56 / 1.88	1.52 / 1.82	1.50 / 1.78	1.48 / 1.75
46	4.05 / 7.21	3.20 / 5.10	2.81 / 4.24	2.57 / 3.76	2.42 / 3.44	2.30 / 3.22	2.22 / 3.05	2.14 / 2.92	2.09 / 2.82	2.04 / 2.73	2.00 / 2.66	1.97 / 2.60	1.91 / 2.50	1.87 / 2.42	1.80 / 2.30	1.75 / 2.22	1.71 / 2.13	1.65 / 2.04	1.62 / 1.98	1.57 / 1.90	1.54 / 1.86	1.51 / 1.80	1.48 / 1.76	1.46 / 1.72
48	4.04 / 7.19	3.19 / 5.08	2.80 / 4.22	2.56 / 3.74	2.41 / 3.42	2.30 / 3.20	2.21 / 3.04	2.14 / 2.90	2.08 / 2.80	2.03 / 2.71	1.99 / 2.64	1.96 / 2.58	1.90 / 2.48	1.86 / 2.40	1.79 / 2.28	1.74 / 2.20	1.70 / 2.11	1.64 / 2.02	1.61 / 1.96	1.56 / 1.88	1.53 / 1.84	1.50 / 1.78	1.47 / 1.73	1.45 / 1.70

Appendix D is reproduced from Snedecor: *Statistical Methods*, Iowa State College Press. Ames, Iowa, by permission of the author and publisher.

The 5 (Roman Type) and 1 (Boldface Type) Percent Points for the Distribution of F*—Concluded

n_1 degrees of freedom (for greater mean square)

n_2	1	2	3	4	5	6	7	8	9	10	11	12	14	16	20	24	30	40	50	75	100	200	500	∞
50	4.03 / 7.17	3.18 / 5.06	2.79 / 4.20	2.56 / 3.72	2.40 / 3.41	2.29 / 3.18	2.20 / 3.02	2.13 / 2.88	2.07 / 2.78	2.02 / 2.70	1.98 / 2.62	1.95 / 2.56	1.90 / 2.46	1.85 / 2.39	1.78 / 2.26	1.74 / 2.18	1.69 / 2.10	1.63 / 2.00	1.60 / 1.94	1.55 / 1.86	1.52 / 1.82	1.48 / 1.76	1.46 / 1.71	1.44 / 1.68
55	4.02 / 7.12	3.17 / 5.01	2.78 / 4.16	2.54 / 3.68	2.38 / 3.37	2.27 / 3.15	2.18 / 2.98	2.11 / 2.85	2.05 / 2.75	2.00 / 2.66	1.97 / 2.59	1.93 / 2.53	1.88 / 2.43	1.83 / 2.35	1.76 / 2.23	1.72 / 2.15	1.67 / 2.06	1.61 / 1.96	1.58 / 1.90	1.52 / 1.82	1.50 / 1.78	1.46 / 1.71	1.43 / 1.66	1.41 / 1.64
60	4.00 / 7.08	3.15 / 4.98	2.76 / 4.13	2.52 / 3.65	2.37 / 3.34	2.25 / 3.12	2.17 / 2.95	2.10 / 2.82	2.04 / 2.72	1.99 / 2.63	1.95 / 2.56	1.92 / 2.50	1.86 / 2.40	1.81 / 2.32	1.75 / 2.20	1.70 / 2.12	1.65 / 2.03	1.59 / 1.93	1.56 / 1.87	1.50 / 1.79	1.48 / 1.74	1.44 / 1.68	1.41 / 1.63	1.39 / 1.60
65	3.99 / 7.04	3.14 / 4.95	2.75 / 4.10	2.51 / 3.62	2.36 / 3.31	2.24 / 3.09	2.15 / 2.93	2.08 / 2.79	2.02 / 2.70	1.98 / 2.61	1.94 / 2.54	1.90 / 2.47	1.85 / 2.37	1.80 / 2.30	1.73 / 2.18	1.68 / 2.09	1.63 / 2.00	1.57 / 1.90	1.54 / 1.84	1.49 / 1.76	1.46 / 1.71	1.42 / 1.64	1.39 / 1.60	1.37 / 1.56
70	3.98 / 7.01	3.13 / 4.92	2.74 / 4.08	2.50 / 3.60	2.35 / 3.29	2.23 / 3.07	2.14 / 2.91	2.07 / 2.77	2.01 / 2.67	1.97 / 2.59	1.93 / 2.51	1.89 / 2.45	1.84 / 2.35	1.79 / 2.28	1.72 / 2.15	1.67 / 2.07	1.62 / 1.98	1.56 / 1.88	1.53 / 1.82	1.47 / 1.74	1.45 / 1.69	1.40 / 1.62	1.37 / 1.56	1.35 / 1.53
80	3.96 / 6.96	3.11 / 4.88	2.72 / 4.04	2.48 / 3.56	2.33 / 3.25	2.21 / 3.04	2.12 / 2.87	2.05 / 2.74	1.99 / 2.64	1.95 / 2.55	1.91 / 2.48	1.88 / 2.41	1.82 / 2.32	1.77 / 2.24	1.70 / 2.11	1.65 / 2.03	1.60 / 1.94	1.54 / 1.84	1.51 / 1.78	1.45 / 1.70	1.42 / 1.65	1.38 / 1.57	1.35 / 1.52	1.32 / 1.49
100	3.94 / 6.90	3.09 / 4.82	2.70 / 3.98	2.46 / 3.51	2.30 / 3.20	2.19 / 2.99	2.10 / 2.82	2.03 / 2.69	1.97 / 2.59	1.92 / 2.51	1.88 / 2.43	1.85 / 2.36	1.79 / 2.26	1.75 / 2.19	1.68 / 2.06	1.63 / 1.98	1.57 / 1.89	1.51 / 1.79	1.48 / 1.73	1.42 / 1.64	1.39 / 1.59	1.34 / 1.51	1.30 / 1.46	1.28 / 1.43
125	3.92 / 6.84	3.07 / 4.78	2.68 / 3.94	2.44 / 3.47	2.29 / 3.17	2.17 / 2.95	2.08 / 2.79	2.01 / 2.65	1.95 / 2.56	1.90 / 2.47	1.86 / 2.40	1.83 / 2.33	1.77 / 2.23	1.72 / 2.15	1.65 / 2.03	1.60 / 1.94	1.55 / 1.85	1.49 / 1.75	1.45 / 1.68	1.39 / 1.59	1.36 / 1.54	1.31 / 1.46	1.27 / 1.40	1.25 / 1.37
150	3.91 / 6.81	3.06 / 4.75	2.67 / 3.91	2.43 / 3.44	2.27 / 3.14	2.16 / 2.92	2.07 / 2.76	2.00 / 2.62	1.94 / 2.53	1.89 / 2.44	1.85 / 2.37	1.82 / 2.30	1.76 / 2.20	1.71 / 2.12	1.64 / 2.00	1.59 / 1.91	1.54 / 1.83	1.47 / 1.72	1.44 / 1.66	1.37 / 1.56	1.34 / 1.51	1.29 / 1.43	1.25 / 1.37	1.22 / 1.33
200	3.89 / 6.76	3.04 / 4.71	2.65 / 3.88	2.41 / 3.41	2.26 / 3.11	2.14 / 2.90	2.05 / 2.73	1.98 / 2.60	1.92 / 2.50	1.87 / 2.41	1.83 / 2.34	1.80 / 2.28	1.74 / 2.17	1.69 / 2.09	1.62 / 1.97	1.57 / 1.88	1.52 / 1.79	1.45 / 1.69	1.42 / 1.62	1.35 / 1.53	1.32 / 1.48	1.26 / 1.39	1.22 / 1.33	1.19 / 1.28
400	3.86 / 6.70	3.02 / 4.66	2.62 / 3.83	2.39 / 3.36	2.23 / 3.06	2.12 / 2.85	2.03 / 2.69	1.96 / 2.55	1.90 / 2.46	1.85 / 2.37	1.81 / 2.29	1.78 / 2.23	1.72 / 2.12	1.67 / 2.04	1.60 / 1.92	1.54 / 1.84	1.49 / 1.74	1.42 / 1.64	1.38 / 1.57	1.32 / 1.47	1.28 / 1.42	1.22 / 1.32	1.16 / 1.24	1.13 / 1.19
1000	3.85 / 6.66	3.00 / 4.62	2.61 / 3.80	2.38 / 3.34	2.22 / 3.04	2.10 / 2.82	2.02 / 2.66	1.95 / 2.53	1.89 / 2.43	1.84 / 2.34	1.80 / 2.26	1.76 / 2.20	1.70 / 2.09	1.65 / 2.01	1.58 / 1.89	1.53 / 1.81	1.47 / 1.71	1.41 / 1.61	1.36 / 1.54	1.30 / 1.44	1.26 / 1.38	1.19 / 1.28	1.13 / 1.19	1.08 / 1.11
∞	3.84 / 6.64	2.99 / 4.60	2.60 / 3.78	2.37 / 3.32	2.21 / 3.02	2.09 / 2.80	2.01 / 2.64	1.94 / 2.51	1.88 / 2.41	1.83 / 2.32	1.79 / 2.24	1.75 / 2.18	1.69 / 2.07	1.64 / 1.99	1.57 / 1.87	1.52 / 1.79	1.46 / 1.69	1.40 / 1.59	1.35 / 1.52	1.28 / 1.41	1.24 / 1.36	1.17 / 1.25	1.11 / 1.15	1.00 / 1.00

Appendix D is reproduced from Snedecor: *Statistical Methods*, Iowa State College Press, Ames, Iowa, by permission of the author and publisher.

References

Abelson, R. P. A note on the Neyman-Johnson technique. *Psychometrika*, 1953, **18**, 213–218.

Aitken, A. C. *Determinants and matrices*. London: Oliver and Boyd, 1956.

Althauser, R. P. Multicollinearity and non-additive regression models. In H. M. Blalock (Ed.), *Causal models in the social sciences*. Chicago: Aldine, 1971.

Anderson, G. J. Effects of classroom social climate on individual learning. *American Educational Research Journal*, 1970, **7**, 135–152.

Anderson, H. E. Regression, discriminant analysis, and a standard notation for basic statistics. In R. B. Cattell (Ed.), *Handbook of multivariate experimental psychology*. Skokie, Ill.: Rand McNally, 1966.

Anderson, N. H. Scales and statistics: Parametric and nonparametric. *Psychological Bulletin*, 1961, **58**, 305–316.

Astin, A. W. "Productivity" of undergraduate institutions. *Science*, 1962, **136**, 129–135.

Astin, A. W. Undergraduate achievement and institutional excellence. *Science*, 1968, **161**, 661–668.

Baker, B. D., Hardyck, C. D., & Petrinovich, L. F. Weak measurement vs. strong statistics: An empirical critique of S. S. Stevens' proscriptions on statistics. *Educational and Psychological Measurement*, 1966, **26**, 291–309.

Bartlett, M. S. Multivariate analysis. *Supplement to the Journal of the Royal Statistical Society*, 1947, **9**, 176–197.

Bates, F., & Douglas, M. L. *Programming language/one*. Englewood Cliffs, N.J.: Prentice-Hall, 1967.

Beaton, A. E. The use of special matrix operators in statistical calculus. Princeton, N.J.: Educational Testing Service, 1964.

Berkowitz, L. Anti-Semitism and the displacement of aggression. *Journal of Abnormal and Social Psychology*, 1959, **59**, 182–187.

517

Berkowitz, L. Social motivation. In G. Lindzey & E. Aronson (Eds.), *Handbook of Social Psychology.* Vol. 3 (2nd ed.) Reading, Mass.: Addison-Wesley, 1969.

Berliner, D. C., & Cahen, L. S. Trait-treatment interaction and learning. In F. N. Kerlinger (Ed.), *Review of research in education*: 1. Itasca, Ill.: Peacock Publishers, 1973, in press.

Blalock, H. M. *Causal inferences in nonexperimental research.* Chapel Hill: University of North Carolina Press, 1964.

Blalock, H. M. Theory building and causal inferences. In H. M. Blalock & A. B. Blalock (Eds.), *Methodology in social research.* New York: McGraw-Hill, 1968.

Blalock, H. M. (Ed.) *Causal models in the social sciences.* Chicago: Aldine, 1971.

Bloom, B. S. Higher mental processes. In R. L. Ebel, V. H. Noll, & R. M. Bauer (Eds.), *Encyclopedia of educational research.* (4th ed.) New York: Macmillan, 1969.

Bock, R. D., & Haggard, E. A. The use of multivariate analysis of variance in behavioral research. In D. K. Whitla (Ed.), *Handbook of measurement and assessment in behavioral sciences.* Reading, Mass.: Addison-Wesley, 1968.

Bohrnstedt, G. W., & Carter, T. M. Robustness in regression analysis. In H. L. Costner (Ed.), *Sociological Methodology 1971.* San Francisco: Jossey-Bass, 1971.

Boneau, C. A. The effects of violations of assumptions underlying the t test. *Psychological Bulletin,* 1960, **57**, 49–64.

Bottenberg, R. A., & Ward, J. H. Applied multiple linear regression. Lackland Air Force Base, Texas: 6570th Personnel Research Laboratory, Aerospace Medical Division, Air Force Systems Command, 1963.

Boudon, R. A new look at correlation analysis. In H. M. Blalock & A. B. Blalock (Eds.), *Methodology in social research.* New York: McGraw-Hill, 1968.

Boyle, R. P. Path analysis and ordinal data. *American Journal of Sociology,* 1970, **75**, 461–480.

Bracht, G. H. Experimental factors related to aptitude-treatment interactions. *Review of Educational Research,* 1970, **40**, 627–645.

Braithwaite, R. B. *Scientific explanation.* Cambridge: Cambridge University Press, 1953.

Brodbeck, M. Logic and scientific method in research on teaching. In N. L. Gage (Ed.), *Handbook of research on teaching.* Skokie, Ill.: Rand McNally, 1963.

Bush, R. R., Abelson, R. P., & Hyman, R. *Mathematics for psychologists*: *Examples and problems.* New York: Social Science Research Council, 1956.

Campbell, D. T. Factors relevant to the validity of experiments in social setting. *Psychological Bulletin,* 1957, **54**, 297–312.

Campbell, D. T., & Stanley, J. C. Experimental and quasi-experimental designs for research. In N. L. Gage (Ed.), *Handbook of research on teaching.* Skokie, Ill.: Rand McNally, 1963.

Cattell, R. B. *Factor analysis.* New York: Harper & Row, 1952.

Citron, A., Chein, I., & Harding, J. Anti-minority remarks; A problem for action research. *Journal of Abnormal and Social Psychology,* 1950, **45**, 99–126.

Cleary, T. A. Test bias: Prediction of grades of negro and white students in integrated colleges. *Journal of Educational Measurement,* 1968, **5**, 115–124.

Cnudde, C. F., & McCrone, D. J. The linkage between constituency attitudes and congressional voting behavior: A causal model. *American Political Science Review,* 1966, **60**, 66–72.

Coats, W. D., & Smidchens, U. Audience recall as a function of speaker dynamism. *Journal of Educational Psychology,* 1966, **57**, 189–191.

Cochran, W. G. Analysis of covariance: Its nature and uses. *Biometrics*, 1957, **13**, 261–281.

Cohen, J. Multiple regression as a general data analytic system. *Psychological Bulletin*, 1968, **70**, 426–443.

Coleman, J. S., Campbell, E. Q., Hobson, C. J., McPartland, J., Mood, A. M., Weinfeld, F. D., & York, R. L. *Equality of educational opportunity*. Washington, D.C.: U.S. Dept. of Health, Education, and Welfare, Office of Education, U.S. Govt. Printing Office, 1966.

Cooley, W. W., & Lohnes, P. R. *Multivariate procedures for the behavioral sciences*. New York: Wiley, 1962.

Cooley, W. W., & Lohnes, P. R. *Multivariate data analysis*. New York: Wiley, 1971.

Creager, J. A. Orthogonal and nonorthogonal methods of partitioning regression variance. *American Educational Research Journal*, 1971, **8**, 671–676.

Cronbach, L. J. *Essentials of psychological testing*. (2nd ed.) New York: Harper & Row, 1960.

Cronbach, L. J. Intelligence? Creativity? A parsimonious reinterpretation of the Wallach-Kogan data. *American Educational Research Journal*, 1968, **5**, 491–511.

Cronbach, L. J., & Furby, L. How should we measure "change" – or should we? *Psychological Bulletin*, 1970, **74**, 68–80.

Cronbach, L. J., & Snow, R. E. Individual differences in learning ability as a function of instructional variables. Report to U.S. Office of Education, 1969.

Cutright, P. National political development: Measurement and analysis. *American Sociological Review*, 1963, **27**, 229–245.

Darlington, R. B. Multiple regression in psychological research and practice. *Psychological Bulletin*, 1968, **69**, 161–182.

Davidson, M. L. Univariate versus multivariate tests in repeated-measures experiments. *Psychological Bulletin*, 1972, **77**, 446–452.

Dewey, J. *Democracy and education*. New York: Macmillan, 1916.

Dewey, J. *How we think*. Boston: D. C. Heath, 1933.

Dixon, W. J. (Ed.), *BMD: Biomedical computer programs*. Berkeley: University of California Press, 1970.

Do teachers make a difference? Washington, D.C.: U.S. Office of Education, 1970.

Dollard, J., Miller, N. E., Doob, L. W., Mowrer, O. H., & Sears, R. R. *Frustration and aggression*. New Haven: Yale University Press, 1939.

Draper, N. R., & Smith, H. *Applied regression analysis*. New York: Wiley, 1966.

Duncan, O. D. Inheritance of poverty or inheritance of race? In D. P. Moynihan (Ed.), *Understanding poverty: Perspectives from the social sciences*. New York: Basic Books, 1969.

Duncan, O. D. Partials, partitions, and paths. In E. F. Borgatta & G. W. Bohrnstedt (Eds.), *Sociological Methodology 1970*. San Francisco: Jossey-Bass, 1970.

Dunnett, C. W. A multiple comparison procedure for comparing several treatments with a control. *Journal of the American Statistical Association*, 1955, **50**, 1096–1121.

Dwyer, P. S. *Linear computations*. New York: Wiley, 1951.

Eckstein, M. A., & Noah, H. J. *Scientific investigations in comparative education*. New York: Macmillan, 1969.

Edwards, A. L. *Expected values of discrete random variables and elementary statistics*. New York: Wiley, 1964.

Edwards, A. L. *Experimental design in psychological research*. (3rd ed.) New York: Holt, Rinehart and Winston, 1968.

Elashoff, J. D. Analysis of covariance: A delicate instrument. *American Educational Research Journal*, 1969, **6**, 383–401.

Ezekiel, M., & Fox, K. A. *Methods of correlation and regression analysis.* (3rd ed.) New York: Wiley, 1959.

Feigl, H., & Brodbeck, M. *Readings in the philosophy of science.* New York: Appleton-Century-Crofts, 1953.

Feldt, L. S. A comparison of the precision of three experimental designs employing a concomitant variable. *Psychometrika*, 1958, **23**, 335–353.

Festinger, L. Informal social communication. *Psychological Review*, 1950, **57**, 271–282.

Fisher, R. A., & Yates, F. *Statistical tables for biological, agricultural and medical research.* (6th ed.) New York: Hafner, 1963.

Frederiksen, N. Factors in in-basket performance. *Psychological Monographs*, 1962, **76** (22, Whole No. 541).

Frederiksen, N., Jensen, O., & Beaton, A. E. Organizational climates and administrative performance. Princeton, N.J.: Educational Testing Service, 1968.

Frederiksen, N., Saunders, D. R., & Wand, B. The in-basket test. *Psychological Monographs*, 1957, **71** (9, Whole No. 438).

Free, L. A., & Cantril, H. *The political beliefs of Americans: A study of public opinion.* New Brunswick, N.J.: Rutgers University Press, 1967.

Freedman, J. L. Involvement, discrepancy, and change. *Journal of Abnormal and Social Psychology*, 1964, **69**, 290–295.

Fruchter, B. *Introduction to factor analysis.* New York: Van Nostrand, 1954.

Fruchter, B. Manipulative and hypothesis-testing factor-analytic experimental designs. In R. B. Cattell (Ed.), *Handbook of multivariate experimental psychology.* Skokie, Ill.: Rand McNally, 1966.

Gagné, R. M. *The conditions of learning.* (2nd ed.) New York: Holt, Rinehart and Winston, 1970.

Games, P. A. Multiple comparisons of means. *American Educational Research Journal*, 1971, **8**, 531–565.

Games, P. A., & Lucas, P. A. Power of the analysis of variance of independent groups on non-normal and normally transformed data. *Educational and Psychological Measurement*, 1966, **26**, 311–327.

Garms, W. I. The correlates of educational effort: A multivariate analysis. *Comparative Education Review*, 1968, **12**, 281–290.

Getzels, J. W., & Jackson, P. W. The teacher's personality and characteristics. In N. L. Gage (Ed.), *Handbook of research on teaching.* Skokie, Ill.: Rand McNally, 1963.

Glass, G. V. Correlations with products of variables: Statistical formulation and implications for methodology. *American Educational Research Journal*, 1968, **5**, 721–728.

Glass, G. V., & Hakstian, A. R. Measures of association in comparative experiments: Their development and interpretation. *American Educational Research Journal*, 1969, **6**, 403–414.

Glass, G. V., & Maguire, T. O. Abuses of factor scores. *American Educational Research Journal*, 1966, **3**, 297–304.

Goldberger, A. S. *Econometric theory.* New York: Wiley, 1964.

Gordon, R. A. Issues in multiple regression. *American Journal of Sociology*, 1968, **73**, 592–616.

Graybill, F. A. *An introduction to linear statistical methods.* Vol. 1. New York: McGraw-Hill, 1961.

Green, B. F. The computer revolution in psychometrics. *Psychometrika*, 1966, **31**, 437–445.

Greenhouse, S. W., & Geisser, S. On methods in the analysis of profile data. *Psychometrika*, 1959, **24**, 95–112.

Guilford, J. P. *Psychometric methods.* (2nd ed.) New York: McGraw-Hill, 1954.

Guilford, J. P. The structure of intellect. *Psychological Bulletin*, 1956, **53**, 267–293.

Guilford, J. P. *The nature of human intelligence.* New York: McGraw-Hill, 1967.

Guttman, L. A new approach to factor analysis: The radex. In P. F. Lazarsfeld (Ed.), *Mathematical thinking in the social sciences.* New York: Free Press, 1954.

Guttman, L. 'Best possible' systematic estimates of communalities. *Psychometrika*, 1956, **21**, 273–285.

Haggard, E. A. Socialization, personality and academic achievement in gifted children. *School Review*, 1957, **65**, 388–414.

Halinski, R. S., & Feldt, L. S. The selection of variables in multiple regression analysis. *Journal of Educational Measurement*, 1970, **7**, 151–157.

Harman, H. *Modern factor analysis.* (2nd ed.) Chicago: University of Chicago Press, 1967.

Harvard Educational Review. Environment, heredity, and intelligence. Reprint series No. 2. Cambridge, Mass., 1969.

Harvey, O. J., Hunt, E. E., & Schroder, H. M. *Conceptual systems and personality organization.* New York: Wiley, 1961.

Hays, W. L. *Statistics for psychologists.* New York: Holt, Rinehart and Winston, 1963.

Heck, D. L. Charts of some upper percentage points of the distribution of the largest characteristic root. *Annals of Mathematical Statistics*, 1960, **31**, 625–642.

Heise, D. R. Affectual dynamics in simple sentences. *Journal of Personality and Social Psychology*, 1969, **11**, 204–213. (a)

Heise, D. R. Problems in path analysis and causal inference. In E. F. Borgatta (Ed.), *Sociological methodology 1969.* San Francisco: Jossey-Bass, 1969. (b)

Heise, D. R. Potency dynamics in simple sentences. *Journal of Personality and Social Psychology*, 1970, **16**, 48–54.

Hempel, C. G. *Aspects of scientific explanation.* New York: Free Press, 1965.

Hemphill, J. K., Griffiths, D. E., & Frederiksen, N. *Administrative performance and personality.* New York: Bureau of Publications, Teachers College Press, Columbia University, 1962.

Herzberg, P. A. The parameters of cross-validation. *Psychometrika*, 1969, **34** (Monogr. Suppl. 16).

Hiller, J. H., Fisher, G. A., & Kaess, W. A computer investigation of verbal characteristics of effective classroom lecturing. *American Educational Research Journal*, 1969, **6**, 661–675.

Hoffman, S. Long road to theory. In S. Hoffman (Ed.), *Contemporary theory in international relations.* Englewood Cliffs, N.J.: Prentice-Hall, 1960.

Holtzman, W. H., & Brown, W. F. Evaluating study habits and attitudes of high school students. *Journal of Educational Psychology*, 1968, **59**, 404–409.

Horst, P. *Factor analysis of data matrices*. New York: Holt, Rinehart and Winston, 1965.

Hotelling, H. The impact of R. A. Fisher on statistics. *Journal of the American Statistical Association*, 1951, **46**, 35–46.

Hummel, T. J., & Sligo, J. R. Empirical comparison of univariate and multivariate analysis of variance procedures. *Psychological Bulletin*, 1971, **76**, 49–57.

Humphreys, L. G., & Ilgen, D. R. Note on a criterion for the number of common factors. *Educational and Psychological Measurement*, 1969, **29**, 571–578.

Jaspen, N. The calculation of probabilities corresponding to values of *z*, *t*, *F*, and chi-square. *Educational and Psychological Measurement*, 1965, **25**, 877–880.

Jensen, A. How much can we boost IQ and scholastic achievement? *Harvard Educational Review*, 1969, **39**, 1–123.

Jessor, R., Graves, T. D., Hanson, R. C., & Jessor, S. L. *Society, personality, and deviant behavior*. New York: Holt, Rinehart and Winston, 1968.

Johnson, P. O., & Fay, L. C. The Johnson-Neyman technique, its theory and application. *Psychometrika*, 1950, **15**, 349–367.

Johnson, P. O., & Jackson, R. W. B. *Modern statistical methods: Descriptive and inferential*. Skokie, Ill.: Rand McNally, 1959.

Johnson, P. O., & Neyman, J. Tests of certain linear hypotheses and their applications to some educational problems. *Statistical Research Memoirs*, 1936, **1**, 57–93.

Jones, E. E., & Gerard, H. B. *Foundations of social psychology*. New York: Wiley, 1967.

Jones, E. E., Rock L., Shaver, K. G., Goethals, G. R., & Ward, L. M. Pattern of performance and ability attribution: An unexpected primacy effect. *Journal of Personality and Social Psychology*, 1968, **10**, 317–340.

Kaplan, A. *The conduct of inquiry*. San Francisco: Chandler, 1964.

Kemeny, J. G. *A philosopher looks at science*. Princeton N.J.: Van Nostrand, 1959.

Kemeny, J. G., Snell, J. L., & Thompson, G. L. *Introduction to finite mathematics*. (2nd ed.) Englewood Cliffs, N.J.: Prentice-Hall, 1966.

Kerlinger, F. N. *Foundations of behavioral research*. New York: Holt, Rinehart and Winston, 1964.

Kerlinger, F. N. Research in education. In R. L. Ebel, V. H. Noll, & R. M. Bauer (Eds.), *Encyclopedia of educational research*. (4th ed.) New York: Macmillan, 1969.

Kerlinger, F. N. A social attitude scale: Evidence on reliability and validity. *Psychological Reports*, 1970, **26**, 379–383.

Kerlinger, F. N. The structure and content of social attitude referents: A preliminary study. *Educational and Psychological Measurement*, 1972, **32**, 613–630.

Kerlinger, F. N. *Foundations of behavioral research*. (2nd ed.) New York: Holt, Rinehart and Winston, 1973.

Kersh, B. Y., & Wittrock, M. C. Learning and discovery: An interpretation of recent research. *Journal of Teacher Education*, 1962, **13**, 461–468.

Khan, S. B. Affective correlates of academic achievement. *Journal of Educational Psychology*, 1969, **60**, 216–221.

Kirk, R. E. *Experimental design: Procedures for the behavioral sciences*. Belmont, California: Brooks/Cole, 1968.

Knief, L. M., & Stroud, J. B. Interrelations among various intelligence, achievement, and social class scores. *Journal of Educational Psychology*, 1959, **50**, 117–120.

Kogan, N., & Wallach, M. A. *Risk taking: A study in cognition and personality.* New York: Holt, Rinehart and Winston, 1964.

Kolb, D. A. Achievement motivation training for underachieving high-school boys. *Journal of Personality and Social Psychology*, 1965, **2**, 783–792.

Koslin, B. L., Haarlow, R. N., Karlins, M., & Pargament, R. Predicting group status from members' cognitions. *Sociometry*, 1968, **31**, 64–75.

Land, K. C. Principles of path analysis. In E. F. Borgatta (Ed.), *Sociological methodology: 1969.* San Francisco: Jossey-Bass, 1969.

Lave, L. B., & Seskin, E. P. Air pollution and human health. *Science*, 1970, **169**, 723–733.

Layton, W. L., & Swanson, E. O. Relationship of ninth grade Differential Aptitude Test scores to eleventh grade test scores and high school rank. *Journal of Educational Psychology*, 1958, **49**, 153–155.

Lee, R. S. Social attitudes and the computer revolution. *Public Opinion Quarterly*, 1970, **34**, 53–59.

Lerner, D. (Ed.), *Cause and effect.* New York: Free Press, 1965.

Li, C. C. *Population genetics.* Chicago: University of Chicago Press, 1955.

Li, C. C. *Introduction to experimental statistics.* New York: McGraw-Hill, 1964.

Li, J. C. R. *Statistical inference.* Ann Arbor: Edwards Brothers, 1964.

Liddle, G. Overlap among desirable and undesirable characteristics in gifted children. *Journal of Educational Psychology*, 1958, **49**, 219–223.

Lindquist, E. F. *Design and analysis of experiments in psychology and education.* Boston: Houghton Mifflin, 1953.

Lohnes, P. R., & Cooley, W. W. *Introduction to statistical procedures: With computer exercises.* New York: Wiley, 1968.

Lord, F. M., & Novick, M. R. *Statistical theories of mental test scores.* Reading, Mass.: Addison-Wesley, 1968.

Lubin, A. The interpretation of significant interaction. *Educational and Psychological Measurement*, 1961, **21**, 807–817.

Lyons, M. Techniques for using ordinal measures in regression and path analysis. In H. L. Costner (Ed.), *Sociological methodology 1971.* San Francisco: Jossey-Bass, 1971.

McClelland, D. C. Toward a theory of motive acquisition. *American Psychologist*, 1965, **20**, 321–333.

McClelland, D. C., Atkinson, J. W., Clark, R. A., & Lowell, E. L. *The achievement motive.* New York: Appleton-Century-Crofts, 1953.

McCracken, D. D. *A guide to Fortran IV programming.* New York: Wiley, 1965.

McGinnis, R. *Mathematical foundations for social analysis.* Indianapolis: Bobbs-Merrill, 1965.

McGuire, C., Hindsman, E., King, F. J., & Jennings, E. Dimensions of talented behavior. *Educational and Psychological Measurement*, 1961, **31**, 3–38.

McNemar, Q. At random: Sense and nonsense. *American Psychologist*, 1960, **25**, 295–300.

McNemar, Q. *Psychological statistics.* (3rd ed.) New York: Wiley, 1962.

Marascuilo, L. A., & Levin, L. R. Appropriate post hoc comparisons for interaction and nested hypotheses in analysis of variance designs: The elimination of type IV errors. *American Educational Research Journal*, 1970, **7**, 397–421.

Mayeske, G. W., Wisler, C. E., Beaton, A. E., Weinfeld, F. D., Cohen, W. M., Okada, T., Proshek, J. M., & Tabler, K. A. *A study of our nation's schools*. Washington, D.C.: U.S. Dept. of Health, Education, and Welfare, Office of Education, 1969.

Miller, R. G. *Simultaneous statistical inference*. New York: McGraw-Hill, 1966.

Miller, W. E., & Stokes, D. E. Constituency influence in congress. *The American Political Science Review*, 1963, **57**, 45–56.

Mitzel, H. Teacher effectiveness. In C. Harris (Ed.), *Encyclopedia of educational research*. (3rd ed.) New York: Macmillan, 1960.

Mood, A. M. Macro-analysis of the American educational system. *Operations Research*, 1969, **17**, 770–784.

Mood, A. M. Partitioning variance in multiple regression analyses as a tool for developing learning models. *American Educational Research Journal*, 1971, **8**, 191–202.

Morrison, D. F. *Multivariate statistical methods*. New York: McGraw-Hill, 1967.

Mosier, C. I. Problems and designs of cross-validation. *Educational and Psychological Measurement*, 1951, **11**, 5–11.

Mullish, H. *Modern programming: Fortran IV*. Waltham, Mass.: Blaisdell, 1968.

Myers, J. L. *Fundamentals of experimental design*. Boston: Allyn and Bacon, 1966.

Moore, M. Aggression themes in a binocular rivalry situation. *Journal of Personality and Social Psychology*, 1966, **3**, 685–688.

Mosteller, F., & Bush, R. R. Selected quantitative techniques. In G. Lindzey (Ed.), *Handbook of social psychology*. Vol. 1. Reading, Mass.: Addison-Wesley, 1954.

Mosteller, F., & Moynihan, D. P. (Eds.), *On equality of educational opportunity*. New York: Vintage Books, 1972.

Nagel, E. Types of causal explanation in science. In D. Lerner (Ed.), *Cause and effect*. New York: Free Press, 1965.

Namboodiri, N. K. Experimental designs in which each subject is used repeatedly. *Psychological Bulletin*, 1972, **77**, 54–64.

Newcomb, T. M. *Social psychology*. New York: Dryden, 1950.

Newton, R. G., & Spurrell, D. J. A development of multiple regression for the analysis of routine data. *Applied Statistics*, 1967, **16**, 51–64. (a)

Newton, R. G., & Spurrell, D. J. Examples of the use of elements for clarifying regression analyses. *Applied Statistics*, 1967, **16**, 165–172. (b)

Neyman, J. R. A. Fisher (1890–1962): An appreciation. *Science*, 1967, **156**, 1456–1460.

Nichols, R. C. Schools and the disadvantaged. *Science*, 1966, **154**, 1312–1314.

Nunnally, J. C. *Psychometric theory*. New York: McGraw-Hill, 1967.

Olkin, J., & Pratt, J. W. Unbiased estimation of certain correlation coefficients. *Annals of Mathematical Statistics*, 1958, **29**, 201–211.

Overall, J. E., & Spiegel, D. K. Concerning least squares analysis of experimental data. *Psychological Bulletin*, 1969, **72**, 311–322.

Pillai, K. C. S. *Statistical tables for tests of multivariate hypotheses*. Manila, Philippines: University of the Philippines, 1960.

Pollack, S. V., & Sterling, T. D. *A guide to PL/I*. New York: Holt, Rinehart and Winston, 1969.

Potthoff, R. F. On the Johnson-Neyman technique and some extensions thereof. *Psychometrika*, 1964, **29**, 241–256.

Powers, R. D., Sumner, W. A., & Kearl, B. E. Recalculation of four adult readability formulas. *Journal of Educational Psychology*, 1958, **49**, 99–105.

Press, S. J. *Applied multivariate analysis*. New York: Holt, Rinehart and Winston, 1972.

Pugh, R. C. The partitioning of criterion score variance accounted for in multiple correlation. *American Educational Research Journal*, 1968, **5**, 639–646.

Quenouille, M. H. *Introductory statistics*. London: Pergamon Press, 1950.

Roe, A., & Siegelman, M. *The origin of interests*. Washington, D.C.: American Personnel and Guidance Association, 1964.

Rokeach, M. *The open and closed mind*. New York: Basic Books, 1960.

Roy, S. N. *Some aspects of multivariate analysis*. New York: Wiley, 1957.

Rulon, P. J., & Brooks, W. D. On statistical tests of group differences. In D. K. Whitla (Ed.), *Handbook of measurement and assessment in behavioral sciences*. Reading, Mass.: Addison-Wesley, 1968.

Ryans, D. Prediction of teacher effectiveness. In C. Harris (Ed.), *Encyclopedia of educational research*. (3rd ed.) New York: Macmillan, 1960.

Sarason, S. B., Davidson, K. S., Lighthall, F. F., Waite, R. R., & Ruebush, B. K. *Anxiety in elementary school children*. New York: Wiley, 1960.

Scannell, D. P. Prediction of college success from elementary and secondary school performance. *Journal of Educational Psychology*, 1960, **51**, 130–134.

Scheffé, H. *The analysis of variance*. New York: Wiley, 1959.

Searle, S. R. *Matrix algebra for the biological sciences*. New York: Wiley, 1966.

Searle, S. R. *Linear models*. New York: Wiley, 1971.

Sherif, M., & Hovland, E. I. *Social judgment*. New Haven: Yale University Press, 1961.

Simon, H. A. Spurious correlations: A causal interpretation. *Journal of the American Statistical Association*, 1954, **49**, 467–479.

Snedecor, G. W., & Cochran, W. G. *Statistical methods*. (6th ed.) Ames, Ia.: Iowa State University Press, 1967.

Stanley, J. C. The influence of Fisher's 'The Design of Experiments' on educational research thirty years later. *American Educational Research Journal*, 1966, **3**, 223–229.

Stewart, D., & Love, W. A general canonical correlation index. *Psychological Bulletin*, 1968, **70**, 160–163.

Tatsuoka, M. M. *Discriminant analysis: The study of group differences*. Champaign, Ill.: Institute for Personality and Ability Testing, 1970.

Tatsuoka, M. M. *Multivariate analysis: Techniques for educational and psychological research*. New York: Wiley, 1971. (a)

Tatsuoka, M. M. *Significance tests: Univariate and multivariate*. Champaign, Ill.: Institute for Personality and Ability Testing, 1971. (b)

Thistlethwaite, D. L., & Wheeler, N. Effects of teacher and peer subcultures upon student aspirations. *Journal of Educational Psychology*, 1966, **57**, 35–47.

Thorndike, R. L. *The concept of over- and underachievement*. New York: Teachers College, Columbia University, Bureau of Publications, 1963.

Thurstone, L. L. *Multiple-factor analysis*. Chicago: University of Chicago Press, 1947.

Thurstone, L. L., & Thurstone, T. G. *Factorial studies of intelligence*. Chicago: University of Chicago Press, 1941.

Tukey, J. W. Causation, regression and path analysis. In O. Kempthorne, T. A. Bancroft, J. W. Gowen, & J. D. Lush (Eds.), *Statistics and mathematics in biology.* Ames, Ia.: Iowa State College Press, 1954.

Turner, M. E., & Stevens, C. D. The regression analysis of causal paths. *Biometrics,* 1959, **15**, 236–258.

Veldman, D. J. *Fortran programming for the behavioral sciences.* New York: Holt, Rinehart and Winston, 1967.

Veldman, D. J., & Peck, R. F. Student teacher characteristics from pupils' viewpoint. *Journal of Educational Psychology,* 1963, **54**, 346–355.

Walberg, H. J. Predicting class learning: An approach to the class as a social system. *American Educational Research Journal,* 1969, **6**, 529–542.

Walker, H. M., & Lev, J. *Statistical inference.* New York: Holt, Rinehart and Winston, 1953.

Wallach, M. A., & Kogan, N. *Modes of thinking in young children.* New York: Holt, Rinehart and Winston, 1965.

Warr, P. B., & Smith, J. S. Combining information about people: Comparisons between six models. *Journal of Personality and Social Psychology,* 1970, **16**, 55–65.

Wilson, A. B. Educational consequences of segregation in a California community. In U.S. Commission on Civil Rights, *Racial isolation in the public schools.* Vol. 2. Washington, D.C.: U.S. Govt. Printing Office, 1967.

Winer, B. J. *Statistical principles in experimental design.* (2nd ed.) New York: McGraw-Hill, 1971.

Wisler, C. E. Partitioning the explained variation in a regression analysis. In G. W. Mayeske et al. *A study of our nation's schools.* Washington, D.C.: U.S. Dept. of Health, Education, and Welfare, Office of Education, 1969.

Wold, H. Causal inference from observational data: A review of ends and means. In M. Wittrock & D. Wiley (Eds.), *The evaluation of instruction: Issues and problems.* New York: Holt, Rinehart and Winston, 1970.

Wold, H., & Jureen, L. *Demand analysis.* New York: Wiley, 1953.

Wolf, R. The measurement of environments. In A. Anastasi (Ed.), *Testing problems in perspective.* Washington, D.C.: American Council on Education, 1966.

Wood, C. G., & Hokanson, J. E. Effects of induced muscular tension on performance and the inverted U function. *Journal of Personality and Social Psychology,* 1965, **1**, 506–510.

Worell, L. Level of aspiration and academic success. *Journal of Educational Psychology,* 1959, **50**, 47–54.

Wright, S. Correlation and causation. *Journal of Agricultural Research,* 1921, **20**, 557–585.

Wright, S. The method of path coefficients. *Annals of Mathematical Statistics,* 1934, **5**, 161–215.

Wright, S. Path coefficients and path regressions: Alternative or complementary concepts. *Biometrics,* 1960, **16**, 189–202. (a)

Wright, S. The treatment of reciprocal interaction, with and without lag, in path analysis. *Biometrics,* 1960, **16**, 423–445. (b)

Youden, W. J. The Fisherian revolution in methods of experimentation. *Journal of the American Statistical Association,* 1951, **45**, 47–50.

Zigler, E., & Child, I. L. Socialization. In G. Lindzey & E. Aronson (Eds.), *Handbook of social psychology.* Vol. 3 (2nd ed.) Reading, Mass.: Addison-Wesley, 1969.

Author Index

A

Abelson, R. P., 258
Althauser, R. P., 415
Anderson, G. J., 403–404, 414, 415
Anderson, H. E., 67, 343
Anderson, N. H., 48
Astin, A. W., 394–396, 399, 409–410, 414, 429, 445, 449

B

Baker, B. D., 48
Bartlett, M. S., 380
Beaton, A. E., 422–424, 429
Berkowitz, L., 370, 435, 449
Berliner, D. C., 49, 240
Blalock, H. M., 15, 16, 327, 446
Bloom, B. S., 49
Bock, R. D., 346, 352
Boneau, C. A., 48
Bracht, G. H., 240
Braithwaite, R. B., 3, 450
Brodbeck, M., 306
Brooks, W. D., 352, 356
Brown, W. F., 5, 391, 392
Bush, R. R., 48

C

Cahen, L. S., 49, 240
Campbell, D. T., 4, 350, 447
Cantril, H., 369–370, 445
Child, I. L., 435
Cnudde, C. F., 327, 328
Coats, W. D., 365
Cochran, W. G., 23, 24, 36, 69

Cohen, J., 109
Coleman, J. S., 5–6, 17, 94, 297, 409, 422, 425–428, 429, 433, 442, 449
Cooley, W. W., 341, 343, 434
Creager, J. A., 298, 304
Cronbach, L. J., 49, 240, 404–405, 414, 415
Cutright, P., 94, 396–397

D

Darlington, R. B., 77, 296, 443
Dewey, J., 49
Dixon, W. J., 291
Dollard, J., 449
Dunnett, C. W., 120
Dwyer, P. S., 61

E

Edwards, A. L., 120
Ezekiel, M., 90, 403

F

Fay, L. C., 258
Festinger, L., 433
Fisher, G. A., 415–417
Fisher, R. A., 23, 215, 260, 350, 351
Fox, K. A., 90, 403
Frederiksen, N., 337, 422–424, 429, 442, 449
Free, L. A., 369–370, 445
Freedman, J. L., 228

Subject Index